Marine Pollution and Microbial Remediation

Milind Mohan Naik
Santosh Kumar Dubey
Editors

Marine Pollution and Microbial Remediation

Springer

Editors
Milind Mohan Naik
Department of Microbiology
Goa University
Taleigao Plateau, Goa, India

Santosh Kumar Dubey
Department of Microbiology
Goa University
Taleigao Plateau, Goa, India

ISBN 978-981-10-1042-2 ISBN 978-981-10-1044-6 (eBook)
DOI 10.1007/978-981-10-1044-6

Library of Congress Control Number: 2016948708

Printed on acid-free paper

This Springer imprint is published by Springer Nature
The registered company is Springer Science+Business Media Singapore Pte Ltd.

This book is dedicated to
Late Shri Mohan Ladu Naik
Retired Police Sub-inspector
Menkurem Bicholim, Goa

Acknowledgement

No good work is created without joining hands, and it will be definitely deluding and improper to say that we could have completed this study single-handed. This effort was not carried out solely by us but was aided by a quite few people whom we wish to thank now because this is the best time and place.

We are thankful to Dr. Satish Shetye, V.C. of Goa University, and V.P. Kamat, Registrar, Goa University, for providing necessary facilities and Prof. Sanjeev Ghadi and Prof. Savita Kerkar for their support. Our sincere thanks go to Prof. Saroj Bhosle, Prof. Irene Furtado, Prof. Sarita Nazareth, Dr. Sandeep Garg, Dr. Lakshangy S. Charya, Dr. Priya D' Costa, Dr. Vishwas Khodse, Dr. Sanika Samant, Dr. Gauri Achari, Dr. Varada Damare, Dr. Trelita de Sousa, Dr. Swapnil Doijad, Dr. S. B. Barbuddhe, Dr. Flory Pereira, Dr. Mahesh Majik, Dr. Teja Gaonkar and Dr. Pranay Morajkar. I express my sincere gratitude to the entire non-teaching staff including Mr. Shashikant Parab, Budhaji, Dominic, Saraswati, Deepa, Narayan and Rudransh for their constant help and support in various ways. We owe our deepest sense of gratitude to our friends Mr. Uday Naik, Amar Degvekar, Kashinath Shetgaonkar and Mrs Pranaya Naik. We also thank Government of India for financial support as SERB-DST young scientist project. We have successfully compiled our creative and thoughtful research work due to genuine concern and painstaking effort of many more well-wishers whose names are not mentioned, but they are still in our heart. So, reward is surely worth for their efforts.

We want to dedicate this book to the late **Shri Mohan Ladu Naik** (retired PSI Menkurem, Goa) and **Mrs Manisha Mohan Naik** (Menkurem).

Contents

About the Editors

Dr. Milind Mohan Naik has a PhD in Microbiology from Goa University. He joined Department of Microbiology in Goa University as assistant professor of marine microbiology in the year 2013. His research efforts have been focused on understanding the metal-resistance mechanism in bacteria from marine and terrestrial environments and their potential applications in bioremediation of polluted environmental sites. He has published over 20 research papers. He is a recipient of SERB-DST Young Scientist project award. He has guided eight postgraduate dissertation projects. He has also worked as scientist 'C' on the 'Malaria Evolution in South Asia' project funded by NIH and University of Washington, USA, in the National Institute of Malaria Research (ICMR). He has worked as scientist in Molbio Diagnostics Pvt. Ltd. in the Research and Development Department in 2012. His research interests are environmental microbiology and nanobiotechnology. He aims to gain knowledge in the research field of marine microbiology and biotechnology with the intent of learning new concepts.

Prof. Santosh Kumar Dubey has a PhD in botany from BHU, Varanasi, Uttar Pradesh. He joined the Department of Microbiology, Goa University, in the year 1996. His research efforts have been focused on understanding the metal and organometal resistance mechanisms in bacteria from marine and terrestrial environments, metagenomics, agriculture microbiology, genetic engineering and molecular biology. He has published 32 research papers in international journals, 9 in national journals and 8 chapters in edited books published by international and national publishers. He has guided 9 PhDs and 30 M. Sc. dissertations. He has five postdoctoral research experiences in several universities of the USA (Clarkson University; University at Buffalo; Roswell Park Cancer Institute, Buffalo, New York; University of Nevada Las Vegas, Las Vegas) and also is a recipient of JSPS fellowship in Japan. He has served as Head, Department of Microbiology, and Chairman, Board of Studies in Microbiology, at Goa University twice. He is programme coordinator of the UGC sponsored innovative programme of M.Sc. in Marine Microbiology in the Department of Microbiology, Goa University. He was awarded a visiting scientistship from the Centre of Marine Environmental Studies, Ehime University, Matsuyama, Japan, and a National Overseas Scholarship (PDF) from the Department of Education M/O Human Resource

Development, Govt. of India. He is a member of the Institutional Biosafety Committee, Goa University, and completed two major projects funded by DBT and DOD, Govt. of India. Currently he is member of the advisory board of UGC for the SAP-DRS programme of the Department of Microbiology and Biotechnology, M.S. University, Baroda, Gujarat.

Assessing Metal Contamination in Recent Creek Sediments Using Fractionation Technique Along Mumbai Coast, India

1

Lina L. Fernandes and G.N. Nayak

Abstract

Three sediment cores from a creek environment in Mumbai, extending from the head to the creek mouth, were studied with respect to elements (Fe, Mn, Cu, Pb, Co, Zn and Cr) along with organic matter (total organic carbon, total phosphorus, total nitrogen) and sediment components (sand, silt, clay). A sequential extraction procedure was also applied to understand the partitioning of trace metals among the different fractions of the sediment. Together with this data, pollution indices were also computed and comparison with numerical sediment quality guidelines carried out. Correlation analysis among the different variables displayed weaker relations of metals with the sediment components, while organic matter and Fe–Mn oxyhydroxides were found to act as important substrates for metal sequestration in the creek region. The fractionation results reveal almost all the elements were associated with the residual fraction, while Mn was high in the bioavailable fraction.

1.1 Introduction

Concerns regarding the trace metals in estuarine environments have changed from the focus of just measuring total metal levels to understanding the processes and controls on metal behaviour. Considering that a metal's availability is affected by its association with one or more of the different sediment components (organic matter, carbonates, sulphides, oxides), total metal concentration provides partial information about its potential interaction between the biotic and abiotic environments. Therefore, to better estimate metal bioavailability, an understanding of the partitioning of a particular metal among these sediment components is necessary. This study addresses the processes and controls on metal bioavailability so that further insight can be gained in the prediction of trace metal uptake in organisms.

L.L. Fernandes (✉)
CSIR- National Institute of Oceanography, Dona Paula, Goa 403 004, India
e-mail: lfernandes@nio.org

G.N. Nayak
Department of Marine Sciences, Goa University, Taleigão, Goa 403 206, India

M.M. Naik, S.K. Dubey (eds.), *Marine Pollution and Microbial Remediation*,
DOI 10.1007/978-981-10-1044-6_1

Thane Creek, in western Mumbai, being surrounded by highly urbanised and industrialised region forms an integral part of Mumbai, subjected to the effects and influences of these developments. Total metal concentrations in sediments of the region have been reported earlier (Sharma et al. 1994; Zingde and Desai 1981; Patel et al. 1975, 1985; Bhosale and Sahu 1991), which found increased level of contaminants such as Cu, Co, Zn, Cr and Ni. In recent years, studies on quality of the creek sediments have been undertaken, and the results indicate that contamination problems concerning heavy metal pollution still exist (Jha et al. 2003; Ram et al. 1998, 2003, 2009). Study carried out by Krishna and Govil (2005) reported the concentration ranges were Cr 177.9–1039 mg/kg (average 521.3 mg/kg), Ni 64.4–537.8 mg/kg (average 183.6 mg/kg), Co 44.8–101.6 mg/kg (average 68.7 mg/kg), Cu 3.10–271.2 mg/kg (average 104.6 mg/kg) and Zn 96.6–763.2 mg/kg (average 191.3 mg/kg). The WHO standard for aquatic life with respect to metals is Fe 200 ppb, Cu 5 ppb, Pb 25 ppb, Zn 30 ppb and Cr 100 ppb. Thus, it is important to understand the geochemical dynamics for better assessing metal transport and its ultimate fate in the creek environment. The total metal concentration in sediments is useful as an indicator of contamination of aquatic environments but inadequate to understand bioavailability, mobility or toxicity of metals to a larger extent (Hooda 2010). Also, as the chemical form determines the toxicity and fate of waterborne metals, it is more appropriate to quantify the different forms of metals than to estimate its total metal concentrations (Davidson et al. 1994; Lima et al. 2001; Jain 2004). Therefore, the overall objectives of the present study are to (i) assess the extent of heavy metal (Fe, Mn, Cu, Pb, Co, Zn and Cr) pollution in the creek sediments and (ii) employ sequential extraction procedure to determine the potential mobility and risk to biota from sediment-associated metals. The present metals were chosen because of their known abundance and toxic effects in environment of highly urbanised and industrialised regions. The research will help to better understand the patterns and pathways of sediment transport and accumulation in this urban-industrial creek setting.

1.2 Materials and Methods

1.2.1 Study Area

Thane Creek is bound by the Ulhas River to the north and Mumbai harbour bay towards the south (Fig. 1.1). The creek lies at longitudes 72°55′ E to 73°00′ E and latitudes 19°00′ N to 19°15′ N. Dense mangroves occur along both the banks of the creek, coupled with heavy urbanisation and industrialisation. Geologically, Mumbai (Thane) region is part of the Deccan Trap that was formed by volcanic effusions at the end of the Cretaceous period (Blasco 1975). The riverine end of the creek is narrow and shallow which gradually broadens and deepens towards the sea. The industrial and domestic waste contributes about 294 mld (million litres per day) and 145–260 mld, respectively, within Thane City limits (TMC-ES 2000). The diversified industries (chemical, pharmaceutical, textile, engineering and major fertiliser complexes), present along the creek banks, release high levels of nitrates and phosphates through their effluents into the creek (Quadros et al. 2009).

1.2.2 Sampling

Three sediment cores were collected, using a hand driven PVC tube of 5.5 cm diameter, from the intertidal regions of Thane Creek (Fig. 1.1), namely, near the creek head (core TC1), middle creek region (core TC5) and near the creek mouth (core TC4). In the laboratory, the cores were subsampled at 2 cm intervals, transferred to labelled polyethylene bags and stored frozen to −20 °C till further analysis.

For sedimentological and geochemical parameters, the sediment samples were dried in an oven at 60 °C. Sediment components were analysed using standard sieve and pipette technique after destruction of organic matter with H_2O_2 (Folk 1974). For geochemical analysis, the

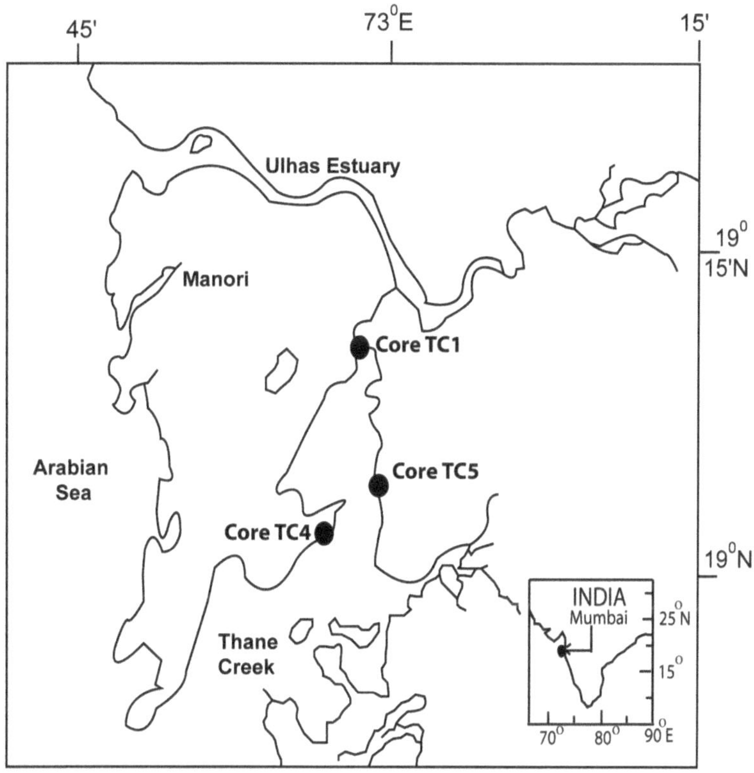

Fig. 1.1 Map showing the sampling locations

sediment samples were finely ground with mortar and pestle. Total organic carbon (TOC) content was determined employing a rapid titration method (in chromic acid environment) (Walkley and Black 1934), while a standard procedure (Grasshoff 1999) was used for total phosphorus (TP) and total nitrogen (TN) in sediment. Mineralogical phases of the sediments (<63 μm fraction) were determined by X-ray diffraction (XRD). Sediment samples for trace metal analysis were digested using a HF–$HClO_4$–HNO_3 total dissolution technique (Jarvis and Jarvis 1985). The digested samples were aspirated for Fe, Al, Zn, Mn, Co, Cr, Cu and Pb with the help of Varian AA 240 FS flame atomic absorption spectrometry (AAS) with an air–acetylene flame for all of the above elements except for Al for which nitrous oxide–acetylene flame was employed at specific wavelengths. Each sample was analysed in triplicate with a relative standard deviation of <5 % for all the metals. Recovery after digestion was checked using a certified reference material (BCSS-1). Percentage recoveries were always over 95 %. For speciation study Tessier's sequential extraction procedure was adopted (Tessier et al. 1979) and the different steps involved are shown in Table 1.1. The five fractions, viz. exchangeable, carbonate bound, Fe–Mn oxide, organic and residual, were determined for Fe, Zn, Mn, Co, Cr, Cu and Pb. The analytical results were statistically processed with correlation analysis to group metals with similar behaviour using the computer software STATISTICA 6.

1.2.3 Enrichment Factor

An enrichment factor was used to differentiate metals from anthropogenic (non-crustal) and geogenic (crustal) sources and to further evaluate the magnitude of contamination in the environment (Feng et al. 2004). Al was used as the reference element:

Table 1.1 Sequential extraction scheme of Tessier et al. (1979)

Step	Fraction	Extraction reagents/ conditions
1.	Exchangeable (F1)	8 ml $MgCl_2$, pH7, 1 h, room temperature, continuous agitation
2.	Acid soluble (carbonates) (F2)	8 ml NaOAc, pH5, 5 h, room temperature, continuous agitation
3.	Reducible (Fe–Mn oxide bound) (F3)	20 ml of 0.04M NH_2OH. HCl in 25 % HOAc, 6 h, 96 ± 3 °C, occasional agitation
4.	Oxidisable (organically bound + sulphide bound) (F4)	3 ml of 0.02M HNO_3 + 5 ml of 30 % H_2O_2, pH 2, 2 h, 85 ± 2 °C, occasional agitation. Add 3 ml of 30 % H_2O_2, repeat 3 h, cool and then add 5 ml 3.2M NH_4OAc in 20 % HNO_3, 0.5 h, room temperature, continuous agitation
5.	Residual (F5)	$HF:HNO_3:HClO_4$ = 7:3:1, dryness, again $HF:HNO_3:HClO_4$, 1 h, 2 ml conc. HCl, dryness, 10 ml 50 % HNO_3, make up to 50 ml with distilled water

$$EF = \left(C_n / C_{\mathrm{Al}}\right)_{\mathrm{sediment}} / \left(C_n / C_{\mathrm{Al}}\right)_{\mathrm{background}}$$

where $(C_n/C_{\mathrm{Al}})_{\mathrm{sample}}$ is the ratio of concentration of the element (C_n) to that of Al; (C_{Al}) in the sediment sample and $(C_n/C_{\mathrm{Al}})_{\mathrm{background}}$ is the same ratio with background value taken of average shale (Turekian and Wedepohl 1961). EF values between 0.5 and 1.5 indicate the metal originates from crustal material or natural processes, whereas EF values greater than 1.5 suggest the presence of an anthropogenic source (Zhang and Liu 2002).

1.2.4 Index of Geoaccumulation (Igeo)

Igeo, proposed by Muller (1979), to calculate metal concentration in sediments by comparing current concentration with undisturbed or crustal sediments levels, was used to quantitatively estimate the metal pollution status of the creek sediments. The Igeo is given by the following equation:

Table 1.2 Geoaccumulation index proposed by Muller (1979)

Pollution intensity	Sediment accumulation	Igeo class
Very strongly polluted	>5	6
Strongly to very strongly polluted	4–5	5
Strongly polluted	3–4	4
Moderately to strongly polluted	2–3	3
Moderately polluted	1–2	2
Unpolluted to moderately polluted	0–1	1
Practically unpolluted	<0	0

$$\mathrm{Igeo} = \log_2\left(C_n / 1.5\mathrm{B}_n\right)$$

where C_n = the measured concentration of the element "n" in the sediment fraction and B_n = geochemical background of element "n", taken from literature (average global shale). The factor 1.5 is used to compensate for possible variations of the reference data due to lithogenic effects. The classification of pollution intensity is shown in Table 1.2.

1.3 Results and Discussion

1.3.1 Sediment Components

In the entire creek region, sand ranges from 0.16 % to 4.77 %, while silt and clay vary from 18.24 % to 44.82 % and 52.80 % to 81.20 % with averages of 0.96 %, 32.44 % and 66.60 %, respectively. From the head to the creek mouth, a gradual decrease in sand and clay percentage and an increase in silt percentage are observed (Table 1.3).

1.3.2 Clay Mineralogy

Clay mineral analysis was performed on selected subsamples of two cores, namely, from the head

Table 1.3 Average concentrations of the different sediment parameters analysed in the cores of the Thane Creek

Variables	Core TC1	Core TC5	Core TC4
Sand (%)	1.52	0.53	0.83
Silt (%)	29.54	33.43	34.34
Clay (%)	68.94	66.04	64.82
TOC (%)	3.32	1.8	1.12
TP (mg/g)	1.11	0.98	0.79
TN (mg/g)	1.46	0.7	1.4
Al (%)	10.65	11.19	11.51
Fe (%)	9.09	8.06	7.06
Mn (ppm)	1518	2025	1884
Cu (ppm)	235	227	242
Pb (ppm)	107	81	99
Co (ppm)	78	44	63
Zn (ppm)	424	138	170
Cr (ppm)	159	201	195

(core TC1) and middle region (core TC5) of the creek. Sample composition was found to be consistent among the selected cores, with smectite (avg. 62.85 %) as the major constituent clay mineral followed by illite (avg. 16.91 %), kaolinite (avg. 10.59 %) and chlorite (avg. 9.65 %). The results are consistent with the study of Bhosale and Sahu (1991), who carried out clay mineral analysis in the creek region and reported the sediment mineralogy was dominated by montmorillonite followed by degraded chlorite, illite and silica to a minor extent. In the creek region, the percentage of smectite ranges from 38.59 to 79.73 %, 7.48 to 28.26 % for illite, 0.94 to 19.97 % for kaolinite and 4.77–20.48 % for chlorite. Thus, the clay fraction is found to be characterised by its mineralogical simplicity, with an absolute predominance of smectite, derived from the weathering of the basalt rocks of the basin. Deccan basalts are the major source for smectite along the western continental shelf of India (Rao and Rao 1995). Clay minerals in the marine environment are found to be largely detrital and widely dispersed (Biscaye 1965). However, in the present case, the bulk mineralogical analyses, in general, show a remarkable consistency in the sediment mineralogy among the sites.

1.3.3 Organic Matter (TOC, TP and TN)

Organic matter concentration, in general, shows a gradual decrease from the head towards the mouth of the creek. TOC ranges from 0.78 to 5.00 % and TP from 0.49 to 1.64 mg/g and TN varies from 0.31 to 2.89 mg/g for the entire creek basin. The average for TOC is 2.08 % and for TP is 0.96 mg/g, while for TN it is 1.19 mg/g. The organic matter content of the cores studied decreases with depth, as would be expected in any soil or sediment profile due to natural decomposition processes.

1.3.4 Metal Geochemistry

In general, of the metals analysed in the sediments, the individual metals followed varying patterns (Table 1.3). In the creek region, Al displays the highest average concentration (avg. 11.12 %), while Co has the lowest average concentration (avg. 62 ppm). The creek head exhibits maximum average concentrations of Fe, Pb, Co and Zn with low average values for Mn and Al. The middle region (TC5) displays highest values of Mn and Cr, along with lowest average values of Pb, Cu, Zn and Co. At the creek mouth (TC4), maximum average concentration of Al and Cu occurs with low average values for Fe. In general, the Al % distribution over the entire creek is found to remain constant. The order of total metal concentrations of sediment samples was found to be Al > Fe > Mn > Zn > Cu > Cr > Pb > Co.

Core TC1, near the creek head, contains more than twice the concentration of metals as compared to the other cores and a reduction in metal content occurs with increasing distance from the creek head. Most of the industries are housed near the head region of the creek. Therefore, during periods of high rainfall and flooding, the core location at TC1 is bound to receive the sediment-loaded water from the industrial and urban areas first and would therefore receive the larger proportion of contamination. The reduced tidal influence and decreased current speed in the inner creek region favour conditions for the settlement of suspended load (NIO 1998). Sediments collected near the mouth of the creek (TC4) reveal a substantial reduction in concentrations of most of the metals as compared to core TC1, as the location of core TC4 is more prone to wave action from the sea, causing the sediments to be more uniformly distributed. Therefore, there is evidence for a distance–concentration relationship in the sediment record, with concentration of metals decreasing with greater distance from the head/input.

1.3.5 Correlation Analysis

In order to understand the processes involved in the distribution of sediment and associated metal in the creek environment, Pearson's correlation analysis at $p<0.05$ was carried out on the different variables studied for all the cores sampled from the creek (Tables 1.4a, 1.4b, and 1.4c). Important factors controlling the abundance of metals in natural environment include organic matter contents and granulometry (Zhang et al. 2007). However, the sediment components are found to exhibit weak or negative correlation as compared to the organic matter which shows better correlations with the metals in the creek region. Many studies have reported a significant positive correlation between the organic matter content and adsorption affinity of heavy metals in sediments (Soares et al. 1999; Lin and Chen 1998). Fe–Mn oxides, on the other hand, seem to associate with most of the metals in core TC5 than cores TC1 and TC4 in which such correlations are seen to a lesser extent. Among the metals, good inter-elemental correlations are observed in all the cores suggesting of similar terrigenous sources or similar mechanisms of transport and accumulation within the sediments. It appears from the correlation analysis that in the creek region, the sediment geochemistry may be characterised by strong associations of metals (M) with three matrices (M–M, M–organic carbon and M–Fe–Mn oxides). Therefore, statistical analysis in general shows that there are three different processes operating in the basin, which help us in understanding the metal behaviour and distribution.

1.3.6 Enrichment Factor and Igeo

Near the head region of the creek (core TC1), Cu, Pb and Co are found to be highly enriched along the entire core length, while the remaining elements show an increase at the upper few cm of the core (Table 1.5). On the other hand in core TC5, sampled from middle region on the eastern bank of the creek, Fe and Zn are below the enriched level, while the remaining elements

Table 1.4a Pearson's correlation coefficients for organic matter and metals in core TI (near creek head) ($p<0.05$, $n=23$)

	TOC	TP	TN	Sand	Silt	Clay	Fe	Mn	Cu	Pb	Co	Zn
TP	−0.14	1.00										
TN	0.30	0.39	1.00									
Sand	0.25	−0.26	−0.20	1.00								
Silt	0.29	−0.23	−0.04	**0.56**	1.00							
Clay	−0.30	0.25	0.06	**−0.66**	**−0.99**	1.00						
Fe	**0.47**	0.36	0.29	0.05	0.09	−0.09	1.00					
Mn	0.35	0.01	**0.48**	0.17	0.38	−0.37	0.30	1.00				
Cu	0.35	0.32	**0.43**	−0.10	0.16	−0.13	0.11	**0.44**	1.00			
Pb	**0.49**	0.22	0.35	−0.05	−0.09	0.09	0.30	0.14	**0.47**	1.00		
Co	−0.25	0.05	**−0.54**	−0.07	−0.04	0.05	0.17	−0.39	0.05	0.16	1.00	
Zn	0.28	−0.19	0.35	−0.01	0.15	−0.13	−0.07	0.37	−0.01	0.02	**−0.60**	1.00
Cr	0.35	0.28	−0.01	0.18	0.27	−0.27	0.29	0.09	**0.60**	**0.54**	**0.51**	−0.36

Table 1.4b Pearson's correlation coefficients for sediment component, organic matter and metals in core TC5 (lower middle creek region) ($p<0.05$, $n=31$)

	TOC	TP	TN	Sand	Silt	Clay	Fe	Mn	Cu	Pb	Co	Zn
TP	−0.25	1.00										
TN	**0.45**	**−0.53**	1.00									
Sand	0.21	0.07	−0.02	1.00								
Silt	−0.20	0.31	0.02	−0.04	1.00							
Clay	0.19	−0.32	−0.02	0.00	−1.00	1.00						
Fe	−0.23	−0.05	0.25	−0.34	0.25	−0.24	1.00					
Mn	0.12	0.00	0.07	**0.36**	−0.15	0.13	0.06	1.00				
Cu	0.02	−0.20	**0.45**	−0.13	0.19	−0.19	**0.69**	**0.42**	1.00			
Pb	**−0.43**	0.35	−0.20	−0.28	**0.49**	**−0.48**	**0.67**	0.13	**0.50**	1.00		
Co	0.02	**0.43**	−0.02	0.16	0.14	−0.15	0.24	**0.46**	0.30	0.32	1.00	
Zn	**0.45**	**−0.44**	**0.74**	−0.28	−0.12	0.14	**0.45**	0.10	**0.63**	−0.04	0.04	1.00
Cr	−0.35	0.35	−0.22	0.01	**0.39**	**−0.39**	**0.42**	0.15	0.31	**0.77**	**0.39**	−0.06

Table 1.4c Pearson's correlation coefficients for sediment component, organic matter and metals in core TC4 (near creek mouth) ($p<0.05$, $n=23$)

	TOC	TP	TN	Sand	Silt	Clay	Fe	Mn	Cu	Pb	Co	Zn
TP	0.00	1.00										
TN	0.41	0.40	1.00									
Sand	0.09	0.37	0.17	1.00								
Silt	0.21	0.33	0.00	**0.43**	1.00							
Clay	−0.21	−0.35	−0.01	**−0.48**	−1.00	1.00						
Fe	0.38	−0.19	0.05	−0.21	−0.21	0.22	1.00					
Mn	0.16	−0.09	−0.07	−0.15	0.01	0.00	0.14	1.00				
Cu	0.01	**0.51**	**0.62**	−0.15	−0.08	0.09	0.04	−0.10	1.00			
Pb	−0.38	**0.52**	0.12	0.27	0.15	−0.16	**−0.52**	−0.18	0.28	1.00		
Co	−0.11	0.25	0.28	−0.19	−0.09	0.10	0.12	0.28	0.26	−0.04	1.00	
Zn	0.28	−0.01	0.31	0.01	−0.08	0.07	0.15	−0.14	0.33	−0.05	−0.08	1.00
Cr	0.35	0.00	0.36	−0.16	−0.21	0.21	0.31	**0.48**	0.14	−0.37	**0.70**	0.11

Table 1.5 Range of enrichment factor for metals from the creek region

EF	Fe	Mn	Cu	Pb	Co	Zn	Cr
TC1	1.20–1.75	1.51–2.33	2.89–4.85	2.86–4.85	2.68–3.85	2.10–4.55	1.15–1.86
TC5	1.02–1.43	1.63–3.19	2.82–4.15	2.24–3.61	1.01–2.57	0.84–1.26	1.26–2.18
TC4	0.79–1.44	1.69–2.97	2.87–4.94	2.34–5.67	1.71–3.28	0.93–1.82	1.26–2.27

point towards an anthropogenic source. Near the creek mouth (TC4), the EF values of Fe and Zn are low, while enrichment is observed for the remaining elements. The distribution obtained suggests that the trace metals may be derived from diverse sources including natural and anthropogenic inputs. Potential sources such as agricultural activities, corrosion of building materials, urban and industrial sewages and atmospheric deposition are probable contributors for the elemental enrichment.

The core sampled near the creek head (core TC1) exhibits unpolluted to moderately polluted Igeo class with respect to Pb, Cu, Zn and Co, while Mn, Fe and Cr remain unpolluted (Table 1.6b). The middle creek region (core TC5) displays moderate pollution with respect to Pb and Mn. In addition to this, in core TC5, moderate to strong pollution is observed with respect to Cu. The remaining elements fall in unpolluted to moderately polluted class. In the core retrieved near the creek mouth (core TC4), Cu, Pb and Co

Table 1.6 Igeo values for Thane cores

Stations	Igeo class	Elements
TC1	1	Cu, Pb, Zn and Co
	0	Mn, Cr, Fe
TC5	3	Cu
	2	Pb and Mn
	1	Fe, Co, Zn and Cr
TC4	2	Cu, Pb and Co
	1	Fe, Mn, Zn and Cr

fall in moderately polluted class, while the remaining elements are in unpolluted to moderately polluted class. In general, it is seen that the creek exhibits moderate pollution with respect to Cu and Pb. The principal anthropogenic sources of Cu include municipal waste and sewage and corrosion of Cu-containing pipelines or fittings. Further Cu salts are used in water supply systems to control biological growth in reservoirs and distribution pipes. The most probable sources of Pb are vehicle exhausts and leaded gasoline. Mn, in the middle creek region, and also Co near creek mouth are observed to fall in moderately polluted class suggesting there might be a risk to the biota.

1.3.7 Sediment Quality Guidelines (SQGs)

Some researchers have used numerical sediment quality guidelines as predictors of contaminants in aquatic sediments (Muwanga 1997). For assessment of ecotoxicological implications of the total metal concentrations in the sediments, SQGs (two sets) developed for aquatic ecosystems (MacDonald et al. 1996; Long et al. 1998) are considered in this study. These sets are defined as (i) the threshold effect level (TEL) and the probable effect level (PEL) and (ii) effect range low (ERL) and effect range median (ERM). ERL and TEL include concentrations below which adverse effects upon sediment-dwelling fauna are unlikely and occur only infrequently. In contrast, at concentrations above ERM and PEL, adverse effects are more likely to occur (Long et al. 1998). SQGs for elements such as Mn, Fe and Co are not available. The mean concentration of heavy metals (Table 1.7) and SQGs of USEPA (Pedersen et al. 1998) when compared indicate that in all the cores, Cu and Cr values are above TEL, PEL and ERL values. For Pb, all the cores exhibit values above TEL but below PEL. In the case of Zn, cores TC1 and TC4 have values higher than TEL and ERL but below PEL, while core TC5 shows values above TEL but below PEL and ERL. All the elements in the study area have values below ERM indicating low risk of adverse effects on organisms. In general, in all the cores collected from the creek region, almost all the elements are found to be above TEL, PEL and ERL but below ERM.

1.3.8 Speciation

Speciation analysis can provide information on heavy metal origin, its form and its biological and physicochemical availability (Das et al. 1995). Speciation analysis was carried out in all the cores. The percentage of metal extracted was calculated from the ratio between the element concentration in each fraction and the sum of concentrations in all fractions. For most of the samples, the sums of the extracted fractions agree within 10 %, with the independently determined total metal concentrations, supporting the overall accuracy of the extraction procedure. The plots of average metal concentration in the different sediment fractions are given in Fig. 1.2.

The distribution of Fe is similar in all the three cores, being dominated by the residual (>80 %) phase (F5), with minor Fe–Mn oxides (F3) and organic matter (F4)-bound fractions. Mn in the all the cores is found to be associated to a considerable extent (70 %) with bioavailable phases. The exchangeable, carbonate, Fe–Mn oxides and organic matter fractions contain 1–9 %, 18–31 %, 34–44 % and 2–12 %, respectively. Most Cu in the sediments is present in the residual fraction, which amount to 80–85 %. Also, some part was extracted mainly in organic and sulphide bound (about 10 %). These results are in accordance to previous results wherein the residual fraction is the most important fraction for Cu, followed by organic fraction (Li et al.

Table 1.7 Screening quick reference table for heavy metals in marine sediment (Buchman 1999)

Element	Near creek head (TC1)	Middle creek (TC5)	Near creek mouth (TC4)	TEL	PEL	ERL	ERM
Fe (%)	9.09	8.06	7.06	–	–	–	–
Mn (ppm)	1518	2025	1884	–	–	–	–
Cu (ppm)	235	228	243	31.6	149	34	270
Pb (ppm)	107	82	100	35.8	128	46.7	218
Co (ppm)	78	44	64	–	–	–	–
Zn (ppm)	424	139	170	121	459	150	410
Cr (ppm)	159	202	196	31.6	111	81	370

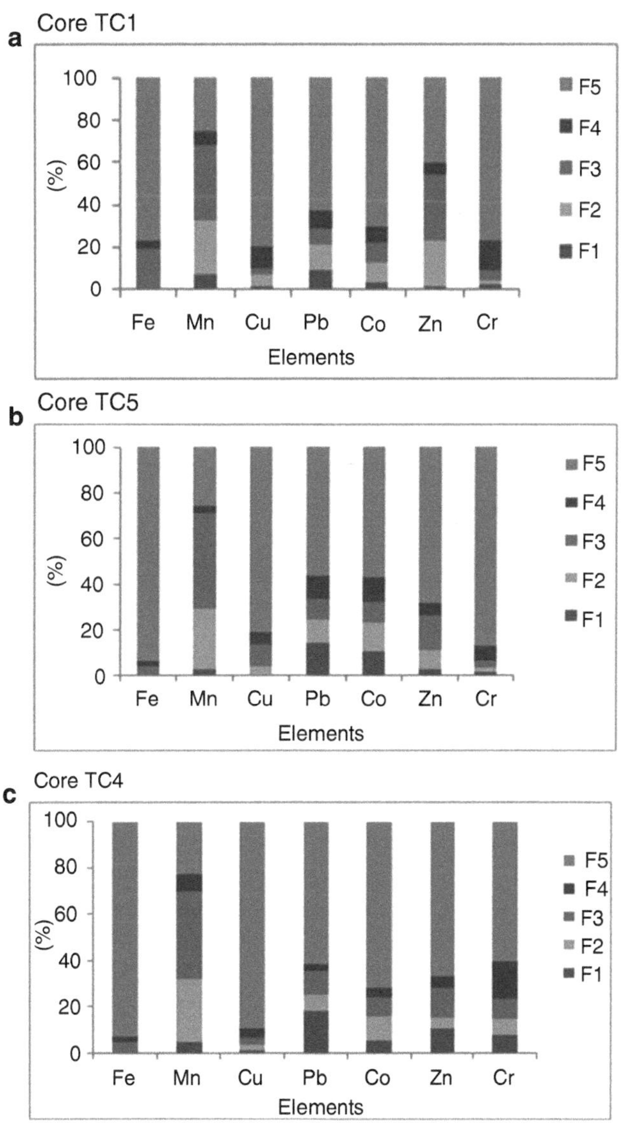

Fig. 1.2 Distribution of metals in the different sediment fractions of (**a**) core TC1, (**b**) core TC5 and (**c**) core TC4

2000, 2001; Ip et al. 2007). Oxidisable fraction, which is the second most dominant Cu host in the sediments of cores TC1 and TC4, consists mainly of organic- and sulphide-bound metals (Tessier et al. 1979; Kersten and Forstner 1989). Although chemical discrimination between these two phases is difficult (Kersten and Forstner 1989), the affinity of Cu for organic particles and coatings is well known (Comber and Gunn 1995).

The highest percentages of Pb contents are in the F5 residual fraction followed by the carbonate-bound fraction (F2) in core TC1 and the exchangeable fractions in cores TC5 and TC4. The distribution patterns of Pb in the samples collected from the study sites are similar. For Co, residual fraction (>60 %) is found dominant. The remaining Co is almost evenly distributed in the other fractions. For Zn, the highest content is in the residual fraction, with nearly comparable percentages in the carbonate and the Fe–Mn oxide fractions. Some authors have proposed that Zn due to its strong bonding with the Fe oxide occurs mainly in the reducible fraction (Fe–Mn oxide) (Carrol et al. 1998; O'Day et al. 1998; Harrington et al. 1998). Further, high stability constant for Zn when combined with Fe–Mn oxides in sediments (Gonzalez et al. 1994) favours its occurrence in the residual and Fe–Mn oxide fractions. For Cr, the residual fraction is always dominant, with the proportion decreasing from 75 % at the inner creek end (TC1) to 60 % near the creek mouth (TC4).

From the results it is evident that different forms of the metals are bound with different strengths. In the three cores, among the metals analysed, the element which showed the highest average percentages in the F1 phase is Pb in core TC1 (8.66 %), core TC5 (14.47 %) and core TC4 (18.21 %). The F2 fraction was highest for Mn in core TC1 (25.19 %), core TC5 (26.33 %) and core TC4 (26.96 %). Higher percentages of Pb and Mn in the exchangeable and carbonate fractions suggest that the metals in the creek sediments are readily more available for exchange and/or release into the creek environment. Singh et al. (2005) and Jain (2004) reported that significant parts of metals introduced by human activity are present in the exchangeable and the carbonate fractions. With respect to the reducible Fe–Mn-associated phase (F3), core TC1 (36.30 %), core TC5 (41.85 %) and core TC4 (38.10 %) show highest amounts of Mn. In the organic fraction (F4), the highest percentage is seen for Cr in cores TC1 (14.46 %) and TC4 (15.98 %), while core TC5 exhibits the presence of Pb (10.42 %). Organic matter plays an important part in controlling the physicochemical behaviours of metals at solid–water interface (Davis 1984; Stumm 1992; Schmitt et al. 2002), with significant influence on the bioavailability, reactivity and mobility of metals in sediments. Metals bound to the reducible phase (Fe–Mn oxides) and organic matter and sulphides are better held by a scavenging effect (i.e. this fraction acts as a sink of metals). Calmano et al. (1993) reported a high proportion of metals strongly bound to the organic–sulphide fraction. Cu (80.05 %) in core TC1 and Fe in core TC5 (93.59 %) and core TC4 (93.12 %) display highest percentage in the residual fraction. Metals associated with residual fractions, being part of the mineral's crystalline structure, remain relatively stable and inert and hence are not easily released into the mobile and bioavailable phases (Tessier et al. 1979; Wong et al. 2007).

The spatio-temporal metal variation results show that at all the sites, the residual fraction is found to be dominant fraction, with contribution of the other fractions to the total concentration only minimal, except for Mn. Therefore, the relative content of the metal in the residual phase can be used as a measure of the contribution of natural sources, and also of the degree of contamination, with a higher percentage indicative of low pollution levels (Singh et al. 2005). For Fe, over 90 % contents are found in the residual fraction. As Fe concentration in the mobile fraction is very low, the sediment may be unpolluted with regard to this metal. The metal distributions in the non-residual phases (F1+F2+F3+F4) are found to vary with the element analysed and the sampling site. The order of mobility (from most to least bioavailable) in metals extracted in the most labile fractions (F1+F2+F3+F4) is Mn (75.05 %) > Zn (59.77 %) > Pb (37.05 %) > Co (29.45 %) > Fe (22.77) > Cr (22.60 %) > Cu (19.95) for core TC1 and Mn (74.43 %) > Pb (43.64 %) > Co (42.78 %) > Zn (31.38 %) > Cu (19.01 %) > Cr (13.01 %) > Fe (6.41 %) for core TC5, while in core TC4, the trend seen is Mn (77.02 %) > Cr (39.40 %) > Pb (38.42 %) > Zn (33.04 %) > Co (28.18 %) > Cu (10.65 %) > Fe (6.88 %). Mn and Zn are present in greater percentage (>50 %) in the labile fractions in samples collected near the creek mouth (TC1) as compared to the other two regions (TC5 and TC4) wherein only Mn is present in higher amounts in the labile fractions. Core TC1 being closer to the industrialised areas is more prone to contaminant input. In general, Mn is found to be the most bioavailable element in the creek region, while Fe is observed to be the least bioavailable element.

Since the chemical form and concentration of the metal govern its bioavailability and toxicity in sediments (Kwon et al. 2001), the high EF values along with higher labile fractions of trace metals can be potential sources for their mobility and bioavailability in the aquatic ecosystems (Harikumar and Jisha 2010). In the creek region, high EF values along with greater percentage of liable fractions are seen only for Mn as compared to the other metals. Hence, if the environmental conditions change, Mn may enter the water column. On the other hand, a major portion of the remaining trace metals studied (50–80 %) are associated with mineral lattices and so are essentially unavailable and least expected to be released into solution under the condition normally encountered in natural waters. Although variations are seen in the metal concentrations, the study indicated no significant deterioration of the sediment quality based on heavy metal pollution in the creek, in spite of the rapid industrial growth in the region over the last few decades.

1.4 Conclusions

Sediment cores analysed, from a creek in Mumbai, showed higher metal concentration near the creek head with decreasing trend towards the creek mouth. Organic matter along with Fe–Mn oxides was found to be the main metal carriers in the region. EF and Igeo pointed towards pollution with respect to Cu and Pb. Based on speciation, most of the elements analysed were found to be in the residual fraction which clearly indicated that these metals were primarily immobile and had or bore the least bioavailability except for Mn. These findings are of great interest since the sediments studied showed relatively high Mn concentrations which can be a highly toxic element for aquatic organisms and fish.

Acknowledgements The first author expresses her sincere thanks to the Department of Science and Technology-SERB, New Delhi for the award of fast track Research Fellowship (SR/FTP/ES-26/2013).

References

Bhosale U, Sahu KC (1991) Heavy metal pollution around the island city of Bombay, India. Part II: distribution of heavy metals between water, suspended particles and sediments in a polluted aquatic regime. Chem Geol 90:285–305

Biscaye PE (1965) Mineralogy and sedimentation of deep sea clay in the Atlantic Ocean and adjacent sea and oceans. Bull Geol Soc Amer 76:803–832

Blasco F (1975) Mangroves of India (Les Mangroves de'Inde) Institute Francias De Pondichery, Travaux de la section. Scientifique et Technique, Tome XIV. Fascicule 1, p 175

Buchman MF (1999) NOAA screening quick reference tables. NOAA HAZMAT Report 99-1. Coastal Protection and Restoration Division, National Oceanic and Atmospheric Administration, Seattle, p 12

Calmano W, Hong J, Förstner U (1993) Binding and remobilisation of heavy metals in contaminated sediment affected by pH and redox potential. Water Sci Tech 28:223–235

Carrol SA, O'Day PA, Piechowski MP (1998) Rock-water interactions, controlling zinc, cadmium and lead

concentrations in surface waters and sediments. Environ Sci Tech 32:956–965
Comber SDW, Gunn AM (1995) The release of trace metals from sewage sludge dispersed at sea. Environ Tech 16:419–432
Das AK, Chakraborty R, Cervera ML, de la Guardia M (1995) Metal speciation in solid matrices. Talanta 42:1007–1030
Davidson CM, Thomas RP, McVey SE, Perala R, Littlejohn D, Ure AM (1994) Evaluation of sequential extraction procedure for the speciation of heavy metals in sediments. Anal Chim Acta 291:277–286
Davis JA (1984) Complexation of trace metals by adsorbed natural organic matter. Geochim et Cosmochim Acta 48:679–691
Feng H, Han X, Zhang W, Yu L (2004) A preliminary study of heavy metal contamination in Yangtze River intertidal zone due to urbanization. Mar Pollut Bull 49:910–915
Folk RL (1974) Petrology of sedimentary rocks. Hemphill, Austin, p 177
Gonzalez MJ, Ramos L, Hernandez LM (1994) Organochlorine and heavy metal residues in the water/sediment system of the Southeast Regional Park in Madrid, Spain. Int J Environ Analy Chem 57:135–150
Grasshoff K (1999) Methods of seawater analysis. Verlag Chemie, Weinheim, p 205
Harikumar PS, Jisha TS (2010) Distribution pattern of trace metal pollutants in the sediments of an urban wetland in the southwest coast of India. Int J Eng Sci Tech 2(5):840–850
Harrington JM, Laforce MJ, Rember WC, Fendorf SE (1998) Phase association and mobilization of iron and trace elements in Coeur d'Alene lake, Idaho. Environ Sci Tech 32:650–656
Hooda PS (2010) Assessing bioavailability of soil trace elements. In: Hooda PS (ed) Trace elements in soil. Willey-Blackwell, Chichester, pp 229–265
Ip CCM, Xi XD, Zhang G, Wai OWH, Li YS (2007) Trace metal distribution in sediments of the Pearl River Estuary and the surrounding coastal area, South China. Environ Pollut 147(2):311–323
Jain CK (2004) Metal fractionation study on bed sediments of River Yamuna, India. Water Res 38(3):569–578
Jarvis IJ, Jarvis K (1985) Rare earth element geochemistry of standard sediments: a study using inductively coupled plasma spectrometry. Chem Geol 53:335–344
Jha SK, Chavan SB, Pandit GG, Sadasivan S (2003) Geochronology of Pb and Hg pollution in a coastal marine environment using global fallout 137-Cs. J Environ Radioact 69:145–157
Kersten M, Forstner U (1989) Speciation of trace elements in sediments. In Batley G (ed) Trace element speciation: analytical methods and problems. Boca Raton, CRC Press, pp 245–317
Krishna AK, Govil PK (2005) Heavy metal distribution and contamination in soils of Thane-Belapur industrial development area, Mumbai, Western India. Environ Geol 47(8):1054–1061
Kwon YT, Lee CW, Ahn BY (2001) Sedimentation pattern and sediments bioavailability in a wastewater discharging area by sequential metal analysis. Microchem J 68(2–3):135–141
Li XD, Shen ZG, Wai OWH, Li YS (2000) Chemical partitioning of heavy metal contaminants in sediments of the Pearl River Estuary. Chem Speciat Bioavail 12(1):17–25
Li XD, Shen ZG, Wai OWH, Li YS (2001) Chemical forms of Pb, Zn and Cu in the sediment profile of the Pearl River Estuary. Mar Pollut Bull 42(3):215–223
Lima MC, Giacomelli MBO, Stupp V, Roberge FD, Barrera PB (2001) Speciation analysis of copper and lead in Tobacco River sediments using the Tessier sequential extraction procedure. Quim Nova 24:734–742
Lin JG, Chen SY (1998) The relationship between adsorption of heavy metal and organic matter in river sediments. Environ Int 24:345–352
Long ER, Field LJ, MacDonald DD (1998) Predicting toxicity in marine sediments with numerical sediment quality guidelines. Environ Toxicol Chem 17(4):714–727
MacDonald DD, Carr RS, Calder FD, Long ER (1996) Development and evaluation of sediment quality guidelines for Florida coastal waters. Ecotoxicology 5:253–278
Muller G (1979) Schwermetalle in den sedimenten des Rheins – Veranderungen seit (1971). Umschau 79:778–783
Muwanga A (1997) Environmental impacts of copper mining at Kilembe, Uganda: a geochemical investigation of heavy metal pollution of drainage waters, stream, sediments and soils in the Kilembe valley in relation to mine waste disposal. Ph.D. dissertation. Universitat Braunschweig, Germany
NIO (1998) Flora and fauna in MbBT area. National Institute of Oceanography, Mumbai, p 412
O'Day PA, Carrol SA, Waychunas GA (1998) Rock-water interactions, controlling zinc, cadmium and lead concentrations in surface waters and sediments, UST ri-state mining district. 1. Molecular identification using X-ray absorption spectroscopy. Environ Sci Tech 32:943–955
Patel B, Mulay CD, Ganguly AK (1975) Radioecology of Bombay harbour- a tidal estuary. Estuar Coast Mar Sci 3:13–42
Patel B, Bangera VS, Patel S, Balaoi MC (1985) Heavy metals in Bombay harbour area. Mar Pollut Bull 16:22–28
Pedersen F, Bjorestad E, Anderson HV, Kjolholt J, Poll C (1998) Characterization of sediments from Copenhagen harbour by use of biotests. Water Sci Tech 37(6–7):233–240

Quadros G, Sukumaran S, Athalye RP (2009) Impact of the changing ecology on intertidal polychaetes in an anthropogenically stressed tropical creek, India. Aqua Ecol 43:977–985

Ram A, Rokade MA, Zingde MD (1998) Mercury in sediments of Ulhas estuary. In: International conference on environmental science (abstract). Regional research laboratory, Trivandrum, p 52

Ram A, Rokade MA, Borole DV, Zingde MD (2003) Mercury in sediments of Ulhas estuary. Mar Pollut Bull 46:846–857

Ram A, Borole DV, Rokade MA, Zingde MD (2009) Diagenesis and bioavailability of mercury in the contaminated sediments of Ulhas estuary, India. Mar Pollut Bull 58:1685–1693

Rao VP, Rao BR (1995) Provenance and distribution of clay minerals in the sediments of the western continental shelf and slope of India. Cont Shelf Res 15:1757–1771

Schmitt D, Taylor HE, Aiken GR, Roth DA, Frimmel FH (2002) Influence of natural organic matter on the adsorption of metal ions onto clay minerals. Environ Sci Tech 36:2932–2938

Sharma P, Borole DV, Zingde MD (1994) 210-Pb geochronology and trace element composition of the sediments in the vicinity of Bombay, west coast of India. Mar Chem 47:227–241

Singh KP, Mohan D, Singh VK, Malik A (2005) Studies on distribution and fractionation of heavy metals in Gomti river sediments – a tributary of the Ganges, India. J Hydrol 312:14–27

Soares HMVM, Boaventura RAR, Machado AASC, Esteves da Silva JCG (1999) Sediments as monitors of heavy metal contamination in the Ave river basin (Portugal): multivariate analysis of data. Environ Pollut 105:311–323

Stumm W (1992) Chemistry of the solid-water interface. Processes at the mineral-water and particle-water interface in natural systems. Wiley, New York

Tessier A, Campbell PGC, Bisson M (1979) Sequential extraction procedure for the speciation of particulate metals. Anal Chem 51:844–851

TMC-ES (2000) Environmental status report prepared by the pollution control cell of Thane Municipal Corporation, p 118

Turekian KK, Wedepohl KH (1961) Distribution of the elements in some major units of the earth's crust. Bull Geol Soc Am 72:175–192

Walkley A, Black JA (1934) The determination of organic carbon by rapid titration method. Soil Sci 37:29–38

Wong CSC, Wu SC, Duzgoren-Aydin NS, Aydin A, Wong MH (2007) Trace metals contamination of sediments in an e-waste processing village in China. Environ Pollut 145:434–442

Zhang J, Liu CL (2002) Riverine composition and estuarine geochemistry of particulate metals in China – weathering features, anthropogenic impact and chemical fluxes. Estuar Coast Shelf Sci 54:1051–1070

Zhang L, Xin Y, Feng H, Jing Y, Ouyang T, Xingtian Y, Liang R, Gao C, Chen W (2007) Heavy metal contamination in western Xiamen Bay sediments and its vicinity, China. Mar Pollut Bull 54:974–982

Zingde MD, Desai BN (1981) Mercury in Thane Creek, Bombay harbour. Mar Pollut Bull 12:237–241

2 Bioremediation of Heavy Metals from Saline Water Using Hypersaline Dissimilatory Sulfate-Reducing Bacteria

Savita Kerkar and Kirti Ranjan Das

Abstract

Salt pans are man-made ecosystems which are fed by the tidal influx of seawater through the estuaries. Most heavy metal contaminants from industries and anthropogenic processes dissolve in water and thus gain entry into the sea. Heavy metals are high-density metallic chemicals that are potentially toxic at low concentrations and present a danger to human and environmental health. The removal of these metals by general physical separation techniques is a crucial issue and chemical treatment is not always environmentally friendly. Biological methods provide an alternative to heavy metal remediation. In the present study, hypersaline dissimilatory sulfate-reducing bacteria (SRB) were found to remediate barium, calcium, cadmium, cobalt, copper, iron, magnesium, molybdenum, zinc, mercury, nickel, and lead metals from saline waters. SRB produce H_2S by utilizing sulfate as electron acceptor, which helps in oxidizing organic matter, and reactive H_2S precipitates dissolved heavy metals as their metal sulfides and thus play an important role in detoxifying saline waters. Among the 11 heavy metals found in the adjoining estuarine seawater, 9 metals were detected in the salt pan water of Ribandar, Goa. Fe, Mn, and Pb were observed in dissolved and particulate form, whereas Hg and Sb were absent. In the salt manufacturing process, the brine starts crystallizing the salt and metal concentrations increase by 10^3 fold in brine and 10^4 in salt crystals. SRB precipitate almost 50 % concentrations of the dissolved metals (from the overlying salt pan water) as their metal sulfides, which gradually get deposited in the underlying salt pan sediments. Hypersaline SRB show optimal sulfate-reducing activity from 80 to 115 psu and are thus potential bioremediators in salt pan ecosystems and in turn have an application in detoxifying industrial effluents containing

S. Kerkar (✉) • K. Ranjan Das
Department of Biotechnology, Goa University,
Goa 403206, India
e-mail: savitakerkar@gmail.com

M.M. Naik, S.K. Dubey (eds.), *Marine Pollution and Microbial Remediation*,
DOI 10.1007/978-981-10-1044-6_2

heavy metals. This study assesses the role of hypersaline SRB strains isolated from salt pans in remediating heavy metal containing saline waters.

2.1 Heavy Metal Influx in Saline Waters

Coastal areas are sites of discharge and accumulation of a range of environmental contaminants due to urbanization and industrialization, which include mining, agriculture, and waste disposal, being the main contributors of metal pollution in estuaries and rivers (Tabak et al. 2005; Ross 1994) which elevate metal concentrations in estuaries (Kumar et al. 2010) and salt crystallizer ponds (Pereira et al. 2013).

Metals, at high concentrations, influence the biochemical activity, growth, and morphology of microbes (Pereira et al. 2012) even at low (5–10 ppm) concentrations. Metals form complexes and combine with inorganic or organic metals and get accumulated in the sediments. Microorganisms use various strategies, like complex formation and extracellular precipitation, reducing metal transport across cell membranes or impermeability (Atlas and Bartha 1997). In some studies, microorganisms have been utilized to remove metal contamination from wastewaters, to separate metals from sediments and soil or to foster metal solubilization for extraction (Lovley and Coates 1997).

The Mandovi estuary of Goa faces a threat of anthropogenic pollution. Consequently the salt pans fed by the estuary would get affected. In the salt pans, metals get concentrated along with the formation of brine. Organisms like Bacteria, Archaea, and Eukarya are known to inhabit and influence the salt pan water and thus the quality of the salt produced. Solar salt obtained from the evaporation of seawater has 86 % sodium chloride and 14 % other trace minerals, viz., SO_4, Mg, Ca, K, HCO_3, Br, Sr, and F (Kerkar and Fernandes 2013).

The Ribandar solar salt pans are fed by the Mandovi estuary and in turn are vulnerable to metal effluent influx from ferromanganese ore mining activity, barge traffic, and sewage disposal, affecting the water and sediment quality in the salt pan and its inhabiting organisms. Solar salt pan is a niche for extremophiles thriving on temperature variation, oxygen availability, solar radiations, pH, nutrient concentration, salinity variation, and water activity. Besides other functions, these extremophilic microbes play a key role in modulating the metal concentrations in the salt produced.

White et al. (1998) have reported comfortable leaching of metal contaminants, viz., Zn, Ni, Mn, Cr, Co, and Cd, from artificially contaminated soil. In wastewater treatment, physiochemical methods, viz, chemical precipitation, carbon absorption, ion exchange, and electrochemistry, are generally used, but still have some disadvantages. When higher concentrations of heavy metals (1–100 mg/l) are present, proportionately the treatment and material cost increases. In some treatment processes, poor selectivity was observed for competitive metal absorption. Biological treatment has an advantage over traditional chemical treatment due to low operational cost, steady effect, and a smooth recovery of the desired metals (Wang et al. 2001; Rehman and Shakoori 2001).

Biosorption and bioaccumulation of metals by microorganisms are probably one of nature's safeguards for reducing metal ion toxicity in the surrounding microbial niche. Potential applications of these phenomena, however, are governed by certain criteria or characteristics of the biosorbent. These include metal affinity, rate of metal uptake, selectivity, temperature tolerance, versatility, and robustness (Eccles 1995).

Bacterial resistance to metals may be due to properties like metal precipitation, metal detoxification, absorption, or accumulation. Bio-absorption involves the cell surface with a complex formation between functional groups like phosphoryl, carbonyl, and hydroxyl present on the cell surface with metal ions. Bioaccumulation involves transport systems and depends on active metabolism of the cell (Mohan and Pittman 2007).

In the Ribandar salt pans of Goa, due to continuous exposure of heavy metals, there is an emergence of metal-tolerant bacterial strains. It was seen that these tolerant bacteria employed specific and multiple mechanisms for detoxification of metals. These metals were thus removed from the overlying water of the salt pans and were found to accumulate in the sediment (Pereira and Kerkar 2014).

The average ranges of metal concentrations recorded in the Ribandar salt pan water, salt crystals, and sediment were listed below in Table 2.1 which shows the concentration ranges of toxic heavy metals such as cadmium, zinc, and lead were well within the permissible limits of 0.001–0.05 ppm, 0.005–5 ppm, and 2–20 ppm in water and 0.03–0.3 ppm, 50–300 ppm, and 2–20 ppm in the sediment, respectively (RSMENR 2002).

Our previous assessment of the concentration of metals in the Ribandar salt pan sediment for all seasons revealed that the metal concentration increases by 52 % during salt harvesting season. Attri and Kerkar (2011) reported the metal concentrations in the Mandovi estuary are 18.3 ± 1.9 % Fe (Attri et al. 2011), 0.19 ± 0.002 % Mn, 36.2 ± 4.2 ppm Co, and 102.3 ± 9.8 ppm Zn.

Attempts were made to precipitate high levels of Co using hypersaline SRB which revealed 90 % of Co was precipitated (where 3 % was by the SRB cells and 87 % by SRA) and hence established the role of SRB in bioremediating Co. It was observed that SRB cells could bioadsorb cobalt minimally; however the actively growing SRB cells actually mediate the bioremediation of Co by utilizing the sulfide metabolized to precipitate Co as cobalt sulfide in the medium.

2.2 Sulfate-Reducing Bacteria

SRB are members of delta subdivision of Proteobacteria. They are strict anaerobes and their permanent habitats being estuarine, marine and salt marsh sediments, saline and hypersaline ponds, and lakes. Due to the high and almost inexhaustible supply of sulfate, SRB are able to produce sulfide in high concentrations which

Table 2.1 Metal concentration in salt pan water, salt crystals, and salt pan sediment

Metals	Dissolved form in salt pan water (in ppm)	Particulate form in salt pan water $\times 10^3$ (in ppm)	Metal concentration in salt crystals $\times 10^4$ (in ppm)	Metal concentration in sediment $\times 10^4$ (in ppm)
Li	–	–	–	0.0044
Mg	1.109	4	1.400	1.3190
Al	0.062	0.359	4.400	3.5245
Ti	0.010	1.0	0.620	0.51
Sc	–	0.001	–	–
V	0.047	0.121	0.013	–
Cr	0.021	0.009	0.018	0.0089
Mn	0.255	0.882	0.064	0.0258
Fe	0.173	2.749	5.900	3.037
Co	–	0.024	0.003	0.002
Ni	0.005	0.041	0.007	0.005
Cu	0.058	0.021	0.005	0.003
Zn	0.136	0.030	0.067	0.004
Sr	0.046	0.120	0.013	0.027
Cd	0.002	–	–	–
Ba	–	–	0.095	0.005
Hg	–	–	–	–
Pb	0.116	0.294	0.002	6.750

precipitate most of the metals present in ionic form into their corresponding metal sulfides. The role of sulfate-reducing bacteria in coastal marine sediments amounts to almost 50 % of the organic material degradation (Jorgensen 1982), and also their involvement in anaerobic turnover of certain metals makes them important as metal contamination detoxifiers.

SRB comprise of anaerobic bacteria that use sulfate as their terminal electron acceptor forming a mixed group which is morphologically and nutritionally diverse. SRB oxidize a range of compounds including fatty acids, organic acids, alcohols, and H_2 as an electron donor and carbon sources. A symbolic expression of SRB metabolism characterized by production of a strong reducing agent, hydrogen sulfide, is able to inhibit growth of other microorganisms present in its environment (Gibson and Suflita 1990). The SRB perform dissimilatory and assimilatory sulfate reduction, with the dissimilatory process far exceeding the assimilatory reduction.

The genera of SRB are generally defined in terms of their morphology rather than physiology. The majority of SRB are reported to stain gram-negative with *Desulfovibrio* being the most encountered genus and *Desulfotomaculum* being the sole gram-positive genus. However gram-staining behavior of SRB is diagnostically unreliable (Boopathy et al. 1998a; Zehnder 1988). Although morphologically diverse, SRB are considered to be physiologically unified. Currently 18 dissimilatory SRB genera are placed into two physiological subgroups. The first group contains *Desulfonema*, *Desulfovibrio*, *Desulfobulbus*, and *Desulfotomaculum* genera as they can utilize ethanol, lactate, and pyruvate or certain fatty acids as carbon and energy sources. The second group contains genera that are specialized in oxidation of acetate and fatty acids such as *Desulfobacter*, *Desulfococcus*, and *Desulfosarcina* (Madigan et al. 1997).

The reduction of sulfate to hydrogen sulfide is an eight-electron reduction reaction:

$$SO_4^{2-} + 8e^- + 8H^+ \langle\!=\!=\!=\rangle$$
$$H_2S + 2H_2O + 2OH^-$$

The reaction proceeds through a number of intermediate stages. The stable sulfate ion is first activated by the enzyme adenosine triphosphate (ATP) sulfurylase to give adenosine phosphosulfate (APS). In dissimilatory sulfate reduction, the sulfate in adenosine phosphosulfate is then reduced to sulfite releasing adenosine monophosphate (AMP). In assimilatory reduction another phosphate molecule adds to APS to form phosphoadenosine phosphosulfate (PAPS), after which the sulfate is reduced. Sulfite is the first product of sulfate reduction in both cases (Madigan et al. 1997). SRB are associated with the systems that are characterized from fouling problems and a pungent smell of H_2S. *Desulfotomaculum nigrificans* and *Desulfovibrio desulfuricans* are the two widely distributed, most common SRB species found in the anaerobic environment. Though SRB favor anaerobic conditions, they still can grow in oxygenated environments, even in slimy deposits where aerobic conditions persist. They also establish themselves in the well water beneath the aerobes like IRB (iron-reducing bacteria) which form biofilms on the surface and use up oxygen, while SRB thrive in the anaerobic condition below the IRB biofilm. The presence of SRB is detected by visualizing yellowish or reddish nodules on metal surfaces and exhibits a black color due to the production of iron sulfide when nodules are broken open. A bright metallic pit on the metal forms on the removal of nodules and releases H_2S when hydrochloric acid is added to it which is characterized by a rotten egg smell.

SRB reduce sulfate via dissimilatory pathway to obtain its energy. It can grow by utilizing miniscule amounts of grease and oil as a nutrient source. Low flow or stagnant water favors its chances of growing. SRB are also considered as biocorrosion agents, as the produced hydrogenase enzyme enables them to use elemental

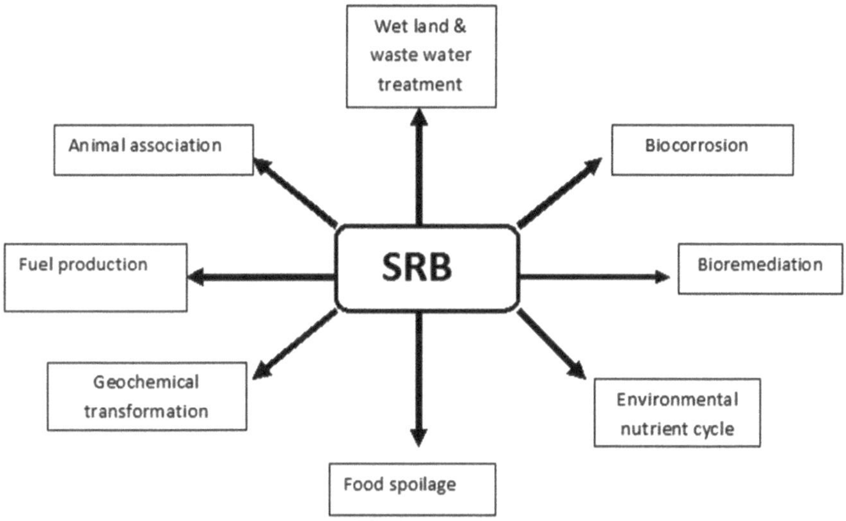

Fig. 2.1 Role of SRB

hydrogen to reduce sulfate and generate H_2S which triggers biocorrosion. Hence iron corrosion through such biological processes occurs very rapidly as compared to normal iron rusting (Fig. 2.1).

2.3 Bioremediation of Heavy Metals

Bioremediation is an ancient technology, dates back to 6000 BC, as evident from compost pile and kitchen middens (NABIR primer 2003), and demonstrates an ancient bioremediation practice by human beings. Bioremediation techniques utilize microbes to remove or convert toxic contaminants present in the environments like water, sediments, soil, and air to a less toxic form. The sewage treatment plant in Sussex, UK, in 1891 (considered to be the first biological treatment system), demonstrates bioremediation process application that started more than 100 years ago (NABIR Primer 2003). The wastewater treatment, using microorganisms to remove heavy metals, is one of the most investigated research areas in the current scenario (Leusch et al. 1995; Kaewsarn 2002; Wu et al. 1996; Xu et al. 2005; Adeniji 2004). There are various strategies being adopted by microorganisms in order to counteract the presence of heavy metals in its surrounding to facilitate bioremediation:

(a) *Biomineralization*: Biomineralization is a process of precipitation of insoluble metal with the interaction of metabolic products of microbes. Such processes result in mineral formation and geochemical depositions. Biomineralization adds value to metal microbe interaction research (Ehrlich 1999; Banfield et al. 2000; White and Gadd 2000).
(b) *Bioaccumulation:* Bioaccumulation is a metal uptake process which requires external energy to enter into the cells and get accumulated. Some physiologically essential metal ions, toxic metals, and radionuclides have been reported to enter into the cell using the energy transport system. For example, K^+ ion uptake is linked with H^+ ion bound to the plasma membrane and ATPase through membrane potential. Such processes get affected by the factors that inhibit energy metabolism of cells. As explained by White and Gadd (1987), absence of substrates, anaerobiosis, low incubation temperature, and respiratory inhibitors like cyanide could affect this metabolism.
(c) *Biosorption*: It is a widely used approach to bioremediate metals and radionuclides, involving passive sequestration of metals by

their interaction with living or dead biological entities. It is effectively used in wastewater treatment (Schiewer and Volesky 2000; Jang et al. 2001).

(d) *Biotransformation*: Microbes can transform toxic metals to less toxic forms by catalyzing them to more volatile or less soluble form. For example, metal precipitation achieved by the microbial reduction of Cr(VI) to Cr(III), Se(VI) to Se(0), V(V) to V(III), Au(III) to Au(0), Pd(II) to Pd(0), and U(VI) to U(IV) has been reviewed by Lloyd (2003).

(e) *Microbially enhanced chemisorption of heavy metals (MECHM)*: This is a class of microbial cell-mediated reactions, which involve the precipitation of metal biomineral called "priming deposit" acting as a nucleation center on which subsequently targeted heavy metals get deposited and promote a targeted metal precipitation reaction (Macaskie et al. 1996). Generally priming deposit is initiated by sulfide or phosphate biomineralization route. SRB adapt such a strategy when Fe is present on a precipitant metal to H_2S produced and then FeS acts as primary deposit and as an adsorbent for target metals (Ellwood et al. 1992; Watson and Ellwood 1994, 1988).

When metal biosorption strategies are not feasible, then a consortium of metal-resistant cells can ensure better remediation, combining biosorption, bioprecipitation, and bioaccumulation. This approach can be used simultaneously in the removal of toxic metals and organic and inorganic pollutants from water. But these living cells have some constraints like sensitivity to extreme pH, high metal concentration, and metabolic energy production requirements limiting their use in bioremediation. Therefore efforts are being made to meet such challenges by metal-resistant microbial strains.

2.4 SRB as Bioremediators of Heavy Metals

The use of microorganisms in metal polluted water remediation was successfully used for the reduction and precipitation of soluble metal sulfates as insoluble sulfides in liquid wastes by using SRB, as they use sulfate as their electron acceptor to produce H_2S, binding with metals to give their metal sulfide. Nowadays this procedure is effectively used for surface water treatment, underground water cleaning up, and even in commercial treatment plants (Fig. 2.2).

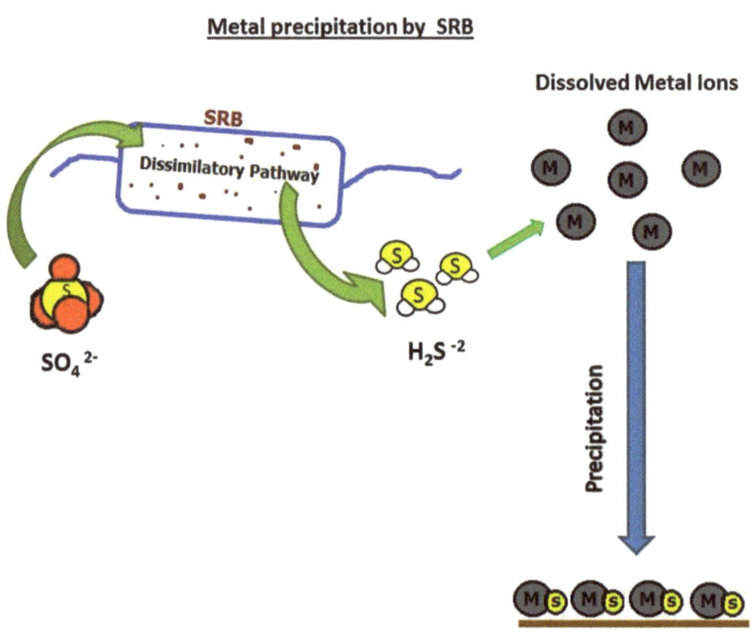

Fig. 2.2 Mechanism of metal precipitation by SRB

Microbiologically produced H_2S is reported to be an effective way to immobilize metals like iron, cadmium, nickel, lead, copper, and zinc in bioprecipitating metals as their insoluble sulfides, e.g., FeS, CdS, NiS, CoS, ZnS, and CuS (White et al. 1998; White and Gadd 2000; Labrenz et al. 2000; Wang et al. 2001; Drzyzga et al. 2002; Valls and Lorenzo 2002; Utgikar et al. 2002; White et al. 2003; Krumholz et al. 2003). Precipitating toxic concentrations of metals as metal sulfides adds value to anoxic wetland and sediment bioremediation (White et al. 1998; Kaksonen et al. 2003; Labrenz and Banfield 2004). Jalali and Baldwin (2000) were able to grow SRB in a solution up to 150 mg L^{-1} of copper and remove copper to levels below 0.1 mg/L. Mining and mineral processing of polymetallic ore in Vromos bay area near the Black Sea coast of southeastern Bulgaria have resulted in the contamination of the surrounding agricultural land with Th, Ra, and U radioactive elements. Other toxic heavy metals (e.g. Co, Cd, and Pb) are also present. Laboratory experiments demonstrated efficient treatment of the soils by using in situ treatment method, where acidified water was used to solubilize the metals, and the SRB in turn immobilized the metals. Real field application of this process gave promising results (Groudev et al. 2001). The sulfate-reducing bacterium, *Desulfovibrio desulfuricans*, was found to be capable in coupling the oxidation of an organic compound with an enzymatic reduction of uranium (VI) to uranium (IV), which precipitated out of solution from groundwater contaminated with uranium (VI) as uraninite (Abdelouas et al. 1999, 2000). SRB are a physiologically important group of microbes that are used extensively in metal bioremediation of water and soil with their produced sulfides to remediate toxic metal contaminants in soil.

Many sulfate-reducing microorganisms can also reduce ferric ions and may preferentially reduce Fe(III) at low electron donor concentrations, typically in the sediments (Coleman et al. 1993; Lovley et al. 1993). Analysis of microbial lipids suggested that sulfate-reducing bacteria belonging to genus *Desulfovibrio* were enriched in zones of ferric reduction in salt pan sediments, suggesting that they might be involved in ferric ion reduction in these environments (Coleman et al. 1993). The biosulfide process described by Rowley et al. (1994) involved the separation of chemical precipitation of sulfide metals from the biological conversion of sulfate to sulfide. The advantage of this process is that the SRB biomass is not exposed to the fluctuating conditions of the wastewater effluent, which means bacterial sensitivity to toxic compounds is eliminated.

2.5 Hypersaline SRB and Metal Bioremediation

Hypersaline environments are those which contain high concentration of salt. These include mainly inland lakes (the Dead Sea, Great Salt Lake, etc.), soda brines, deep sea brines, polynias, and marine salt pans. SRB are omnipresent and are hyperactive in ecosystems with high sulfate concentrations and are considered to be one of the oldest forms of bacterial life on earth. Hypersaline environments support the growth of salt-loving organisms and these halophilic organisms can tolerate such environments which limit the growth of other organisms. Halophilic organisms produce a variety of stable and unique biomolecules during their growth that may be useful in various biotechnological applications (Kerkar 2004). Literature reveals that halophiles possess excellent metal-scavenging capability. Metal-tolerant strains of bacteria from hypersaline niches were found to tolerate various metals at higher concentrations and can be considered as a potential candidate for metal removal from wastes. In hypersaline environments like solar lakes and salt pans, SRB communities carry out extremely active sulfate reduction with a temporal, spatial, and functional separation of activities among the species present. SRB maintain an osmotically isotonic cytoplasm to cope up with the outside medium osmolarity, as their survival strategy in order to thrive in hypersaline environments (Kerkar 2004). We have detected Desulfobacteraceae at salt concentrations of 475 g/l which indicates the existence of a unique oxidizing SRB with an inorganic osmoadaptation strategy or the presence of spatial microniches of lower salt concentrations in the salt pan. Harithsa et al. (2002) assessed the tolerance of $HgCl_2$ and $Pb(NO_3)_2$ salts using three mesohaline SRB (HSR1, HSR4, HSR14) at 95 psu with concentrations ranging from 50,100 to 500 µg/ml. Growth and SRA were the assessed parameters. Growth of HSR1 strain was 80 % at 100–200 µg/ml of mercury, while sulfate-reducing activity decreased to 60 % as compared to the control. HSR14 could grow normally at 200 µg/ml of $HgCl_2$, but SRA was inhibited by 60 %. In the presence of 500 µg/ml of $Pb(NO_3)_2$, HSR4 growth was stimulated by 160 % and SRA by 170 % as compared to the control. There is a possibility that some hypersaline SRB strains can tolerate heavy metals more efficiently than their mesohaline counterparts and thus can be a better candidate for metal bioremediation (Kerkar and Lokabharathi 2007, 2011).

2.6 Measurement of SRA and Its Variation with Salinity

In hypersaline anoxic sediments (Oren 1999; Skyring 1987; Ollivier et al 1994) and hypersaline microbial mats (Canfield and Des Marais 1993; Caumette et al 1994), sulfate reduction governed by bacteria is of great ecological and biochemical importance. Changes in biological lability and the amount of organic matter that undergoes decomposition show a large variation in the sulfate reduction rate in marine sediments (Goldhaber and Kaplan 1974). In the marine coastal ecosystem, SRA contributes almost 50 % of organic carbon turnover in the sediments, while the total sediment respiration rate is estimated to be 2.5–5.5 g cm^{-2} day^{-1}. Therefore SRB play an important role in hypersaline ecosystems like solar salt pans of Goa.

To ascertain the sulfate reduction rate in this complex system, multiple methods were carried out to determine an integrated and comparative estimate of SRA. Four different methods were followed:

1. Spectroscopic method for fatty acid-amended sediment
2. Radio isotope ^{35}S method
3. Monthly measurement of increase in natural sulfide content
4. Monthly assessment of decrease in natural sulfate content

All the estimations were carried out during the peak salt manufacturing seasons from 0 to 2, 2 to 5, and 5 to 10 cm depth of sediment cores and the values were expressed as ng g^{-1} h^{-1}.

From fatty acid-amended sediment method, SRA at three depths were 2929, 1379, and 1342 ng g^{-1} h^{-1}, while ^{35}S method at 85 psu was found to be 3,713,721 and 200 ng g^{-1} h^{-1} but 330 psu. SRA varied from 108, 25, and 13 ng g^{-1} h^{-1}, respectively, along the depth. SRA based on increase in natural sulfide was found to be 65, 9, and 33 ng g^{-1} h^{-1}, while natural decrease in sulfate was measured to be 6.2, 0.1, and 3.5 ng g^{-1} h^{-1}, respectively, at the three depths. Results showed a trend of depth-wise decrease in SRA measured by all four methods and higher values were obtained for surficial sediment at 0–2 cm depth (Fig. 2.3).

SRA measurements by fatty acid amended and ^{35}S methods were comparable, while values at 330 psu were very low. Thus 330 psu sediment was amended with 10 mM of fatty acid cocktail which helped in measuring SRA by ^{35}S within 24 h. To overcome auto-oxidation of sulfide and other artifacts due to long incubation time, short-period incubation is generally recommended. SRA measured by fatty acid-amended technique (3190 nM cm^{-3} day^{-1}) are comparable to the tracer technique (2050 nM cm^{-3} day^{-1}) which was 1.3 times higher in SRA measured by ^{35}S method (85 psu) at surficial sediments, which were four orders higher from the values given by Li et al. (1999) of 0.5 nM ml^{-1} day^{-1}.

The rise in sulfide concentration revealed an SRA of 65 ng g^{-1} h^{-1}, which was two orders lower than the values of above two methods, probably because of the generated sulfides that rapidly get oxidized in the system chemically or biologically in their natural conditions. Moreover H_2S easily escapes out of the system. SRA based on decrease

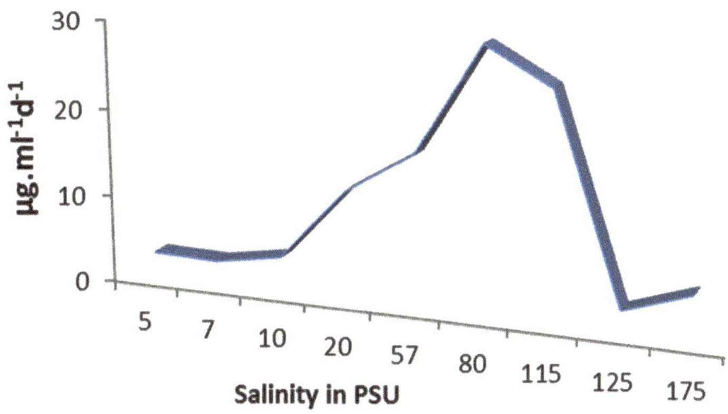

Fig. 2.3 Effect of salinity on sulfate reduction activity

in sulfate concentration level yielded a very low value of 6.2 ng g^{-1} h^{-1}, which is one order lower than the sulfide increase value. Natural sulfide and sulfate concentration analyses showed a 100-fold difference between the sulfide formed and the increase in sulfate concentration. Such a variation in a chemically stable compound could be explained through biological oxidation in the natural system and is much faster than biological reduction. Salinity plays a key role in the process as 85 psu salinity yielded higher SRA than 330 psu. Thus we measured SRA on different salinity gradients to obtain an optimal value for SRA. It was found that salinity enhances SRA as presented in Fig. 2.4. Increase in SRA was observed from 20 to 115 psu. Maximum activity was observed at 80 psu; however the activity drastically decreased at 125 psu. Highest values of 29.96 and 25.87 μg $ml^{-1}day^{-1}$ were obtained for 80 and 115 psu suggesting it to be the optimal salinity to foster higher SRA. We observed the sulfate reduction rate is somewhat related to the optimal salinity range of 80–115 psu as the values are significantly higher (Fig. 2.3).

From the obtained data, it could be a promising alternative to use a consortium of hypersaline SRB from Ribandar salt pan in heavy metal remediation as they resist and precipitate more heavy metals than its mesohaline counterpart, thus enabling them as a better candidate for saline wastewater treatment.

2.7 Effect of Metals on SRA

To access the impact of certain metals influencing activity of hypersaline SRB, lead (Pb) at concentrations of 0, 100–500 psu, was used at incubation periods of 7 and 14 days with different carbon sources (individually) like formate, acetate, lactate, butyrate, etc. Formate was found to stimulate the growth and respiration after 14 days. Similarly when selected strains were grown with mercury (Hg) at 0–500 ppm. it was found that the activity was inhibited. However formate and lactate stimulated the activity up to 100 ppm of Hg and SRA was observed up to 400 ppm of Hg. These experiments indicated that different metals influence the SRA, whereas the presence of low concentrations of specific carbon sources with the respective metal stimulates the SRA. Therefore addition or the presence of certain carbon sources in an environment contaminated with metals stimulates the SRA and thus could be used to enhance the rate of metal bioremediation, i.e., in the formation of metal sulfides, and to increase the level of tolerance to higher concentrations of the metal. The higher the SRA, the higher the production of

Fig. 2.4 Metal remediation from industrial effluent

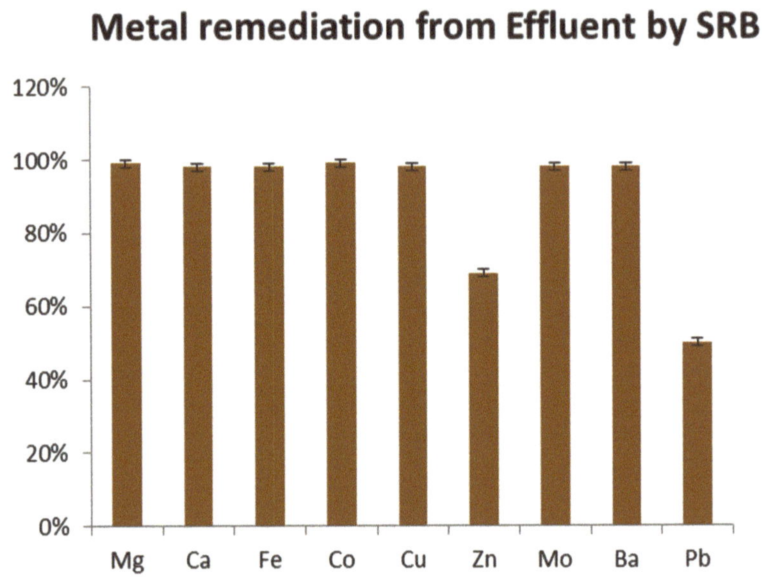

H_2S which ultimately acts upon the heavy metals and precipitates them as metal sulfide, thus increasing bioremediation efficiency.

2.8 Bioremediation of Industrial Effluent with SRB

Organisms in hypersaline ecosystems can tolerate high metal concentrations in their surrounding and adapt various strategies for survival in such extreme conditions. Thus SRB from such environments are considered to be ideal in mitigating problems with environments having high metal contaminations. This is due to their adaptability to extreme saline condition and metal precipitation capabilities. Researchers worldwide have surveyed the heavy metal contents of the sediments from rivers, salt pans, bays, lagoons, and harbors and mostly detected sulfur in the deposits of heavy metals and attributed it to the role of SRB. H_2S gets discharged into the environment as the final product of sulfate respiration via SRB which are responsible for precipitation of metal ions as less soluble metal sulfides. The H_2S production is generally considered to be the main reaction that governs the metal sulfide precipitation (Amacher et al. 1993).

Modified Hatchikians media (1972) of 100 psu were prepared by dissolving NaCl in seawater to isolate hypersaline SRB from the Ribandar salt pans. SRB tolerating higher concentration of Pb, Hg, Ni, and Co were used in bioremediation resulting in almost 100 % heavy metal precipitation from industrial effluents with a salinity ranging from 8 to 90 psu and containing various metals.

Bioprecipitation of nine metals (viz., Mg, Ca, Fe, Co, Cu, Zn, Mo, Ba, Pb) using mix consortia (three strains) of potential SRB out of 15 different metals in varying concentration (listed in Table 2.2) from an industrial effluent was achieved over an incubation period of 17 days. Experimental results show that the consortium was capable of precipitating out nine metals at 50–99 % concentrations.

Tolerance of SRB to environment stress factors like salinity and temperature in the salt pans and their capability of reducing toxic metals at neutral to alkaline pH has an added advantage over traditional metal remediation techniques requiring acidic condition for metal removal. This process of using hypersaline SRB for bioremediation of metals is clearly a more attractive option as the anaerobic waste treatment systems are advantageous as they do not require oxygen and mixing as the aerobic counterparts.

Table 2.2 Different metal concentrations in industrial effluent

Metals in effluent	Mg	Al	Ca	Ti	V	Cr	Mn	Fe	Co	Ni	Cu	Zn	Cd	Ba	Pb
Concentration in mg/L)	3.41	2.54	12.53	0.20	22.17	0.13	7.98	33.35	0.06	0.75	11.30	76.20	3.39	0.04	0.06

2.9 Conclusion

Hypersaline SRB develop tolerance and their metabolic activity gets stimulated in the presence of metals in its surrounding environment by virtue of its survival strategy in this natural habitat bearing higher metal concentrations. Since SRA is the main functional process in SRB which controls the bioremediation of heavy metals from saline waters, stimulating the sulfate-reducing activity would enhance the formation of high levels of sulfides which in turn would lead to the rapid precipitation of metal sulfides and thus remediate the heavy metals from the surrounding waters. Saline waters have limited types and low concentrations of carbon sources. Hypersaline sulfate-reducing bacteria have a tremendous potential and application in bioremediating heavy metals from saline waters, and their activity can be enhanced on supplementing with low concentrations of a specific carbon source according to the strain's requirement. Depending upon the type of metal contamination, individual hypersaline strains with specific metal tolerance levels or mixed consortia with multiple metal tolerance and high SRA could be used to bioremediate metal-contaminated saline waters.

References

Abdelouas A, Lutze W, Nuttall HE (1999) Uranium contamination in the subsurface: characterization and remediation. In: Burns PC, Finch R (eds) Uranium: mineralogy, geochemistry and the environment, vol 38, Reviews in mineralogy., pp 433–473

Abdelouas A, Lutze W, Gong W et al (2000) Biological reduction of uranium in groundwater and subsurface soil. Sci Total Environ 250:21–35

Adeniji A (2004) EPA report: bioremediation of arsenic, chromium, lead, and mercury

Amacher MC, Brown RW, Kotuby-Amacher J, Willis A (1993) Adding sodium hydroxide to study metal removal in a stream affected by acid mine drainage. USDA-FS, Research Paper INT- 465

Atlas RM, Bartha R (1997) Microbial ecology: fundamentals and applications. The Benjamin/Cumming Publishing Company, Redwood City

Attri K, Kerkar S (2011) Seasonal assessment of heavy metal pollution in tropical mangrove sediments (Goa, India). J Ecobiotechnol 3:9–15

Attri K, Kerkar S, LokaBharathi PA (2011) Ambient iron concentration regulates the sulfate reducing activity in the mangrove swamps of Divar, Goa, India. Estuar Coast Shelf Sci 95:156–164

Banfield JF, Welch SA, Zhang H, Ebert TT, Penn RL (2000) Crystal growth and microstructural evolution of FeOOH biomineralization products. Science 289:751–754

Boopathy R, Gurgas M, Ullian J, Manning JF (1998) Metabolism of explosive compounds by sulfate reducing bacteria. Curr Microbiol 37:127–131

Canfield DE, Des Marais DJ (1993) Biogeochemical cycles of carbon, sulfur, and free oxygen in a microbial mat. Geochim Cosmochim Acta 57:3971–3984

Caumette P, Matheron R, Raymond K, Relexans JC (1994) Microbial mats in the hypersaline ponds of Mediterranean salterns (Salins-de-Giraud, France). FEMS Microbiol Ecol 13:273–286

Coleman ML, Hedrick BD et al (1993) Reduction of Fe(III) in sediments by sulfate reducing bacteria. Nature 361:436–438

Drzyzga O, El Mamouni R, Agathos SN, Gottschal JC (2002) Dehalogenation of chlorinated ethenes and immobilization of Nickel in anaerobic sediment column under sulfidogenic conditions. Environ Sci Technol 36:2630–2635

Eccles H (1995) Removal of heavy metals from effluent streams – why select a biological process? Int Biodeterior Biodegrad 5:5–16

Ehrlich HL (1999) Microbes as geologic agents: their role in mineral formation. Geomicrobiol J 16:135–153

Ellwood DC, Hill MJ, Watson JHP (1992) Pollution control using microorganisms and magnetic separation. In: Fry JC, Gadd GM, Herbert RA, Jones CW, Watson-Craik IA (eds) Microbial control of pollution. Soc Gen Microbiol, symposium no.48. Cambridge university press, Cambridge, pp 89–112

Gibson SA, Suflita JM (1990) Anaerobic degradation of 1,4,5-trichlorophenoxyacetic acid in samples from a methanogenic aquifer: stimulation by short-chain organic acid and alcohols. Appl Environ Microbiol 53:254–260

Goldhaber MB, Kaplan IR (1974) The sulfur cycle. In: Goldberg ED (ed) The sea, vol 5, Marine Chemistry. Wiley, New York, pp 469–655

Groudev S, Georgiev P, Spasova I, Komnitsas K (2001) Bioremediation of a soil contaminated with radioactive elements. Hydrometallurgy 59:311–318

Harithsa S, Kerkar S, Loka Bharathi PA (2002) Mercury and lead tolerance in hypersaline sulfate-reducing bacteria. Mar Pollut Bull 44:726–732

Hatchikian EC (1972) Mechanism d'oxido-reduction chez les bacteries sulfato-reductrices. Thesis Marseilles

Jalali K, Baldwin SA (2000) The role of sulphate reducing bacteria in copper removal from aqueous sulphate solutions. Water Res 34:797–806

Jang A, Kim SM, Kim SY, Lee SG, Kim IS (2001) Effect of heavy metal (Cu, Pb, and Ni) on the compositions of EPS in biofilms. Water Sci Technol 43(6):59–66

Jorgensen BB (1982) Mineralization of organic matter in the sea bed: the role of sulphate reduction. Nature 296:643–645

Kaewsarn P (2002) Biosorption of copper(II) from aqueous solutions by pre-treated biomass of marine algae Padina sp. Chemosphere 47(10):1081–1085

Kaksonen AH, Riekkola-vanhanen M-L, Puhakka JA (2003) Optimization of metal sulphide precipitation in fluidized-bed treatment of acidic wastewater. Water Res 37:255–266

Kerkar S (2004) Ecology of hypersaline microorganisms. In: Ramaiah N (ed) Marine microbiology facets & opportunities. NIO Goa, India, pp 53–67

Kerkar S, Fernandes MS (2013) A comparative assessment of Goan natural solar salt and its adequacy in iodine content and nutritive value. Int Food Res J 20:2317–2321

Kerkar S, LokaBharathi PA (2007) Stimulation of sulphate reducing activity at salt – saturation in the salterns of Ribandar, Goa. India. Geomicrobiol J 24:101–110

Kerkar S, LokaBharathi PA (2011) G model revisited: seasonal changes in kinetics of sulphate reducing activity in the saltern of Ribandar, Goa, India. Geomicrobiology 28:187–197

Krumholz LR et al (2003) Immobilization of cobalt by sulphate reducing bacteria in subsurface sediments. Geomicrobiol J 20:61–72

Kumar CSR, Joseph MM, Kumar TRG, Renjith KR et al (2010) Spatial variability and contamination of heavy metals in the inter-tidal systems of a tropical environment. Int J Environ Res 4(4):691–700

Labrenz M, Banfield JF (2004) Sulfate-reducing bacteria-dominated biofilms that precipitate ZnS in a subsurface circumneutral-pH mine drainage system. Microb Ecol 47:205–217

Labrenz M, Druschel GK, Ebert TT, Gilbert B et al (2000) Formation of sphalerite (ZnS) deposits in natural biofilms of sulfate-reducing bacteria. Science 290(5497):1744–1747

Leusch A, Holan ZR, Volesky B (1995) Biosorption of heavy metals (Cd, Cu, Ni, Pb, Zn) by chemically-reinforced biomass of marine algae. J Chem Technol Biotechnol 62:279–288

Li JH, Purdy KJ, Takii S, Hayashi H (1999) Seasonal changes in ribosomal RNA of sulphate-reducing bacteria and sulphate reducing activity in a freshwater lake sediment. FEMS Microb Ecol 28:31–39

Lloyd JR (2003) Microbial reduction of metals and radionuclides. FEMS Microbiol Rev 27:411–425

Lovley DR, Coates JD (1997) Bioremediation of metal contamination. Curr Opin Biotechnol 8:285–289

Lovley DR, Roden EE, Phillips EJP, Woodward JC (1993) Enzymatic iron and uranium reduction by sulphate reducing bacteria. Mar Geol 113:41–53

Macaskie LE, Lloyd JR, Thomas RAP, Tolley MR (1996) The use of microorganism for the remediation of solutions contaminated with actinide elements, other nuclides and organic contaminants generated by nuclear fuel cycle activities. Nucl Energy 35:257–271

Madigan MT, Martinko JM, Parker J (eds) (1997) Biology of microorganisms, 8th edn. Prentice Hall Upper Saddle River Press, London

Mohan D, Pittman CU (2007) Arsenic removal from water/wastewater using adsorbents—a critical review. J Hazard Mater 142:1–53

Natural and Accelerated Bioremediation Research (NABIR) (2003) Program, Office of Biological and Environmental Research, Office of Science, U.S. Department of Energy. What is Bioremediation. p 9

Ollivier B, Caumette P, Garcia JL, Mah RA (1994) Anaerobic bacteria from hypersaline environments. Microbiol Rev 58(1):27–38

Oren A (1999) Bioenergetic aspects of halophilism. Microbiol Mol Biol Rev 63(2):334–348

Pereira F, Kerkar S (2014) Metal detoxification in hypersaline environments. In: 4th National Seminar on Pollution in Urban Industrial Environment, pp 29–37

Pereira F, Krishnan KP, Sinha RK, Kerkar S (2012) Insights on metal-microbe interactions in Bacillus sp. and Chromohalobacter sp. from a solar saltern. J Ecobiotechnol 4:14–24

Pereira F, Kerkar S, Krishnan KP (2013) Bacterial response to dynamic metal concentrations in the surface sediments of a solar saltern (Goa, India). Environ Monit Assess 185:3625–3636

Rehman A, Shakoori AR (2001) Heavy metal resistance *chlorella spp.* isolated from tannery effluents and their role in remediation of hexavalent chromium in industrial wastewater. Bull Environ Contam Toxicol 66:542–546

Ross SM (1994) Retention, transformation and mobility of toxic metals in soils. In: Ross SM (ed) Toxic metals in soil-plant systems. Wiley, Chichester, pp 63–152

Rowley MV, Warkentin DD, Yan VT, Piroshco BM (1994) The biosulfide process: integrated biological/chemical acid mine drainage treatment- results of laboratory piloting. Paper presented at the international Land Reclamation and Mine Drainage conference and the Third international conference on the Abatement of Acidic Drainage, Pittsburgh, PA, 24–29 April 1994

RSMENR (2002) Rivers State Ministry of Environment and Natural Resources.Interim guidelines and Standards on environmental pollution control and management, pp 39–45

Schiewer S, Volesky B (2000) Biosorption processes for heavy metal removal. In: Lovley DR (ed) Environmental microbe – metal interactions. ASM Press, Washington, DC, pp 329–357

Skyring GW (1987) Sulphate reduction in coastal ecosystems. Geomicrobiol J 5:295–374
Tabak HH, Lens P, Van Hullebusch ED, Dejonghe W (2005) Developments in bioremediation of soils and sediments polluted with metals and radionuclides-1. Microbial processes and mechanisms affecting bioremediation of metal contamination and influencing metal toxicity and transport. Rev Environ Sci Biotechnol 4(3):115–156
Utgikar V, Harmon S, Chaudhary N, Tabak H, Govind R, Haines J (2002) Inhibition of sulphate reducing bacteria by metal sulfide formation in bioremediation of acid mine drainage. Environ Toxicol 17:40–48
Valls M, de Lorenzo V (2002) Exploiting the genetic and biochemical capacities of bacteria for remediation of heavy metal pollution. FEMS Microbiol Rev 26:327–338
Wang J, Tao Y, Zhou JT, Gong XY (2001) Biosorption of chromium(VI) ions from aqueous solution by a novel bacterial exopolymers. Technol Water Treat 27(3):145–147
Watson JHP, Ellwood DC (1994) Biomagnetic separation and extraction process for heavy metals from solution. Miner Eng 7:1017–1028
Watson JHP, Ellwood DC (1988) A biomagnetic separation process for the removal of heavy metal ions from solution, Paper presented at the international conference on control of environmental problems from metal mines, Roros, 20–24 June 1988
White C, Gadd GM (1987) Inhibition of H+ efflux and K+ uptake and induction of K+ efflux in yeast by heavy metals. Toxicity Assess 2:437–447
White C, Gadd GM (2000) Copper accumulation by sulphate reducing bacterial biofilm. FEMS Microbiol Lett 183:313–318
White C, Sharman AK, Gadd GM (1998) An integrated microbial process for the bioremediation of soil contaminated with toxic metals. Nat Biotechnol 16(6):572–575
White C, Dennis JS, Gadd GM (2003) A mathematical process model for cadmium bioprecipitation by sulphate reducing bacterial biofilm. Biodegradation 14:139–151
Wu QJ, Song YL, Li FD (1996) Studies of microbial treatment process for wastewater with high concentration of Cr(VI). Technol Water Treat 22(3):165–167
Xu Y, Xiao H, Sun S (2005) Study on anaerobic treatment of wastewater containing hexavalent chromium. J Zhejiang Univ Sci 6B:574–579
Zehnder JBA (ed) (1988) Biology of anaerobic microorganisms. Wiley, New York

3 Lead- and Mercury-Resistant Marine Bacteria and Their Application in Lead and Mercury Bioremediation

Milind M. Naik and S.K. Dubey

Abstract

With rapid industrialisation, enormous amounts of industrial waste including heavy metals have accumulated in marine environments over several decades and require special attention. Untreated wastes from mining, metal refining industries, battery manufacturing industries, sewage sludge, power plants and waste incineration plants often contain substantially high levels of lead (Pb) and mercury (Hg); when dumped into marine and estuarine waters, these pose serious threat to environmental biota and urgently need to be removed from polluted marine/estuarine sites. Lead and mercury are non-bioessential, persistent and hazardous heavy metal pollutants of environmental concern. Bioremediation of heavy metals using Pb- and Hg-resistant bacteria has become a potential alternative to the existing technologies for the removal and/or recovery of toxic Pb and Hg from waste waters before releasing it into marine/estuarine water bodies for environmental safety. Various strategies through which marine/estuarine bacteria resist high concentrations of lead/mercury include efflux mechanisms, extracellular sequestration, biosorption, precipitation, reduction, volatilisation, alteration in cell morphology, enhanced siderophore production, altered permeability, demethylation and intracellular bioaccumulation. These unique characteristics of marine/estuarine bacteria proved to be an ideal tool in bioremediation of lead and mercury from contaminated marine and estuarine environmental sites.

3.1 Introduction

Metals that have a specific gravity greater than 5 or density more than 5 g/cm^3 are termed as heavy metals, e.g. mercury, lead, copper and cadmium (Gadd 1992; Dash and Das 2012). The International Union of Pure and Applied

M.M. Naik (✉) • S.K. Dubey
Department of Microbiology, Goa University, Taleigao Plateau, Goa, India
e-mail: milindnaik4@gmail.com

M.M. Naik, S.K. Dubey (eds.), *Marine Pollution and Microbial Remediation*,
DOI 10.1007/978-981-10-1044-6_3

Chemistry (IUPAC) technical report has eliminated the term "heavy metal", and the alternative name proposed is "toxic metal" (Duffus 2002), but still the term "heavy metal" is used extensively in scientific world. Contamination of marine and freshwater bodies due to release of "toxic metals" lead and mercury poses serious threat to natural biota including humans (Nies 1999; Fernandes and Beiras 2001; Dirilgen 2011; Dash and Das 2012). Untreated wastes from mining, metal refining industries, car battery manufacturing industries, sewage sludge, hydroelectric power plants, paper and pulp industries and waste incineration plants often contain substantially high levels of Pb and Hg when dumped into marine/estuarine waters, pose serious threat to environmental biota (Naik et al. 2012d, 2013; De et al. 2014). It is interesting to mention that even Antarctic Ocean water, which is considered relatively more pristine than any other ocean water, is also contaminated with heavy metals due to anthropogenic activities (Bonner 1984).

Lead and mercury do not have any biological function and are toxic to cells in a variety of ways which include DNA damage, oxidative damage to proteins and lipids and binding to essential proteins and enzymes (Nies 1999; Asmub et al. 2000; Hartwig et al. 2002; Naik et al. 2012a, 2013; De et al. 2014). However, the most common route of exposure to mercury and lead is by eating marine fish containing methylmercury and lead (Eisler 1988; Nascimento and Chartone-Souza 2003; De et al. 2014). Due to rapid industrialisation, enormous amounts of industrial waste containing heavy metals such as (mercury and lead) have accumulated in marine and terrestrial environments over several decades and require special attention. Therefore, the US Environmental Protection Agency (EPA) has included lead and mercury in the list of the most hazardous inorganic wastes (Cameron 1992).

Lead and mercury are mutagenic and teratogenic metals causing severe deleterious effects on human beings such as neurodegenerative impairment, renal failure, reproductive damage, neurological diseases and cancer (Kumagai and Nishimura 1978; Lam et al. 2007; Naik et al. 2012b; Naik and Dubey 2013b). Bioavailability of metals is an important factor for metal toxicity since soluble metals can readily penetrate cell membranes (Roane 1999). Therefore, metal immobilisation or detoxification strategies are applied by microbes to counteract toxic effect of heavy metals. Some natural microbial population that possesses a variety of protective mechanisms can survive and colonise at very high concentrations of toxic lead and mercury without any impact on their growth and metabolism. This unique characteristic of heavy metal-resistant microbes including bacteria makes them an ideal tool for bioremediation of metal-contaminated marine and terrestrial sites. Various strategies through which they resist high concentrations of heavy metals include efflux mechanism, alteration in cell morphology, siderophore production, biosorption, precipitation, volatilisation, reduction, demethylation, oxidation, extracellular sequestration, reduced permeability and intracellular bioaccumulation (Borremans et al. 2001; Barkay et al. 2003; Naik et al. 2012a, 2013; Naik and Dubey 2013b; De et al. 2014). Bioremediation processes are cost effective, highly efficient and environment friendly as compared to physico-chemical methods for removal of toxic metals; therefore, over the last few decades, attention has been focused towards exploiting marine microbes for heavy metal bioremediation. Understanding the mechanism by which marine and estuarine bacteria sequester/detoxify/biotransform lead and mercury to protect themselves from their toxic effects on physiological processes is crucial to the development of microbial processes for their concentration, removal and recovery from industrial effluents, sewage, marine sediments and marine/estuarine waters.

In the present chapter, we are focusing on lead- and mercury-resistant marine and estuarine bacteria and how we can employ them to bioremediate marine and estuarine sites highly contaminated with lead and mercury. Here we will stress on various resistant mechanisms employed by marine and estuarine bacteria to resist very high concentrations of lead and mercury and exploitation of these resistant mechanisms to clean up and biomonitor marine and estuarine contaminated sites.

3.2 Lead and Mercury Pollution in Marine and Estuarine Environments

3.2.1 Mercury

Mercury is the sixteenth rarest element on earth; however, concentrations of global mercury has increased approximately threefold due to various anthropogenic activities, and the world's oceans are the major reservoirs for its deposition (Mason et al. 2012; De et al. 2014). Various anthropogenic sources of mercury are burning of coal and petroleum products, dental fillings, use of mercurial fungicides in agriculture, paper making industry and mercury catalysts in industries. Inorganic and organic mercury compounds such as elemental mercury (Hg^0), mercuric mercury (Hg^{+2}), methylmercuric chloride (MMC) and dimethylmercury are discharged into marine environment through untreated sewage and industrial effluents (Wang et al. 2004; Dash and Das 2012; De et al. 2014).

According to some recent models on the flow of mercury through the environment, it is suggested that natural sources account for about 10 % of the estimated 5500–8900 tons of mercury currently being released to the atmosphere from all sources (De et al. 2014). Minamata disease which causes severe neurological disorder and was discovered in Minamata Bay, Japan, in 1956 is the first record of severe mercury poisoning in people who consumed marine fish and shellfish from mercury-contaminated marine waters. Thousands of people where affected and 887 were killed (Nascimento and Chartone-Souza 2003). In Minamata Bay, very high level of mercury was reported which caused a serious neurological disorder in humans referred as "Minamata disease". The level of total mercury in seawater of Minamata Bay ranged from 56 to 285 ng/L and 2.1–506 ng/L (Kumagai and Nishimura 1978). Interestingly, surface sediment sample from semi-enclosed bay, "Gunnekleivfjorden" from Southwest Norway, contained mercury ranging from 90 to 350 ppm (Skei 1978). The mercury concentrations analysed along Indian coasts at (i) Mormugao (15°24″35′ N, 73°48″2′ E; Hg concentration 152–456 ng/l in water and 53–194 ng/g dry sediment), (ii) Gopalpur (19°18″12′ N, 84°57″55′ E; Hg concentration 2–117 ng/l in water and 72–128 ng/g dry sediment) and (iii) Chennai (13°6″40′ N, 80°18″3′ E; Hg concentration 100–2,100 ng/l in water and 237–338 ng/g dry sediment) were found to be very high (De et al. 2007). Methylmercury is the most toxic among all the forms of mercury, affecting the immune system, altering the genetic system and causing damage to the nervous system including coordination and the senses of touch, taste and sight (De et al. 2014). Mercury causes toxicity by binding to the sulfhydryl groups of enzymes and proteins, thereby inactivating crucial cell functions. Generally, mercury accumulates upwards through aquatic food chains, so that organisms at higher trophic levels have higher mercury concentrations (Nascimento and Chartone-Souza 2003). Mercury is of environmental concern because it biomagnifies in the food chain by up to seven orders of magnitude, resulting in high concentrations in top predators such as fish and polar bears (Hintelmann 2010; Sonne 2010). Europe and North America are reducing their contribution towards the global mercury burden, but the emission rates in Asia are increasing at frightening rates (Dash and Das 2012) and need special attention.

3.2.2 Lead

Lead (Pb), obtained mainly from galena ore (PbS), is known to mankind for last 7,000 years, and its poisoning has been reported for at least 2,500 years (Nriagu 1978; Eisler 1988). Lead is well known to inhibit haem biosynthesis, causes serious neurodegenerative diseases and reproductive impairment, interferes with kidney function, possesses carcinogenic properties and, when blood level exceeds 70 μg/dl, results in coma and death (Naik et al. 2012b). Lead is known to cause damage to DNA, protein and lipid and to also replace essential metal ions such as Zn, Ca and Fe from enzymes (Nies 1999; Asmub et al. 2000; Naik et al. 2013). Lead has broad range of applications in various industries viz. petroleum, electronics, battery, paints, ceramics, stained glass, biocide preparation and ammunitions with annual global demand of refined lead exceeding 87 lakh tonnes (Naik et al. 2012c, 2013). Lead is a non-bioessential persistent environmental pollutant with half-life of approximately 5,000 years and biomagnifies through the trophic levels and accumulates at high concentrations in top predators such as fish and polar bears. Environmental lead has increased more than 1,000-fold over the past three decades as a result of extensive anthropogenic activities (Naik et al. 2013). The WHO has recommended <10 μg/L lead as safe permissible level in drinking water (Watt et al. 2000). Wastes from industries, sewage sludge, power plants and incineration plants contain substantial amounts of toxic lead which are finally discharged into marine and estuarine environment which cause serious damage to marine micro- and macrofauna.

The analysis of heavy metals in coastal waters and sediments from Tianjin Bohai Bay, China, revealed that Pb and Zn were the main heavy metal pollutants in the coastal waters of the bay. High levels of Pb and Zn appeared especially near the estuary, indicating that river discharge was the main pollution source. Analysis of data for the period 1987–2004 indicated that Pb pollution in coastal waters was due to river discharge before 2001. Lead levels did not decrease after 2001 when annual run-off levels declined, indicating that Pb pollution by atmospheric deposition had increased due to the use of leaded petrol in motor cars. Pb, Zn and Cd were the dominant polluting elements in superficial sediments from Tianjin Bohai Bay, with levels in excess of the corresponding upper limits of environmental background values (Meng et al. 2008; Naik et al. 2012d). Dissolved Pb, ranging from 5–15 ng/kg in open-ocean surface waters up to 50 ng/kg in highly polluted coastal waters, enters marine food chains at the 10^{-8} g Pb/g level by adsorption onto surfaces of algae which are consumed by herbivores. Dissolved Pb and biological productivity tend to be inversely correlated in surface ocean waters (Burnett et al. 1980). Sediment sample of Mandovi Estuary of Goa revealed appreciable levels of lead ranging from 4.5 to 46.5 μg g^{-1} at different source points (Alagarsamy 2006). The pollution load index (PLI) for Pb, Fe, Mn, Zn, Cu, Co and Cr for Divar sediments (Mandovi Estuary) was far greater (i.e. 1.65–2.19) than that of Tuvem (Chapora Estuary) (0.91–1.3) reflecting the intensity of anthropogenic inputs into the ecosystem due to transport of ferromanganese ore along the Mandovi River (Attri and Kerkar 2011). Estuarine sediment around the North Irish Sea, UK, was reported to be contaminated with 52–207 μg/g lead (Smith and Orford 1989). Lead concentrations in surface sediments in a near-shore environment, Jurujuba Sound, Southeast Brazil, were found to be contaminated with 64–174 μg/g lead (Neto et al. 2000). Lead concentration in estuarine sediments along the western coast of Mauritius was accounted to be 27 mg/Kg (Ramessur 2004). Lead in marine/estuarine environment is a persistent environmental pollutant, slowly accumulates and results in biomagnification in the food chain and is referred as cumulative poison, and there is pressing need to remove lead from marine contaminated sites.

3.3 Lead- and Mercury-Resistance Mechanisms in Marine and Estuarine Bacteria

3.3.1 Mercury Resistance Mechanisms

Unusual rise in mercury-resistant bacteria (MRB) in coastal waters and sediments of India from year 1997 to 2003 was reported by Ramaiah and De (2003). The mercury-resistant bacteria (MRB) that belong to *Pseudomonas*, *Proteus*, *Xanthomonas*, *Alteromonas*, *Aeromonas* and

Enterobacteriaceae were isolated from west coast of India and were also found to grow in the presence of different toxicants, viz. heavy metals (Pb, Cd), pesticides, phenol, formaldehyde, formic acid and trichloroethane (De et al. 2003). Mercury-resistant marine bacteria are found to be very promising in dealing with mercury and most of other highly toxic heavy metals and xenobiotics (De et al. 2006). Occurrence of large fractions of mercury-resistant bacteria in the Bay of Bengal was reported by De and Ramaiah (2006). *Bacillus* isolates from Minamata Bay sediment were selected for the ability to volatilise mercury from a range of organomercurials (including methylmercury) (Nakamura and Silver 1994). The higher abundance of the Gram-negative *merA* gene in the Seine Estuary mudflats indicates a relationship between the degree of anthropogenic pollution and the abundance of the *merA* gene in the mudflat sediments (Ramond et al. 2008). Five different types of resistance mechanisms (Fig. 3.1) have been reported in bacteria for mercury which include (i) reduced uptake of mercuric ions, (ii) demethylation of methylmercury followed by conversion to mercuric sulfide compounds, (iii) enzymatic reduction of Hg^{+2} to Hg^0, (iv) sequestration of methylmercury and (v) intracellular bioaccumulation of mercury by bacteria (Barkay et al. 2003).

(i) *Reduced uptake of mercuric ions*: In *Enterobacter aerogenes*, resistance to mercury is by reduction in the cellular permeability to Hg^{2+} ions which is believed eventual due to the expression of a couple of plasmid-encoded proteins (Phung 1996). Since bacteria have reduced permeability to mercury, it can tolerate very high concentration of mercury in contaminated environment.

(ii) *Demethylation of methylmercury followed by conversion to mercuric sulfide compounds:* The best studied resistance system in marine bacteria is primarily based on clustered genes in an operon (i.e. *mer*). Genes conferring resistances to mercury compounds are clustered in an operon in most known naturally occurring systems. Tn*MERI1* was found as the first mercury resistance transposon identified from Gram-positive bacteria and was harboured by a Minamata Bay sediment-isolated bacterial strain which was designated as *Bacillus megaterium* MB1. The *mer* operon encoded by Tn*MERI1* has *merR*, *merT*, *merP*, *merA* and *merB* genes, which code for metal-specific activator-repressor, transporting, extracellular metal ion binding, mercuric reductase and organomercurial lyase, respectively (Huang et al. 2010).

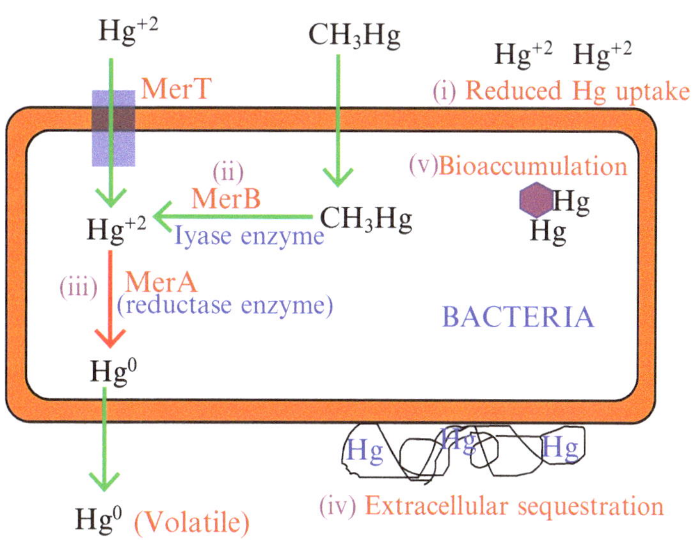

Fig. 3.1 Mercury resistance mechanisms in marine/estuarine bacteria. (i) Reduced uptake of mercuric ions, (ii) demethylation of methylmercury by enzyme organomercurial lyase, (iii) enzyme mercuric reductase responsible for reduction of Hg^{+2} to Hg^0, (iv) extracellular sequestration of methylmercury and Hg^{+2} and (v) intracellular bioaccumulation of mercury

In *mer* operon, determinants *RTPCDAB* in marine bacteria are located on plasmid or transposons and also can be found in chromosomes (Nascimento and Chartone-Souza 2003; Barkay et al. 2003). The *mer* operon of a mercury-resistant *Pseudoalteromonas haloplanktis* strain was isolated from Minamata Bay, Japan (Lohara et al. 2001). The mercury-resistant *mer* operon is also reported in marine gliding *flavobacterium*, *Tenacibaculum discolor* 9A5 (Allen et al. 2013), whereas a diversity of mercury resistance determinants among *Bacillus* strains were isolated from sediment of Minamata Bay (Narita et al. 2003).

(iii) *Enzymatic reduction of* Hg^{+2} *to* Hg^0

mer Operon

merR: *merR* is a regulatory gene in *mer* operon. Mer R, the metalloregulatory protein, has high affinity for Hg^{+2}. In the absence of Hg^{+2}, it binds to the promoter region of *mer* operon and represses the transcription of structural genes from *mer* operon (TPCDAB) responsible for mercury resistance. Mer R protein activates transcription of structural genes whose product requires mercury resistance, in the presence of inducing concentration of Hg^{+2} (Lohara et al. 2001; Allen et al. 2013; Nascimento and Chartone-Souza 2003; Barkay et al. 2003; Huang et al. 2010).

merD: A secondary regulatory protein also binds the same operator-promoter region as MerR, although very weakly.

merT: Product of *merT* gene is required for transport of Hg^{+2} inside bacterial cell where Hg^{+2} is acted upon by mercuric reductase (encoded by *merA*) and reduced to Hg^0.

merB: It confers resistance to organomercurials such as methylmercury (CH_3-Hg). It hydrolyses C-H by coding an enzyme called organomercurial lyase, which transforms CH_3-Hg to less toxic Hg^{+2}. Hg^{+2} is acted upon by mercuric reductase (encoded by *merA*) and reduced to Hg^0 (volatile form).

merA: It encodes enzyme mercuric reductase responsible for reduction of Hg^{+2} (toxic) to less toxic volatile form Hg^0.

merP: MerP protein acts as an extracellular metal ion-binding protein and can also act as free radical scavenger. MerP protein possesses a highly conserved domain with two cysteine residues for mercuric ion binding.

merC: The exact role of MerC protein is not known, but mutating *merC* does not have any effect on mercury resistance.

(iv) *Extracellular sequestration of* Hg^{+2} *and methylmercury*: The exopolymer (EPS) present in the biofilm of marine bacteria has a specific relation with heavy metals; mercury binds with EPS and gets entrapped extracellularly. The immobilised mercury outside bacterial cell is relatively toxic, and therefore bacteria producing EPS can tolerate very high concentration of mercury present in its immediate environment. Mercury entrapped into EPS can be detoxified by activity of *mer* operon (Anthony 2014), i.e. *mer B* encodes enzyme called organomercurial lyase, which transforms CH_3-Hg to less toxic Hg^{+2}, and *merA* encodes enzyme mercuric reductase responsible for reduction of Hg^{+2} (toxic) to less volatile form Hg^0.

(v) *Mercury bioaccumulation by bacteria*: Although methylmercury is more toxic than Hg^{2+}, in some bacteria, methylmercury is characterised to be the less toxic form. Methylation has been reported in bacteria from water, soil and sediments and is both plasmid and chromosomally encoded (Barkay et al. 2003; Miller et al. 2005). In *Desulfovibrio desulfuricans*, the methylation of mercury exists as a duo-step process which elaborates the transfer of a methyl group from methyltetrahydrofolate to methylcobalamin to Hg.

(vi) *Mercury methylation*: Marine mercury-resistant bacteria (Gram negative) isolated from Bay of Bengal along Odisha coast were found negative for *mer* operon. These bacterial isolates were found to resist very high amount of mercury (MIC 50 ppm) by bioaccumulating mercury intracellularly (Sinha 2012).

3.3.1.1 Detoxification of Toxic Chemicals and Heavy Metals by Mercury-Resistant Bacteria

De et al. (2007, 2008) have reported that marine mercury-resistant bacteria exposed to polluted environments such as coastal areas can tolerate, detoxify or biotransform a variety of other toxicants. Several mercury-resistant marine bacteria (*Bacillus pumilus*, *Alcaligenes faecalis*, *Brevibacterium iodinum* and *Pseudomonas aeruginosa*) from the coastal waters of India were found to biotransform heavy metals, viz. cadmium, lead and xenobiotics like polychlorinated biphenyls and tributyltin. A *Pseudomonas aeruginosa* strain CH07 aerobically degraded 14 toxic polychlorinated biphenyls including congeners with five or more chlorine atoms on the biphenyl ring and was also equally efficient in degrading more than 54 % tributyltin. These bacteria offer great biotechnological opportunities in bioremediation of marine waters contaminated with different xenobiotics and heavy metals.

3.3.2 Lead-Resistance Mechanisms

Various lead-resistant mechanisms employed by marine/estuarine lead-resistant bacteria include (i) efflux mechanism, (ii) intracellular lead bioaccumulation, (iii) extracellular sequestration, (iv) cell surface biosorption and (v) precipitation (Fig. 3.2).

(i) *Efflux mechanism*
Heavy metal-resistant marine bacteria possess ATPase-mediated efflux pump which maintains intracellular heavy metal homeostasis by effluxing excessive heavy metals outside the cells (Naik et al. 2013; Naik et al. 2013). Soft metal transporting P_{IB}-type ATPases are group of proteins involved in transport of heavy metals outside the cell membrane and governing bacterial heavy metal resistance (Nies and Silver 1995; Rensing et al. 1999; Coombs and Barkay 2004). Lead-resistant bacterial isolates *Pseudomonas stutzeri* M-9 and *Vibrio harveyi* M-11 isolated from Zuari Estuary showed efflux-mediated lead resistance (Naik et al. 2013). *Pseudomonas stutzeri* M-9 and *Vibrio harveyi* M-11 exhibited resistance up to 0.8 mM and 1.2 mM lead nitrate, respectively. Nested PCR clearly demonstrated presence of *pbrA* gene (amplicon size, 750 bp) belonging to P-type ATPase family on chromosomal genome, and 5.4±0.7- and 7.9±0.9-fold expression of *pbrA* gene in *Pseudomonas stutzeri* strain M-9 and *Vibrio harveyi* strain M-11 when grown in TMM amended with 0.5 mM lead nitrate confirmed efflux-mediated lead resis-

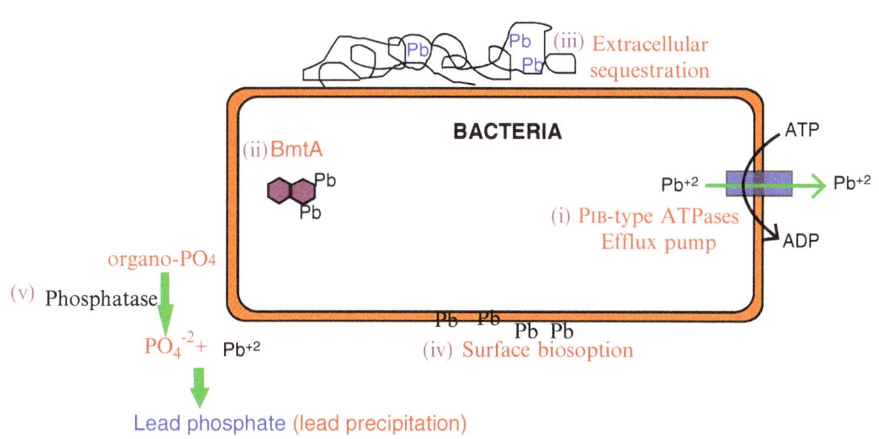

Fig. 3.2 Lead resistance mechanisms in marine/estuarine bacteria. (i) Efflux pump, (ii) intracellular lead bioaccumulation by bacterial metallothionein, (iii) extracellular sequestration in EPS, (iv) cell surface biosorption to negative groups (carboxyl, hydroxyl) and (v) lead precipitation into insoluble form catalysed by phosphatase enzyme

tance in both bacterial isolates. These estuarine bacterial isolates possess *pbrA* gene encoding P-type ATPase for lead resistance and *mdrL* gene for multidrug resistance via efflux pumps, suggesting possible contamination of Zuari Estuary with heavy metals/antibiotics. The *Ralstonia metallidurans* CH34 complete operon *pbrUTRABCD*, conferring efflux-mediated lead resistance, has already been sequenced (Borremans et al. 2001). Lead(II) resistance in *Cupriavidus metallidurans* CH34: interplay between plasmid and chromosomally located functions (Taghavi et al. 2009). *pbrUTRABCD* operon was found to be responsible for lead resistance in *Cupriavidus metallidurans* CH34.

(ii) *Intracellular bioaccumulation*

Intracellular metal bioaccumulation and homeostasis in cell cytosol involve the low molecular weight, cysteine-rich metallothioneins which range from 3.5 to 14 kDa (Hamer 1986). Metallothioneins play an important role in immobilisation of toxic heavy metals, thereby protecting bacterial metabolic processes catalysed by enzymes (Blindauer et al. 2002; Liu et al. 2003). *P. aeruginosa* strain WI-1 isolated from Mandovi Estuary possesses bacterial metallothionein (BmtA) to alleviate lead (Pb^{+2}) toxicity (Naik et al. 2012a). *P. aeruginosa* strain WI-1 resists 0.6 mM lead nitrate by bioaccumulating 26.5 mg lead/g dry weight of cells intracellularly. SDS-PAGE analysis confirmed lead-induced bacterial metallothionein with molecular weight 11 kDa, which corresponds to the predicted *bmtA* gene. Thus, estuarine bacteria possessing metallothioneins are an ideal tool for bioremediation of heavy metal-contaminated environmental sites.

(iii) *Extracellular sequestration*

A metal immobilisation strategy is applied by microbes to counteract toxic effects of heavy metals by secreting extracellular high molecular weight biopolymers referred as exopolysaccharides (EPSs). Bacterial EPS and its possible role in bioaccumulation of Cu and Pb in a marine food chain were investigated using a partially purified and chemically characterised EPS isolated from *Marinobacter* sp. (Bhaskar and Bhosle 2006). In the marine bacterium *P. aeruginosa* CH07, lead was entrapped in EPS indicating it as a possible resistance mechanism (De et al. 2007, 2008). Lead ions could interact with carboxyl, hydroxyl and amide groups and glucuronic acid from different chains of the polyanionic EPS produced by bacteria which acted as an electrostatic bridge between them producing a mesh of polymer large enough to sequester very high levels of lead. Therefore, exopolysaccharide producing lead-resistant bacterial strains may serve as a potential bioremediative agent (lead biosorbent) in lead-contaminated environmental sites (Naik et al. 2012b, 2013).

(iv) *Surface biosorption*

This surface biosorption of lead is due to various negatively charged chemical groups present on the bacterial cell surface, viz. the carboxyl group of the peptidoglycan serves as main metal binding site at the cell wall of Gram-positive bacteria, whereas phosphate groups contribute significantly in the case of Gram-negative bacteria (Gadd and White 1993; Naik et al. 2013). Shamim et al. (2013) reported that lead-resistant *Aeromonas caviae* strain KS-1 isolated from Mandovi Estuary,Goa, India, can resist lead up to 1.4 mM, and significant biosorption of lead (8 %) on the cell surface of this isolate was clearly revealed by scanning electron microscopy coupled with energy-dispersive X-ray spectroscopy.

(v) *Bioprecipitation*

Soluble (bioavailable) lead is toxic to bacteria; therefore, resistant bacteria precipitate toxic lead as insoluble complexes to reduce their bioavailability and toxicity (Naik et al. 2013). Three pleiotropic, quorum sensing-defective *Vibrio harveyi* mutants were observed to precipitate soluble Pb^{+2} as an insoluble compound. The compound was purified and subjected to

X-ray diffraction and elemental analyses. These assays identified the precipitated compound as $Pb_9(PO4)_6$, an unusual and complex lead phosphate salt that is produced synthetically at 200 °C (Mire et al. 2004). This reaction is catalysed by phosphatase enzyme. Sulfate-reducing bacteria (SRBs) are anaerobic heterotrophic bacteria in marine environment which also precipitate heavy metals as insoluble metal sulfides, e.g. ZnS, PbS, CdS and CuS (Chamberlain et al. 1988). Lead-resistant *Bacillus iodinium* GP13 and *Bacillus pumilus* S3 precipitate lead as lead sulfide (PbS) (De et al. 2008).

3.3.2.1 Co-resistance and Cross Resistance

Several mercury-resistant marine bacteria isolated from the coastal waters of India were evaluated for their ability to biotransform heavy metals like cadmium and lead, e.g. *Pseudomonad aeruginosa* strain CH07 (De et al. 2007). Mercury-resistant marine bacterial isolates *Pseudomonas*, *Proteus*, *Xanthomonas*, *Alteromonas*, *Aeromonas* and *Enterobacteriaceae* were also found to tolerate very high concentrations of lead and cadmium (De et al. 2003). Lead-resistant *P. aeruginosa* strain WI-1 isolated from Mandovi Estuary possesses bacterial metallothionein (BmtA) to alleviate Pb^{+2} toxicity (Naik et al. 2012a). *P. aeruginosa* strain WI-1 resists 0.6 mM lead nitrate by bioaccumulating 26.5 mg lead/g dry weight of cells intracellularly. *P. aeruginosa* strain WI-1 also showed cross-tolerance to cadmium, mercury and tributyltin chloride (TBTCl) along with resistance to multiple antibiotics (Naik et al. 2012a). The two bacterial strains *P. stutzeri* M-9 and *V. harveyi* M-11 isolated from Zuari Estuary resist lead nitrate up to 0.8 and 1.2 mM by efflux mechanism (Naik et al. 2013). *P. stutzeri* M-9 and *V. harveyi* M-11 were also found resistant to cadmium and mercury along with multiple antibiotics. MIC of antibiotics for lead-resistant bacterial strain M-9 was 50 μg/disc (ampicillin), 30 μg/disc (chloramphenicol), 10 μg/disc (norfloxacin), 25 μg/disc (co-trimoxazole) and 30 μg/disc (cephalexin), whereas for strain M-11 it was 30 μg/disc (nalidixic acid), 50 μg/disc (ampicillin), 15 μg/disc (erythromycin), 30 μg/disc (chloramphenicol), 30 μg/disc (cephalexin) and 25 μg (co-trimoxazole).

Metal contamination has been reported to function as a selective agent in the proliferation of antibiotic resistance (Baker-Austin et al. 2006). The co-selection mechanisms of antibiotic and metal resistance include co-resistance (different resistance determinants present on the same genetic element) and cross resistance (the same genetic determinant responsible for resistance to multiple antibiotics and heavy metals). Co-resistance occurs when the genes specifying resistant phenotypes are located together on the same genetic element such as plasmid transposons or integron (Chapman 2003). There is growing concern that metal contamination in marine environment may co-select multiple drug-resistant pathogens. Therefore biomonitoring and bioremediation of marine and estuarine environment is a desperate need.

3.4 Conclusion

Over last three decades, attention has been focused towards exploiting marine/estuarine microbes for lead and mercury bioremediation. Marine and estuarine bacteria possess different lead and mercury resistance mechanisms; therefore, it is necessary and highly desirable to characterise such marine bacterial isolates with reference to their biochemical and genetic mechanism of resistance. Lead- and mercury-resistant bacterial strains possessing various lead-resistant mechanisms such as efflux mechanism, extracellular sequestration, biosorption, precipitation, demethylation, reduction, volatilisation, demethylation and intracellular bioaccumulation discussed in this chapter may serve as potential biotechnological agents for bioremediation of lead- and mercury-contaminated marine and estuarine environmental sites. Valuable properties already present in certain bacterial strains can be combined or improved through state-of-the-art genetic engineering tools. Marine/estuarine bacteria containing genetic determinants on plasmid, transposon or chromosomal DNA which are responsible for lead and mercury resistance can be used to genetically engineer marine bacteria to bioremediate very high amount of lead and mercury. Construction of whole-cell bacterial bioreporters by exploiting resistant mechanisms present in bacteria and using recombinant DNA technology represents a convenient testing method for quantifying the bioavailability of heavy metals including lead and mercury in environmental samples (Hynninen et al. 2010). Therefore, resistant determinant present on *pbr* operon and *mer* operon can be exploited to construct whole-cell bioreporters to monitor picogram levels of lead/mercury in marine and estuarine sites.

Application of genetically engineered microorganisms (GEMs) in bioremediation has received a great deal of attention but has largely been confined to laboratory environment. Their practical impact and delivery under field conditions need to be evaluated. Legislations and biosafety norms should be strictly adhered to in this regard before employing these bacteria in metal bioremediation under field conditions. Every nation should implement strict laws to monitor mercury/lead pollution in their respective territories so that this problem can be minimised. Reclamation of lead-/mercury-polluted environments using the marine/estuarine microbes has been an effective, affordable and ecofriendly technological solution.

Acknowledgements The authors would like to acknowledge Dr. Satish Shetye, Vice Chancellor of Goa University, for providing necessary facilities. Dr. Milind Naik also thank Government of India for financial support as SERB-DST young scientist project.

References

Alagarsamy R (2006) Distribution and seasonal variation of trace metals in surface sediments of the Mandovi estuary, west coast of India. Estuar Coast Shelf Sci 67:333–339

Allen RC, Tu YK, Navarez MJ, Bobbs AS, Friesen JW, Lorsch JR, Voet JG, Hamlett NV (2013) The mercury resistance (mer) operon in marine gliding flavobacterium, tenacibaculum discolor 9A5. FEMS Microbiol Ecol 83:135–148

Anthony E (2014) Bioremediation of mercury by biofilm forming mercury resistant marine bacteria. Dissertation, NIT, Rourkela

Asmub M, Mullenders LHF, Hartwig A (2000) Interference by toxic metal compounds with isolated zinc finger DNA repair proteins. Toxicol Lett 112:227–231

Attri K, Kerkar S (2011) Seasonal assessment of heavy metal pollution in tropical mangrove sediments (Goa, India). J Ecobiotechnol 3:9–15

Baker-Austin C, Wright MS, Stepanauskas R, McArthur JV (2006) Co-selection of antibiotics and metal resistance. Trends Microbiol 14:176–182

Barkay T, Miller SM, Summers AO (2003) Bacterial mercury resistance from atoms to ecosystem. FEMS Microbiol Rev 27:355–384

Bhaskar PV, Bhosle NB (2006) Bacterial extracellular polymeric substances (EPS) a carrier of heavy metals in the marine food-chain. Environ Int 32:192–198

Blindauer CA, Harrison MD, Robinson AK, Parkinson JA, Bowness PW, Sadler PJ, Robinson NJ (2002) Multiple bacteria encode metallothioneins and SmtA-like fingers. Mol Microbiol 45:1421–1432

Bonner WN (1984) Conservation and the Antarctic. In: Laws RM (ed) Antarctic ecology, vol II. Academic, London, pp 821–850

Borremans B, Hobman JL, Provoost A, Brown NL, van der Lelie D (2001) Cloning and functional analysis of the pbr lead resistance determinant of *Ralstonia metallidurans* CH34. J Bacteriol 183:5651–5658

Burnett M, Ng A, Settle D, Patterson CC (1980) Impact of man on coastal marine ecosystems. In: Branica M, Konrad Z (eds) Lead in the marine environment (pp 7–13). Proceedings of the International experts discussion on lead occurrence, fate and pollution in the ,marine environment, Rovinj, Yugoslavia, 18–22 October 1977. doi:10.1016/B978-0-08-022960-7.50006-3

Cameron RE (1992) Guide to site and soil description for hazardous waste characterization. Metals, 250. (Environmental Protection Agency EPA/600/4–91/029)

Chamberlain AHL, Simmonds SE, Garn BJ (1988) Marine 'copper-tolerant' sulphate reducing bacteria and their effects on 90/10 copper-nickel (CA 706). Int Biodeter 24:213–219

Chapman JS (2003) Disinfectant resistance mechanisms, cross resistance, and co-resistance. Int Biodeterior Biodegradation 51:271–276

Coombs JM, Barkay T (2004) Molecular evidence for the evolution of metal homeostasis genes by lateral gene transfer in bacteria from the deep terrestrial subsurface. Appl Environ Microbiol 70:1698–1707

Dash HR, Das S (2012) Bioremediation of mercury and the importance of bacterial *mer* genes. Int Biodeterior Biodegradation 75:207–213

De J, Ramaiah N (2006) Occurrence of large fractions of mercury-resistant bacteria in the Bay of Bengal. Curr Sci 9:368–371
De J, Ramaiah N, Mesquita A, Verlekar XN (2003) Tolerance to various toxicants by marine bacteria highly resistant to mercury. Mar Biotechnol 5:185–193
De J, Sarkar A, Ramaiah N (2006) Bioremediation of toxic substances by mercury resistant marine bacteria. Ecotoxicology 15:385–389
De J, Ramaiah N, Bhosle NB, Garg A, Vardanyan L, Nagle VL, Fukami K (2007) Potential of mercury-resistant marine bacteria for detoxification of chemicals of environmental concern. Microbes Environ 22:336–345
De J, Ramaiah N, Vardanyan L (2008) Detoxification of toxic heavy metals by marine bacteria highly resistant to mercury. Mar Biotechnol 10:471–477
De J, Dash HR, Das S (2014) Mercury pollution and bioremediation—a case study on biosorption by a mercury-resistant marine bacterium. In: Das S (ed) Microbial biodegradation and bioremediation, pp 137–166. DOI: 10.1016/B978-0-12-800021-2.00006-6
Dirilgen N (2011) Mercury and lead: assessing the toxic effects on growth and metal accumulation by Lemna minor. Ecotoxicol Environ Saf 74:48–54
Duffus JH (2002) Heavy metals a meaningless term? IUPAC technical report. Pure Appl Chem 74:793–807
Eisler R (1988) Lead hazards to fish, wildlife, and invertebrates: a synoptic review. US Fish Wildl Serv Biol Rep 85:1–14
Fernandes N, Beiras R (2001) Combined toxicity of dissolved mercury with copper, lead and cadmium on embryogenesis and early larval growth of the *Paracentrotus lividus* sea urchin. Ecotoxicology 10:263–271
Gadd GM (1992) Microbial control of heavy metal pollution. In: Fry JC, Gadd GM, Herbert RA, Jones CW, Watson-Craiki IA (eds) Microbial control of pollution. Cambridge University Press, Cambridge, pp 59–88
Gadd GM, White C (1993) Microbial treatment of metal pollution- a working biotechnology. Trends Biotechnol 11:353–359
Hamer DH (1986) Metallothioneins. Ann Rev Biochem 55:913–951
Hartwig A, Asmuss M, Ehleben I, Herzer U, Kostelac D, Pelzer A, Schwerdtle T, Burkle A (2002) Interference by toxic metal ions with DNA repair processes and cell cycle control: molecular mechanisms. Environ Health Perspect 110:797–799
Hintelmann H (2010) Organomercurials: their formation and pathways in the environment. Met Ions Life Sci 7:365–401
Huang C-C, Chien M-F, Lin K-H (2010) Bacterial mercury resistance of Tn*MER11* and its' application in bioremediation. In: Hamamura N, Suzuki S, Mendo S, Barroso CM, Iwata H, Tanabe S (eds) Interdisciplinary studies on environmental chemistry—biological responses to contaminants. Terrapub, Tokyo, pp 23–29
Hynninen A, Tonismann K, Virta M (2010) Improving the sensitivity of bacterial bioreporters for heavy metals. Bioeng Bugs 1:132–138
Kumagai M, Nishimura H (1978) Mercury distribution in seawater in Minamata Bay and the origin of particulate mercury. J Oceanogr Soc Jpn 34(2):50–56
Lam TVP, Agovino XN, Roche L (2007) Linkage study of cancer risk among lead exposed workers in New Jersey. Sci Total Environ 372:455–462
Liu T, Nakashima S, Hirose K, Uemura Y, Shibasaka M, Katsuhara M, Kasamo K (2003) A metallothionein and CPx-ATPase handle heavy-metal tolerance in the filamentous cyanobacterium *Oscillatoria brevis*. FEBS Lett 542:159–163
Lohara K, Liyama R, Nakamura K, Silver S, Sakai M, Takeshita M, Furukawa K (2001) The *mer* operon of a mercury-resistant *Pseudoalteromonas haloplanktis* strain isolated from Minamata Bay, Japan. Appl Microbiol Biotechnol 56:736–741
Mason RP, Choi AL, Fitzgerald WF, Hammerschmidt CR, Lamborg CH, Soerensen AL, et al (2012) Mercury biogeochemical cycling in the ocean and policy implications. Environ Res. http://dx.doi.org/10.1016/j.envres.2012.03.013
Meng W, Qin Y, Zheng B, Zhang L (2008) J Environ Sci 20:814–819
Miller LR, Patel B, Dong A, Fiedler D, Falkowski M, Zelikova JSM (2005) NmerA, the metal binding domain of mercuric ion reductase, removes Hg^{2+} from proteins, delivers it to the catalytic core, and protects cells under glutathione depleted conditions. Biochemistry 44:11402–11416
Mire CE, Tourjee JA, O'Brien WF, Ramanujachary KV, Hecht GB (2004) Lead precipitation by *Vibrio harveyi*: evidence for novel quorum-sensing interactions. Appl Microbiol Biotechnol 70:855–864
Naik MM, Dubey SK (2013) Lead resistant bacteria: lead resistance mechanisms, their applications in lead bioremediation and biomonitoring. Ecotoxicol Environ Saf 98:1–7
Naik MM, Pandey A, Dubey SK (2012a) *Pseudomonas aeruginosa* strain WI-1 from Mandovi estuary possesses metallothionein to alleviate lead toxicity and promotes plant growth. Ecotoxicol Environ Saf 79:129–133. Elsevier
Naik MM, Pandey A, Dubey SK (2012b) Biological characterization of lead- enhanced exopolysaccharide produced by a lead resistant *Enterobacter cloacae* strain P2B. Biodegradation 23:775–783. Springer
Naik MM, Shamim K, Dubey SK (2012c) Biological characterization of lead resistant bacteria to explore role of bacterial metallothionein in lead resistance. Curr Sci 103:426–429. Current Science Association, in collaboration with the Indian Academy of Sciences
Naik MM, Pandey A, Dubey SK (2012d) Bioremediation of metals mediated by marine bacteria. In: Satyanarayana T et al (eds) Microorganisms in environmental management. doi:10.1007/978-94-007-2229-3_29

Naik MM, Dubey SK, Khanolkar D, D'Costa B (2013) P-type ATPase and MdrL efflux pump-mediated lead and multi-drug resistance in estuarine bacterial isolates. Curr Sci 105:1366–1372
Nakamura K, Silver S (1994) Molecular analysis of mercury-resistant *Bacillus* isolates from sediment of Minamata Bay, Japan. Appl Environ Microbiol 60:4596–4599
Narita M, Chiba K, Nishizawa H, Ishii H, Huang CC, Kawabata Z, Silver S, Endo G (2003) FEMS Microbiol Lett 223:73–82
Nascimento AMA, Chartone-Souza E (2003) Operon *mer*: bacterial resistance to mercury potential for bioremediation of contaminated environments. Genet Mol Res 2:92–101
Neto JAB, Smith BJ, McAllister JJ (2000) Heavy metal concentrations in surface sediments in a nearshore environment, Jurujuba Sound, Southeast Brazil. Environ Pollut 109:1–9
Nies NH (1999) Microbial heavy metal resistance. Appl Microbiol Biotechnol 51:730–750
Nies DH, Silver S (1995) Ion efflux systems involved in bacterial metal resistances. J Ind Microbiol Biotechnol 14:186–199
Nriagu JO (1978) The biogeochemistry of lead in the environment. Part B. Biological effects. Elsevier/North Holland Biomedical Press, Amsterdam, p 397
Phung LT (1996) Bacterial mercury resistance proteins. Encycl Metalloproteins:209–217
Ramaiah N, De J (2003) Unusual rise in mercury resistant bacteria in coastal environs. Microb Ecol 45:444–454
Ramessur R (2004) Statistical comparison and correlation of zinc and lead in estuarine sediments along the western coast of Mauritius. Environ Int 30:1039–1044
Ramond J-B, Berthe T, Lafite R, Deloffre J, Ouddance B, Petit F (2008) Relationships between hydrosedimentary processes and occurrence of mercury-resistant bacteria (*merA*) in estuary mudflats (Seine, France). Mar Pollut Bull 56:1168–1176
Rensing C, Ghosh M, Rosen BP (1999) Families of soft-metal-ion transporting ATPases. J Bacteriol 181:5891–5897
Roane TM (1999) Lead resistance in two bacterial isolates from heavy metal-contaminated soils. Microb Ecol 37(3):218–224
Shamim K, Naik MM, Pandey A, Dubey SK (2013) Isolation and identification of *Aeromonas caviae* strain KS-1 as TBTC and lead resistant estuarine bacteria. Environ Monit Assess 185:5243–5249
Sinha R (2012) Bioaccumulation of mercury in marine bacteria: a novel approach of mercury remediation. Dissertation, NIT Orisa. pp 34–42
Skei JM (1978) Serious mercury contamination of sediments in a Norwegian semi-enclosed bay. Mar Pollut Bull 9:191–193
Smith BJ, Orford JD (1989) Scales of pollution in estuarine sediment around the North Irish Sea. In: Sweeney JC (ed) The Irish sea, a resource at risk. Special publication no.3. Geographical Society of Ireland. pp 107–115
Sonne C (2010) Health effects from long-range transported contaminants in Arctic top predators: an integrated review based on studies of polar bears and relevant model species. Environ Int 36:461–491
Taghavi S, Lesaulnier C, Monchy S, Wattiez R, Meargey M, van der Lelie D (2009) Lead(II) resistance in *Cupriavidus metallidurans* CH34: interplay between plasmid and chromosomally-located functions. Antonie Van Leeuwenhoek 96:171–182
Wang Q, Kim D, Dionysiou DD, Sorial GA, Timberlake D (2004) Sources and remediation for mercury contamination in aquatic systems-a literature review. Environ Pollut 131:323–336
Watt GCM, Britton A, Gilmour HG, Moore MR, Murray GD, Robertson SJ (2000) Public health implications of new guidelines for lead in drinking water: a case study in an area with historically high water lead levels. Food Chem Toxicol 38:73–79

Microbial Remediation of Organometals and Oil Hydrocarbons in the Marine Environment

4

Andreia Cruz*, Ana Julia Cavaleiro*, Ana M.S. Paulo, António Louvado, M. Madalena Alves, Adelaide Almeida, and Ângela Cunha

Abstract

Marine environments are exposed to pollution that mostly results from human activities. Organometals and oil hydrocarbons are among the most hazardous pollutants. In surface waters and along the water column, these compounds are more easily degraded than in sediments, especially under anoxic conditions, where they are highly persistent. Due to their negative impact in living organisms, decontamination of polluted marine sites with minimum collateral impacts is imperative. Bioremediation strategies, benefiting from the ability of aerobic and anaerobic microorganisms to degrade organometals or oil hydrocarbons to simpler and less toxic derivatives, represent an alternative to traditional physicochemical decontamination methods. Different bioremediation strategies have been applied in marine environments, including monitored natural recovery, biostimulation, bioaugmentation and phytoremediation. Individual microbial agents or mixed microbial consortia able to remediate these pollutants in marine environments have been identified, and the most relevant mechanisms of biodegradation of pollutants are characterised.

This chapter provides an overview on microbial bioremediation of organometals and oil hydrocarbons in marine environments, focusing on the bioremediation concept, microbial aerobic/anaerobic agents, metabolic pathways and genetic determinants involved in the degradation/transformation processes while highlighting the importance of microbial consortia and their applications. A critical analysis of the advantages and limitations of microbial remediation and a perspective on future developments are also provided.

*Both authors contributed equally to this chapter.

A. Cruz (✉) • A. Louvado • A. Almeida • Â. Cunha
Biology Department & Centre for Environmental and Marine Studies (CESAM), University of Aveiro, Campus de Santiago, 3810-193 Aveiro, Portugal
e-mail: andreiacruz@gmail.com

A.J. Cavaleiro • A.M.S. Paulo • M.M. Alves
Centre of Biological Engineering, University of Minho, Campus de Gualtar, 4710-057 Braga, Portugal

M.M. Naik, S.K. Dubey (eds.), *Marine Pollution and Microbial Remediation*,
DOI 10.1007/978-981-10-1044-6_4

4.1 Introduction

One of the major concerns of the industrialised world is the high level of marine pollution with hazardous compounds, such as organometals and oil hydrocarbons. These pollutants can reach the marine environment by different sources, mostly associated with anthropogenic activities like industrial and municipal wastewater disposal, the use of pesticides or fertilisers, sewage sludge, run-off of landfill leachates and harbour activities (Hoch 2001; Du et al. 2014; Perelo 2010) (Fig. 4.1).

The main source of organometals, in particular organotin compounds (OT), is their use as a component of antifouling paints employed as coating in the hull of ships and/or boats and in nautical equipment (Hoch 2001). Hydrocarbons reach marine environments as a consequence of biosynthesis by aquatic or terrestrial organisms, but its main sources are petrochemical industries; incomplete combustion of fuels; post-depositional transformation of biogenic precursors or diffusing from the mantle, oil spills, petroleum source rocks or reservoirs; and circulation of ships and tank washings (Perelo 2010).

These contaminants may be discharged directly in coastal waters or in open sea. In areas where chronic inputs occur, pollutants can be transferred to deeper areas, due to their adsorption to suspended particulates, and/or tend to accumulate in bottom sediments (Gadd 2000; Louvado et al. 2015; Rainbow 1995). Sediments represent the major sink for these compounds (Huang et al. 2013; Rahmanpour et al. 2014; Maanan et al. 2015; Antizar-Ladislao 2008; Oliveira et al. 2015). On the other hand, when resuspended, particulate-bound contaminants may be remobilised into the water column, becoming more bioavailable (Roberts 2012).

Organometals and hydrocarbons have immediate and long-term effects on marine

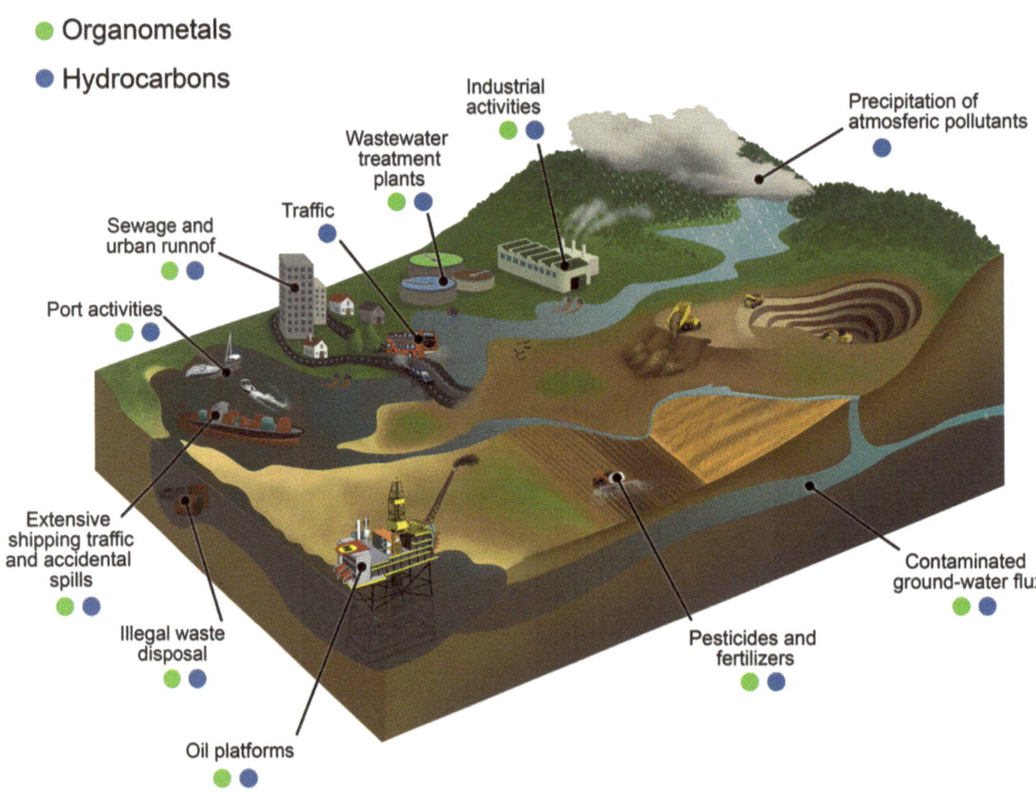

Fig. 4.1 Sources of organometals and oil hydrocarbons pollution into the marine environment

organisms, from bacteria to mammals and from the molecular to the community levels. Some of these compounds have the potential to be bioaccumulated and biomagnified along aquatic trophic webs (Cooney and Wuertz 1989; Cruz et al. 2015; McGenity et al. 2012b; O'Brien and Keough 2014; Zhang et al. 2015). Considering the magnitude and the persistence of the ecotoxicological effects, efficient and sustainable approaches for remediation are required. Conventional methods have been used to remediate pollution in marine environments. The common approaches to remove organometals are sorption, ion exchange, precipitation and electrochemical techniques (Du et al. 2014). In turn, the remediation of hydrocarbons has been performed through physical adsorption with activated carbon, hydrophobic meshes and nanomaterials (Han et al. 2015; Nafees et al. 2013; Lin et al. 2015; Sabir 2015), chemical dispersants (Kujawinski et al. 2011) and oxidants (Usman et al. 2012). Most of these methods are expensive, are labour demanding and can generate secondary materials or by-products that must often be disposed as hazardous compounds (Lee et al. 2012; Du et al. 2014). Thus, as the use of bioremediation, other alternatives are being considered.

The concept of bioremediation has been described as the process of using the metabolic potential of microorganisms to immobilise, degrade, transform or remove hazardous components from the environment (Watanabe 2001). In fact, microorganisms play important roles in the environmental fate of toxic compounds through a multiplicity of physical, chemical and biological mechanisms (Gadd 2004).

Bioremediation techniques can be classified as *ex situ* and *in situ*. *Ex situ* approaches imply the removal of the contaminated material from the natural site to be treated elsewhere. On the other hand, *in situ* technologies are performed at the contaminated site (Boopathy 2000). Strategies for the *in situ* bioremediation of marine environment include (i) monitored natural recovery (MNR) using the "self-healing" capacities of the indigenous microbial population in combination with naturally occurring physical and chemical processes (Yu et al. 2005); (ii) biostimulation as the metabolic activation of the indigenous population by manipulating the factors that affect microbial growth, as, for instance, the supply of nutrients (such as the addition of N and P to counteract nutrient limitation) or surfactants that enhance dispersion and bioavailability of the pollutant (Zhang and Lo 2015; Yu et al. 2005); (iii) bioaugmentation as the introduction of natural or engineered species that enhance biodegradation/transformation of specific contaminants (Cruz et al. 2014); and (iv) phytoremediation as the use of plants and algae for the degradation and removal of contaminants from the environment (Iwamoto and Nasu 2001; Czako et al. 2006; Cruz-Uribe and Rorrer 2006). Combinations of bioremediation strategies, for example, phytoremediation and biostimulation, may also be applied (Bianchi et al. 2010; Moreira et al. 2011; Oliveira et al. 2015).

Low cost is one of the major advantages of bioremediation relatively to conventional strategies. Additionally, it is a non-invasive approach with minimum undesirable effects in the ecosystem (Perelo 2010). However, bioremediation has some drawbacks that can limit the success of its application, such as the relatively long periods of time involved and the higher uncertainty of results, in comparison with conventional methods, because of the multiplicity of environmental factors affecting the biological processes (Perelo 2010; Boopathy 2000). At the laboratory scale, bioremediation of organometals and hydrocarbons has shown to be efficient, even for cleaning up contaminated samples presenting low toxic concentrations, which would be impracticable with conventional remediation techniques. At environmental scales, few examples of hydrocarbon bioremediation were reported (Rocchetti et al. 2014).

This chapter provides an overview of the main microbial agents involved in the bioremediation of organometals and oil hydrocarbons in marine environments, metabolic pathways, transformation processes and associated genes and presents a critical analysis of the efficiency and constraints to the practical implementation of bioremediation protocols.

4.2 Organometals

4.2.1 Organometal Contamination in Marine Environments

Organometals are generally defined as compounds having at least one metal–carbon (normally metal–alkyl) polarised bond, i.e. where the carbon is more electronegative than the metal (Craig 2003). These compounds can occur in the environment by being naturally formed there or associated to anthropogenic inputs (Gadd 1993).

Organometals formed by arsenic, mercury, tin and lead are used in a wide variety of industrial processes due to their biocidal properties (Craig 2003; Frache and Rivaro 2000). Within organometals, organotins (OT) are the most widely used, reaching a worldwide production estimated at 50,000 tons; thus, their impact in the marine environment is highly relevant (Cruz et al. 2015; Ayanda et al. 2012). OT conform to the general formula $R_{(4-n)}SnX_n$ with $n = 0 - 3$, in which R is any organic alkyl or aryl group and X is an anionic species, namely, halide, oxide, chloride and hydroxide (Rudel 2003). The properties of OT vary significantly, depending mainly upon the number and nature of the R groups, but also upon the type of the ligand (X) (Hoch 2001). The toxicity of the OT can be more influenced by the alkyl substitutes than the anionic substitutes (Antizar-Ladislao 2008). In general, inorganic tin is nontoxic, whereas trisubstituted compounds have maximum toxicological activity (Sekizawa et al. 2003).

The solubility of OT in water decreases with increasing number and length of the organic substitutes, but it also depends on the particular X (Hoch 2001). Water solubility of most OT is low and also dependent on pH, ionic strength and temperature. Data on solubility for tributyltin chloride (TBT-Cl) range from 5 to 50 mg L^{-1}, whereas for dibutyltin chloride (DBT-Cl_2) is as high as 92 mg L^{-1} (Rudel 2003). Due to low water solubility, OT strongly bind to suspended material and inorganic sediment particles (Laughlin Jr et al. 1986).

Tributyltin (TBT) was, during decades, the most used organometal. International Maritime Organization (IMO) called for a worldwide treaty that bans the application of TBT-based paints starting 1 January 2003 and total prohibition by 1 January 2008 (EC 2002; IMO 2001). As a consequence of partially prohibiting the use of TBT in several developed countries, the recovery of affected gastropod and oyster populations has been observed (Alzieu 1998; Evans et al. 2000; Evans 1999). However, the use of TBT as biocide in antifouling paints is still permitted in IMO non-member countries (Sousa et al. 2013). Also the ineffectiveness of the alternative products has led to the illegal use of TBT, resulting in environmental concentrations that remain high enough to motivate concern (Okoro et al. 2011; Santillo et al. 2001; Barroso and Moreira 2002; Santos et al. 2002).

The major deleterious effect of TBT is imposex, a phenomenon characterised by the superimposition of male sexual characteristics (vas deferens, penis) on female gastropods (Smith 1971; Barroso et al. 2002), caused by endocrine disruption. However, negative effects of TBT are not restricted to invertebrates. TBT inhibits the growth and metabolic activity of bacteria (Gadd 2000; Cruz et al. 2012) and marine microalgae (Gadd 2000), causes apoptosis in marine sponges (Batel et al. 1993), decreases neonate survival in crustaceans (Takeuchi et al. 2001), causes alterations on shell growth in molluscs (Alzieu 2000) and causes masculinisation in fish females (Shimasaki et al. 2003) and inhibition of hepatic cytochrome P450 and ethoxyresorufin-O-deethylase (EROD) and penthoxyresorufin-O-depenthylase (PROD) in marine mammals (Kim et al. 1998). Additionally, TBT shows a substantial potential for bioaccumulation and biomagnification through the food chains, thus being of particular concern in long-lived biota (Murata et al. 2008).

Dafforn et al. (2011) revised the global levels of TBT in marine environments and reported that in the water column, the maximum values (3.2 µg

L^{-1}) were observed in Singapore in 2000, and in sediments, the maximum value (89 μg g^{-1}) was registered in Australia, in 2006. The hotspots of TBT are the marinas and commercial harbours in developed countries, where the contamination in the surface water and sediments is correlated with the intensity of shipping or boating activities (Dafforn et al. 2011).

In the aquatic environment, TBT is quickly removed from the water column and accumulated in the sediment compartment due to the high specific gravity (near 1.2 kg L^{-1} at 20 °C) (Landmeyer et al. 2004), low solubility (less than 10 mg L^{-1} at 20 °C and pH 7.0) (Fent 1996) and log K_{ow} values near 4.4 at pH 8. Additionally, TBT is ionisable and exhibits a pK_a acidity constant of 6.25 (Antizar-Ladislao 2008). TBT can be degraded by debutylation to form dibutyltin (DBT), monobutyltin (MBT) and ultimately inorganic tin (Cooney 1995; Gadd 2000; Antizar-Ladislao 2008). Considering that TBT and its degradation products are frequently associated with marine pollution, these compounds will be addressed in this chapter as a model organometal contaminant, and a particular focus will be given to their properties and bioremediation strategies.

Rates of TBT degradation may be influenced by several biotic and abiotic factors, such as concentration of dissolved/suspended organic matter, salinity, temperature, pH, light and the nature and density of the microbial communities (Dubey and Roy 2003). However, the degradation process occurs primarily through biotic processes, and the efficiency of TBT biodegradation will depend mainly on the bioavailability of this compound (Sakultantimetha et al. 2011).

In the marine environment, the half-life of TBT can be different depending on the ecological compartment. In the water column, TBT can be rapidly degraded with a half-life of several days (Gadd 2000). In sediments, the rate of TBT debutylation is dependent upon microbial activity (Sternberg et al. 2010), and in deep sediments, the degradation rates are much lower than in the water column, being the half-life within 1.9–3.8 years (Gadd 2000).

Aerobic conditions facilitate the degradation of TBT, which takes about 1–3 months to be degraded, in comparison with anaerobic conditions, that result in higher persistence of the compound (6 months to 8.7 years) (Sternberg et al. 2010).

A microcosm set-up, simulating a temperate shallow estuarine, was used to determine the persistence and behaviour of TBT. Radiolabelled TBT was introduced in a microcosm containing estuarine water and sediment. The experiment was conducted during 40 days, and TBT and its degradation products were monitored during this period. TBT rapidly disseminated to the three compartments of the microcosm, being more than 60 % of the TBT and of its metabolites found in the sediment. In the water column, the TBT half-life was 2.55 days for the first 11 days and then slowed to 13.4 days. The concentration of TBT adsorbed on suspended particles was three orders of magnitude higher than that in the dissolved form (Dai et al. 1998).

4.2.2 Bioremediation of Organotin Compounds (OT) in Marine Environments

In the water column as in sediments, OT degradation is predominantly performed by microorganisms such as bacteria, fungi and microalgae (Table 4.1).

Although the biodegradation of TBT in marine environments is long known, the mechanistic reactions involved are still not completely understood. It was suggested that the enzyme homogentisate 1,2-dioxygenase is involved in the aerobic degradation of butyltins in microalgae, via successive dealkylations (Tsang et al. 1999; Lee et al. 1989).

Recently, two strains with TBT-degradative capability, *Klebsiella pneumoniae* SD9 and *Alcaligenes faecalis* SD9, showed enhanced siderophore production, in the presence of TBT (Khanolkar et al. 2015a, b). Moreover, *Alcaligenes faecalis* SD9 increased the production of exopolysaccharide (EPS) when exposed

Table 4.1 OT-degrading microorganisms isolated from marine environment

	Agent	Reference
Fungi	*Tramates versicolor*	Barug (1981)
	Chaetomium globosum	
	Coniophora puteana	Dubey and Roy (2003)
	Coriolus versicolor	
	Cunninghamella elegans	Bernat and Dlugonski (2006)
Microalgae	*Pavlova lutheri*	Saint-Louis et al. (1994)
	Skeletonema costatum	Lee et al. (1989)
	Nannochloropsis oculata	Taha et al. (2009)
	Dunaliella parva	
	Leptocylindrus danicus	Xie et al. (2011)
	Amphidinium carterae	
Bacteria	*Pseudomonas aeruginosa*	Barug (1981)
	Alcaligenes faecalis	
	Pseudomonas diminuta	Kawai et al. (1998)
	Pseudomonas aeruginosa USS25	Roy and Bhosle (2006)
	Pseudomonas stutzeri DN2	Khanolkar et al. (2014)
	Aeromonas molluscorum Av27	Cruz et al. (2007)
	Enterobacter cloacae TISTR1971	Sakultantimetha et al. (2010)
	Alcaligenes faecalis SD5	Khanolkar et al. (2015b)
	Klebsiella pneumoniae SD9	Khanolkar et al. (2015a)

to TBT (Khanolkar et al. 2015b). Thus, in these studies, authors concluded that siderophore and EPS might be involved in TBT degradation processes. Nonetheless, information on microbial degradation processes is still scarce.

Some few studies were performed with algae and fungi (Barug 1981; Dubey and Roy 2003; Bernat and Dlugonski 2006; Lee et al. 1989; Saint-Louis et al. 1994; Taha et al. 2009; Xie et al. 2011), but the vast majority is focused on bacteria.

TBT degradation rates and the changes in the bacterial community were evaluated during 150 days in microcosm experiments with sediments from the Mekong River in Vietnam (Suehiro et al. 2006). The initial concentration of TBT was of 1.0–1.4 $\mu g\ g^{-1}$ (dry weight) and decreased to 0.6 $\mu g\ g^{-1}$ (dry weight) at the end of the experiment. The identifications of the TBT-degrading strains were not achieved, but denaturing gradient gel electrophoresis (DGGE) and bacterial number profiles indicated that the community was well adapted to TBT pollution in the Mekong River sediment and could efficiently contribute for TBT degradation (Suehiro et al. 2006).

At a larger scale, a mesocosm experiment during which TBT and its degradation products were monitored was performed during 278 days. Radiolabelled TBT (590 ± 20 ng L^{-1}) was added to a 13 m^3 marine enclosure with near-natural water column and benthos. The degradation rate was estimated as 0.20 per day for 15 days, decreasing afterwards to 0.10 per day. Two thirds of the TBT present in the water was degraded to DBT that was further converted to MBT, and one third of the TBT was degraded directly to MBT (Adelman et al. 1990).

A 150-day microcosm approach was used to evaluate the ability of *Aeromonas molluscorum* Av27, a TBT-resistant/degrading bacterium, to enhance TBT degradation in estuarine sediments. At the end of the experiment, 28 % of the TBT had been degraded to DBT and MBT. TBT degradation was significantly enhanced by *A. molluscorum* Av27. The characterisation of the structural diversity of bacterial community indicated that *Proteobacteria* was the predominant phylum, followed by *Bacteroidetes*. At the beginning of the experiment, the relative abundance of different classes could be ordered as *Gammaproteobacteria* (55 %) > *Deltaproteobacteria*(34 %)>*Alphaproteobacteria* (6 %) > unclassified *Proteobacteria* (5 %) (Cruz et al. 2014). At the end of the experiment, the abundance of *Deltaproteobacteria* and *Alphaproteobacteria* increased. *Deltaproteobacteria* includes sulphur- and metal-reducing bacteria (Webster et al. 2010), and *Alphaproteobacteria* includes members that are sensitive to hydrocarbons (LaMontagne et al. 2004; Polymenakou et al. 2005; Sun et al. 2013), which may have contributed to the degradation of pollutants over time. Although some changes in bacterial community structure were noticed, the relative abundance of the phylotypes was generally maintained (Cruz et al. 2014).

The biological degradation of TBT in harbour sediments under aerobic and anaerobic conditions and upon land deposition demonstrated that TBT degradation was faster under aerobic conditions, as above mentioned, and it increased with increasing temperature. Using 320 μg kg^{-1} of TBT, it was observed that TBT half-life decreased from 90 to 4 weeks when temperature raised from 5 to 55 °C, in aerobic conditions. After 7 months, at 55 °C, in aerobic conditions, the degradation was completed, being the concentration of TBT below the detection limit (1 μg kg^{-1} dw) (Brandsch et al. 2001).

Experiments with sediments from various Japanese coastal areas, involving successive enrichments with 100 μg L^{-1} of TBT with 21-day intervals, were conducted in order to select TBT-degrading bacteria. The ratio of TBT degradation varied between 0 and 72.7 %. Two microorganisms with high homology with *Halomonas aquamarina* and *Halomonas alimentaria* were isolated and identified, and the ability to degrade 60 μg L^{-1} and 97 μg L^{-1} of TBT after 21 days of incubation was demonstrated (Hamada-Sato et al. 2002).

The effect of bioaugmentation and the combined action of coadjuvants to enhance bioavailability was confirmed in microcosm experiments. Under natural attenuation control conditions, the half-life of TBT was 578 days confirming its high persistence in the sediment. Different experimental conditions consisting in the improvement of bioavailability (manipulation of salinity, surfactant addition and sonication), the stimulation of biological activity (aeration and temperature manipulation), the inoculation with a TBT-resistant bacterium (*Enterobacter cloacae*) and nutrient addition (succinate, glycerol and L-arginine) were tested. The shorter half-life estimates were 4–5 days, after bioavailability improvement and inoculation of *E. cloacae*, both with or without nutrient addition (Sakultantimetha et al. 2011).

As shown in the examples given above, some of the bioremediation processes described are attributed to bacteria consortia, usually associated with the indigenous bacterial community. By the many experiments performed, it has been reinforced that the use of microbial consortia offers considerable advantages over the use of pure cultures.

4.3 Oil Hydrocarbons

4.3.1 Oil Hydrocarbon Contamination in Marine Environments

Oil demand is increasing worldwide, largely due to development countries, averaging 94.0 million barrels per day in 2015 (International Energy Agency 2015). The intensive exploitation of petroleum sources, and the vast range of activities developed by the oil industry, increases the risk of accidental oil spills. Oil-derived pollutants enter marine environments through leaks during drilling, transport and storage, discharge of water at offshore oil exploration and combustion of fossil fuels (Notar et al. 2001; Guitart et al. 2007, 2010). Chronic contamination of harbour sediments from shipping activities, fuel oil spills and run-offs is also frequent (Coates et al. 1997). In addition, oil hydrocarbons are introduced in the marine ecosystems by natural sources such as oil seepages that represent 47 % of the oil released to marine environments (National Academy of

Science 2002; Kvenvolden and Cooper 2003). Despite the significant contribution of natural sources, anthropogenic oil spills are still the major cause for the fast and uncontrolled release of high quantities of oil in specific areas, causing severe ecological and economical damages. In 1989, the Exxon Valdez oil spill in Alaska caused the release of 41–144,000 m^3 of crude throughout 2100 km of coastline, affecting surface water, sediments, wetland, beaches and other coast features and caused the death of microorganisms, plants and animals, including birds, cetaceans, sea otters and seals (Carson et al. 1992; Peterson et al. 2003). During the Gulf War (1990–1991), more than 10 million barrels (approx. 1.6 million m^3) of crude oil were spilled in the Arabian Gulf, and the impacts on marshlands and mud tidal flats persisted for over a decade (Bejarano and Michel 2010). In 2010, the Deepwater Horizon oil spill in the Gulf of Mexico released 4.9 million barrels (approx. 0.8 million m^3) of crude oil that impacted salt marshes (Silliman et al. 2012) and animals (Barron 2012).

At sublethal concentrations (chronic contamination), hydrocarbons induce growth inhibition in sensitive prokaryote groups (Vázquez and Rial 2014) and significant changes in the structure and function of bacterial and fungal communities (Bik et al. 2012). They affect embryogenesis and growth in invertebrates (Bellas et al. 2013; Vieira and Guilhermino 2012), cause oxidative stress in benthic communities (Coelho et al. 2015), induce phototoxicity and metabolic disorders, impair development in fish, particularly in early growth stages (Incardona et al. 2013; Lucas et al. 2014), cause immunotoxicity and adrenal and lung disease in marine vertebrates, including mammals (Barron 2012; Schwacke et al. 2013), and reduce growth and photosynthetic activity in algae and macrophytes (Lewis and Pryor 2013).

Crude oil is a very complex mixture of organic compounds, mainly composed by aliphatic and aromatic hydrocarbons, resins and asphaltenes. Aliphatic hydrocarbons may be saturated or unsaturated (with double bonds in the carbon chain) and can appear with linear, branched or cyclic structures. Aromatic hydrocarbons possess one or more organic rings with delocalised electrons. The different physical and chemical properties of the oil components, namely, solubility, volatility and tendency to adsorb to particles or sediments (Table 4.2), greatly influence bioavailability, biodegradation and, subsequently, oil composition after its release in the environment.

Short-chain alkanes, monoaromatics and low molecular polycyclic aromatic hydrocarbons (PAHs), which have a higher solubility and vapour pressure (Table 4.2), will dissolve and evaporate easily. Afterwards, long-chain and alkylated alkanes will be biodegraded, leaving a remnant composed of high molecular weight (HMW) PAH, resins and other more recalcitrant pollutants. The rate of biological degradation is lower for more complex hydrocarbons, which are more resistant to biodegradation (Atlas 1995).

4.3.2 Hydrocarbon Degradation by Aerobic Marine Bacteria

Marine oil-degrading bacteria tend to be metabolically more specialised than their terrestrial counterparts (Yakimov et al. 2007). Some hydrocarbon-degrading bacteria actually prefer other carbon sources, but obligate hydrocarbon-degrading bacteria, also designated as hydrocarbonoclastic bacteria, are also present in the environment and will flourish after pollution (Yakimov et al. 2007). Hydrocarbon-degrading capacity is found in a wide range of *taxa* but is particularly frequent among α and γ-*Proteobacteria* (Yakimov et al. 2007). Aerobic alkane biodegradation can be performed by facultative anaerobic or obligate aerobic bacteria (Yakimov et al. 2007). Facultative anaerobic bacteria have a versatile metabolism and will not degrade alkanes if other more attractive carbon sources are available (Rojo 2009). Some examples of alkane-degrading bacterial *genus* are *Alcanivorax*, *Oleiphilus*, *Oleispira* and *Thalassolituus* (Yakimov et al. 2007). PAH-degrading isolates are widespread among different bacteria phyla, being related to *Proteobacteria*

Table 4.2 Physical and chemical properties of selected hydrocarbons

Group[a]	Name	Molecular formula	Molecular weight (g mol^{-1})	Solubility (mg L^{-1})[b]	Vapour pressure (Pa)[b]	Log K_{ow}[c] (–)	EPA carcinogenic classification[d]
Aliphatics	Hexane	C_6H_{14}	86.2	13.0	17,600	3.3	D
	Hexadecane	$C_{16}H_{34}$	226.4	4.5×10^{-5}	<10	8.3	D
	Cyclohexane	C_6H_{12}	84.2	43.0	10,300	3.2	D
BTEX	Benzene	C_6H_6	78.1	2.0×10^{-2}	12,700	2.1	A
	Toluene	C_7H_8	92.1	6.0×10^{-3}	2,920	2.7	D
	Ethylbenzene	C_8H_{10}	106.2	1.0×10^{-3}	950	3.2	D
	Xylene	C8H10	106.2	2.0×10^{-3}	787	3.0–3.2	D
PAH	Naphthalene	$C_{10}H_8$	128.2	31.7	11	3.3	C
	Phenanthrene	$C_{14}H_{10}$	178.2	1.3	2×10^{-2}	4.5	D
	Pyrene	$C_{16}H_{10}$	202.3	0.1	6×10^{-4}	5.3	D

[a]Aliphatic hydrocarbons, monoaromatic (BTEX benzene, toluene, ethylbenzene and xylene) or polycyclic aromatic hydrocarbons (PAH)
[b]Aqueous solubility and vapour pressure at 20 °C
[c]K_{ow} = octanol–water partition coefficient
[d]EPA carcinogenic classification (all classes described): A-human carcinogenic, B1 and B2-probable human carcinogenic, C-possible human carcinogenic, D-not classifiable as to human carcinogenicity, E-evidence of noncarcinogenicity for humans
Adapted from Weelink et al. (2010), McGenity et al. (2012b)

(α, β and γ classes), *Actinobacteria*, *Cyanobacteria*, *Bacteroidetes* and *Firmicutes*. Bacteria belonging to the genera *Cycloclasticus*, *Pseudomonas* (γ-*Proteobacteria*) and some genera of the *Sphingomonadaceae* family (γ-*Proteobacteria*, *Novosphingomonas*, *Sphingomonas* and *Sphingobium*) are frequently referred to as PAH degraders (Dyksterhouse et al. 1995; Kasai et al. 2002).

Before a pollution event, oil-degrading bacteria have a low relative abundance in the undisturbed community (Vila et al. 2010). In response to an oil spill, the pioneer oil-degrading populations will bloom, and with time, the bacterial community will increase in diversity, as secondary strains feed on bacterial metabolites and on the less toxic oil intermediates (Vila et al. 2010; Yakimov et al. 2007). In the aftermath, the community is reported to return to its initial structure (Kasai et al. 2001; Yang et al. 2014). Shifts in the metabolic and structural diversity of the community are directly related with the pool of hydrocarbons (Ringelberg et al. 2001). A field work conducted after the Deepwater Horizon oil spill revealed that bacterial communities developing immediately after the accident were dominated by *Oceanospirillaceae* and pseudomonads. However, in a subsequent phase, when the proportion of alkanes and cycloalkanes was the highest, the relative abundance of PAH degraders like *Colwellia*, *Cycloclasticus* and *Pseudoalteromonas* increased. *Methylomonas* and *Methylophaga* responding to methane in the presence of high molecular weight organic compounds persisted for weeks after the accident (Dubinsky et al. 2013).

Pseudomonas species have widely documented as hydrocarbon degraders, and the genes related with the capacity to degrade different hydrocarbons are associated with plasmids that may be transmitted between cells (Friello et al. 2001). This was the basis for the patented construction of super-degrader strains of *P. aeruginosa* carrying multiple degradative plasmids (Chakrabarty 1992).

4.3.2.1 Pathways of Aerobic Hydrocarbon Degradation

Aerobic biodegradation of alkanes and aromatic hydrocarbons can occur within the same bacterial species, although some species are specialised in the utilisation of one or the other type of hydrocarbons (Rojo 2009; Yakimov et al. 2007). In both processes, the reactions occur intracellularly, though the uptake of the pollutants is limited by their low bioavailability. To overcome this, PAH- and alkane-degrading bacteria adhere to oil droplets and secrete biosurfactants that enhance emulsification and assist uptake (Ron and Rosenberg 2002). Alkanes and PAH are very stable molecules that present a reduced reactivity. Thus, their biodegradation is unattractive since a large amount of energy is needed to activate these compounds (Rojo 2009). In aerobic catalysis, alkanes and PAH activation occurs in the first step of biodegradation through oxygenation of the compound by monooxygenases in the case of alkanes and by dioxygenases for PAH (although monooxygenases are also reported). These reactions make the organics more reactive in future biodegradation steps (Rojo 2009). Alkane oxygenation can occur at terminal or subterminal positions in the carbon chain. In both cases, alkane will form a fatty acid that will be degraded by a β-oxidation pathway. A schematic representation of the aerobic pathway of degradation PAH is presented in Fig. 4.2. The di- or monooxygenation of the molecule is catalysed by the α-subunit of an aromatic ring enzyme and results in a metabolite with one or two hydroxyl radicals. This intermediate ring structure is cleaved by intradiol or extradiol oxygenases, yielding protocatechuates and catechol intermediates (Jiménez et al. 2004; Nzila 2013; Seo et al. 2009). Frequently, PAH with benzenic rings sequentially form dihydriols with n-1 aromatic rings until reaching catechol that, by ring cleavage, is converted to tricarboxylic acid cycle (TCA) intermediates (Nzila 2013). In alkane and PAH biodegradation, the downstream steps are more common among bacteria, while the oxygenation reaction is less widespread and considered the rate-limiting step (de Lorenzo 2008). The genes that encode these essential enzymes are frequently located in mobile genetic elements (i.e. transposons and plasmids) that are prone to horizontal transfer, which, in some cases, may occur between strains phylogenetically distant (broad-host range) (van Beilen et al. 2001; Jutkina et al. 2011; Ben Said et al. 2008). These elements may remain after the recovery period and will quickly respond to a new pollution event (Yang et al. 2014).

4.3.3 Hydrocarbon Degradation by Anaerobic Marine Bacteria

In the water column or in surface waters, oil bioremediation occurs mainly due the activity of aerobic microorganisms. However, a significant amount of water-soluble or particle-associated hydrocarbons may settle on the seafloor, reaching deep-sea sediments where oxygen is scarce (Kimes et al. 2014). Oil spills may also impact the shoreline, affecting beaches and salt marshes (Brundrett et al. 2015). In the fine-grained intertidal sediments, oxygen diffusion is limited, and these environments are generally anoxic (Head and Swannell 1999; McGenity 2014).

Anaerobic biodegradation of aliphatic and aromatic hydrocarbons has been demonstrated in laboratory enrichment cultures developed from marine sediments and was reported to occur *in situ* in several marine environments (Coates et al. 1997; Weelink et al. 2010; Jaekel et al. 2013). Biodegradation at *in situ* conditions was also verified in mesocosms artificially contaminated with naphthalene (Massias et al. 2003; Suárez-Suárez et al. 2011) or crude oil (Massias et al. 2003; Suárez-Suárez et al. 2011) which were installed on the seabed of undisturbed areas. Although anaerobic oil biodegradation is widespread, natural attenuation in anoxic marine sediments is slow, and oil persists for long periods of time (Kolukirik et al. 2011).

Pure cultures of anaerobic Archaea and Bacteria able to couple hydrocarbon degradation to sulphate or Fe(III) reduction have been isolated from hydrocarbon-rich marine environments, namely, contaminated sediments and hydrocarbon seeps, but no isolate obtained from these environments was reported using nitrate as electron acceptor (Mbadinga et al. 2011; Weelink et al. 2010).

The majority of the isolated alkane-degrading anaerobic bacteria retrieved from marine environments are sulphate reducers affiliated with the

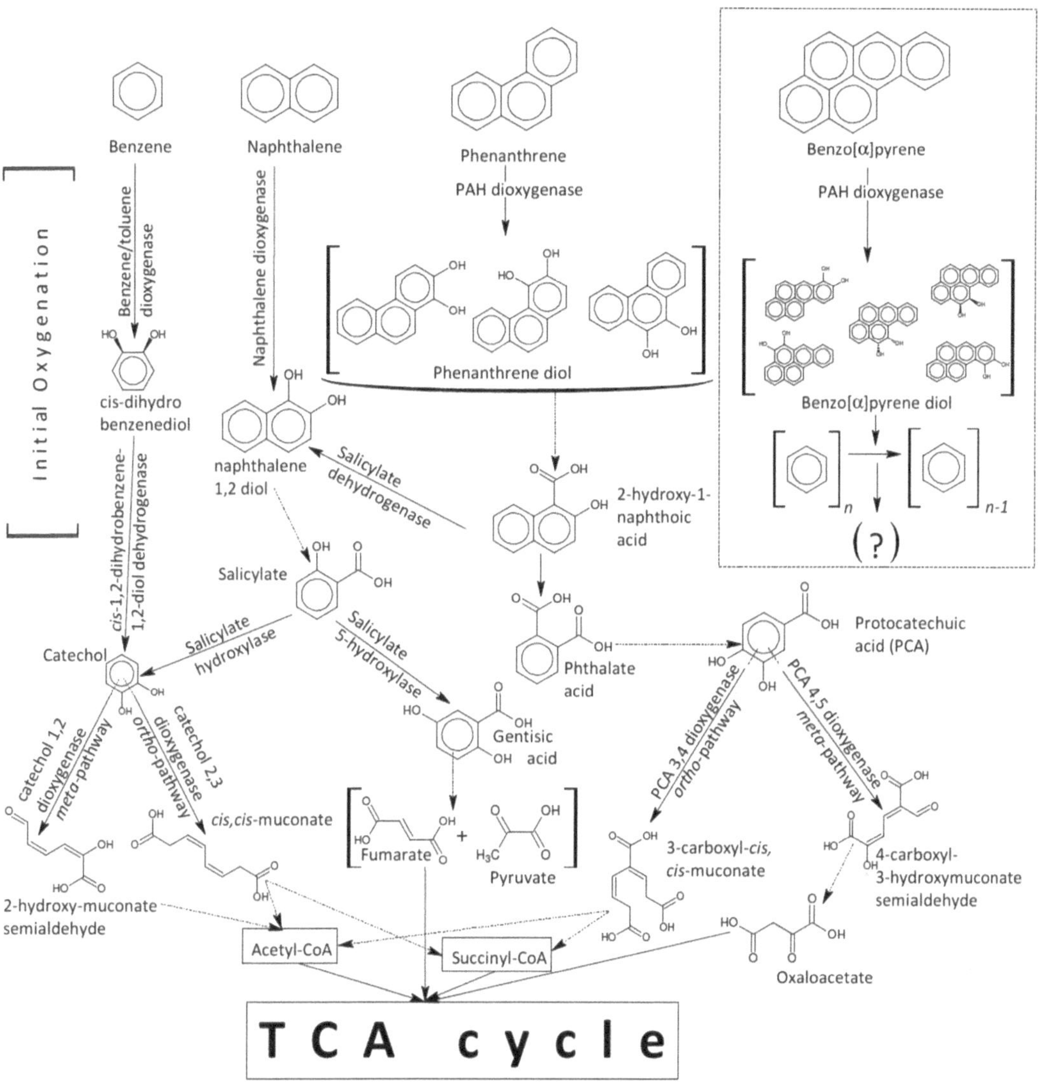

Fig. 4.2 Schematic representation of pathways of aerobic degradation of aromatic hydrocarbons

family *Desulfobacteraceae* within the *Deltaproteobacteria* (Table 4.3). Among them, the ability to grow on long-chain *n*-alkanes (C12–C20) was described in *Desulfococcus oleovorans* Hxd3, *Desulfatibacillum aliphaticivorans* CV2803T, *Desulfatibacillum alkenivorans* AK-01 and strain Pnd3, from the *Desulfococcus/Desulfatibacillum* cluster (Mbadinga et al. 2011). Strain BuS5, clustering within *Desulfosarcina/Desulfococcus* clade, is only able to grow on hydrocarbon gases propane and butane (Kniemeyer et al. 2007). Novel cold-adapted propane- and butane-degrading enrichment cultures were recently reported by Jaekel et al. (2013), all affiliated with the *Desulfosarcina/Desulfococcus* cluster. The crucial role of this microbial clade on anaerobic alkane biodegradation was also highlighted by Kleindienst et al. (2014) using stable isotope probing (SIP) techniques.

Only one thermophilic axenic culture capable of alkane biodegradation was reported, *Desulfothermus naphthae* TD3^T (Rueter et al. 2004). This microorganism was retrieved from sediments of the hydrothermal vent Guaymas

Table 4.3 Overview of bacterial axenic cultures isolated from marine environments capable of hydrocarbon degradation coupled to sulphate reduction

	Class	Source	Substrate			Motility	Spores	Optimal T	pH	Salinity ($g\ L^{-1}$)	Ref.
			Alkanes	Alkenes	Aromatics			(°C)	Optimum	(Range/optimal)	
Desulfococcus oleovorans Hxd3	δ-prot	Oil–water separator	C12–C20	C16	n.d	N	N	28–30	n.d.	20[a]	Aeckersberg et al. (1991)
Desulfatibacillum aliphaticivorans CV2803^T	δ-prot	Contaminated marine sediment	C13–C18	C7–C23	–	N	N	28–35	7.5	6–45/24	Cravo-Laureau et al. (2004)
Desulfatibacillum alkenivorans AK-01	δ-prot	Contaminated estuarine sediment	C13–C18	C15, C16	n.d	N	N	26–28	6.9–7.0	1–60/10	So and Young (1999)
Strain Pnd3	δ-prot	Marine sediments	C14–C17	C16	n.d	N	n.d.	30	n.d.	20[b]	Aeckersberg et al. (1998)
Strain BuS5	δ-prot	Marine hydrocarbon seeps	C3–C4	n.d.	n.d	n.r	n.d.	28	6.9	26[b]	Kniemeyer et al. (2007)
Desulfothermus naphthae TD3^T	δ-prot	Sediments from hydrothermal vent	C6–C14	–	–	Y	n.d.	60–65	6.5–6.8	21	Rueter et al. (2004)
Desulfobacula toluolica Tol2	δ-prot	Marine sediments	–	n.d.	To	Y	n.d.	28	7.0–7.1	20	Rabus et al. (1993)
Desulfotignum toluenicum H3	δ-prot	Oil reservoir model column	–	n.d.	To	N	n.d.	34	7.2	5-55/15	Ommedal and Torsvik (2007)
Strain oXyS1	δ-prot	Oil-contaminated sediments	–	n.d.	To, oXy	n.r.	n.d.	32	7.5	26[b]	Harms et al. (1999)
Strain mXyS1	δ-prot	Oil-contaminated sediments	–	n.d.	To, mXy	n.r.	n.d.	30	7.2	26[b]	Harms et al. (1999)

δ-prot Deltaproteobacteria, *n.d.* not determined, *n.r.* not reported, *Y* yes, *N* no, *Be* benzene, *Nap* naphthalene, *To* toluene, *mXy m*-xylene, *oXy o*-xylene, – = no growth observed
[a]No growth in freshwater;
[b]Growth was observed at this value, but the range and optimum conditions were not evaluated

Basin and confirmed to use C6 to C14 alkanes as substrates (Table 4.3). Another group of microorganisms that degrade aliphatic hydrocarbons is composed by Archaea from the genus *Archaeoglobus* and the order *Thermococcales. Archaeoglobus*-related species are not known to utilise alkanes, but *Archaeoglobus fulgidus* strain VC-16 reduces thiosulphate coupled to the anaerobic oxidation of n-alkenes (Beeder et al. 1994).

Among the monoaromatic hydrocarbons, benzene is the most recalcitrant to anaerobic biodegradation. Benzene oxidation has been observed in sediments and in microcosms, but only recently isolates capable of benzene biodegradation were described (e.g. *Dechloromonas* and *Azoarcus* spp.*)*. However, these isolates use nitrate as terminal electron acceptor, and none was isolated from the marine environment (Weelink et al. 2010). Several sulphidogenic bacterial isolates belonging to the *Deltaproteobacteria* class were reported as capable of growing on toluene. Four were retrieved from marine sediments: *Desulfobacula toluolica, Desulfotignum toluenicum,* strain oXyS1 (98.7 % similarity with *Desulfosarcina ovata*) and strain mXyS1 (86.9 % similarity with *Desulfococcus multivorans*) (Table 4.3).

Pure archaeal and bacterial cultures able to use Fe(III) as electron acceptor in the degradation of aromatic hydrocarbons have also been reported. *Ferroglobus placidus* is a hyperthermophilic Archaea isolated from a hydrothermal vent sediment, capable of benzene degradation at 85 °C coupled to Fe(III) reduction (Holmes et al. 2011).

Oil biodegradation under methanogenic conditions was first demonstrated in highly enriched cultures able to convert long-chain alkanes to methane and CO_2 (Zengler et al. 1999). Recent studies provided strong evidence on the phylogeny of the microorganisms involved in these reactions. *Smithella/Syntrophus* spp. from the *Syntrophaceae* family, within *Deltaproteobacteria* class, have been associated to alkane biodegradation, in syntrophy with hydrogenotrophic methanogens (Zengler et al. 1999; Gray et al. 2011; Siddique et al. 2012). Although these reactions are endergonic under standard conditions, as shown by the positive Gibbs free energy change (Dolfing et al. 2008), they become energetically favourable when the hydrogen partial pressure is low, which is generally achieved by the activity of methanogenic or sulphidogenic partners. Anaerobic degradation of benzene under methanogenic conditions was also demonstrated in microcosms (Weiner and Lovley 1998).

4.3.3.1 Pathways of Anaerobic Hydrocarbon Degradation

Usually, the microorganisms involved in aromatic hydrocarbons biodegradation are not able to degrade aliphatics, and the reverse is also true. Several metabolic pathways of anaerobic hydrocarbon biodegradation have been proposed, all involving biochemical reactions that differ completely from the ones described for aerobic hydrocarbon catabolism.

The first mechanism described was shown to occur during toluene biodegradation. Activation of the aromatic hydrocarbon occurs by addition of the methyl group to the double bond of fumarate, yielding benzylsuccinate, which is then further converted to benzoyl-CoA. This compound was previously identified as an important intermediate in the degradation of several aromatic hydrocarbons, resulting from a carbon–carbon reaction that does not require any co-substrate. The enzyme involved in toluene activation, benzylsuccinate synthase (*bssABC*), presents a type of glycine radical protein. The *bss* pathway was found in phylogenetically diverse bacteria-degrading toluene under nitrate-, Fe(III)- and sulphate-reducing conditions, and *bss* genes were also identified in a methanogenic toluene-degrading culture, as well as in marine sediments (Weelink et al. 2010; von Netzer et al. 2013). Other pathways for toluene activation have also been proposed, namely, hydroxylation of the methyl group via a dehydrogenase to benzyl alcohol or hydroxylation to cresol, although the latter is considered a minor pathway (Weelink et al. 2010; Widdel and Rabus 2001).

A mechanism similar to the *bss* pathway was proposed for the activation of alkanes by anaerobic bacteria, despite the higher energy required for C–H bond cleavage in alkanes comparatively to toluene. In this case, fumarate addition at the subterminal position C-2 of alkanes yields an alkyl-succinate and is catalysed by the glycyl radical enzyme alky-succinate synthase, encoded by the *assABC* genes (Widdel and Rabus 2001; Grossi et al. 2008). An alternative pathway for anaerobic alkane biodegradation was suggested that involves carboxylation of the alkyl chain at the C-3 position and removal of the C-1 and C-2 carbon atoms, to produce a one carbon shorter fatty acid (Grossi et al. 2008; Kimes et al. 2014). The possibility of different activation mechanisms being performed by different microorganisms, or under different redox conditions (e.g. sulphate-reducing or methanogenic), was proposed (Aitken et al. 2013).

Due to the high stability of benzene, anaerobic degradation of this compound is difficult, and the mechanisms of activation and further degradation of benzene are still unknown. Proposed pathways for benzene activation are (i) anaerobic hydroxylation, with the formation of an alcohol; (ii) direct carboxylation yielding benzoate; and (iii) methylation to toluene. These reactions proceed towards the formation of central metabolites such as benzoyl-coA, which is further degraded to CO_2 (Weelink et al. 2010).

4.3.4 Bioremediation of Oil Hydrocarbons

The capacity of microbes to remove large quantities of oil is limited, even in areas with continuous natural seepage where hydrocarbon-degrading microbial populations are positively selected (e.g. the Gulf of Mexico) (Ramirez-Llodra et al. 2011). Several environmental constraints affect the activity of oil-degrading microorganisms, namely, temperature, pH, salinity, water activity, hydrostatic pressure and availability of resources such as carbon and energy sources, electron acceptors and inorganic nutrients (Head et al. 2014). Bioremediation attempts to counter these limitations and enhance biodegradation of pollutants are mainly accomplished through biostimulation, by adding nutrients, alternative electron acceptors or co-substrates, or through bioaugmentation, using natural or engineered microbes.

4.3.4.1 Nutrient Supply

Oil biodegradation depends on the growth of hydrocarbon-degrading microorganisms, in which N and P typically account for approximately 15 % and 1 % of the biomass dry weight, respectively (Head et al. 2014). Due to the high carbon content of hydrocarbons, N and P are generally limiting in oil-contaminated environments. Thus, additional N and P supplies can improve hydrocarbon biodegradation (Head et al. 2006). The addition of 23:2 N/P garden fertiliser significantly enhanced the efficiency of bioremediation of coastal habitats following the Exxon Valdez oil spill (Bragg et al. 1994). Biostimulation with the oleophilic fertiliser S200 after the Prestige oil spill, in 2002, increased the bioavailability and stimulated the degradation of high molecular weight n-alkanes and alkylated PAH (Jiménez et al. 2007). The requirements of N and P are lower for anaerobic microorganisms, which typically have lower biomass yields than aerobes, e.g. 2–10 % in methanogenic hydrocarbon-degrading consortia compared to the 50 % growth yield expected for aerobic heterotrophs (Gray et al. 2011).

Kolukirik et al. (2011) assessed the hydrocarbon-degrading activity in marine sediments from the Halic Bay (Istanbul), a highly anaerobic deep sludge, with hydrocarbon concentrations in the range of 4000–6000 and 1500–3000 mg L^{-1} for aromatics and aliphatics, respectively. These sediments were incubated in anaerobic microcosms prepared with the same TOC/N/P ratio of the sediment porewater (1000/5/1) and in microcosms where N and P concentrations were gradually increased. The lowest TOC/N/P ratio studied (i.e. 1,000/40/6) induced a 92 % improvement in hydrocarbons removal, along with a ninefold increase in biogas production, comparatively to the natural Halic Bay conditions. This strategy resulted in a hydrocarbon degradation rate around 700 μg g^{-1} L^{-1} day^{-1} which, although still much lower than the aerobic degradation rates, represents significant enhancement of anaerobic microbial activity. In nutrient biostimulation, the low cost and stock availability of inorganic hydrophilic fertilisers favour their usage in emergency situations. However, dispersion hinders their effectiveness in open ocean and in tidal zones. Thus, the continuous application of exogenous N is required in order to maintain a high biodegradation rate (Nikolopoulou et al. 2007). The quantity applied should be balanced (Singh et al. 2014), since the excessive use of fertilisers worsens eutrophication of aquatic ecosystems (Haehnel et al. 2014) and can reduce degradation due to microbial inhibition (Singh et al. 2014) and O_2 depletion (Macaulay and Rees 2014). Alternative strategies, such as the use of oleophilic and slow-release fertilisers (Nikolopoulou et al. 2007) or the positive selection of endogenous N_2-fixing bacteria (Chronopoulou et al. 2013; Toccalino et al. 1993), have been conjectured. Water-soluble nutrients coated with an oleophilic degradable compound will be less dispersed and maintain higher bioavailability to attached oil-degrading bacteria (Nikolopoulou et al. 2007), also providing an additional carbon source. Dinitrogen fixation is encountered in many hydrocarbonoclastic bacteria in undisturbed and polluted environments (Al-Mailem et al. 2010, 2013; Dashti et al. 2015; Sorkhoh et al. 2010; Thavasi et al. 2006), and naphthalene adduction has been shown to promote dinitrogen fixation (Hanson et al. 2012). In general, high carbon and low N conditions promote dinitrogen fixation (Karl et al. 2002; Sorkhoh et al. 2010) in association with hydrocarbon biodegradation.

4.3.4.2 Electron Acceptors

When oil biodegradation is limited by poor oxygen mass transfer, mechanical aeration increases hydrocarbon removal. Genovese et al. (2014) reported improved hydrocarbon biodegradation (up to 98 %) in mesocosm experiments in which anaerobic marine sediments contaminated with petroleum were oxygenated during 3 months. If the use of mechanical techniques to enhance oxygen mass transfer is not feasible, alternative electron acceptors can be used (e.g. manganese, iron(III) and sulphate). Sulphate, nitrate and a mixture of sulphate+nitrate enhanced biodegradation of two-, three- and four-ring PAH from the 16 PAH listed as priority pollutants by the US Environmental Protection Agency (Lu et al. 2012), although five- and six-ring PAH showed the lowest biodegradation rates under these redox conditions. Successful improvement in anaerobic bioremediation of phenanthrene was also accomplished when gypsum ($CaSO_4 \cdot 2H_2O$) and nitrocellulose were used as slow-release electron acceptors (Tang et al. 2005).

Bacterial guilds able to use different electron acceptors are likely to co-occur within sediments. Thus, supplementation with a cocktail of electron acceptors may be an efficient strategy for enhancing oil metabolism in mixed microbial communities. In laboratory microcosms inoculated with wetland sediments of Barataria–Terrebonne estuary (the USA), a mixture of different electron acceptors, i.e. sulphate, nitrate and bicarbonate, yielded extensive biodegradation of diesel fuel No. 2, relatively to the addition of single electron acceptors (Boopathy 2003). Under mixed electron acceptor conditions, 99 % removal of diesel fuel No. 2 was achieved within 510 days, while for the same period, only 62 % and 27 % were degraded upon of sulphate addition and by natural attenuation, respectively. Siegert et al. (2011) reported an increase in the rate of methane production from hexadecane in microcosms amended with ferrihydrite, comparatively to the ones supplemented with sulphate (87 ± 2 and 38 ± 7 nmol cm^{-3} day^{-1}, respectively). The positive effect of ferrihydrite was possibly related with the growth of *Methanosarcina*-like Archaea, which, once triggered by the Fe(III) addition, can boost methane production (Siegert et al. 2011). The potential of chlorate to support oil mineralisation was also demonstrated by Brundrett et al. (2015). In this study, mesocosms were prepared with oiled salt marsh sediments from an area impacted by the Deepwater Horizon oil spill and studied over a 4-month period. Crude oil degradation in mesocosms amended with chlorate was not significantly different ($p > 0.05$) from that in the aerated mesocosms, both presenting similar reduction (66 ± 2 %) in the alkane/hopane ratio. Chlorate was rapidly exhausted in both replicates during the first 30 days, and successive chlorate additions were consumed at similar rates. These authors suggest the use of chlorate to promote *in situ* oil mineralisation for the bioremediation of contaminated salt marshes.

4.3.4.3 Addition of Co-substrates

Amendments with organic co-substrates have also been tested in laboratory batch assays. The presence of co-substrates may enhance the growth of hydrocarbon-degrading microorganisms or stimulate complex microbial communities, facilitating the degradation of the more stable compounds present in the oil mixture. The presence of an exploitable carbon and energy source can also stimulate the degradation of a second nongrowth compound (e.g. hydrocarbons) that otherwise would not be attacked by the microorganisms, in a process called co-metabolism.

Incubation of contaminated marine harbour sediments with acetate or lactose under anaerobic conditions significantly improved hydrocarbon biodegradation (Dell'Anno et al. 2012), indicating that bioremediation strategies which sustain high bacterial diversity may be more efficient than those selecting specific *taxa*. The positive effects of acetate or methanol as co-substrates on the anaerobic biodegradation of total petroleum hydrocarbons (TPH, C_{10}–C_{40}, >1000 mg kg^{-1} dry weight) in marine sediments were also shown by Zhang and Lo (2015).

4.3.4.4 Surfactants

A reduced oil–water interface limits oil biodegradation. Through the application of surfactants, the interfacial tension is reduced, and oil–water emulsification increases, thereby increasing the cell-oil contact area and, if all other conditions permit, enhance biodegradation (Hazen et al. 2010). Application of dispersants can reduce oil droplets to 10–100 μm diameter and significantly shorten biodegradation half-time, even without the use of fertilisers (Prince et al. 2013). A number of dispersants are listed in the National Contingency Plan produced by the EPA (United States Environmental Protection Agency 2015). The effect of a microemulsion composed by Tween 80, 1-pentanol, linseed oil and deionised water, on TPH biodegradation in nitrate-induced bioremediation, was investigated in sediment column assays reported by Zhang et al. (2014). After 6 weeks of treatment, higher TPH removal efficiencies were attained with the combined use of nitrate and microemulsion, compared with the use of nitrate alone (30 % and 8 %, respectively). The choice of surfactants should take into consideration the associated costs, effectiveness and toxicity (Megharaj et al. 2011). The increase of pollutant dispersion by surfactant addition will lead to higher mortality and higher bioconcentration of PAH in macrofauna (Milinkovitch et al. 2011). Toxicity of some surfactants per se towards oil-degrading bacteria can actually inhibit natural biodegradation (Louvado et al. 2012). Additionally, surfactants can increase eutrophication (García et al. 2009), and similarly to fertilisers, the use of chemical surfactants may be unacceptable in some sensitive environments (Haehnel et al. 2014). Therefore, biosurfactants have been conjectured as environment-friendly alternatives (Joo and Kim 2013). Surface-active lipopeptide production was documented on marine bacteria, namely, strains of *Bacillus subtilis* (Tareq et al. 2015; Dusane et al. 2011), *Bacillus mojavensis* (Ma et al. 2012) and *Brevibacterium luteolum* (Vilela et al. 2014). Some marine bacteria also produce surface-active glycolipids (Tareq et al. 2015). The production of biosurfactants under anaerobic conditions by halotolerant facultative anaerobic microorganisms has been reported, e.g. *Bacillus mojavensis* strain JF-2 (former *Bacillus licheniformis* strain JF-2) and *Bacillus* strain SP018 (McInerney et al. 1990; Pfiffner et al. 1986), and presents high potential as a bioremediation strategy in anaerobic marine environments. However, the direct application of semi-purified biosurfactants in environmental remediation is extremely expensive. The use of biosurfactant-producing oil-degrading microorganisms or consortia could be a more sustainable approach.

4.3.4.5 Bioaugmentation

Bioaugmentation can be beneficial if the natural community is incapable of biodegradation or if, even possessing that ability, biodegradation occurs slowly. However, this approach is not always successful, and in many cases it is also considered too expensive for field bioremediation of oil-impacted sites (Megharaj et al. 2011). The maladaptation of the bioaugmenting strains to the new environment diminishes their capacity to outcompete the natural strains that initiate degradation immediately after spill (Ron and Rosenberg 2014). Often, logistics difficult a prompt intervention and, in that case, natural communities have a time advantage. Compared to the bioaugmentation performed with pure cultures, the use of active hydrocarbon-degrading sediments may be an interesting strategy (Head and Swannell 1999).

4.3.5 Consortia of Marine Bacteria for Enhanced Hydrocarbon Bioremediation

Oil is a complex mixture of compounds, from which almost all constituents are biodegradable. Complete oil biodegradation can only be achieved by mixed communities composed by microorganisms specialised in the degradation of alkanes, monoaromatics and PAH, among other components (Yakimov et al. 2007; Tang et al. 2010). Furthermore, a single strain is often either incapable of completely degrading a specific pollutant or does it very slowly (Festa et al. 2013; Wang et al. 2008; Tixier et al. 2002). In the lack of the necessary enzymatic capacities, more toxic intermediate metabolites may accumulate, and inhibition of biodegradation will likely occur (Festa et al. 2013; Tixier et al. 2002). In this situation, a more successful biodegradation is achieved by a consortium of strains (Cui et al. 2014; Festa et al. 2013; Gallego et al. 2013; HuiJie et al. 2011; Mao et al. 2012; Nzila 2013. Vallero 2010). In a synergistic consortium, metabolically complementary bacteria work on different steps along the degradation pathway, and this avoids the building-up of inhibiting metabolites (Bouchez et al. 1999). While some strains are directly involved in the biodegradation process, others are indirectly beneficial (Ron and Rosenberg 2014). Secondary strains can increase PAH bioavailability through cell hydrophobicity and biosurfactant production (Pedetta et al. 2013) and provide essential growth factors (Pedetta et al. 2013; Sorensen et al. 2005; Singh et al. 2014). Consortia can also involve microalgae (Tang et al. 2010) and fungi (Boonchan et al. 2000). The microalgae *Scenedesmus obliquus* is incapable of oil biodegradation but promoted it when added as an axenic culture to a four-strain bacterial consortium composed by obligate alkane and PAH degraders (Tang et al. 2010). Algae can indirectly stimulate oil biodegradation through the release of oxygen and EPS that promote adsorption of hydrophobic pollutants and attachment of hydrocarbonoclastic bacteria, besides providing an additional carbon source (McGenity et al. 2012a). Reciprocally, algae can benefit from CO_2, dinitrogen and iron produced by the bacteria, and most importantly, they can benefit from removal of toxic hydrocarbon (McGenity et al. 2012a).

In N-depleted microcosms, the PAH-degrading genus *Cycloclasticus* was favoured, while the alkane-degrading genus *Alcanivorax* thrives under N supplementation (Singh et al. 2014). In this case, nutrient supplementation may favour the biodegradation of the more abundant and less toxic alkanes over the least abundant and more toxic PAHs. The understanding of the complexity of interactions established within hydrocarbon-degrading microbial communities, as well as the understanding of the benefits that can come from these interactions, is still in its infancy. The unveiling of the black box of microbial ecology of hydrocarbon-contaminated marine habitats can help to optimise the application of biostimulation strategies and fulfil absent metabolic roles by bioaugmentation (Röling and van Bodegom 2014).

4.4 Critical and Future Perspectives

Bioremediation of hydrocarbons and organometals is likely to gain applicability in the future, since these pollutants can be biodegraded to simpler and, in general, less toxic compounds. A diversity of microorganisms has been identified as capable of degrading or removing these from different marine ecosystems, and a multiplicity of metabolic pathways has been characterised. Moreover, the high potential of microorganisms, either isolated or in consortia, for bioremediation is currently accepted, and several bioremediation agents are listed in the EPA National Contingency Plan (United States Environmental Protection Agency 2015).

Despite the significant advances from the last decades on the major mechanisms and microbial agents of organometals and hydrocarbon biodegradation, bioremediation of these compounds in marine environments is still in its infancy. Laboratory-scale experiments, pilot-scale tests and some few full-scale remediation trials have been conducted as an attempt to demonstrate the potential of bioremediation for the clean-up of polluted marine environments and simultaneously overcome the economic, technical and logistic difficulties of this process. These goals are yet to be fully achieved and intensive research on bioremediation is still needed.

Bioremediation is generally not adopted as the primary strategy for treating environmental contamination mainly because of the difficulties in predicting the outcome of bioremediation processes. Physical, chemical and biological factors significantly affect the results of bioremediation, and for that reason, the extrapolations from laboratory-scale experiments to the field are still associated with a high degree of uncertainty. Moreover, the responses of complex microbial communities, which include several uncharacterised organisms, in natural environments or under different environmental conditions, are not easily predictable. Difficulties in confirming the efficacy of bioremediation at field scales and the concerns raised by the introduction of allochthonous microorganisms are also constraints to the practical application of these strategies. In fact, it has been proposed that EPA should actually ban bioremediation agents that can result in the introduction of non-native microorganisms (Prince William Sound Regional Citizens' Advisory Council 2015). Notwithstanding, bioremediation can be highly efficient and sometimes the only possible strategy for the removal of recalcitrant pollutants.

Future research should consider the forms and possible fate of each contaminant throughout the whole bioremediation process, by following the interactions between different contaminants and covering the biological processes occurring since the release from the original sources up to the formation of end products and/or sinking in marine environmental compartments. This overall tracking should not disregard the evaluation of impacts and risks of the contaminants, as well as of their degradation/transformation products, on environment and human health.

Other approaches to optimise the bioremediation processes are being pursued. Considering the effectiveness of genetically engineered microorganisms (GEM) in other biotechnological research areas, the applicability and adaptability of GEM or designed consortia, able to cope with multiple adverse/stress conditions, may represent a significant development in bioremediation of pollutants in marine environments. For that, GEM need be proven safe and the difficulties with public acceptance should be overcome.

It is consensual that the design and implementation of bioremediation approaches and the predictive analysis of the process outcome represent complex multidisciplinary tasks. However, bioremediation is an indispensable tool for preserving or restoring environmental quality in the framework of sustainability, and the need for efficient protocols justifies keeping bioremediation as a priority topic in the research agenda.

Acknowledgments The authors thank the financial support from the European Regional Development Fund (ERDF), through the Operational Programme Thematic Factors of Competitiveness (COMPETE), and the Portuguese Foundation for Science and Technology (FCT) in the frame of project FCOMP-010124-FEDER-027917 (FCT: PTDC/AAG-TEC/3428/2012). The authors also thank the FCT strategic funding of UID/BIO/04469/2013 unit, COMPETE 2020 (POCI-01-0145-FEDER-006684), and Project RECI/BBB-EBI/0179/2012 (FCOMP-01-0124-FEDER-027462). Research of AC was funded by the postdoctoral grant (BPD/UI88/2886/2013), from the project "Sustainable Use of Marine Resources" – MARES (CENTRO-07-ST24-FEDER-002033), funded by QREN, Mais Centro- Programa Operacional Regional do Centro e União Europeia/Fundo Europeu de Desenvolvimento Regional. AL was funded by a PhD grant SFRH/BD/86447/2012 funded by FCT. Research of AJC was supported by ERC grant (project 323009).

References

Adelman D, Hinga KR, Pilson MEQ (1990) Biogeochemistry of butyltins in an enclosed marine ecosystem. Environ Sci Technol 24(7):1027–1032

Aeckersberg F, Bak F, Widdel F (1991) Anaerobic oxidation of saturated hydrocarbons to CO2 by a new type of sulfate-reducing bacterium. Arch Microbiol 156(1):5–14

Aeckersberg F, Rainey FA, Widdel F (1998) Growth, natural relationships, cellular fatty acids and metabolic adaptation of sulfate-reducing bacteria that utilize long-chain alkanes under anoxic conditions. Arch Microbiol 170(5):361–369

Aitken C, Jones D, Maguire M, Gray N, Sherry A, Bowler B, Ditchfield A, Larter S, Head I (2013) Evidence that crude oil alkane activation proceeds by different mechanisms under sulfate-reducing and methanogenic conditions. Geochim Cosmochim Acta 109:162–174

Al-Mailem DM, Sorkhoh NA, Salamah S, Eliyas M, Radwan SS (2010) Oil-bioremediation potential of Arabian Gulf mud flats rich in diazotrophic hydrocarbon-utilizing bacteria. Int Biodeter Biodegr 64(3):218–225

Al-Mailem DM, Eliyas M, Radwan SS (2013) Oil-bioremediation potential of two hydrocarbonoclastic, diazotrophic Marinobacter strains from hypersaline areas along the Arabian Gulf coasts. Extremophiles 17(3):463–470

Alzieu C (1998) Tributyltin: case study of a chronic contaminant in the coastal environment. Ocean Coast Manag 40(1):23–36

Alzieu C (2000) Impact of tributyltin on marine invertebrates. Ecotoxicology 9(1–2):71–76

Antizar-Ladislao B (2008) Environmental levels, toxicity and human exposure to tributyltin (TBT)-contaminated marine environment. A review. Environ Int 34(2): 292–308

Atlas RM (1995) Petroleum biodegradation and oil spill bioremediation. Mar Pollut Bull 31(4):178–182

Ayanda OS, Fatoki OS, Adekola FA, Ximba BJ (2012) Fate and remediation of organotin compounds in seawaters and soils. Chem Sci Trans 1(3):470–481

Barron MG (2012) Ecological impacts of the deepwater horizon oil spill: implications for immunotoxicity. Toxicol Pathol 40(2):315–320

Barroso CM, Moreira MH (2002) Spatial and temporal changes of TBT pollution along the Portuguese coast: inefficacy of the EEC directive 89/677. Mar Pollut Bull 44(6):480–486

Barroso C, Moreira M, Bebianno M (2002) Imposex, female sterility and organotin contamination of the prosobranch *Nassarius reticulatus* from the Portuguese coast. Mar Ecol Prog Ser 230:127–135

Barug D (1981) Microbial degradation of bis(tributyltin) oxide. Chemosphere 10(10):1145–1154

Batel R, Bihari N, Rinkevich B, Dapper J, Schacke H, Schroder HC, Muller WEG (1993) Modulation of organotin-induced apoptosis by the water pollutant methyl mercury in a human lymphoblastoid tumor cell line and a marine sponge. Mar Ecol Prog Ser Oldendorf 93(3):245–251

Beeder J, Nilsen RK, Rosnes JT, Torsvik T, Lien T (1994) *Archaeoglobus fulgidus* isolated from hot North Sea oil field waters. Appl Environ Microbiol 60(4):1227–1231

Bejarano AC, Michel J (2010) Large-scale risk assessment of polycyclic aromatic hydrocarbons in shoreline sediments from Saudi Arabia: environmental legacy after twelve years of the Gulf war oil spill. Environ Pollut 158(5):1561–1569

Bellas J, Saco-Álvarez L, Nieto Ó, Bayona JM, Albaigés J, Beiras R (2013) Evaluation of artificially-weathered standard fuel oil toxicity by marine invertebrate embryogenesis bioassays. Chemosphere 90(3): 1103–1108

Ben Said O, Goñi-Urriza MS, El Bour M, Dellali M, Aissa P, Duran R (2008) Characterization of aerobic polycyclic aromatic hydrocarbon-degrading bacteria from Bizerte lagoon sediments, Tunisia. J Appl Microbiol 104(4):987–997

Bernat P, Dlugonski J (2006) Acceleration of tributyltin chloride (TBT) degradation in liquid cultures of the filamentous fungus *Cunninghamella elegans*. Chemosphere 62(1):3–8

Bianchi V, Masciandaro G, Ceccanti B, Doni S, Iannelli R (2010) Phytoremediation and bio-physical conditioning of dredged marine sediments for their re-use in the environment. Water Air Soil Pollut 210(1–4): 187–195

Bik HM, Halanych KM, Sharma J, Thomas WK (2012) Dramatic shifts in benthic microbial eukaryote communities following the Deepwater Horizon oil spill. PLoS One 7(6):e38550

Boonchan S, Britz M, Stanley G (2000) Degradation and mineralization of high-molecular-weight polycyclic aromatic hydrocarbons by defined fungal-bacterial cocultures. Appl Environ Microbiol 66:1007–1019

Boopathy R (2000) Factors limiting bioremediation technologies. Bioresour Technol 74(1):63–67

Boopathy R (2003) Anaerobic degradation of No. 2 diesel fuel in the wetland sediments of Barataria-Terrebonne estuary under various electron acceptor conditions. Bioresour Technol 86(2):171–175

Bouchez M, Blanchet D, Bardin V, Haeseler F, Vandecasteele J-P (1999) Efficiency of defined strains and of soil consortia in the biodegradation of polycyclic aromatic hydrocarbon (PAH) mixtures. Biodegradation 10(6):429–435

Bragg JR, Prince RC, Harner EJ, Atlas RM (1994) Effectiveness of bioremediation for the Exxon Valdez oil spill. Nature 368:413–418

Brandsch R, Nowak KE, Binder N, Jastorff B (2001) Investigations concerning the sustainability of remediation by land deposition of tributyltin contaminated harbour sediments. J Soils Sediment 1(4):234–236

Brundrett M, Horita J, Anderson T, Pardue J, Reible D, Jackson WA (2015) The use of chlorate, nitrate, and perchlorate to promote crude oil mineralization in salt marsh sediments. Environ Sci Pollut 22:15377–15385

Carson RT, Mitchell RC, Hanemann WM, Kopp RJ, Presser S, Ruud PA (1992) A contingent valuation study of lost passive use values resulting from the Exxon Valdez oil spill. University Library of Munich, Germany

Chakrabarty A (1992) Microorganisms having multiple compatible degradative energy-generating plasmids and preparation thereof 1980. Biotechnology 24:535–545

Chronopoulou P-M, Fahy A, Coulon F, Païssé S, Goñi-Urriza M, Peperzak L, Acuña Alvarez L, McKew BA, Lawson T, Timmis KN, Duran R, Underwood GJC, McGenity TJ (2013) Impact of a simulated oil spill on benthic phototrophs and nitrogen-fixing bacteria in mudflat mesocosms. Environ Microbiol 15(1):242–252

Coates JD, Woodward J, Allen J, Philp P, Lovley DR (1997) Anaerobic degradation of polycyclic aromatic hydrocarbons and alkanes in petroleum-contaminated marine harbor sediments. Appl Environ Microbiol 63(9):3589–3593

Coelho FJRC, Cleary DFR, Rocha RJM, Calado R, Castanheira JM, Rocha SM, Silva A, Simões MMQ, Oliveira V, Lillebo AI (2015) Unraveling the interactive effects of climate change and oil contamination on laboratory-simulated estuarine benthic communities. Global Chang Biol 21(5):1871–1886

Cooney JJ (1995) Organotin compounds and aquatic bacteria: a review. Helgol Mar Res 49(1):663–677

Cooney JJ, Wuertz S (1989) Toxic effects of tin-compounds on microorganisms. J Ind Microbiol 4(5):375–402

Craig P (2003) Organometallic compounds in the environment, 2nd edn. Wiley, Chichester

Cravo-Laureau C, Matheron R, Cayol J-L, Joulian C, Hirschler-Réa A (2004) *Desulfatibacillum aliphaticivorans* gen. nov., sp. nov., an n-alkane-and n-alkene-degrading, sulfate-reducing bacterium. Int J Syst Evol Microbiol 54(1):77–83

Cruz A, Caetano T, Suzuki S, Mendo S (2007) *Aeromonas veronii*, a tributyltin (TBT)-degrading bacterium isolated from an estuarine environment, Ria de Aveiro in Portugal. Mar Environ Res 64:639–650

Cruz A, Oliveira V, Baptista I, Almeida A, Cunha A, Suzuki S, Mendo S (2012) Effect of tributyltin (TBT) in the metabolic activity of TBT-resistant and sensitive estuarine bacteria. Environ Toxicol 27(1):11–17

Cruz A, Henriques I, Sousa ACA, Baptista I, Almeida A, Takahashi S, Tanabe S, Correia A, Suzuki S, Anselmo AM (2014) A microcosm approach to evaluate the degradation of tributyltin (TBT) by *Aeromonas molluscorum* Av27 in estuarine sediments. Environ Res 132:430–437

Cruz A, Anselmo AM, Suzuki S, Mendo S (2015) Tributyltin (TBT): a review on microbial resistance and degradation. Crit Rev Environ Sci Technol 45(9):970–1006

Cruz-Uribe O, Rorrer GL (2006) Uptake and biotransformation of 2, 4, 6-trinitrotoluene (TNT) by microplantlet suspension culture of the marine red macroalga *Portieria hornemannii*. Biotechnol Bioeng 93(3): 401–412

Cui Z, Xu G, Gao W, Li Q, Yang B, Yang G, Zheng L (2014) Isolation and characterization of *Cycloclasticus* strains from Yellow Sea sediments and biodegradation of pyrene and fluoranthene by their syntrophic association with *Marinobacter* strains. Int Biodeter Biodegr 91:45–51

Czako M, Feng X, He Y, Liang D, Marton L (2006) Transgenic *Spartina alterniflora* for phytoremediation. Environ Geochem Health 28(1–2):103–110

Dafforn KA, Lewis JA, Johnston EL (2011) Antifouling strategies: history and regulation, ecological impacts and mitigation. Mar Pollut Bull 62:453–465

Dai S, Huang G, Chen C (1998) Fate of 14C-labeled tributyltin in an estuarine microcosm. Appl Organomet Chem 12:585–590

Dashti N, Ali N, Eliyas M, Khanafer M, Sorkhoh NA, Radwan SS (2015) Most hydrocarbonoclastic bacteria in the total environment are diazotrophic, which highlights their value in the bioremediation of hydrocarbon contaminants. Microbes Environ 30(1):70–75

de Lorenzo V (2008) Systems biology approaches to bioremediation. Curr Opin Biotechnol 19(6):579–589

Dell'Anno A, Beolchini F, Rocchetti L, Luna GM, Danovaro R (2012) High bacterial biodiversity increases degradation performance of hydrocarbons during bioremediation of contaminated harbor marine sediments. Environ Pollut 167:85–92

Dolfing J, Larter S, Head I (2008) Thermodynamic constraints on methanogenic crude oil biodegradation. ISME J 2:442–452

Du J, Chadalavada S, Chen Z, Naidu R (2014) Environmental remediation techniques of tributyltin contamination in soil and water: a review. Chem Eng J 235:141–150

Dubey SK, Roy U (2003) Biodegradation of tributyltins (organotins) by marine bacteria. Appl Organomet Chem 17(1):3–8

Dubinsky EA, Conrad ME, Chakraborty R, Bill M, Borglin SE, Hollibaugh JT, Mason OU, M Piceno Y, Reid FC, Stringfellow WT (2013) Succession of hydrocarbon-degrading bacteria in the aftermath of the Deepwater Horizon oil spill in the Gulf of Mexico. Environ Sci Technol 47(19):10860–10867

Dusane DH, Pawar VS, Nancharaiah YV, Venugopalan VP, Kumar AR, Zinjarde SS (2011) Anti-biofilm potential of a glycolipid surfactant produced by a tropical marine strain of *Serratia marcescens*. Biofouling 27(6):645–654

Dyksterhouse SE, Gray JP, Herwig RP, Lara JC, Staley JT (1995) *Cycloclasticus pugetii* gen. nov., sp. nov., an aromatic hydrocarbon-degrading bacterium from marine sediments. Int J Syst Bacteriol 45:116–123

EC (2002) Official Journal of the European Community 2002, Commission Directive 2002/62/EC of 9 July 2002. vol LI 83. European Community

Evans SM (1999) Tributyltin pollution: the catastrophe that never happened. Mar Pollut Bull 38(8):629–636

Evans SM, Birchenough AC, Brancato MS (2000) The TBT ban: out of the frying pan into the fire? Mar Pollut Bull 40(3):204–211

Fent K (1996) Ecotoxicology of organotin compounds. Crit Rev Toxicol 26(1):3–117

Festa S, Coppotelli BM, Morelli IS (2013) Bacterial diversity and functional interactions between bacterial strains from a phenanthrene-degrading consortium obtained from a chronically contaminated-soil. Int Biodeter Biodegr 85:42–51

Frache R, Rivaro P (2000) Occurence, pathways and bioaccumulation of organometallic compounds in marine environments. In: Gianguzza A, Ezio P, Sammartano S (eds) Chemical processes in marine environments, vol Part III. Environmental science. Springer, Berlin, pp 201–211

Friello DA, Mylroie JR, Chakrabarty AM (2001) Use of genetically engineered multi-plasmid microorganisms for rapid degradation of fuel hydrocarbons. Int Biodeterior Biodegrad 48(1):233–242

Gadd GM (1993) Microbial formation and transformation of organometallic and organometalloid compounds. FEMS Microbiol Rev 11(4):297–316

Gadd GM (2000) Microbial interactions with tributyltin compounds: detoxification, accumulation, and environmental fate. Sci Total Environ 258(1–2):119–127

Gadd GM (2004) Microbial influence on metal mobility and application for bioremediation. Geoderma 122(2–4):109–119

Gallego S, Vila J, Tauler M, Nieto J, Breugelmans P, Springael D, Grifoll M (2013) Community structure and PAH ring-hydroxylating dioxygenase genes of a marine pyrene-degrading microbial consortium. Biodegradation 25:543–556

García MT, Campos E, Marsal A, Ribosa I (2009) Biodegradability and toxicity of sulphonate-based surfactants in aerobic and anaerobic aquatic environments. Water Res 43(2):295–302

Genovese M, Crisafi F, Denaro R, Cappello S, Russo D, Calogero R, Santisi S, Catalfamo M, Modica A, Smedile F (2014) Effective bioremediation strategy for rapid *in situ* cleanup of anoxic marine sediments in mesocosm oil spill simulation. Front Microbiol 5:(Article 162)

Gray ND, Sherry A, Grant RJ, Rowan AK, Hubert CRJ, Callbeck CM, Aitken CM, Jones DM, Adams JJ, Larter SR (2011) The quantitative significance of *Syntrophaceae* and syntrophic partnerships in methanogenic degradation of crude oil alkanes. Environ Microbiol 13(11):2957–2975

Grossi V, Cravo-Laureau C, Guyoneaud R, Ranchou-Peyruse A, Hirschler-Réa A (2008) Metabolism of n-alkanes and n-alkenes by anaerobic bacteria: a summary. Org Geochem 39(8):1197–1203

Guitart C, García-Flor N, Bayona JM, Albaigés J (2007) Occurrence and fate of polycyclic aromatic hydrocarbons in the coastal surface microlayer. Mar Pollut Bull 54(2):186–194

Guitart C, García-Flor N, Miquel JC, Fowler SW, Albaigés J (2010) Effect of the accumulation of polycyclic aromatic hydrocarbons in the sea surface microlayer on their coastal air-sea exchanges. J Mar Syst 79(1–2):210–217

Haehnel J, Jeschek J, Schulz-Bull DE (2014) Quantitative determination of microbial oil degradation and of oil absorption by a new oil-binding system in a Baltic Sea mesocosm experiment. Int Oil Spill Conf Proc 2014(1):1059–1072

Hamada-Sato N, Asuka T, Kobayashi T, Imada C, Mizuishi K, Takeuchi M, Watanabe E (2002) Degradation of tributyltin by marine microorganisms in sediment collected from Tokyo Bay. Fish Sci 68(No. Sup 1):643–644

Han Z, Sani B, Akkanen J, Abel S, Nybom I, Karapanagioti HK, Werner D (2015) A critical evaluation of magnetic activated carbon's potential for the remediation of sediment impacted by polycyclic aromatic hydrocarbons. J Hazard Mater 286:41–47

Hanson BT, Yagi JM, Jeon CO, Madsen EM (2012) Role of nitrogen fixation in the autecology of *Polaromonas naphthalenivorans* in contaminated sediments. Environ Microbiol 14(6):1544–1557

Harms G, Zengler K, Rabus R, Aeckersberg F, Minz D, Rosselló-Mora R, Widdel F (1999) Anaerobic oxidation of o-xylene, m-xylene, and homologous alkylbenzenes by new types of sulfate-reducing bacteria. Appl Environ Microbiol 65(3):999–1004

Hazen TC, Dubinsky EA, DeSantis TZ, Andersen GL, Piceno YM, Singh N, Jansson JK, Probst A, Borglin SE, Fortney JL (2010) Deep-sea oil plume enriches indigenous oil-degrading bacteria. Science 330(6001):204–208

Head IM, Swannell RPJ (1999) Bioremediation of petroleum hydrocarbon contaminants in marine habitats. Curr Opin Biotechnol 10(3):234–239

Head IM, Jones DM, Roling WFM (2006) Marine microorganisms make a meal of oil. Nat Rev Microbiol 4(3):173–182

Head IM, Gray ND, Larter SR (2014) Life in the slow lane; biogeochemistry of biodegraded petroleum containing reservoirs and implications for energy recovery and carbon management. Front Microbiol 5(Article 566):1–23

Hoch M (2001) Organotin compounds in the environment – an overview. Appl Geochem 16(7–8):719–743

Holmes DE, Risso C, Smith JA, Lovley DR (2011) Anaerobic oxidation of benzene by the hyperthermophilic archaeon *Ferroglobus placidus*. Appl Environ Microbiol 77(17):5926–5933

Huang L, Pu X, Pan J-F, Wang B (2013) Heavy metal pollution status in surface sediments of Swan Lake lagoon and Rongcheng Bay in the northern Yellow Sea. Chemosphere 93(9):1957–1964

HuiJie L, CaiYun Y, Yun T, GuangHui L, TianLing Z (2011) Using population dynamics analysis by DGGE to design the bacterial consortium isolated from mangrove sediments for biodegradation of PAHs. Int Biodeter Biodegr 65(2):269–275

IMO (2001) International convention on the control of harmful anti-fouling systems on ships AFS/CONF/26. International Maritime Organization, London

Incardona JP, Swarts TL, Edmunds RC, Linbo TL, Aquilina-Beck A, Sloan CA, Gardner LD, Block BA, Scholz NL (2013) Exxon Valdez to deepwater horizon: comparable toxicity of both crude oils to fish early life stages. Aquat Toxicol 142:303–316

International Energy Agency (2015) Oil market report. Accessed on 30 Apr 2015

Iwamoto T, Nasu M (2001) Current bioremediation practice and perspective. J Biosci Bioeng 92(1):1–8

Jaekel U, Musat N, Adam B, Kuypers M, Grundmann O, Musat F (2013) Anaerobic degradation of propane and butane by sulfate-reducing bacteria enriched from marine hydrocarbon cold seeps. ISME J 7(5): 885–895

Jiménez J, Miñambres B, García J, Díaz E (2004) Genomic insights in the metabolism of aromatic compounds in *Pseudomonas*. In: Ramos J-L (ed) Pseudomonas. Springer, US, pp 425–462

Jiménez N, Viñas M, Bayona JM, Albaiges J, Solanas AM (2007) The Prestige oil spill: bacterial community dynamics during a field biostimulation assay. Appl Microbiol Biotechnol 77(4):935–945

Joo M, Kim J (2013) Characteristics of crude oil biodegradation by biosurfactant-producing bacterium *Bacillus subtilis* JK-1. J Korean Soc Appl Biol Chem 56(2):193–200

Jutkina J, Heinaru E, Vedler E, Juhanson J, Heinaru A (2011) Occurrence of plasmids in the aromatic degrading bacterioplankton of the baltic sea. Genes (Basel) 2(4):853–868

Karl D, Michaels A, Bergman B, Capone D, Carpenter E, Letelier R, Lipschultz F, Paerl H, Sigman D, Stal L (2002) Dinitrogen fixation in the world's oceans. In: Boyer E, Howarth R (eds) The nitrogen cycle at regional to global scales. Springer, Netherlands, pp 47–98

Kasai Y, Kishira H, Syutsubo K, Harayama S (2001) Molecular detection of marine bacterial populations on beaches contaminated by the Nakhodka tanker oil-spill accident. Environ Microbiol 3(4):246–255

Kasai Y, Kishira H, Harayama S (2002) Bacteria belonging to the genus *Cycloclasticus* play a primary role in the degradation of aromatic hydrocarbons released in a marine environment. Appl Environ Microbiol 68:5625–5633

Kawai S, Kurokawa Y, Harino H, Fukushima M (1998) Degradation of tributyltin by a bacterial strain isolated from polluted river water. Environ Pollut 102(2–3):259–263

Khanolkar DS, Naik MM, Dubey SK (2014) Biotransformation of tributyltin chloride by *Pseudomonas stutzeri* strain DN2. Braz J Microbiol 45(4):1239–1245

Khanolkar D, Dubey SK, Naik MM (2015a) Biotransformation of tributyltin chloride to less toxic dibutyltin dichloride and monobutyltin trichloride by *Klebsiella pneumoniae* strain SD9. Int Biodeterior Biodegrad 104:212–218

Khanolkar D, Dubey SK, Naik MM (2015b) Tributyltin chloride (TBTCl)-enhanced exopolysaccharide and siderophore production in an estuarine *Alcaligenes faecalis* strain. Arch Environ Contam Toxicol 68:612–621

Kim GB, Nakata H, Tanabe S (1998) *In vitro* inhibition of hepatic cytochrome P450 and enzyme activity by butyltin compounds in marine mammals. Environ Pollut 99(2):255–261

Kimes NE, Callaghan AV, Suflita JM, Morris PJ (2014) Microbial transformation of the Deepwater Horizon oil spill-past, present, and future perspectives. Front Microbiol 5(Article 603):1–11

Kleindienst S, Herbst F-A, Stagars M, von Netzer F, von Bergen M, Seifert J, Peplies J, Amann R, Musat F, Lueders T (2014) Diverse sulfate-reducing bacteria of the *Desulfosarcina/Desulfococcus clade* are the key alkane degraders at marine seeps. ISME J 8(10):2029–2044

Kniemeyer O, Musat F, Sievert SM, Knittel K, Wilkes H, Blumenberg M, Michaelis W, Classen A, Bolm C, Joye SB (2007) Anaerobic oxidation of short-chain hydrocarbons by marine sulphate-reducing bacteria. Nature 449(7164):898–901

Kolukirik M, Ince O, Ince BK (2011) Increment in anaerobic hydrocarbon degradation activity of halic bay sediments via nutrient amendment. Microb Ecol 61(4):871–884

Kujawinski EB, Kido Soule MC, Valentine DL, Boysen AK, Longnecker K, Redmond MC (2011) Fate of dispersants associated with the Deepwater Horizon oil spill. Environ Sci Technol 45(4):1298–1306

Kvenvolden KA, Cooper CK (2003) Natural seepage of crude oil into the marine environment. Geo-Mar Lett 23(3–4):140–146

LaMontagne MG, Leifer I, Bergmann S, Van De Werfhorst LC, Holden PA (2004) Bacterial diversity in marine hydrocarbon seep sediments. Environ Microbiol 6(8):799–808

Landmeyer JE, Tanner TL, Watt BE (2004) Biotransformation of tributyltin to tin in freshwater river-bed sediments contaminated by an organotin release. Environ Sci Technol 38(15):4106–4112

Laughlin RB Jr, French W, Guard HE (1986) Accumulation of bis (tributyltin) oxide by the marine mussel *Mytilus edulis*. Environ Sci Technol 20(9):884–890

Lee RF, Valkirs AO, Seligman PF (1989) Importance of microalgae in the biodegradation of tributyltin in estuarine waters. Environ Sci Technol 23(12):1515–1518

Lee SE, Chung JW, Won HS, Lee DS, Lee YW (2012) Removal of methylmercury and tributyltin (TBT) using marine microorganisms. Bull Environ Contam Toxicol 88:239–244

Lewis M, Pryor R (2013) Toxicities of oils, dispersants and dispersed oils to algae and aquatic plants: review and database value to resource sustainability. Environ Pollut 180:345–367

Lin X, Lu F, Chen Y, Liu N, Cao Y, Xu L, Wei Y, Feng L (2015) One-step breaking and separating emulsion by tungsten oxide coated mesh. ACS Appl Mater Inter 7(15):8108–8113

Louvado A, Coelho FJRC, Domingues P, Santos AL, Gomes NCM, Almeida A, Cunha A (2012) Isolation of surfactant-resistant Pseudomonads from the estuarine surface microlayer. J Microbiol Biotechnol 22(3):283–291

Louvado A, Gomes NC, Simões MM, Almeida A, Cleary DF, Cunha A (2015) Polycyclic aromatic hydrocarbons in deep sea sediments: microbe-pollutant interactions in a remote environment. Sci Total Environ 526:312–328

Lu X-Y, Li B, Zhang T, Fang HHP (2012) Enhanced anoxic bioremediation of PAHs-contaminated sediment. Bioresour Technol 104:51–58

Lucas J, Perrichon P, Nouhaud M, Audras A, Leguen I, Lefrancois C (2014) Aerobic metabolism and cardiac activity in the descendants of zebrafish exposed to pyrolytic polycyclic aromatic hydrocarbons. Environ Sci Pollut R 21(24):13888–13897

Ma Z, Wang N, Hu J, Wang S (2012) Isolation and characterization of a new iturinic lipopeptide, mojavensin A produced by a marine-derived bacterium *Bacillus mojavensis* B0621A. J Antibiot 65(6):317–322

Maanan M, Zourarah B, Sahabi M, Maanan M, Le Roy P, Mehdi K, Salhi F (2015) Environmental risk assessment of the Moroccan Atlantic continental shelf: the role of the industrial and urban area. Sci Total Environ 511:407–415

Macaulay BM, Rees D (2014) Bioremediation of Oil Spills: A Review of Challenges for Research Advancement. Ann Environ Sci 8:9–37

Mao J, Luo Y, Teng Y, Li Z (2012) Bioremediation of polycyclic aromatic hydrocarbon-contaminated soil by a bacterial consortium and associated microbial community changes. Int Biodeter Biodegr 70:141–147

Massias D, Grossi V, Bertrand J-C (2003) *In situ* anaerobic degradation of petroleum alkanes in marine sediments: preliminary results. Compt Rendus Geosci 335(5):435–439

Mbadinga SM, Wang L-Y, Zhou L, Liu J-F, Gu J-D, Mu B-Z (2011) Microbial communities involved in anaerobic degradation of alkanes. Int Biodeterior Biodegrad 65(1):1–13

McGenity TJ (2014) Hydrocarbon biodegradation in intertidal wetland sediments. Curr Opin Biotechnol 27:46–54

McGenity T, Folwell B, McKew B, Sanni G (2012a) Marine crude-oil biodegradation: a central role for interspecies interactions. Aquat Biosyst 8(1):10

McGenity TJ, Folwell BD, McKew BA, Sanni GO (2012b) Marine crude-oil biodegradation: a central role for interspecies interactions. Aquatic Biosyst 8:10–28

McInerney MJ, Javaheri M, Nagle DP Jr (1990) Properties of the biosurfactant produced by *Bacillus licheniformis* strain JF-2. J Ind Microbiol 5(2–3):95–101

Megharaj M, Ramakrishnan B, Venkateswarlu K, Sethunathan N, Naidu R (2011) Bioremediation approaches for organic pollutants: a critical perspective. Environ Int 37(8):1362–1375

Milinkovitch T, Kanan R, Thomas-Guyon H, Le Floch S (2011) Effects of dispersed oil exposure on the bioaccumulation of polycyclic aromatic hydrocarbons and the mortality of juvenile *Liza ramada*. Sci Total Environ 409(9):1643–1650

Moreira ITA, Oliveira OMC, Triguis JA, dos Santos AMP, Queiroz AFS, Martins CMS, Silva CS, Jesus RS (2011) Phytoremediation using *Rizophora mangle* L. in mangrove sediments contaminated by persistent total petroleum hydrocarbons (TPH's). Microchem J 99(2):376–382

Murata S, Takahashi S, Agusa T, Thomas NJ, Kannan K, Tanabe S (2008) Contamination status and accumulation profiles of organotins in sea otters (*Enhydra lutris*) found dead along the coasts of California, Washington, Alaska (USA), and Kamchatka (Russia). Mar Pollut Bull 56:641–649

Nafees M, Waseem A, Khan AR (2013) Comparative study of laterite and bentonite based organoclays: Implications of hydrophobic compounds remediation from aqueous solutions. Sci World J Article ID 681769

National Academy of Science (2002) Oil in the sea III: inputs, fates and effects. The National Academies Press, Washington, D C
Nikolopoulou M, Pasadakis N, Kalogerakis N (2007) Enhanced bioremediation of crude oil utilizing lipophilic fertilizers. Desalination 211(1–3):286–295
Notar M, Leskovšek H, Faganeli J (2001) Composition, distribution and sources of polycyclic aromatic hydrocarbons in sediments of the gulf of Trieste, Northern Adriatic Sea. Mar Poll Bull 42(1):36–44
Nzila A (2013) Update on the co-metabolism of organic pollutants by bacteria. Environ Poll 178:474–482
O'Brien AL, Keough MJ (2014) Ecological responses to contamination: a meta-analysis of experimental marine studies. Environ Pollut 195:185–191
Okoro HK, Fatoki OS, Adekola HA, Ximba BJ, Snyman RG (2011) Sources, environmental levels and toxicity of organotin in marine environment – a review. Asian J Chem 23(2):473–482
Oliveira V, Gomes NCM, Almeida A, Silva AMS, Silva H, Cunha Â (2015) Microbe-assisted phytoremediation of hydrocarbons in estuarine environments. Microb Ecol 69(1):1–12
Ommedal H, Torsvik T (2007) *Desulfotignum toluenicum* sp. nov., a novel toluene-degrading, sulphate-reducing bacterium isolated from an oil-reservoir model column. Int J Syst Evol Microbiol 57(12):2865–2869
Pedetta A, Pouyte K, Herrera Seitz MK, Babay PA, Espinosa M, Costagliola M, Studdert CA, Peressutti SR (2013) Phenanthrene degradation and strategies to improve its bioavailability to microorganisms isolated from brackish sediments. Int Biodeter Biodegr 84:161–167
Perelo LW (2010) Review: *in situ* and bioremediation of organic pollutants in aquatic sediments. J Hazard Mater 177(1–3):81–89
Peterson CH, Rice SD, Short JW, Esler D, Bodkin JL, Ballachey BE, Irons DB (2003) Long-term ecosystem response to the Exxon Valdez oil spill. Science 302(5653):2082–2086
Pfiffner S, McInerney M, Jenneman G, Knapp R (1986) Isolation of halotolerant, thermotolerant, facultative polymer-producing bacteria and characterization of the exopolymer. Appl Environ Microbiol 51:1224–1229
Polymenakou P, Bertilsson S, Tselepides A, Stephanou E (2005) Bacterial community composition in different sediments from the Eastern Mediterranean Sea: a comparison of four 16S ribosomal DNA clone libraries. Microb Ecol 50(3):447–462
Prince William Sound Regional Citizens' Advisory Council (2015) Prince William sound regional citizens' advisory council comments on 40 CFR parts 110 and 300, National Contingency Plan Subparts A and J. Alaska
Prince RC, McFarlin KM, Butler JD, Febbo EJ, Wang FCY, Nedwed TJ (2013) The primary biodegradation of dispersed crude oil in the sea. Chemosphere 90(2):521–526
Rabus R, Nordhaus R, Ludwig W, Widdel F (1993) Complete oxidation of toluene under strictly anoxic conditions by a new sulfate-reducing bacterium. Appl Environ Microbiol 59(5):1444–1451
Rahmanpour S, Ghorghani NF, Ashtiyani SML (2014) Heavy metal in water and aquatic organisms from different intertidal ecosystems, Persian Gulf. Environ Monit Assess 186(9):5401–5409
Rainbow PS (1995) Biomonitoring of heavy metal availability in the marine environment. Mar Pollut Bull 31(4):183–192
Ramirez-Llodra E, Tyler PA, Baker MC, Bergstad OA, Clark MR, Escobar E, Levin LA, Menot L, Rowden AA, Smith CR, Van Dover CL (2011) Man and the last great wilderness: human impact on the deep sea. PLoS One 6(8):e22588
Ringelberg DB, Talley JW, Perkins EJ, Tucker SG, Luthy RG, Bouwer EJ, Fredrickson HL (2001) Succession of phenotypic, genotypic, and metabolic community characteristics during *in vitro* bioslurry treatment of polycyclic aromatic hydrocarbon-contaminated sediments. Appl Environ Microbiol 67(4):1542–1550
Roberts DA (2012) Causes and ecological effects of resuspended contaminated sediments (RCS) in marine environments. Environ Int 40:230–243
Rocchetti L, Fonti V, Beolchini F, Dell'Anno A (2014) Bioremediation of contaminated marine sediments: examples of successful applications. Waste Manag Environ VII 180:335
Rojo F (2009) Degradation of alkanes by bacteria. Environ Microbiol 11(10):2477–2490
Röling WFM, van Bodegom PM (2014) Toward quantitative understanding on microbial community structure and functioning: a modeling-centered approach using degradation of marine oil spills as example. Front Microbiol 5:125
Ron EZ, Rosenberg E (2002) Biosurfactants and oil bioremediation. Curr Opin Biotechnol 13(3):249–252
Ron EZ, Rosenberg E (2014) Enhanced bioremediation of oil spills in the sea. Curr Opin Biotechnol 27(0):191–194
Roy U, Bhosle S (2006) Microbial transformation of tributyltin chloride by *Pseudomonas aeruginosa* strain USS25 NCIM-5224. Appl Organomet Chem 20(1):5–11
Rudel H (2003) Case study: bioavailability of tin and tin compounds. Ecotoxicol Environ Saf 56(1):180–189
Rueter P, Rabus R, Wilkest H, Aeckersberg F, Rainey FA, Jannasch HW, Widdel F (2004) Anaerobic oxidation of hydrocarbons in crude oil by new types of sulphate-reducing bacteria. Nature 372(6505):455–458
Sabir S (2015) Approach of cost-effective adsorbents for oil removal from oily water. Crit Rev Environ Sci Technol 45(17):1916–1945
Saint-Louis R, Pelletier E, Marsot P, Fournier R (1994) Distribution and effects of tributyltin chloride and its degradation products on the growth of the marine alga *Pavlova lutheri* in continuous culture. Water Res 28(12):2533–2544

Sakultantimetha A, Keenan HE, Beattie TK, Aspray TJ, Bangkedphol S, Songsasen A (2010) Acceleration of tributyltin biodegradation by sediment microorganisms under optimized environmental conditions. Int Biodeterior Biodegrad 64(6):467–473

Sakultantimetha A, Keenan HE, Beattie TK, Bangkedphol S, Cavoura O (2011) Bioremediation of tributyltin contaminated sediment: degradation enhancement and improvement of bioavailability to promote treatment processes. Chemosphere 83(5):680–686

Santillo D, Johnston P, Langston WJ (2001) Tributyltin (TBT) antifoulants: a tale of ships, snails and imposex. In: Harremoes P, Gee D, MacGarvin M et al (eds) Late lessons from early warnings: the precautionary principle 1896–2000. European Environmental Agency, Copenhagen, pp 135–148

Santos MM, Hallers-Tjabbes CCT, Santos AM, Vieira N (2002) Imposex in *Nucella lapillus*, a bioindicator for TBT contamination: re-survey along the Portuguese coast to monitor the effectiveness of EU regulation. J Sea Res 48(3):217–223

Schwacke LH, Smith CR, Townsend FI, Wells RS, Hart LB, Balmer BC, Collier TK, De Guise S, Fry MM, Guillette LJ Jr (2013) Health of common bottlenose dolphins (*Tursiops truncatus*) in Barataria Bay, Louisiana, following the Deepwater Horizon oil spill. Environ Sci Technol 48(1):93–103

Sekizawa J, Suter II G, Birnbaum L (2003) Integrated human and ecological risk assessment: a case study of tributyltin and triphenyltin compounds. Hum Ecol Risk Assess 9(1):325–342

Seo J-S, Keum Y-S, Li QX (2009) Bacterial degradation of aromatic hydrocarbons. Int J Environ Res Public Health 6:278–309

Shimasaki Y, Kitano T, Oshima Y, Inoue S, Imada N, Honjo T (2003) Tributyltin causes masculinization in fish. Environ Toxicol Chem 22(1):141–144

Siddique T, Penner T, Klassen J, Nesbo C, Foght JM (2012) Microbial communities involved in methane production from hydrocarbons in oil sands tailings. Environ Sci Technol 46(17):9802–9810

Siegert M, Cichocka D, Herrmann S, Grundger F, Feisthauer S, Richnow H-H, Springael D, Kruger M (2011) Accelerated methanogenesis from aliphatic and aromatic hydrocarbons under iron-and sulfate-reducing conditions. FEMS Microbiol Lett 315(1):6–16

Silliman BR, van de Koppel J, McCoy MW, Diller J, Kasozi GN, Earl K, Adams PN, Zimmerman AR (2012) Degradation and resilience in Louisiana salt marshes after the BP-Deepwater Horizon oil spill. Proc Natl Acad Sci 109(28):11234–11239

Singh AK, Sherry A, Gray ND, Jones DM, Bowler BF, Head IM (2014) Kinetic parameters for nutrient enhanced crude oil biodegradation in intertidal marine sediments. Front Microbiol 5:(Article 160)

Smith BS (1971) Sexuality in the American mud snail *Nassarius obsoletus*. Proc Malacolog Soc London 39:377–378

So CM, Young LY (1999) Isolation and characterization of a sulfate-reducing bacterium that anaerobically degrades alkanes. Appl Environ Microbiol 65(7):2969–2976

Sorensen SJ, Bailey M, Hansen LH, Kroer N, Wuertz S (2005) Studying plasmid horizontal transfer *in situ*: a critical review. Nat Rev Micro 3(9):700–710

Sorkhoh NA, Ali N, Dashti N, Al-Mailem DM, Al-Awadhi H, Eliyas M, Radwan SS (2010) Soil bacteria with the combined potential for oil utilization, nitrogen fixation, and mercury resistance. Int Biodeter Biodegr 64(3):226–231

Sousa ACA, Pastorinho MR, Takahashi S, Tanabe S (2013) Organotin compounds from snails to humans. In: Lichtfouse ESJ, Robert D (eds) Environmental chemistry for a sustainable world: pollutant diseases, remediation and recycling, vol 4. Springer, Switzerland, pp 215–275

Sternberg R, Gooding M, Hotchkiss A, LeBlanc G (2010) Environmental-endocrine control of reproductive maturation in gastropods: implications for the mechanism of tributyltin-induced imposex in prosobranchs. Ecotoxicology 19(1):4–23

Suárez-Suárez A, López-López A, Tovar-Sánchez A, Yarza P, Orfila A, Terrados J, Arnds J, Marqués S, Niemann H, Schmitt-Kopplin P (2011) Response of sulfate-reducing bacteria to an artificial oil-spill in a coastal marine sediment. Environ Microbiol 13(6):1488–1499

Suehiro F, Kobayashi T, Nonaka L, Tuyen BC, Suzuki S (2006) Degradation of tributyltin in microcosm using Mekong River sediment. Microb Ecol 52(1):19–25

Sun MY, Dafforn KA, Johnston EL, Brown MV (2013) Core sediment bacteria drive community response to anthropogenic contamination over multiple environmental gradients. Environ Microbiol 15(9):2517–2531

Taha HM, Said HA, Abbas NH, Khaleafa AFM (2009) Biosorption and biodegradation of the antifouling compound tributyltin (TBT) by microalgae. Am Eurasian J Sci Res 4(1):1–6

Takeuchi I, Takahashi S, Tanabe S, Miyazaki N (2001) *Caprella* watch: a new approach for monitoring butyltin residues in the ocean. Mar Environ Res 52(2):97–113

Tang YJ, Carpenter S, Deming J, Krieger-Brockett B (2005) Controlled release of nitrate and sulfate to enhance anaerobic bioremediation of phenanthrene in marine sediments. Environ Sci Technol 39(9):3368–3373

Tang X, He L, Tao X, Dang Z, Guo C, Lu G, Yi X (2010) Construction of an artificial microalgal-bacterial consortium that efficiently degrades crude oil. J Hazard Mater 181(1–3):1158–1162

Tareq FS, Lee H-S, Lee Y-J, Lee JS, Shin HJ (2015) Ieodoglucomide C and ieodoglycolipid, new glycolipids from a marine-derived bacterium *Bacillus licheniformis* 09IDYM23. Lipids 50(5):513–519

Thavasi R, Jayalakshmi S, Balasubramanian T, Banat IM (2006) Biodegradation of crude oil by nitrogen fixing

marine bacteria *Azotobacter chroococcum*. Res J Microbiol 1(5):401–408
Tixier C, Sancelme M, Aıt-Aıssa S, Widehem P, Bonnemoy F, Cuer A, Truffaut N, Veschambre H (2002) Biotransformation of phenylurea herbicides by a soil bacterial strain, *Arthrobacter* sp. N2: structure, ecotoxicity and fate of diuron metabolite with soil fungi. Chemosphere 46(4):519–526
Toccalino PL, Johnson RL, Boone DR (1993) Nitrogen limitation and nitrogen fixation during alkane biodegradation in a sandy soil. Appl Environ Microbiol 59(9):2977–2983
Tsang CK, Lau PS, Tam NFY, Wong YS (1999) Biodegradation capacity of tributyltin by two *Chlorella* species. Environ Pollut 105(3):289–297
United States Environmental Protection Agency (2015) U.S. Environmental protection agency national contingency plan product schedule. EPA,Washington, DC
Usman M, Faure P, Hanna K, Abdelmoula M, Ruby C (2012) Application of magnetite catalyzed chemical oxidation (Fenton-like and persulfate) for the remediation of oil hydrocarbon contamination. Fuel 96:270–276
Vallero DA (2010) Applied microbial ecology. Environmental biotechnology. Academic, San Diego
van Beilen JB, Panke S, Lucchini S, Franchini AG (2001) Analysis of *Pseudomonas putida* alkane-degradation gene clusters and flanking insertion sequences: evolution and regulation of the *alk* genes. Microbiology 147:1621–1630
Vázquez JA, Rial D (2014) Inhibition of selected bacterial growth by three hydrocarbons: mathematical evaluation of toxicity using a toxicodynamic equation. Chemosphere 112:56–61
Vieira LR, Guilhermino L (2012) Multiple stress effects on marine planktonic organisms: influence of temperature on the toxicity of polycyclic aromatic hydrocarbons to *Tetraselmis chuii*. J Sea Res 72:94–98
Vila J, Nieto J, Mertens J, Springael D, Grifoll M (2010) Microbial community structure of a heavy fuel oil-degrading marine consortium: linking microbial dynamics with polycyclic aromatic hydrocarbon utilization. FEMS Microbiol Ecol 73:349–362
Vilela WFD, Fonseca SG, Fantinatti-Garboggini F, Oliveira VM, Nitschke M (2014) Production and properties of a surface-active lipopeptide produced by a new marine *Brevibacterium luteolum* strain. Appl Biochem Biotechnol 174(6):2245–2256
von Netzer F, Pilloni G, Kleindienst S, Krüger M, Knittel K, Gründger F, Lueders T (2013) Enhanced gene detection assays for fumarate-adding enzymes allow uncovering of anaerobic hydrocarbon degraders in terrestrial and marine systems. Appl Environ Microbiol 79:543–552
Wang B, Lai Q, Cui Z, Tan T, Shao Z (2008) A pyrene-degrading consortium from deep-sea sediment of the West Pacific and its key member *Cycloclasticus* sp. P1. Environ Microbiol 10(8):1948–1963
Watanabe K (2001) Microorganisms relevant to bioremediation. Curr Opin Biotechnol 12(3):237–241
Webster G, Rinna J, Roussel EG, Fry JC, Weightman AJ, Parkes RJ (2010) Prokaryotic functional diversity in different biogeochemical depth zones in tidal sediments of the Severn Estuary, UK, revealed by stable-isotope probing. FEMS Microbiol Ecol 72(2):179–197
Weelink SAB, van Eekert MHA, Stams AJM (2010) Degradation of BTEX by anaerobic bacteria: physiology and application. Rev Environ Sci Biol Technol 9(4):359–385
Weiner JM, Lovley DR (1998) Rapid benzene degradation in methanogenic sediments from a petroleum-contaminated aquifer. Appl Environ Microbiol 64(5):1937–1939
Widdel F, Rabus R (2001) Anaerobic biodegradation of saturated and aromatic hydrocarbons. Curr Opin Biotechnol 12(3):259–276
Xie Y, Su R, Zhang L, Wang X A study on biosorption and biodegradation of tributyltin by two red tide microalgae. In: International Conference on Remote Sensing, Environment and Transportation Engineering (RSETE), 2011, Nanjing, China, 24–26 June 2011. pp 7331–7334
Yakimov M, Timmis K, Golyshin P (2007) Obligate oil-degrading marine bacteria. Curr Opin Biotechnol 18:257–266
Yang T, Nigro LM, Gutierrez T, D'Ambrosio L, Joye SB, Highsmith R, Teske A (2014) Pulsed blooms and persistent oil-degrading bacterial populations in the water column during and after the Deepwater Horizon blowout. Deep Sea Res II (in press) http://dx.doi.org/10.1016/j.dsr2.2014.01.014i
Yu KSH, Wong AHY, Yau KWY, Wong YS, Tam NFY (2005) Natural attenuation, biostimulation and bioaugmentation on biodegradation of polycyclic aromatic hydrocarbons (PAHs) in mangrove sediments. Mar Pollut Bull 51(8):1071–1077
Zengler K, Richnow HH, Rosselló-Mora R, Michaelis W, Widdel F (1999) Methane formation from long-chain alkanes by anaerobic microorganisms. Nature 401(6750):266–269
Zhang Z, Lo IMC (2015) Biostimulation of petroleum-hydrocarbon-contaminated marine sediment with co-substrate: involved metabolic process and microbial community. Appl Microbiol Biotechnol 99:5683–5696
Zhang Z, Zheng G, Lo IMC (2014) Enhancement of nitrate-induced bioremediation in marine sediments contaminated with petroleum hydrocarbons by using microemulsions. Environ Sci Pollut 22:8296–8306
Zhang G, Pan Z, Wang X, Mo X, Li X (2015) Distribution and accumulation of polycyclic aromatic hydrocarbons (PAHs) in the food web of Nansi Lake, China. Environ Monit Assess 187(4):1–12

5 Extracellular Polysaccharide Production by Bacteria as a Mechanism of Toxic Heavy Metal Biosorption and Biosequestration in the Marine Environment

P.V. Bramhachari
and Ganji Purnachandra Nagaraju

Abstract

Marine environments are one of the most diverse environments owing to their vast natural resource of imperative functional molecules. Interestingly, marine bacteria offer a great diversity of polysaccharides which could play an important role in biotechnology and industry. Among the various bioactive compounds, marine exopolymers are attracting major interest and attention due to their structural and functional diversity. Bacterial exopolysaccharides (EPSs) contain ionizable functional groups, which enable them to bind and sequestrate toxic heavy metal ions. Due to their biodegradability and safety of the environment, biosorption of heavy metals by these biopolymers has attracted considerable attention as promising alternatives capable to compete with expensive, inefficient and conventional technologies, including chemical precipitation, adsorption on activated carbon, membrane separations, ion exchange and solvent extraction methods. This review particularly emphasizes on utilization of marine bacteria in the field of bioremediation and understanding the mechanism behind acquiring the characteristic feature of adaptive responses. Fundamental insights regarding metals in relation to metal-binding proteins/peptides for immobilization, information regarding genetic engineering for enzymes

P.V. Bramhachari (✉)
Department of Biotechnology and Botany, Krishna University, Machilipatnam 521001, Andhra Pradesh, India
e-mail: veerabramha@gmail.com

G.P. Nagaraju
Department of Hematology and Medical Oncology, Winship Cancer Institute, Emory University, Atlanta, GA 30322, USA

M.M. Naik, S.K. Dubey (eds.), *Marine Pollution and Microbial Remediation*,
DOI 10.1007/978-981-10-1044-6_5

involved in metal transformation and strategies that can be employed to overcome the bottlenecks associated with microbial-based remediation are highlighted in this review. The important engineering properties based on structural characteristics such as adsorption, biodegradability and hydrophilicity/hydrophobicity of EPS matrix are also discussed. A thorough understanding of microbes that produce exopolysaccharides for metal biosequestration and biosorption would solve several problems in bioremediation process.

5.1 Introduction

Majority of the marine microbial world remain unexploited due to the enormous size of the marine biosphere which offers affluent flora and fauna, signifying a vast natural resource of essential functional commercial grade products. Among the various bioactive compounds, marine exopolymers are attracting principal interest attributable to their structural and functional diversity which tends to find abundant applications in bioremediation (Satpute et al. 2010). Marine environments are one of the most stressful environments due to their unstable nature of temperature, pH, salinity and sea surface temperature and are more correctly adapted to the adverse conditions, hence possessing complex characteristic features of adaptation (Mancuso Nichols et al. 2005). Therefore, the bacteria isolated from the marine environments are believed to be better utilized in bioremediation of heavy metals, hydrocarbons and many other recalcitrant compounds and xenobiotics through production of extracellular polymeric substances (EPSs). Heavy metal contamination in coastal and marine environments is receiving increasingly serious threat to naturally stressed marine ecosystems and moreover humans directly rely on marine resources for food, industry and recreation. Heavy metals are introduced to coastal and marine environments through a variety of sources and activities including sewage and industrial effluents, brine discharges, coastal modifications and oil pollution. The seriousness of heavy metal contamination is further compounded by the fact that they are by and large water-soluble, nondegradable, vigorous-oxidizing agents and are strongly bonded to many biochemicals inhibiting their functions. Many heavy metal ions and metalloids are well known to be toxic or otherwise carcinogenic to humans (Fu and Wang 2011). Heavy metals can contribute to degradation of marine ecosystems by reducing species diversity and profusion through accumulation of metals in living organisms and food chains (Hosono et al. 2011). Therefore, there is a critical need to study the heavy metal resistance mechanism in marine bacteria with a comprehensive approach to combat and manage heavy metal pollution in the marine environment.

In recent years, marine microbial EPSs have fascinated additional attentions, particularly for those originated from marine bacteria (Jouault et al. 2004). Nevertheless, there is a growing interest in isolating novel bacteria producing EPSs from marine environments with potential applications (Arena 2004). Exopolysaccharides (EPSs) excreted by majority of marine microorganisms are known to produce EPS for cryoprotection, halotolerance, attachment to substrate, nutrient uptake, and formation of colonies and biofilms (Decho 1990; Hoagland et al. 1993; Decho et al. 2010). Due to scores of ecological roles, EPSs are expected to be present at a significant level in aquatic systems, primarily as colloidal organic matter. Exopolymeric substances (EPSs) are yet inadequately explored organic ligands that are heterogeneous, polyfunctional macromolecules containing uronic acids, neutral sugars, mono- and polysaccharides, amino acids, and proteins (Mancuso Nichols et al. 2005; Verdugo et al. 2004; Hassler et al. 2011). Interestingly, these EPSs are primarily composed of polysaccharides and glycoproteins, in addition to charged functional groups and amphiphilic, adsorptive, and adhesive properties (Quigley et al. 2002; Gutiérrez et al. 2007), and provide a substantial pool of organic carbon available to serve as a natural ligand source for numerous other molecules, together with trace metals (Bhaskar et al. 2005). Extracellular polymeric substances (EPSs) are produced by a wide range of microorganisms from both terrestrial and marine environments. EPSs differ among organisms and producing conditions in carbohydrate chain length and branching, composition, type of sugar linkages, and presence of additional chemical groups such as sulfates, proteins, lipids, and even nucleic acids (Ruas-Madiedo and De Los Reyes-Gavilán 2005; Mancuso Nichols et al. 2005). Different bacterial strains produce EPS of diverse chemical composition and structure (Lemoine et al. 1997; Mancuso Nichols et al. 2005); nevertheless a single strain can also produce more than one kind of EPS (Schiano Moriello et al. 2003). Similarly, the type and amount of EPS produced by a bacterial strain can be customized by exposure to certain environmental conditions such as salinity (Vyrides and Stuckey 2009), temperature (Mancuso Nichols et al. 2005a, b, c), or presence of heavy metals (Guibaud et al. 2005; Naik et al. 2012). Therefore, the term EPS not only refers to a single chemically defined molecule but also encompasses a complex mixture of diverse polysaccharides and other ancillary compounds. In nature, bacterial EPSs serve a range of biological functions, viz., as a buffer material for biodegradable compounds and trace metal nutrients or as part of a protective structures against predators, desiccation, salinity, and cytotoxic compounds and toward high or low temperatures (Nicolaus et al. 2010; Poli et al. 2010).

The metal-binding properties of microbial EPS are well investigated (Geesey et al. 1988) and widely employed in bioremediation of heavy metals (Loaec et al. 1997). In the marine environment, binding of heavy metals by microbial EPS (dissolved or particulate) may comprise broad implications on its dynamics with superior binding capacity for metals than any other known sorbents (Quigley et al. 2002) and form multiple complexes with metal ions. Since EPSs are usually the primary barrier of microbial cells that directly contact and interact with metals in aqueous environment, they are of vital importance through not only protecting the interior microbial cells (Ozturk et al. 2009) but also for geochemical cycling and remediation of metals in natural environments (Wu et al. 2013). Some metals are indispensable to many physiological processes (e.g., copper and iron), while others are nonessential to the cell life (e.g., cadmium and lead). Conversely, in excess, all metals are deleterious to the cells (Ishaque et al. 2006; Valls and de Lorenzo 2002; Lewinson et al. 2009). Several protection mechanisms have already been described for bacterial cells to avoid/minimize the toxicity of heavy metals. These strategies can crop up extracellularly, for example, the sequestration of metal ions on outer cell surface. However, these strategies are necessitated to be pooled with other cellular and biochemical mechanisms to achieve homeostasis, such as exporting the metals through P-type ATPases and detoxify the metal cations by efflux and/or the binding of these cations to metallothioneins, preventing free toxic metals (Nies 2003; Baptista and Vasconcelos 2006).

Therefore, to better engineer such metal biosorption processes, it is essential to have profound understanding on the biological mechanisms of metal adsorption by microbial EPS. The discovery of an effective EPS-producing marine microorganism could therefore enhance the use of environmentally biodegradable EPS molecules in industry and could reduce the dependence on biohazardous, nondegradable synthetic polymers. In these perspectives, marine microbial polysaccharides show impending source of macromolecules with high-value additions from the marine environment (Pomin 2014). Thus a thorough understanding of marine microbes that produce EPSs for metal sequestration and detoxification would result in economic production with several applications in bioremediation process. This review summarizes fundamental insights regarding the marine microbial diversity, extracellular and intracellular metal detoxification strategies, and chemical, enzymatic, and genetic modification of bacterial EPS. It also evaluates the biotechnological applications of exopolysaccharides and future prospects of both naturally occurring and genetically modified bacterial strains in bioremediation and biomonitoring of hazardous metals present in metal-polluted environment for improved detoxification and degradation of toxic pollutants.

5.2 Marine Microbial Diversity Producing EPS

Several EPS-producing marine strains were investigated, which led to the discovery and isolation of novel biomacromolecules (Finore et al. 2014). Marine bacterial exopolysaccharides have been the subject of interest in numerous reviews (Guezennec 2002; Nazarenko et al. 2003; Mancuso Nichols et al. 2005b; Laurienzo 2010; Satpute et al. 2010; Freitas et al. 2011; Jouault and Delbarre-Ladrat 2014; Pomin 2014). Most of the marine derived EPS are bacterial (*viz.* mesophilic, heterotrophic, psychrophilic), archaeal (thermophilic and halophilic) been shown to produce EPS (Rinker and Kelly 2000; Schiano Moriello et al. 2003; Marx et al. 2009; Nicolaus et al. 2010; Poli et al. 2011; Finore et al. 2014).

Mancuso Nichols et al. (2004) depicted the production of EPS by the marine strains *Pseudoalteromonas* CAM025 and CAM036 isolated in Antarctica seawater and sea ice. Most recently, few more strains from Arctic sea ice were also demonstrated to produce EPS with cryoprotective effect (Marx et al. 2009; Liu et al. 2013). The halophilic strains *Halomonas maura* (Arias et al. 2003), *Halomonas ventosae* (Martinez-Canovas et al. 2004; Mata et al. 2006), *Halomonas alkaliantarctica* (Poli et al. 2007), *Hahella chejuensis* (Poli et al. 2010) and the archaeal halophilic *Haloferax mediterranei* (Parolis et al. 1996) isolated from hypersaline environments were shown to produce EPS; a few of them are sulfated (Poli et al. 2010). Marine thermophilic anaerobes *Sulfolobus*, *Thermococcus* and *Thermotoga* were described to produce EPS (Vanfossen et al. 2008). *Thermococcus litoralis* produces an EPS which contains sulfate and phosphorus substituents (Rinker and Kelly 2000); the archaea *Sulfolobus solfataricus* were shown to produce a sulfated polysaccharide (Nicolaus et al. 1993). *Geobacillus* sp., *Bacillus thermodenitrificans*, and *B. licheniformis* thermophilic strains were isolated from shallow marine hydrothermal vents of Vulcano Island (Italy) and the polysaccharides they produced were analyzed (Poli et al. 2010).

Several bacteria belonging to gammaproteobacteria *Alteromonadales* or *Vibrionales* orders isolated from the deep-sea *Polychaeta* annelids *Alvinella pompejana* and *A. caudata* tissues were also shown to produce different EPS: HYD1545 (Vincent et al. 1994), HYD1644 (Dubreucq et al. 1996), HYD721 (Rougeaux et al. 1999a), HYD657 (Cambon-Bonavita et al. 2002) and HE800 (Raguenes et al. 1997; Rougeaux et al. 1999b). The most convincing evidence also came from another gammaproteobacterium isolated from a deep-sea hydrothermal vent shrimp (*A. macleodii* subsp. *fijiensis biovar medioatlantica*) and was shown to produce the EPS MS907 (Raguenes et al. 2003), whereas *Alteromonas macleodii* sp. *fijiensis* producing the pyruvated EPS ST716 (Raguenes et al. 1996; Rougeaux et al. 1998) and *Alteromonas infernus* producing the sulfated EPS GY785 were isolated from hydrothermal fluids (Raguenes et al. 1997; Roger et al. 2004). In another study, *Pseudoalteromonas* strains produced diverse EPS: SM9913 isolated from deep-sea sediments produces an acetylated EPS (Qin et al. 2007) and SM20310 from the arctic sea ice produces a complex α-mannan (Liu et al. 2013). *Pseudomonas*, *Alteromonas*, *Paracoccus* and *Vibrio* sp. bacteria producing EPS were explored from an extreme marine ecosystem. Among them, *Paracoccus zeaxanthinifaciens subsp. payriae* and *Vibrio* sp. RA29 are shown to produce sulfated polysaccharides and *Vibrio* sp. MO 245 produced a biopolymer very similar to EPS produced by *Vibrio diabolicus* (Guézennec et al. 2011). Several researchers supported the fact that chemical and structural diversity of these macromolecules confirms the high value of marine environment as a source of exciting chimiodiversity.

5.3 Extracellular Biosequestration of Metals by Marine EPS

The adsorption of heavy metals by EPS is non-metabolic and energy independent and can be caused by interaction between metal cations and negative charge of acidic functional groups (COOH, OH, C=O, PO_4, NH_2) of EPS as shown in Fig. 5.1. Here the EPSs are recommended as surface-active agents for heavy metal removal because of their extensive capacity. Similarly, a cell-bound polysaccharide produced by the marine bacterium *Zoogloea* spp. is also known to act as an adsorbent of metal ions like cadmium, ferrous, lead, and chromium (Kong et al. 1998). Anionic polysaccharides have a strong affinity for metal cations (Rendleman 1978); for instance, alginate, a naturally occurring polysaccharide invented solely of uronic acids, was shown to stabilize Fe (III) (Sreeram et al. 2004). For instance, in seawater, the Fe (III) can coordinate with carboxyl groups of these anionic polysaccharides (Gyurcsik and Nagy 2000; Nagy et al. 2003), with greater stability, while two or more of the carboxyl groups are able to interact with a single Fe (III) cation (Rendleman 1978). In this context, sorption of heavy metals based on metal-binding capacities of biological materials can be an alternative solution (Gadd 2009; Park et al. 2010; Vieira and Volesky 2010). Among numerous biosorbents (e.g., yeast, seaweed, bacteria, fungi), bacterial exopolysaccharides (EPSs) have been effectively and successfully used in heavy metal removal studies (Comte et al. 2006; Loaec et al. 1997, 1998). Moreover, EPSs have ionizable functional groups like carboxyl, acetate, hydroxyl, amine, phosphate, and more rarely sulfate groups, which are potential binding sites for the sequestration of heavy metals (Fig. 5.1) and show metal binding (Liu and Fang 2002). It is admitted that metal sorption implies a physico-chemical interaction between the metal cation and functional groups based on physical sorption,

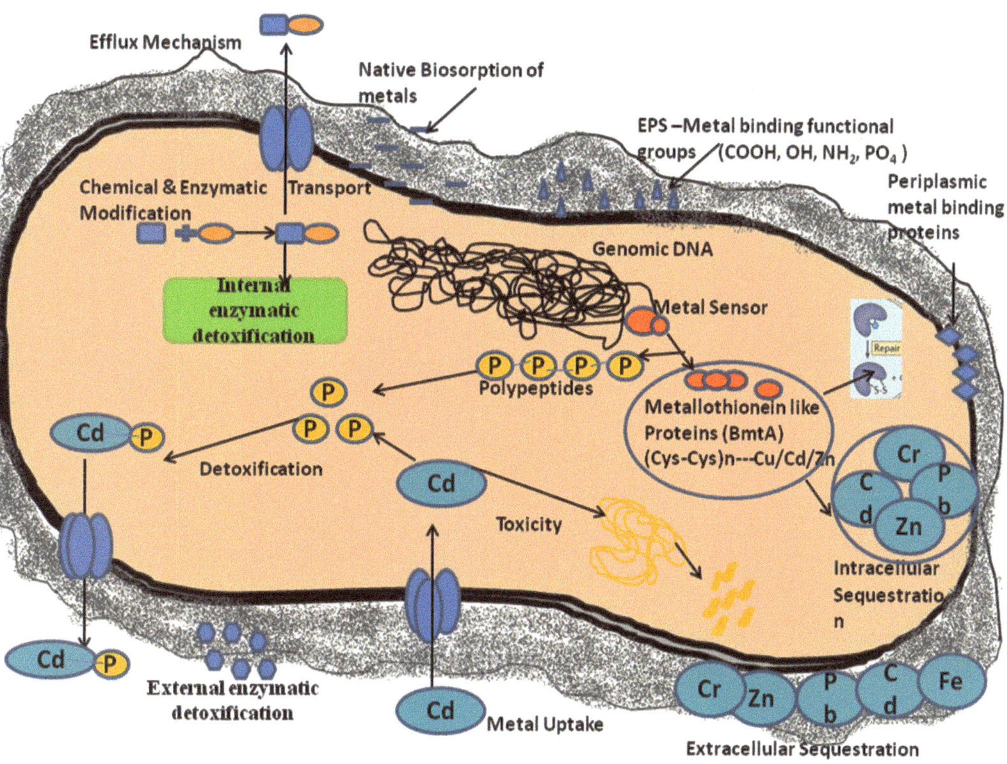

Fig. 5.1 Mechanism of metal biosorption and biosequestration in the marine bacteria

ion exchange, complexation, and/or precipitation (Gadd and White 1989). Moreover, metal sorption also depends on external factors like pH, metal concentration, or biomass concentration.

Extracellular high molecular weight biopolymers secreted by marine bacterial cells, referred to as exopolysaccharides (EPS), consist of macromolecules such as polysaccharides, proteins, nucleic acids, humic substances, lipids, and other non-polymeric constituents of low molecular weight (Bramhachari et al. 2007). These exopolysaccharides are chemically diverse and are mostly acidic heteropolysaccharides with functional groups such as hydroxyl, carboxyl, amides, and phosphoryl which exhibit high affinity toward heavy metals (Bramhachari and Dubey. 2006; Bramhachari et al. 2007; Bhaskar and Bhosle 2006; Braissant et al. 2007). Bacterial EPSs play a key role in initial attachment of cells to different substrata, cell-to-cell aggregation protection against desiccation, and resistance to harmful exogenous materials (Decho 1990; Iyer et al. 2004; Pal and Paul 2008). It is generally assumed that diverse microbial biopolymers were shown to possess potential to bind heavy metals, including lead, with different degree of specificity and affinity (De et al. 2008; Bhaskar and Bhosle 2006; Pal and Paul 2008; Naik et al. 2012). Bacterial EPS and its possible role in bioaccumulation of Cu and Pb in marine food chains were investigated using a partially purified and chemically characterized EPS isolated from *Marinobacter sp*. (Bhaskar and Bhosle 2006). The exopolymer binding process is important in the downward transport of metals in the ocean environment (Decho 1990). In the marine bacterium *P. aeruginosa* CH07, lead was entrapped in EPS indicating a possible resistance mechanism (De et al. 2007, 2008). EPSs are high molecular weight polyanionic polymers which bind metals by electrostatic interaction between metal cation and negatively charged components of EPS resulting in metal immobilization within the exopolymeric matrix (Roane 1999; van Hullebusch et al. 2003).

Marine sulfur-reducing bacteria were shown to produce EPS and formed complex with heavy metals like Ni(II), Cr(VI), and Mo(IV). Interestingly, the predominance of uronic acid in EPS suggests that high molybdenum content was detected in the EPS than other metals and it was also assumed to form stable complex with uronic acids (Beech and Cheung 1995). Interestingly, Shah et al. (2000) reported that the extracellular polysaccharide produced by a marine *Cyanothece* sp. ATCC 51142 contained 2-methylglucose, mannuronic acid, and sulfated mannose, linked by β-1,4-linkage and 3,6-linkages, and was exploited toward metal removal from solutions. In another study, *Enterobacter cloacae*, an exopolysaccharide-producing marine bacterium, was tested for its Cr (VI) tolerance and chelation. X-ray fluorescence (XRF) spectroscopy analysis of both the biomass and the exopolysaccharide revealed that 60–70 % chromium was accumulated by this bacterium (Iyer et al. 2004). In the case of marine environments, several studies were also performed on macroscopic characterization of heavy metal uptake and concentration, retention, and release of radioactive materials by aquatic organisms (Iyer et al. 2005; Sakamoto et al. 2008). For instance, the exopolysaccharide produced by the marine bacterium *Enterobacter cloacae* was reported to have excellent chelating properties with respect to 65 % of cadmium, 20 % copper, and 8 % cobalt at 100 mg/L of metal concentration (Iyer et al. 2005). In the marine environment, the EPS is also capable of binding lead and could potentially be used in water treatment plants (Mancuso Nichols et al. 2005). Interactions between cations and polysaccharides also conferred both algal and bacterial EPS with the potential to adsorb heavy metal contaminants such as Cd^{2+} (Guibaud et al. 2005), Pb^{2+} (Comte et al. 2008), and Hg^{2+} (Zhang et al. 2012), which can then be incorporated into the food chain (Bhaskar and Bhosle 2006). Disconcerting concentrations of these heavy metals were found in the Arctic marine food web (Campbell et al. 2005); EPS from sea-ice organisms may play an imperative role in the fate of these contaminants. Exceptional interest is the dependence of heavy metal adsorption to EPS on properties such as salinity, pH, and Ca^{2+} concentrations (Comte et al.

2008) that underwent seasonal changes in sea-ice brines. The ecological implications of metal-binding properties of bacterial EPS and its role in the bioaccumulation of pollutants in the marine food chain were investigated using a partially purified microbial EPS from *Marinobacter sp.* This EPS was shown to selectively bind more amount of Cu^{2+} per mg (182 nmol mg^{-1}) of EPS than Pb^{2+}(13 nmol mg^{-1}) (Bhaskar and Bhosle 2006).

Purified EPS from marine *Pseudoalteromonas* sp. was explored for its role in Fe chemical speciation, solubility, as well as bioavailability for two keystone Southern Ocean phytoplankton microbial strains. Strikingly, the combined effect of EPS on Fe was shown to increase the sequestrating time of bioavailable Fe in the euphotic zone, therefore possibly sustaining and controlling primary productivity (Hassler et al. 2011). In a most recent report on the biofilm-forming marine bacterial isolates, *Pseudomonas putida* SP-1, capable of volatilizing 89 % of mercury, has confirmed the efficiency of mercury-resistant marine bacteria in bioremediation (Zhang et al. 2012). It is noteworthy that the proteomic responses of the marine bacterium *Pseudomonas fluorescens* BA3SM1 revealed cellular adaptations to Cd, Zn, and Cu by inducing defense mechanisms, i.e., biofilm formation and modification of envelope properties to increase the extracellular metal biosorption (Poirier et al. 2013). More recently, Morcillo and colleagues (2014) investigated the biosorption and biomineralization of radioactive U(VI) associated with the marine bacterium *Idiomarina loihiensis* MAH1. Interestingly, the U(VI) speciation in sea-water samples was shown to be more intricate, by forming different complexes; however, the removal of uranium was demonstrated due to biosorption to extracellular polysaccharides (EPS) and cell wall components as evident from TEM analysis. Biomass and extracellular polymeric substances secreted by biofilm-forming marine bacterium *Pseudomonas aeruginosa* JP-11 effectively removed 58.760 ± 10.62 % and 29.544 ± 8.02 % of Cd, respectively (Chakraborty and Das 2014). In another study, a marine exopolysaccharide-producing bacteria *Bacillus thuringiensis* PW-05 was found to volatilize mercury efficiently and also could resist higher concentrations of $CdCl_2$, $ZnSO_4$, $PbNO_3$, and Na_2HAsO_4 (Dash et al. 2014). Removal of Na^+ by binding with exopolymeric substances was demonstrated in *Rhodopseudomonas palustris* strains TN114 and PP803. Interestingly, the principal component from EPS of strain PP803 was identified as a polysaccharide (≈18 kDa) mainly composed of galacturonic acid (Nunkaew et al. 2015). Strikingly, *Alcaligenes faecalis* was shown to enhance exopolysaccharides and siderophore production, when cells were exposed to toxic levels of organometal (TBTCl), suggesting their involvement in conferring resistance to this antifouling biocide and degradative capability, respectively (Khanolkar et al. 2015). Extracellular polymeric substances (EPSs) produced by marine bacteria play a fundamental role in different stages of biofilm formation, maturation, and maintenance. The influence of loosely bound and tightly bound EPS, extracted from marine *Pseudomonas* NCIMB 2021, on the electrochemical behavior of 70 Cu–30 Ni (wt %) alloy in static artificial seawater (ASW) and on the chemical composition of oxide layers was studied. Interestingly, Bautista and colleagues (2015) recently demonstrated a metal corrosion inhibition effect and a low oxide layer thickness are shown with loosely bound EPS.

5.3.1 Marine Microbial Surface-Active Molecules

Marine microbial surface-active molecules possess several physiological roles of biosurfactants including the increasing of the surface area and bioavailability of hydrophobic water-insoluble substrates, heavy metal binding, bacterial pathogenesis, quorum sensing, and biofilm formation (Ron and Rosenberg 1999; Flemming and Wingender 2010). Microbial surface-active metabolites, called biosurfactants (polysaccharide moiety attached to lipid and/or protein), are such metal-complexing agents that were previously reported to be successful in the bioremediation of heavy metal-contaminated environments (Mulligan et al. 2001; Singh and Cameotra 2004).

There are many reasons that make biosurfactants promising alternative agents for remediation purposes. These are their less toxic nature (Poremba et al. 1991), better environmental compatibility, and biodegradability (Georgiou et al. 1992). Most biosurfactants reported till date are obtained from microorganisms of terrestrial origin. Nonetheless, the marine environment that forms the majority of the earth's surface serves as a huge catalog of microbial population producing a variety of active metabolites and molecules. In the quest of these molecules, Das et al. (2008) reported another marine bacterium with heavy metal sequestration potential and removal of lead and cadmium (Das et al. 2009). There is a growing evidence to implicate uronic acids of EPS in conferring these macromolecules with an ability to interface with hydrophobic organic chemicals, such as hydrocarbons (Gutierrez et al. 2008, 2009). Strikingly, Gutierrez and colleagues (2008) demonstrated an emulsifying and metal ion-binding activity of a glycoprotein exopolymer produced by marine *Pseudoalteromonas* sp. strain TG12. Amino acids and peptides are also frequently found associated with marine bacterial EPS, which conferred amphiphilic characteristics to these polymers (Gutierrez et al. 2009; Wolfaardt et al. 1999). Scores of marine microorganisms such as *Acinetobacter* sp., *Arthrobacter* sp., *Pseudomonas* sp., *Halomonas* sp., *Myroides* sp., *Corynebacteria* sp., *Bacillus* sp., and *Alteromonas* sp. were studied for production of bioemulsifiers and exopolysaccharides (Bouchotroch et al. 2000; Pepi et al. 2005; Satpute et al. 2010; Iyer et al. 2006). Hexavalent chromium reduction and trivalent chromium tolerance of marine *Bacillus* sp. MTCC 5514 are known to be mediated by its extracellular enzyme, chromate reductase, and its biosurfactant (Gnanamani et al. 2010). An excellent alternative for enhanced metal bioremediation is the use of microbial biomolecules such as microbial surfactants, and extracellular polymers suggested the increase in efficiency of metal reduction/sequestration by microorganisms to achieve field bioremediation of metal-contaminated sites (Singh and Cameotra 2004).

5.4 Intracellular Metal Sequestration

5.4.1 Biosorption of Heavy Metals

Increasing contamination of marine environment with heavy metals has become a major global concern. Several techniques, such as adsorption, chemical precipitation, chemical oxidation and reduction, ion exchange, and evaporative recovery, were used for removal and recovery of valuable or toxic metals from wastewater (Singh et al. 1998). Since these physicochemical methods are incredibly expensive, the applications of biological methods are emerging as a viable alternative (Muraleedharan et al. 1991). Interestingly, several algae, bacteria, fungi, mosses and macrophytes and several higher plants were previously employed for metal recovery from water systems (Singh et al. 1998). In an attempt to find more economically viable biological methods, much interest has been focused on finding novel marine microbes able to bind large amounts of metals rapidly. It is interesting to note that *Vibrio harveyi*, a normal inhabitant of the marine environment, is reported to possess the potential for bioaccumulation of cadmium up to 23.3 mg Cd^{2+}/g of dry cells (Abd-Elnaby et al. 2013). In line with that, Von Canstein et al. (2002) reported a consortium of marine bacteria to efficiently remove mercury in a bioreactor in disturbance-independent mechanism. Likewise, a new combination of genetic systems in marine bacteria for the potential degradation capability of phenol and heavy metals was also described (El-Deeb 2009). The effect of cell wall-associated extracellular polymeric substances (EPS) of a Gram-negative bacterium *Shewanella oneidensis* strain MR-1 on Zn(II) and Pb(II) adsorption was investigated using an amalgamation of titration/batch uptake studies, surface complexation modeling, and attenuated total reflectance–Fourier transform infrared (ATR-FTIR) spectroscopy. Interestingly, this study demonstrated a greater amount of EPS in enhanced Zn(II) and Pb(II) uptake. Attributable to its dissimilatory metal-reducing activity, *S. oneidensis* MR-1 was chosen as a target of marine bioremediation research (Ha et al. 2010). Correspondingly, certain purple non-sulfur marine bacterial isolates, e.g., *Rhodobium marinum* and *Rhodobacter sphaeroides*, have also been found to possess the potential of removing heavy metals like copper, zinc, cadmium, and lead from the contaminated environments either by biosorption or biotransformation (Panwichian et al. 2011). Cao and his colleagues (2011) compared U(VI) immobilization using cells with bound EPS (bEPS) and cells with minimal EPS and demonstrated that (i) EPS from *Shewanella* sp. HRCR-1 biofilms contributed significantly to U(VI) immobilization, particularly at low initial U(VI) concentrations, through both sorption and reduction. Therefore, the marine bacteria have been designated for assessing marine pollution through tolerance (Das et al. 2007) and biosorption of heavy metals (Das et al. 2009). Biosorption capabilities and extracellular polysaccharides (EPSs) were demonstrated in marine *Bacillus cereus* CURB-4 strain and were shown to produce more EPS (71 ± 0.6 %) in the presence of Cu (II) ions (Rajaram et al. 2013).

5.4.2 Small Metallothionein Proteins

Microorganisms have evolved several resistance mechanisms to withstand the toxic effects of heavy metals and organometals. One of the common mechanisms is induction of specific metal-binding proteins facilitating the sequestration/bioaccumulation of toxic metals inside the cell. These well-studied metal-binding proteins are referred to as metallothioneins (MTs). Intracellular metal bioaccumulation and homeostasis in cell cytosol involves the low-molecular-weight, cystein-rich metallothioneins which range from 3.5 to 14 kDa (Hamer 1986). Metallothioneins play a significant role in immobilization of toxic heavy metals in that way protecting bacterial metabolic processes catalyzed by enzymes (Fig. 5.1) (Blindauer et al. 2002). Several cyanobacterial and bacterial strains were reported to encode metallothioneins for maintaining cytosolic metal homeostasis such as *Synechococcus* PCC 7942 (SmtA), *Anabaena* PCC 7120 (SmtA), *Oscillatoria brevis* (BmtA), *Pseudomonas aeruginosa* (BmtA) and *Pseudomonas putida* (BmtA) (Blindauer et al. 2002). *P. aeruginosa* strain WI-1 isolated from Mandovi estuary possesses bacterial metallothionein (BmtA) to alleviate lead toxicity (Naik et al. 2012). Typically, the metals for which metallothioneins have the highest affinity are Cd, Pb, Hg, Cu and Zn but also bind to Ni with lower affinity. The heterotrophic marine bacterium, *Vibrio alginolyticus*, interestingly showed production of extracellular copper-binding compounds when exposed to copper in a sea-water medium (Schreiber et al. 1990). Proteomic responses of the marine oxic sediment bacterium *Pseudomonas fluorescens* to Zn, Cu, and Cd metals demonstrated de novo synthesis of metal responsive proteins, indicative of metal-stimulated synthesis, particularly for Cu and Cd (Poirier et al. 2008). Recent genome-mining studies suggest that marine *Synechococcus* sp. WH8102 have metal uptake systems for zinc and proteins that utilize zinc as a cofactor. Interestingly, a mechanism for zinc sequestration was particularly efficient and the expression of SYNW2224, a putative porin protein, was shown upregulated during growth in zinc-depleted conditions (Barnett et al. 2014).

5.5 Biotechnological Applications of Marine Exopolysaccharides

Microbial exopolysaccharides are of significant commercial value, since they embody an interesting alternative to the plant and macro algal exopolysaccharides traditionally used in the food, textile, painting, cosmetic, paper, pharmaceutical and oil industries. Contemporary commercial production of microbial polysaccharides is built on fermentation of heterotrophic bacteria. Marine bacteria offer a great diversity of polysaccharides which could play an important role in biotechnology and industry as well as in future development of cell therapy and regenerative medicine among other applications. Mancuso Nichols et al. (2005)

isolated several EPS-producing strains from the Antarctic sea ice and seawater and found that mannose represented the most abundant neutral sugar in the EPSs. However, the ecological role of mannose in these EPSs is still unknown. The structure of this EPS is different from that of EPSs secreted by other marine bacteria. For instance, EPS produced by *V. diabolicus* contains equal amounts of uronic acids and hexosamines (hyaluronic acid like) and was shown to enhance bone healing in rats (Zanchetta et al. 2003). Kim and Yim (2007) showed that the EPS produced by the Antarctic bacterium *Pseudoalteromonas arctica* KOPRI 21653 could improve the freeze-thaw survival ratio of *E. coli* and suggested that it may have biotechnological potential as a cryo-protection agent. In another study by Liu et al. (2013), a *Pseudoalteromonas* strain SM20310, screened from Arctic sea ice, produced a highly complex mannan. This EPS had a cryoprotective effect and could also improve high-salinity tolerance, a characteristic feature beneficial for the strain to adapt to sea-ice environment. Although several known marine bacteria were shown to produce structural and functionally diverse EPSs, few of the produced EPSs from *Salipiger mucosus* strain A3T, a halophilic species belonging to *Alphaproteobacteria*, are of immense biotechnological importance, so the investigation for novel EPSs that have innovative applications is still of potential interest (Llamas et al. 2010).

Marine bacteria have become increasingly popular for producing novel EPS molecules. The presence of charged EPS from the Arctic marine psychrophile has intricate implications for the dynamics of carbonates in the sea-ice environment given that these charged polysaccharides, unlike non-ionic polysaccharides, have well-known effects on the precipitation of carbonates (Hardikar and Matijević 2001). In the case of cold-adapted marine organisms, EPS was shown to increase the stability and half-life of a cold-active extracellular aminopeptidase isolated from subzero Arctic marine sediment bacteria *C. psychrerythraea* 34H (Huston et al. 2004). Mannose is a main component of many EPSs from cold marine environment strains like *Pseudoalteromonas arctica* KOPRI 21653, *Pseudoalteromonas* strain SM20310 and *Pseudoalteromonas haloplanktis* TAC 125, isolated from Antarctic seawater and has a mannan and phosphomannan structure (Corsaro et al. 2004; Kim and Yim 2007; Liu et al. 2013).

5.6 Strategies for Enzymatic Modifications of EPS

The marine biodiversity has previously shown a great potential in providing new biocatalysts (Trincone 2010). Quite a lot of enzymes have become promising biotechnological tools for in vitro synthesis or in vivo biosynthesis engineering. Conversely, the in silico genomic data would also consent to identify new enzymatic tools to modify microbial glycopolymers. Most importantly, the depolymerizing enzymes can be used in in vitro depolymerization process but also as tools to study the chemical structure. Carbohydrate sulfotransferases and other enzymes grafting substituents such as acetate, as well as enzymes catalyzing their removal, may be used in vitro for binding or for elimination of substituents which are of great importance for the final bioactivity of the molecule. Identifying both the polysaccharide chemical structure and genetic cluster for the biosynthesis would broaden the knowledge of the glycopolymer enzymatic function in biosynthetic pathway by means of genetic knockouts.

Several research studies proved that exopolysaccharides can be modified by the enzymatic action of transferases and hydrolases which possess the capability to add or remove functional groups such as acetyls, pyruvyls (Marzocca et al. 1991), glyceryls (Kuo et al. 1986), succinyls (Reuber and Walker 1993), lactyls (Maalej et al. 2014), or a combination of these, leading to variations in polymer surface electrostatics and solubility. Additionally, epimerization can drastically alter the structural conformation of polysaccharides, affecting polymer interactions within the biofilm (Steigedal et al. 2008). Some of these modifications were earlier studied with respect to their importance in bacterial virulence, pathogenesis, biofilm formation, or symbiosis (Ridout et al. 1997), as well as their commercial utility in the food and cosmetic industries. Despite this wealth of knowledge, there remain a number of unresolved questions regarding the biological implications of EPS modifications in marine bacteria as well. It is interesting to note that glycosaminoglycans (GAGs) are a group of polymers that are typically composed of a disaccharide repeat unit containing an amino sugar and a hexuronic acid (Laurent and Fraser 1992; Esko and Lindahl 2001; DeAngelis 2002). GAGs were initially thought to exist only in the animal kingdom, where they serve essential biological functions; however, there was an emergence of GAG-like polymers among few prokaryote strains (DeAngelis et al. 2002). Prokaryotic GAGs are typically less complex than their eukaryotic counterparts due to an absence of chemical modifications such as sulfation on the structures (Raedts et al. 2011). The identification of the bacteria glucuronyl C5-epimerase has proven elusive (Raedts et al. 2011). However, an enzyme (RED65_08024) from the marine bacterium *Bermanella marisrubi* that shares 37 % sequence similarity with the human glucuronyl C5-epimerase was characterized and shown in vitro to convert GlcA to IdoA in de-sulfated mouse HS (Raedts et al. 2013). This glucuronyl C5-epimerase represents the first prokaryotic protein capable of generating IdoA residues and only identified epimerase that can function on bacterial polysaccharides after post polymerization, besides AlgG and AlgE1-7. Unfortunately, the EPS produced by *B. marisrubi* has not been characterized, so its target remains unknown. Therefore, increasing list of identified modifications in marine bacteria will allow future efforts to focus on linking these modifications to specific biosynthetic genes and biofilm phenotypes.

5.7 Strategies for Chemical Modifications of EPS

Interestingly, chemical modifications are widely demonstrated in terrestrial microbes (D'ayala et al. 2008; Laurienzo 2010; Senni et al. 2011): acid hydrolysis (Colliec et al. 1994), radical depolymerization (Nardella et al. 1996), N-deacetylation (Zou et al. 1998) and sulfation (Guezennec et al. 1998). However, several drawbacks such as lack of control and regioselectivity, use of organic solvents, and nonhomogeneous conditions are identified (Al-Horani and Desai 2010). More recently, ionic liquids were also used for cellulose sulfation in homogeneous media (Gericke et al. 2011). Only few reports have been dealt with chemical modifications on marine EPS (D'ayala et al. 2008). The most convincing evidence for the low molecular weight over sulfated derivatives of the EPS GY785 from the deep-sea bacterium *A. infernus* and EPS HE800 from *V. diabolicus* has been obtained by depolymerization, acid hydrolysis, or free-radical reaction followed by sulfation with sulfur trioxide pyridine complex (Jouault et al. 2001; Senni et al. 2011). Nevertheless, there are no reports of chemically modified EPS used in bioremediation.

5.8 Genetic Manipulation in Marine Bacteria to Enhance Metal Detoxification

In order to increase the bioremediation potential and/or metabolic activity of any bacteria, insertions of certain functional genes are highly indispensable into their genome. This phenomenon can be achieved by insertion of new genes into the genomic complexion, insertion of new plasmid, alteration of metabolic pathways like transport and chemotaxis and, most importantly, adaption of features toward the environmental conditions (Pieper and Reineke 2000). However, limited reports are available to date for the genetic manipulation of marine bacteria to achieve a goal of enhanced bioremediation (Valls and de Lorenzo 2002). An interesting result came from a study with the insertion of *bmtA* gene coding for metallothionein into suitable vector and its transformation into marine bacteria was conducted and successfully employed in highly metal-contaminated environments (Chen et al. 1999). These genetically modified marine bacteria expressing extremely high amount of metallothionein can be employed to bioremediate marine sites highly contaminated with heavy metals. These hyper-metal accumulating bacteria maintain metal homeostasis by reducing metal bioavailability. Expression of metallothioneins on the cell surface by fusion with cell surface proteins improves the bioaccumulation capacity of bacteria (Chen et al. 1999). Similarly, *Pseudoalteromonas haloplanktis*, possessing a shuttle plasmid-encoding suppressor for amber mutation, was used for genetic manipulation in bioremediation studies (Kivela et al. 2008). Bacteria possessing plasmid with *merA* gene responsible for converting toxic form of mercury to nontoxic form may be transformed into marine bacteria for better application in field conditions of bioremediation of mercury (De Rore et al. 1994). *Deinococcus radiodurans*, the most radio-resistant marine organism, was modified genetically to metabolize toluene and ionic form of mercury from nuclear wastes (Brim et al. 2000). Cheung and Gu (2003) demonstrated chromate ($CrO4^{2-}$) reduction in marine sulfate-reducing bacteria. In a recent report, Stahl et al. (2015) elucidated the function of two CzcCBA pumps encoded by *Marinobacter adhaerens* HP15's genome during exposure to cadmium, zinc and cobalt by generating single and double knockout mutants. Modification of bacterial isolates which can overproduce EPS and biosurfactants may be a good strategy for bioremediation of significant amount of heavy metals. Microbial enzymes can be modified to increase their kinetics to reductively precipitate heavy metal ions or increase efficiency of enzyme to effectively detoxify metals (Karigar and Rao 2011).

5.9 Future Prospects in Marine EPS Research

The important roles of marine microbial EPS in metal biosorption have been widely documented. However, the underlying mechanisms of the metal–EPS interactions and engineering of practical metal biosorption processes are largely unclear. In fact, the complexity in the EPS composition and the diversity in metal species in the vast marine biosphere make it difficult to explore into metal–EPS interactions and microbial-based metal biosorption. Furthermore, the details regarding the genes and proteins involved in the polysaccharide modifications remain largely unavailable. Characterization of these genes and proteins will probably provide details on to how the levels and types of modifications are regulated under different conditions. Therefore, to gain an in-depth understanding of the metal detoxification process and promote its practical implementation, future research efforts in the above directions are warranted.

5.10 Concluding Remarks

Marine bacteria offer a great diversity of polysaccharides which could play an important role in biotechnology and industry as well as in future development. Fundamental studies on marine bacterial exopolysaccharides are relatively few in number compared to other polysaccharides. There are indeed many gaps in our knowledge on fundamental questions such as the structure, physicochemical properties and biosynthesis. Thus, research is urgently wanted on these aspects. The data currently available indicate that marine bacterial exopolysaccharides have considerable biotechnological importance; nevertheless, further studies are highly indispensable to evaluate the feasibility of their practical application. This may limit the practical application of marine bacterial exopolysaccharides as a metal-chelating agent. Studies on interactions between metals and EPS produced by bacteria have established the phenomenon of biosorption as one of the vital processes in bioremediation of toxic heavy metal-contaminated systems, the components of EPS as a whole playing the central role (Watanabe et al. 2003). Therefore, further research work is warranted to determine whether marine bacterial exopolysaccharides have high and selective affinities for metal ions in a given natural system. However, the structural complexity of the EPS contributed by its major components has made it complicated to elucidating the functional role and virtual contribution of each EPS component in sequestration and biosorption of metal(s). Therefore, in-depth understanding will abet in engineering the EPSs with enhanced characteristics of metal sorption and detoxification system for effective bioremediation of environmental concern. Despite the fact that marine microorganisms are better adapted to rapidly changing environmental conditions, little has been known regarding the mechanism of resistance to the deleterious environment. Hence, the research in this aspect will be ready to lend a hand in understanding the genetic mechanism of the metal resistance. Some modifications in their genetic system may endow with useful, high-potential, and more efficient bacterial entity for enhanced bioremediation.

Acknowledgments Dr.P.V. Bramhachari gratefully acknowledges DST-SERB, Government of India for financial support under the Grant No: SR/FT/LS-109-2011

References

Abd-Elnaby H, Abou-Elela GM, El-Sersy NA (2013) Cadmium resisting bacteria in Alexandria Eastern Harbor (Egypt) and optimization of cadmium bioaccumulation by Vibrio harveyi. Afr J Biotechnol 10(17):3412–3423

Al-Horani RA, Desai UR (2010) Chemical sulfation of small molecules—advances and challenges. Tetrahedron 66(16):2907–2918

Arena A (2004) Exopolysaccharides from marine thermophilic bacilli induce a Th1 cytokine profile in human PBMC. Eur Congr Clin Microbiol Infect Dis 14:1317

Arias S, Del Moral A, Ferrer MR, Tallon R, Quesada E, Bejar V (2003) Mauran, an exopolysaccharide produced by the halophilic bacterium *Halomonas maura*, with a novel composition and interesting properties for biotechnology. Extremophiles 7:319–326

Baptista MS, Vasconcelos MT (2006) Cyanobacteria metal interactions: requirements, toxicity, and ecological implications. Crit Rev Microbiol 32(3):127–137

Barnett JP, Scanlan DJ, Blindauer CA (2014) Identification of major zinc-binding proteins from a marine cyanobacterium: insight into metal uptake in oligotrophic environments. Metallomics 6(7):1254–1268

Bautista BET, Wikieł AJ, Datsenko I, Vera M, Sand W, Seyeux A, Marcus P (2015) Influence of extracellular polymeric substances (EPS) from *Pseudomonas* NCIMB 2021 on the corrosion behaviour of 70Cu–30Ni alloy in seawater. J Electroanal Chem 737:184–197

Beech IB, Cheung CWS (1995) Interactions of exopolymers produced by sulphate reducing bacteria with metal ions. Int Biodeterior Biodegrad 35:59–72

Bhaskar PV, Bhosle NB (2006) Bacterial extracellular polymeric substances (EPS) a carrier of heavy metals in the marine food-chain. Environ Int 32:192–198

Bhaskar PV, Grossart HP, Bhosle NB, Simon M (2005) Production of macroaggregates from dissolved exopolymeric substances (EPS) of bacterial and diatom origin. FEMS Microbiol Ecol 53(2):255–264

Blindauer CA, Harrison MD, Robinson AK, Parkinson JA, Bowness PW, Sadler PJ, Robinson NJ (2002) Multiple bacteria encode metallothioneins and SmtA-like zinc fingers. Mol Microbiol 45(5):1421–1432

Bouchotroch S, Quesada E, Izquierdo I, Rodriguez M, Bejar V (2000) Bacterial exopolysaccharides produced by newly discovered bacteria belonging to the genus *Halomonas*, isolated from hypersaline habitats in Morocco. J Ind Microbiol Biotechnol 24:374–378

Braissant O, Decho AW, Dupraz C, Glunk C, Przekop KM, Visscher PT (2007) Exopolymeric substances of sulfate-reducing bacteria: interactions with calcium at alkaline pH and implication for formation of carbonate minerals. Geobiology 5:401–411

Bramhachari PV, Dubey SK (2006) Isolation and characterization of exopolysaccharide produced by *Vibrio harveyi* strain VB23. Lett Appl Microbiol 43(5):571–577

Bramhachari PV, Kavi Kishor PB, Ramadevi R, Kumar R, Rao BR, Dubey SK (2007) Isolation and characterization of mucous exopolysaccharide produced by *Vibrio furnissii* VB0S3. J Microbiol Biotechnol 17:44–51

Brim H, McFarlan SC, Fredrickson JK, Minton KW, Zhai M, Wackett LP, Daly MJ (2000) Engineering Deinococcus radiodurans for metal remediation in radioactive mixed waste environments. Nat Biotechnol 18:85–90

Cambon-Bonavita MA, Raguenes G, Jean J, Vincent P, Guezennec J (2002) A novel polymer produced by a bacterium isolated from a deep-sea hydrothermal vent polychaete annelid. J Appl Microbiol 93:310–315

Campbell LM, Norstrom RJ, Hobson KA, Muir DC, Backus S, Fisk AT (2005) Mercury and other trace elements in a pelagic Arctic marine food web (Northwater Polynya, Baffin Bay). Sci Total Environ 351:247–263

Cao B, Ahmed B, Kennedy DW, Wang Z, Shi L, Marshall MJ, Beyenal H (2011) Contribution of extracellular polymeric substances from *Shewanella* sp. HRCR-1 biofilms to U (VI) immobilization. Environ Sci Technol 45(13):5483–5490

Chakraborty J, Das S (2014) Characterization and cadmium-resistant gene expression of biofilm-forming marine bacterium *Pseudomonas aeruginosa* JP-11. Environ Sci Pollut Res 21(24):14188–14201

Chen W, Bruhlmann F, Richins RD, Mulchandani A (1999) Engineering of improved microbes and enzymes for bioremediation. Curr Opin Biotechnol 10:137–141

Cheung KH, Gu JD (2003) Reduction of chromate ($CrO4^{2-}$) by an enrichment consortium and an isolate of marine sulfate reducing bacteria. Chemosphere 52:1523–1529

Colliec S, Boisson-vidal C, Jozefonvicz J (1994) A low molecular weight fucoidan fraction from the brown seaweed *Pelvetia canaliculata*. Phytochemistry 35(3):697–700

Comte S, Guibaud G, Baudu M (2006) Biosorption properties of extracellular polymeric substances (EPS) resulting from activated sludge according to their type: soluble or bound. Process Biochem 41(4):815–823

Comte S, Guibaud G, Baudu M (2008) Biosorption properties of extracellular polymeric substances (EPS) towards Cd, Cu and Pb for different pH values. J Hazard Mater 151(1):185–193

Corsaro MM, Lanzetta R, Parrilli E, Parrilli M, Tutino ML, Ummarino S (2004) Influence of growth temperature on lipid and phosphate contents of surface polysaccharides from the Antarctic bacterium *Pseudoalteromonas haloplanktis* TAC 125. J Bacteriol 186(1):29–34

d'Ayala GG, Malinconico M, Laurienzo P (2008) Marine derived polysaccharides for biomedical applications: chemical modification approaches. Molecules 13(9):2069–2106

Das S, Shanmugapriya R, Lyla PS, Khan SA (2007) Heavy metal tolerance of marine bacteria—an index of marine pollution. Nat Acad Sci Lett (India) 30:279–284

Das P, Mukherjee S, Sen R (2008) Improved bioavailability and biodegradation of a model polyaromatic hydrocarbon by a biosurfactant producing bacterium of marine origin. Chemosphere 72:1229–1234

Das P, Mukherjee S, Sen R (2009) Biosurfactant of marine origin exhibiting heavy metal remediation properties. Bioresour Technol 100(20):4887–4890

Dash HR, Mangwani N, Das S (2014) Characterization and potential application in mercury bioremediation of highly mercury-resistant marine bacterium *Bacillus thuringiensis* PW-05. Environ Sci Pollut Res 21(4):2642–2653

De Rore H, Top E, Houwen F, Mergcay M, Verstraete W (1994) Evolution of heavy metal-resistant transconjugants in a soil environment with a concomitant selective pressure. FEMS Microbiol Ecol 14:263–273

De J, Ramaiah N, Bhosle NB, Garg A, Vardanyan L, Nagle VL, Fukami K (2007) Potential of mercury resistant marine bacteria for detoxification of chemicals of environmental concern. Microbes Environ 22:336–345

De J, Ramaiah N, Vardanyan L (2008) Detoxification of toxic heavy metals by marine bacteria highly resistant to mercury. Mar Biotechnol 10(4):471–477

DeAngelis PL (2002) Evolution of glycosaminoglycans and their glycosyltransferases: implications for the extracellular matrices of animals and the capsules of pathogenic bacteria. Anat Rec 268(3):317–326

DeAngelis PL, Gunay NS, Toida T, Mao W-J, Linhardt RJ (2002) Identification of the capsular polysaccharides of Type D and F *Pasteurella multocida* as unmodified heparin and chondroitin, respectively. Carbohydr Res 337:1547–1552

Decho AW (1990) Microbial exopolymer secretions in ocean environments: their role(s) in food webs and marine processes. In: Barnes M (ed) Oceanogr Mar Biol Annu Rev. Aberdeen University Press, Aberdeen, pp 73–153

Decho AW, Visscher RS, Norman PT (2010) Quorum sensing in natural environments: emerging views from microbial mats. Trends Microbiol 18:73–80

Dubreucq G, Domon B, Fournet B (1996) Structure determination of a novel uronic acid residue isolated from the exopolysaccharide produced by a bacterium originating from deep sea hydrothermal vents. Carbohydr Res 290(2):175–181

El-Deeb B (2009) Natural combination of genetic systems for degradation of phenol and resistance to heavy metals in phenol and cyanide assimilating bacteria. Malay J Microbiol 5:94–103

Esko JD, Lindahl U (2001) Molecular diversity of heparan sulfate. J Clin Invest 108(2):169–173

Finore I, Di Donato P, Mastascusa V, Nicolaus B, Poli A (2014) Fermentation technologies for the optimization of marine microbial exopolysaccharide production. Mar Drugs 12:3005–3024

Flemming HC, Wingender J (2010) The biofilm matrix. Nat Rev Microbiol 8:623–633

Freitas F, Alves VD, Reis MAM (2011) Advances in bacterial exopolysaccharides: from production to biotechnological applications. Trends Biotechnol 29:388–398

Fu F, Wang Q (2011) Removal of heavy metal ions from wastewaters: a review. J Environ Manag 92:407–418

Gadd GM (2009) Biosorption: critical review of scientific rationale, environmental importance and significance for pollution treatment. J Chem Technol Biotechnol 84(1):13–28

Gadd GM, White C (1989) Removal of thorium from simulated acid process streams by fungal biomass. Biotechnol Bioeng 33(5):592–597

Geesey GG, Jang L, Jolley JG, Hankins MR, Iwaoka T, Griffiths PR (1988) Binding of metal ions by extracellular polymers of biofilm bacteria. Water Sci Technol 20(11–12):161–165

Georgiou G, Lin SC, Sharma MM (1992) Surface-active compounds from microorganisms. Biotechnology 10:60–65

Gericke M, Liebert T, Seoud OAE, Heinze T (2011) Tailored media for homogeneous cellulose chemistry: ionic liquid/co-solvent mixtures. Macromol Mater Eng 296(6):483–493

Gnanamani A, Kavitha V, Radhakrishnan N, Rajakumar GS, Sekaran G, Mandal AB (2010) Microbial products (biosurfactant and extracellular chromate reductase) of marine microorganism are the potential agents reduce the oxidative stress induced by toxic heavy metals. Colloids Surf B: Biointerfaces 79(2):334–339

Guezennec J (2002) Deep-sea hydrothermal vents: a new source of innovative bacterial exopolysaccharides of biotechnological interest? J Ind Microbiol Biotechnol 29:204–208

Guezennec J, Ortega-Morales O, Raguenes G, Geesey G (1998) Bacterial colonization of artificial substrate in the vicinity of deep-sea hydrothermal vents. FEMS Microbiol Ecol 26(2):89–99

Guézennec J, Moppert X, Raguénès G, Richert L, Costa B, Simon-Colin C (2011) Microbial mats in French Polynesia and their biotechnological applications. Process Biochem 46:16–22

Guibaud G, Comte S, Bordas F, Dupuy S, Baudu M (2005) Comparison of the complexation potential of extracellular polymeric substances (EPS), extracted from activated sludge and produced by pure bacterial strains, for cadmium, lead and nickel. Chemosphere 59:629–638

Gutiérrez T, Mulloy B, Black K, Green DH (2007) Glycoprotein emulsifiers from two marine *Halomonas* species: chemical and physical characterization. J Appl Microbiol 103:1716–1727

Gutierrez T, Shimmield T, Haidon C, Black K, Green DH (2008) Emulsifying and metal ion binding activity of a glycoprotein exopolymer produced by *Pseudoalteromonas* sp. strain TG12. Appl Environ Microbiol 74:4867–4876

Gutierrez T, Morris G, Green DH (2009) Yield and physicochemical properties of EPS from *Halomonas* sp. strain TG39 identifies a role for protein and anionic residues (sulfate and phosphate) in emulsification of n-hexadecane. Biotechnol Bioeng 103:207–216

Gyurcsik B, Nagy L (2000) Carbohydrates as ligands: coordination equilibria and structure of the metal complexes. Coord Chem Rev 203:81–149

Ha J, Gélabert A, Spormann AM, Brown GE (2010) Role of extracellular polymeric substances in metal ion complexation on *Shewanella oneidensis*: batch uptake, thermodynamic modeling, ATR-FTIR, and EXAFS study. Geochim Cosmochim Acta 74(1):1–15

Hamer DH (1986) Metallothionein1, 2. Annu Rev Biochem 55(1):913–951

Hardikar VV, Matijević E (2001) Influence of ionic and nonionic dextrans on the formation of calcium hydroxide and calcium carbonate particles. Colloids Surf A Physicochem Eng Asp 186(1):23–31

Hassler CS, Alasonati E, Nichols CM, Slaveykova VI (2011) Exopolysaccharides produced by bacteria isolated from the pelagic Southern Ocean—role in Fe binding, chemical reactivity, and bioavailability. Mar Chem 123(1):88–98

Hoagland KD, Rosowski JR, Gretz MR, Roemer SC (1993) Diatom extracellular polymeric substances: function, fine structure, chemistry, and physiology. J Phycol 29:537–566

Hosono T, Su C, Delinom R, Umezawa Y, Toyota T, Kaneko S, Taniguchi M (2011) Decline in heavy metal contamination in marine sediments in Jakarta Bay, Indonesia due to increasing environmental regulations. Estuar Coast Shelf Sci 92:297–306

Huston AL, Methe B, Deming JW (2004) Purification, characterization, and sequencing of an extracellular cold-active aminopeptidase produced by marine psychrophile *Colwellia psychrerythraea* strain 34H. Appl Environ Microbiol 70(6):3321–3328

Ishaque AB, Johnson L, Gerald T, Boucaud D, Okoh J, Tchounwou PB (2006) Assessment of individual and combined toxicities of four non-essential metals (As, Cd, Hg and Pb) in the microtox assay. Int J Environ Res Public Health 3:118–120

Iyer A, Mody K, Jha B (2004) Accumulation of hexavalent chromium by an exopolysaccharide producing marine *Enterobacter cloacae*. Mar Pollut Bull 49(11):974–977

Iyer A, Mody K, Jha B (2005) Biosorption of heavy metals by a marine bacterium. Mar Pollut Bull 50(3):340–343

Iyer A, Mody K, Jha B (2006) Emulsifying properties of a marine bacterial exopolysaccharide. Enzym Microb Technol 38(1):220–222

Jouault SC, Chevolot L, Helley D, Ratiskol J, Bros A, Sinquin C, Fischer AM (2001) Characterization, chemical modifications and in vitro anticoagulant properties of an exopolysaccharide produced by Alteromonas infernus. Biochim Biophys Acta Gen Subj 1528(2):141–151

Jouault SC, Delbarre-Ladrat C (2014) Marine-derived bioactive polysaccharides from microorganisms. In: Brahmachari G (ed) Natural bioactive molecules. Impacts and prospects. Narosa Publishing House, New Delhi, pp 5.1–5.21

Jouault SC, Zanchetta P, Helley D, Ratiskol J, Sinquin C, Fischer AM et al (2004) Microbial polysaccharides of marine origin and their potential in human therapeutics. Pathol Biol 52:127–130

Karigar CS, Rao SS (2011) Role of microbial enzymes in the bioremediation of pollutants: a review. Enzym Res 2011:1–11

Khanolkar D, Dubey SK, Naik MM (2015) Tributyltin Chloride (TBTCl)-enhanced exopolysaccharide and siderophore production in an estuarine *Alcaligenes faecalis* strain. Arch Environ Contam Toxicol 68(4):612–621

Kim SJ, Yim JH (2007) Cryoprotective properties of exopolysaccharide (P-21653) produced by the Antarctic bacterium, *Pseudoalteromonas arctica* KOPRI 21653. J Microbiol (Seoul Korea) 45(6):510–514

Kivela HM, Madonna S, Krupovic M, Tutino ML, Bamford JKH (2008) Genetics for Pseudoalteromonas

provides tools to manipulate marine bacterial virus PM2. J Bacteriol 190:1298–1307
Kong JY, Lee HW, Hong JW, Kang YS, Kim JD, Chang MW, Bae SK (1998) Utilization of a cell-bound polysaccharide produced by the marine bacterium *Zooglea* sp.: new biomaterial for metal adsorption and enzyme immobilization. J Mar Biotechnol 6:99–103
Kuo M-S, Mort AJ, Dell A (1986) Identification and location of l-glycerate, an unusual acyl substituent in gellan gum. Carbohydr Res 156:173–187
Laurent TC, Fraser JR (1992) Hyaluronan. FASEB J 6:2397–2404
Laurienzo P (2010) Marine polysaccharides in pharmaceutical applications: an overview. Mar Drugs 8:2435–2465
Lemoine J, Chirat F, Wieruszeski JM, Strecker G, Neeser N, Favre JR (1997) Structural characterization of the exocellular polysaccharides produced by *Streptococcus thermophilus* SFi39 and SFi12. Appl Environ Microbiol 63:3512–3518
Lewinson O, Lee AT, Rees DC (2009) A P-type ATPase importer that discriminates between essential and toxic transition metals. Proc Natl Acad Sci U S A 106:4677–4682
Liu H, Fang HH (2002) Characterization of electrostatic binding sites of extracellular polymers by linear programming analysis of titration data. Biotechnol Bioeng 80(7):806–811
Liu S-B, Chen X-L, He H-L, Zhang X-Y, Xie B-B, Yu Y et al (2013) Structure and ecological roles of a novel exopolysaccharide from the Arctic Sea Ice bacterium *Pseudoalteromonas* sp. strain SM20310. Appl Environ Microbiol 79:224–230
Llamas I, Mata JA, Tallon R, Bressollier P, Urdaci MC, Quesada E, Béjar V (2010) Characterization of the exopolysaccharide produced by *Salipiger mucosus* A3T, a halophilic species belonging to the Alphaproteobacteria, isolated on the Spanish mediterranean seaboard. Mar Drugs 8(8):2240–2251
Loaec M, Olier R, Guezennec J (1997) Uptake of lead, cadmium and zinc by a novel bacterial exopolysaccharide. Water Res 31:1171–1179
Loaec M, Olier R, Guezennec J (1998) Chelating properties of bacterial exopolysaccharides from deep-sea hydrothermal vents. Carbohydr Polym 35(1):65–70
Maalej H, Boisset C, Hmidet N, Buon L, Heyraud A, Nasri M (2014) Purification and structural data of a highly substituted exopolysaccharide from *Pseudomonas stutzeri* AS22. Carbohydr Polym 112:404–411
Mancuso Nichols CA, Garon S, Bowman JP, Raguenes G, Guezennec J (2004) Production of exopolysaccharides by Antarctic marine bacterial isolates. J Appl Microbiol 96:1057–1066
Mancuso Nichols CA, Guezennec J, Bowman JP (2005a) Bacterial exopolysaccharides from extreme marine environments with special consideration of the southern ocean, sea ice, and deep-sea hydrothermal vents. Mar Biotechnol 7(4):253–271
Mancuso Nichols C, Bowman JP, Guezennec J (2005b) Effects of incubation temperature on growth and production of exopolysaccharides by an Antarctic sea ice bacterium grown in batch culture. Appl Environ Microbiol 71:3519–3523
Mancuso Nichols C, Garon Lardière S, Bowman JP, Nichols PD, Gibson JAE, Guézennec J (2005c) Chemical characterization of exopolysaccharides from Antarctic marine bacteria. Microb Ecol 49:578–589
Martinez-Canovas MJ, Quesada E, Llamas I, Béjar V (2004) *Halomonas ventosae* sp. nov., a moderately halophilic, denitrifying, exopolysaccharide-producing bacterium. Int J Syst Evol Microbiol 54:733–737
Marx JG, Carpenter SD, Deming JW (2009) Production of cryoprotectant extracellular polysaccharide substances (EPS) by the marine psychrophilic bacterium *Colwellia psychrerythraea* strain 34H under extreme conditions. Can J Microbiol 55:63–72
Marzocca MP, Harding NE, Petroni EA, Cleary JM, Ielpi L (1991) Location and cloning of the ketal pyruvate transferase gene of *Xanthomonas campestris*. J Bacteriol 173:7519–7524
Mata JA, Béjar V, Llamas I, Arias S, Bressollier P et al (2006) Exopolysaccharides produced by the recently described halophilic bacteria *Halomonas ventosae* and *Halomonas anticariensis*. Res Microbiol 157:827–835
Morcillo F, Gonza'lez-Mun˜oz MT, Reitz T, Romero-Gonzalez ME, Arias JM et al (2014) Biosorption and Biomineralization of U (VI) by the marine bacterium *Idiomarina loihiensis* MAH1: effect of background electrolyte and pH. PLoS ONE 9(3):e91305
Moriello VS, Lama L, Poli A, Gugliandolo C, Maugeri TL, Gambacorta A, Nicolaus B (2003) Production of exopolysaccharides from a thermophilic microorganism isolated from a marine hot spring in flegrean areas. J Ind Microbiol Biotechnol 30(2):95–101
Mulligan CN, Yong CN, Gibbs BF (2001) Heavy metal removal from sediments by biosurfactants. J Hazard Mater 85:111–125
Muraleedharan TR, Iyengar L, Venkobachar C (1991) Biosorption: alternative for metal removal and recovery. Curr Sci 61:379–385
Nagy L, Yamaguchi T, Yoshida K (2003) Application of EXAFS and XANES methods in coordination chemistry of carbohydrates and their derivatives. Struct Chem 14:77–84
Naik MM, Pandey A, Dubey SK (2012) *Pseudomonas aeruginosa strain WI-1* from Mandovi estuary possesses metallothionein to alleviate lead toxicity and promotes plant growth. Ecotoxicol Environ Saf 79:129–133
Nardella A, Chaubet F, Boisson-Vidal C, Blondin C, Durand P, Jozefonvicz J (1996) Anticoagulant low molecular weight fucans produced by radical process and ion exchange chromatography of high molecular weight fucans extracted from the brown seaweed Ascophyllum nodosum. Carbohydr Res 289:201–208

Nazarenko EL, Komandrova NA, Gorshkova RP, Tomshich SV, Zubkov VA, Kilcoyne M et al (2003) Structures of polysaccharides and oligosaccharides of some gram-negative marine proteobacteria. Carbohydr Res 338:2449–2457

Nicolaus B, Manca MC, Ramano I, Lama L (1993) Production of an exopolysaccharide from two thermophilic archaea belonging to the genus *Sulfolobus*. FEMS Microbiol Lett 109:203–206

Nicolaus B, Kambourova M, Oner ET (2010) Exopolysaccharides from extremophiles: from fundamentals to biotechnology. Environ Technol 31:1145–1158

Nies DH (2003) Efflux-mediated heavy metal resistance in prokaryotes. FEMS Microbiol Rev 27(2–3):313–339

Nunkaew T, Kantachote D, Nitoda T, Kanzaki H, Ritchie RJ (2015) Characterization of exopolymeric substances from selected *Rhodopseudomonas palustris* strains and their ability to adsorb sodium ions. Carbohydr Polym 115:334–341

Ozturk S, Aslim B, Suludere Z (2009) Evaluation of chromium (VI) removal behaviour by two isolates of *Synechocystis* sp. in terms of exopolysaccharide (EPS) production and monomer composition. Bioresour Technol 100:5588–5593

Pal A, Paul AK (2008) Microbial extracellular polymeric substances: central elements in heavy metal bioremediation. Indian J Microbiol 48:49–64

Panwichian S, Kantachote D, Wittayaweerasa B, Mallavarapu M (2011) Removal of heavy metals by exopolymeric substances produced by resistant purple non sulphur bacteria isolated from contaminated shrimp ponds. Electron J Biotechnol 14. ISSN: 0717-3458

Park D, Yun YS, Park JM (2010) The past, present, and future trends of biosorption. Biotechnol Bioprocess Eng 15(1):86–102

Parolis H, Parolis LAS, Boan IF, Rodriguez-Valera F, Widmalm G, Manca MC et al (1996) The structure of the exopolysaccharide produced by the halophilic archaeon *Haloferax mediterranei* strain R4 (ATCC 33500). Carbohydr Res 295:147–156

Pepi M, Cesàro A, Liut G, Baldi F (2005) An Antarctic psychrotrophic bacterium *Halomonas* sp. ANT-3b, growing on n-hexadecane, produces a new emulsifying glycolipid. FEMS Microbiol Ecol 53:157–166

Pieper DH, Reineke W (2000) Engineering bacteria for bioremediation. Curr Opin Biotechnol 11(3):262–270

Poirier I, Jean N, Guary JC, Bertrand M (2008) Responses of the marine bacterium *Pseudomonas fluorescens* to an excess of heavy metals: physiological and biochemical aspects. Sci Total Environ 406:76–87

Poirier I, Hammann P, Kuhn L, Bertrand M (2013) Strategies developed by the marine bacterium *Pseudomonas fluorescens* BA3SM1 to resist metals: a proteome analysis. Aquat Toxicol 128:215–232

Poli A, Esposito E, Orlando P, Lama L, Giordano A, De Appolonia F et al (2007) *Halomonas alkaliantarctica* sp. nov., isolated from saline lake Cape Russell in Antarctica, an alkalophilic moderately halophilic, exopolysaccharide-producing bacterium. Syst Appl Microbiol 30:31–38

Poli A, Anzelmo G, Nicolaus B (2010) Bacterial exopolysaccharides from extreme marine habitats: production, characterization and biological activities. Mar Drugs 8:1779–1802

Poli A, Di Donato P, Abbamondi GR, Nicolaus B (2011) Synthesis, production, and biotechnological applications of exopolysaccharides and polyhydroxyalkanoates by Archaea. Archaea 2011:693253

Pomin VH (2014) Marine medicinal glycomics. Front Cell Infect Microbiol 4:5

Poremba K, Gunkel W, Lang S, Wagner F (1991) Marine biosurfactants, III. Toxicity testing with marine microorganisms and comparison with synthetic surfactants. Z Naturforsch 46c:210–216

Qin G, Zhu L, Chen X, Wang PG, Zhang Y (2007) Structural characterization and ecological roles of a novel exopolysaccharide from the deep-sea psychrotolerant bacterium *Pseudoalteromonas* sp. SM9913. Microbiology 153:1566–1572

Quigley MS, Santschi PH, Hung CC, Guo L, Honeyman BD (2002) Importance of acid polysaccharides for 234Th complexation to marine organic matter. Mar Chem 76:27–45

Raedts J, Kengen SWM, van der Oost J (2011) Occurrence of L-iduronic acid and putative D-glucuronyl C5-epimerases in prokaryotes. Glycoconj J 28:57–66

Raedts J, Lundgren M, Kengen SWM, Li J-P, van der Oost J (2013) A novel bacterial enzyme with D-glucuronyl C5-epimerase activity. J Biol Chem 288:24332–24339

Raguenes G, Pignet P, Gauthier G, Peres A, Christen R, Rougeaux H et al (1996) Description of a new polymer-secreting bacterium from a deep-sea hydrothermal vent, *Alteromonas macleodii subsp. fijiensis*, and preliminary characterization of the polymer. Appl Environ Microbiol 62:67–73

Raguenes GH, Peres A, Ruimy R, Pignet P, Christen R, Loaec M et al (1997) *Alteromonas infernus* sp. nov., a new polysaccharide-producing bacterium isolated from a deep-sea hydrothermal vent. J Appl Microbiol 82:422–430

Raguenes G, Cambon-Bonavita MA, Lohier JF, Boisset C, Guezennec J (2003) A novel, highly viscous polysaccharide excreted by an *Alteromonas* isolated from a deep-sea hydrothermal vent shrimp. Curr Microbiol 46:448–452

Rajaram R, Banu JS, Mathivanan K (2013) Biosorption of Cu (II) ions by indigenous copper-resistant bacteria isolated from polluted coastal environment. Toxicol Environ Chem 95(4):590–604

Rendleman JA (1978) Metal-polysaccharide complexes-part 1. Food Chem 3:47–79

Reuber TL, Walker GC (1993) Biosynthesis of succinoglycan, a symbiotically important exopolysaccharide of *Rhizobium meliloti*. Cell 74:269–280

Ridout MJ, Brownsey GJ, York GM, Walker GC, Morris VJ (1997) Effect of o-acyl substituents on the functional behaviour of Rhizobium meliloti succinoglycan. Int J Biol Macromol 20(1):1–7

Rinker KD, Kelly RM (2000) Effect of carbon and nitrogen sources on growth dynamics and exopolysaccharide production for the hyperthermophilic archaeon *Thermococcus litoralis* and bacterium *Thermotoga maritima*. Biotechnol Bioeng 69:537–547

Roane TM (1999) Lead resistance in two bacterial isolates from heavy metal contaminated soils. Microb Ecol 37:218–224

Roger O, Kervarec N, Ratiskol J, Colliec-Jouault S, Chevolot L (2004) Structural studies of the main exopolysaccharide produced by the deep-sea bacterium *Alteromonas infernus*. Carbohydr Res 339:2371–2380

Ron EZ, Rosenberg E (1999) Natural roles of biosurfactants. Environ Microbiol 3:229–236

Rosenberg E, Ron EZ (1999) High-and low-molecular-mass microbial surfactants. Appl Microbiol Biotechnol 52(2):154–162

Rougeaux H, Talaga P, Carlson RW, Guezennec J (1998) Structural studies of an exopolysaccharide produced by *Alteromonas macleodii* subsp. *fijiensis* originating from a deep-sea hydrothermal vent. Carbohydr Res 312:53–59

Rougeaux H, Guezennec J, Carlson RW, Kervarec N, Pichon R, Talaga P (1999a) Structural determination of the exopolysaccharide of *Pseudoalteromonas* strain HYD 721 isolated from a deep-sea hydrothermal vent. Carbohydr Res 315:273–285

Rougeaux H, Kervarec N, Pichon R, Guezennec J (1999b) Structure of the exopolysaccharide of *Vibrio diabolicus* isolated from a deep-sea hydrothermal vent. Carbohydr Res 322:40–45

Ruas-Madiedo P, De Los Reyes-Gavilán CG (2005) Invited review: methods for the screening, isolation, and characterization of exopolysaccharides produced by lactic acid bacteria. J Dairy Sci 88(3):843–856

Sakamoto N, Kano N, Imaizumi H (2008) Biosorption of uranium and rare earth elements using biomass of algae. Bioinorg Chem Appl 1–8. doi:10.1155/2008/706240

Satpute SK, Banat IM, Dhakephalkar PK, Banpurkar AG, Chopade BA (2010) Biosurfactants, bioemulsifiers and exopolysaccharides from marine microorganisms. Biotechnol Adv 28:436–450

Schiano Moriello V, Lama L, Poli A, Gugliandolo C, Maugeri TL, Gambacorta A, Nicolaus B (2003) Production of exopolysaccharides from a thermophilic microorganism isolated from a marine hot spring in flegrean areas. J Ind Microbiol Biotechnol 30:95–101

Schreiber DR, Millero FJ, Gordon AS (1990) Production of an extracellular copper-binding compound by the heterotrophic marine bacterium *Vibrio alginolyticus*. Mar Chem 28(4):275–284

Shah V, Ray A, Garg N, Madamwar D (2000) Characterization of the extracellular polysaccharide produced by a marine cyanobacterium, *Cyanothece* sp. ATCC 51142, and its exploitation toward metal removal from solutions. Curr Microbiol 40(4):274–278

Senni K, Pereira J, Gueniche F, Delbarre-Ladrat C, Sinquin C, Ratiskol J ... Colliec-Jouault S (2011) Marine polysaccharides: a source of bioactive molecules for cell therapy and tissue engineering. Marine Drug 9(9):1664–1681

Singh P, Cameotra SS (2004) Enhancement of metal bioremediation by use of microbial surfactants. Biochem Biophys Res Commun 319:291–297

Singh S, Pradhan S, Rai LC (1998) Comparative assessment of Fe^{3+} and Co^{+} biosorption by field and laboratory-grown *Microcystis*. Process Biochem 33(495):504

Stahl A, Pletzer D, Mehmood A, Ullrich MS (2015) Marinobacter adhaerens HP15 harbors two CzcCBA efflux pumps involved in zinc detoxification. Antonie Van Leeuwenhoek 108(3):649–658

Steigedal M, Sletta H, Moreno S, Maerk M, Christensen BE, Bjerkan T et al (2008) The *Azotobacter vinelandii* AlgE mannuronan C-5-epimerase family is essential for the *in vivo* control of alginate monomer composition and for functional cyst formation. Environ Microbiol 10:1760–1770

Trincone A (2010) Potential biocatalysts originating from sea environments. J Mol Catal B Enzym 66(3):241–256

Valls M, de Lorenzo V (2002) Exploiting the genetic and biochemical capacities of bacteria for the remediation of heavy metal pollution. FEMS Microbiol Rev 26:327–338

Van Hullebusch ED, Zandvoort MH, Lens PNL (2003) Metal immobilization by biofilms: mechanisms and analytical tools. Rev Environ Sci Biotechnol 2:9–33

Vanfossen AL, Lewis DL, Nichols JD, Kelly RM (2008) Polysaccharide degradation and synthesis by extremely thermophilic anaerobes. Ann NY Acad Sci 1125:322–337

Verdugo P, Alldredge AL, Azam F, Kirchman DL, Passow U, Santschi PH (2004) The role of marine gel-phase on carbon cycling in the ocean. Mar Chem 92:65–66

Vieira RH, Volesky B (2010) Biosorption: a solution to pollution? Int Microbiol 3(1):17–24

Vincent P, Pignet P, Talmont F, Bozzi L, Fournet B, Guezennec J (1994) Production and characterization of an exopolysaccharide excreted by a deep-sea hydrothermal vent bacterium isolated from the polychaete annelid *Alvinella pompejana*. Appl Environ Microbiol 60:4134–4141

Von Canstein H, Li Y, Leonhäuser J, Haase E, Felske A, Deckwer WD, Wagner-Döbler I (2002) Spatially oscillating activity and microbial succession of mercury-reducing biofilms in a technical-scale bioremediation system. Appl Environ Microbiol 68(4):1938–1946

Vyrides I, Stuckey DC (2009) Adaptation of anaerobic biomass to saline conditions: Role of compatible solutes and extracellular polysaccharides. Enzym Microb Technol 44(1):46–51

Watanabe M, Kawahara K, Sasaki K, Noparatnaraporn N (2003) Biosorption of cadmium ions using a photosynthetic bacterium, *Rhodobacter sphaeroides* S and a marine photosynthetic bacterium. *Rhodobacter sphaeroides* S and their biosorption kinetics. J Biosci Bioeng 95(4):374–378

Wolfaardt GM, Lawrence JR, Korber DR (1999) Function of EPS. In: Microbial extracellular polymeric substances. Springer, Berlin, pp 171–200

Wu Y, Xia L, Yu Z, Shabbir S, Kerr PG (2013) *In situ* bioremediation of surface waters by periphytons. Bioresour Technol 151:367–372

Zanchetta P, Lagarde N, Guezennec J (2003) A new bone-healing material: a hyaluronic acid-like bacterial exopolysaccharide. Calcif Tissue Int 72(1):74–79

Zhang W, Chen L, Liu D (2012) Characterization of a marine-isolated mercury-resistant *Pseudomonas putida* strain SP1 and its potential application in marine mercury reduction. Appl Microbiol Biotechnol 93(3):1305–1314

Zou W, Laferriere CA, Jennings HJ (1998) Oligosaccharide fragments of the type III group B streptococcal polysaccharide derived from *S. pneumoniae* type 14 capsular polysaccharide by a chemoenzymatic method. Carbohydr Res 309(3):297–301

Biosurfactant: A Promising Approach Toward the Remediation of Xenobiotics, a Way to Rejuvenate the Marine Ecosystem

6

Siddhardha Busi and Jobina Rajkumari

Abstract

Xenobiotics – petroleum hydrocarbons, dyes, heavy metals, pesticides, and antifoulants like tributyltin (TBT) which are dispersed in the marine environment through various activities – have caused ecological and social catastrophes. Pesticides, metals, dyes, etc. used in agriculture and other industrial purposes eventually reach the sea through streams and rivers. The presence of these xenobiotics in the water bodies may lead to many serious health consequences as xenobiotics like heavy metals are mutagenic and carcinogenic in nature. Heavy metals are usually discharged from power generation industries, and some like mercury are an active component in many pesticides. Synthetic dyes and dry cleaning fluid like tetrachloroethylene used largely in dyeing and printing industries constitute one of the major organic pollutants. Polycyclic hydrocarbons and crude oil sludge are discharged into the marine environment as a result of offshore oil drilling, overflow from oil tanker, and ship accidents and as by-products of coal treatment. As a result of increasing public awareness on environmental pollution and the need for sustainable development, various research programs have been initiated for development of technologies to manage the xenobiotics. The use of surface-active biomolecules (biosurfactant) synthesized by microorganisms in bioremediation has been proposed in recent years and is gaining prominence due to their high potential in mobilization and solubilization of various pollutants. Biosurfactants such as rhamnolipid, sophorolipid, surfactin, and alasan produced by diverse species of Pseudomonas, Candida, Bacillus, and Acinetobacter, respectively, have been widely studied and reported for the remediation of hydrocarbons (e.g., phenanthrene, fluoranthene, pyrene, hexadecane, kerosene, and 1-methyl naphthalene) and heavy metals, e.g.,

S. Busi (✉) • J. Rajkumari
Department of Microbiology, Pondicherry University,
Puducherry 605014, India
e-mail: siddhardha.busi@gmail.com

M.M. Naik, S.K. Dubey (eds.), *Marine Pollution and Microbial Remediation*,
DOI 10.1007/978-981-10-1044-6_6

cadmium and pesticides (endosulfan and beta-cypermethrin). Moreover, *Bacillus subtilis* and *Bacillus licheniformis* are well-known producers of biosurfactant employed in oil recovery and bioremediation. The biosurfactant seem to enhance the biodegradation of oils and lipids as a result of its emulsifying property. Owing to its biodegradability and low toxicity, they might serve as promising tool for the use in bioremediation of both organic and inorganic pollutants. This chapter provides an overview on the advances in the application of microbial biosurfactant in rejuvenating marine ecosystem and in combating the issues of xenobiotics.

6.1 Introduction

Marine environment is the largest habitat covering approximately 70 % of the total earth surface. Oceans play a major role in regulating earth's climate and harbor a huge diversity of living organisms. The marine ecosystems have been acting as a sink for large amounts of toxic wastes/xenobiotics, which even in minute quantities threatens the ocean communities. Rapid industrialization and extraction of natural resources to feed the growing urbanization have resulted in large-scale environmental pollution. Xenobiotics, chemicals which are foreign to life and persistent organic pollutants, are now ubiquitous in the environment, raising its risk to public health and its surrounding environment at an alarming rate. Pollution of marine ecosystem has endangered the aquatic flora and fauna. The uptake and accumulation of toxic chemicals like pesticides in the food chain and its biomagnification lead to lethal behavioral effect on animals and contribute to the rise in global epidemic of cancer and other chronic diseases. Environmental pollution of the marine environment is one of the most important problems in the context of human health and havoc to the ecosystem. The need of the hour is to develop novel, eco-friendly, and cost-effective way to revive the polluted environment and restore the marine ecosystem.

6.2 Xenobiotics in Marine Ecosystem

Xenobiotics refer to chemical compounds like synthetic pesticides, herbicides, and industrial pollutants which are foreign to the environment. Unlike the organic pollutants that are biodegradable, these pollutants, e.g., heavy metals, dyes, nuclear wastes, etc. are extremely stable making them persistent in the environment. Xenobiotics released mainly from industries like chemical and pharma, oil spills, nuclear explosions, and agricultural practices reach the marine environment through land runoffs and as industrial effluents. The major environmental threats include residual compounds like pesticides, petroleum hydrocarbons, different dyes, paints, heavy metals, radioactive substances, fertilizer, insecticides, etc.

Petroleum Hydrocarbons Petroleum represents one of the most important industries today without which modern life will come to a halt. The world depends on petroleum and petroleum products for its economic development and technological advances. Although the petroleum plants and crude oil refineries provide vital fuel like gasoline and power to the society, they are also responsible for the global pollution of oceans with large amount of hazardous waste from refineries. Poisoning of sea by man-made disasters is one of the urgent and serious environmental problems all over the world. The pollutions are caused by spilling of crude oil from tankers during transportation, collisions, and leakage of oil and gas on the ocean floor from underground pipelines. On April 20, 2010, the world witnessed the largest accidental oil spill in the Gulf of Mexico, also known as BP oil spill. Following the explosion and sinking of deep water horizon oil rig off the coast of Louisiana (USA), the open ducts continued to discharge oil (estimated 3.19 million barrel) into the ocean for 87 days until finally plugged on July 2010 (British 2010). Another incident of such oil dispersing tragedy includes the Nigeria delta. In the hydrocarbon family, polycyclic aromatic hydrocarbons (PAHs), e.g., naphthalene, phenanthrene, biphenyl, anthracene, and pyrene are the most toxic pollutants and major constituent of crude oil. PAHs are the large group of organic compounds formed by the hydrogen and carbon atoms with two or more fused benzene rings. These are the hydrophobic compounds and insoluble in water. More than 1000 types of PAHs compounds are present in the environment with differing in the number and position of aromatic rings. PAH ingested by marine organisms passes through food chain and ultimately reaches man. Most of the PAHs are carcinogenic, mutagenic, and teratogenic to many organisms including mammals. 3,4-Benzopyrene and its derivatives are well-known carcinogen. Oil/hydrocarbon waste being hydrophobic in nature forms slick layer on the surface of water bodies which blocks the passage of sunlight and prevents gas exchange, respiration, and photosynthesis of phytoplankton communities (Yeung et al. 2011; Asimiea and Sam-Wobo 2011). Marine organisms are affected due to the shortage of dissolved oxygen and fail to cope with the polluted environment. The oil spill has the physical, psychological, genotoxic, and endocrine effects in human. It also causes disorientation, lesions, and deformities in fishes, sea turtles, dolphins, and sea birds.

Dyes and Paints Dyes are persistent xenobiotics, and their presence even in minute amount reduces penetration of sunlight, affecting photosynthetic activity in aquatic life (Abdelkader et al. 2011; Elaziouti et al. 2011). Textile and paper printing industries commonly use azo dyes, anthraquinone, and phthalocyanine dyes. The degradation of these dyes results in the production of aromatic amines, which may also be carcinogenic and mutagenic. Anti-biofouling agents in paint like tributyltin (TBT) are also considered harmful.

Insecticides and Pesticides A large number of insecticides, nematocides, herbicides, fungicides, and pesticides are used on a daily basis in agriculture. Indiscriminate use of chemicals like methyl bromide, chlorinated organics, endosulfan, nitrophenols, organophosphorus compounds, synthetic pyrethroids, morpholine, carbamates, etc. severely contaminates the groundwater, and the runoff flows to rivers and ultimately reaches the ocean (Lalithakumari 2011). The demand and use of beta-cypermethrin (beta-CP), a synthetic pyrethroid (SP) insecticide, have been considerably increased due to its high insecticidal efficiency and with the restrictions and ban on import and export of organophosphate and carbamate pesticides. However, beta-CP poses serious threat to the survival of aquatic fishes and invertebrates. It may even cause serious health consequences like reproductive, neurotoxicity, and developmental disorders in humans (Zhang et al. 2011). The use of high amount of nitrogen- and phosphorous-rich fertilizers causes algal bloom and creates a zone of hypoxia with loss of species diversity in sea. The main problem associated with these chemicals is bioconcentration and biomagnification. Lipophilic insecticides like dichlorodiphenyltrichloroethane (DDT) accumulate into the fatty tissues of fishes and pass on to successive trophic levels. DDT causes birth defects and cancer in humans and also drastically affects the bird population.

Heavy Metals Heavy metals like mercury, arsenic, lead, cadmium, and zinc are highly toxic and accumulate in the organisms leading to biomagnifications. This increases their toxicity to the subsequent tropic levels. These heavy metals are highly carcinogenic. Arsenic and lead administered above the permissible level in agriculture causes groundwater poisoning. Copper used in marine antifouling paints was found to be lethal to marine organisms. Chronic exposure to cadmium is responsible for the "Itai-itai" disease in Japan which is characterized by renal dysfunction, severe osteoporosis, and anemia in women (Mulligan et al. 2001). Mercury is a potent neurotoxin and is responsible for the Minamata disease in Japan (Mulligan et al. 2001). In children, Mercury and lead poisoning causes brain damage and behavioral disturbances. For example, the methylated mercury present in the sea organisms (fish, shellfish, etc.) enters into the human body by food which adversely affects mankind. The effects of heavy metal toxicity are due to their ability in interrupting the process of transcription and translation of proteins, removal of metal ion from biomolecules, and thus deactivating them (Varsha et al. 2011).

6.3 Bioremediation

Several physical and chemical methods have been employed in the treatment or removal of xenobiotics. The physicochemical methods are expensive and often result in undesirable toxic by-products. To overcome these problems, eco-friendly technique like bioremediation has been employed. Microbial bioremediation relies on either removal or degradation of toxic pollutants present in environment into less toxic or nontoxic forms. The process takes place through intracellular accumulation either by microbes or by enzymatic transformation. Many microorganisms naturally possess the ability to degrade, transform, or chelate various toxic chemicals, but these processes are relatively slow. Polycyclic aromatic hydrocarbons (PAHs), crude oil sludge, dyes, and pesticides that pollute the marine environment are toxigenic, mutagenic, and carcinogenic. The low hydrophobicity of these compounds limits their degradation by microorganisms, which is a common problem encountered in bioremediation of contaminated sites. In such scenario, microbial surfactants increase the bioavailability of these hydrophobic compounds, thereby enhancing the solubility of pollutants and making it susceptible to microbial attack. Some of the isolated biosurfactant increases the solubility of hydrocarbons like pyrene while other contributes to an enhanced solubilization of phenanthrene or fluorene. Bordoloi and Konwar (2009) reported enhanced solubility and metabolism of petroleum hydrocarbons in presence of biosurfactant obtained from different *P. aeruginosa* strains (Bordoloi and Konwar 2009). The effect of biosurfactant on the biodegradation of crude oil by various marine bacterial isolates like *Bacillus megaterium*, *Corynebacterium kutscheri*, and *P. aeruginosa* has been studied (Thavasi et al. 2011).

6.4 Introduction to Biosurfactant

Biosurfactants are surface-active compounds that reduce the surface tension between interface of liquid- liquid or liquid-solid. Structurally, all biosurfactant are amphiphilic in nature, i.e., they consist of hydrophilic head and hydrophobic tail. The hydrophilic moiety comprises of a carbohydrate, amino acid, cyclic peptide, phosphate, carboxylic acid, or alcohol with the hydrophobic group of long-chain fatty acids, hydroxyl fatty acids, or α-alkyl-β-hydroxy fatty acids (Mulligan et al. 2001). Biosurfactant possesses different chemical structures and exhibits different properties. They can be lipopeptides, glycolipids, neutral lipids, or fatty acids. They are mainly classified into two classes: low-molecular-weight surface-active agents (biosurfactant) and high-molecular-weight surface-active agents (bioemulsifiers) (Rosenberg and Ron 1999). Microbial surfactants have been increasingly gaining attention due to its interesting and diverse properties such as low toxicity, biodegradability, eco-friendly nature, possibility of large-scale production, selectivity, and optimal activity at extreme conditions of temperatures and pH.

6.5 Biosurfactant-Producing Microorganisms

A wide range of bacteria, fungi, and yeast produces biosurfactant either intracellularly or extracellularly. The genus *Pseudomonas* is well known for the production of rhamnolipids; *Bacillus subtilis* produces a lipopeptide, surfactin (Ron and Rosenberg 2001; Mulligan 2005). *Nocardia amarae* (Moussa et al. 2006), *Saccharomyces lipolytica* CCT- 0913 (Lima and Alegre 2009), *Candida bombicola* (Casas and García-Ochoa 1999; Casas et al. 1997; Williams 2009), *Candida lipolytica* (Sarubbo et al. 2007; Rufino et al. 2007), *Candida ishiwadae* (Thanomsub et al. 2004), and *Candida batistae* (Konishi et al. 2008) are among the most common biosurfactant-producing yeasts and fungi.

Marine microorganisms have gained attention particularly for their biocompatibility and bioremediation potential of the sea polluted by crude oil. High diversity of biosurfactant-producing marine microorganisms offers great potential in bioremediation of petrochemical waste in oceanic environments. The marine microorganisms produce four major classes of biosurfactants such as glycolipids; lipopeptides and lipoproteins; fatty acids, neutral lipids, and phospholipids; and polymeric biosurfactants (Desai and Banat 1997). Hassanshahian (2014) isolated biosurfactant-producing bacteria of genera *Shewanella*, *Vibrio*, *Gallaecimonas*, *Brevibacterium*, *Psychrobacter*, and *Pseudomonas* from samples collected from the coastline of the Persian Gulf. Maneerat and Phetrong (2007) reported biosurfactant-producing marine bacteria, *Vibrio parahaemolyticus*, *Bacillus subtilis*, *Micrococcus luteus*, *Myroides* sp., *Acinetobacter anitratus*, and *B. pumilus*, isolated from oil spill-polluted seawater along Thailand coast. Marine bacteria *Alcaligenes* sp. has been reported to produce glycolipid. Zinjarde and Pant (2002) also reported a marine bacterium, *Yarrowia lipolytica*, for the production of polymeric biosurfactant with the ability to emulsify alkanes and crude oil.

It is believed that microbes secrete biosurfactant into the culture medium to support their growth by facilitating the translocation of insoluble substrates across cell membranes (Zinjarde and Pant 2002). These amphiphilic compounds provide access to hydrophobic substrates by reducing surface tension and increase in the area of contact of insoluble compounds, thus enhancing the mobility, bioavailability, and biodegradation of xenobiotics (Abraham 1998).

The structure, size, and composition of biosurfactant synthesized are influenced by the microorganism, medium composition, and culture conditions (Lang and Wullbrandt 1999; Franzetti et al. 2008). The carbon sources used for biosurfactant production include petroleum hydrocarbons, carbohydrates, oil sludge, lactic whey and distiller wastes, starchy substrates, renewable resources, industrial or municipal wastewater, etc. (Lang and Wullbrandt 1999; Kosaric 1992; Gautam and Tyagi 2006) (Table 6.1).

6.6 Classification of Biosurfactants

Surfactants, diverse group of surface-active compounds, are mainly classified by their chemical structure and molecular weight. Depending on the molecular weight, microbial surfactants can be classified into two main classes, biosurfactant and bioemulsans. Biosurfactants include molecules with low molecular mass, like surfactin and rhamnolipid having a molecular weight of 1036 and 802 Da, respectively (Mulligan and Gibbs 1990). On the other hand, bioemulsans include high-molecular-weight polymers like emulsan and alasan with an average molecular weight of about 1000 kDa and 1 MDa, respectively (Kim et al. 1997; Navon-Venezia et al. 1995).

Due to their amphiphilic nature, the biosurfactant and emulsifiers alter the facial interface of contaminant by different mechanisms, resulting in enhanced bioremediation. High-molecular-weight bioemulsifiers are tested as enhancers of hydrocarbon biodegradation in bioremediation of marine environment. These emulsifiers have a great potential in stabilizing emulsions between

Table 6.1 Different type of biosurfactant and their microbial source

Class/type of biosurfactant	Microorganisms	Reference(s)
Glycolipids		
Rhamnolipids	*Pseudomonas fluorescens*, *Pseudomonas* sp., *Pseudomonas aeruginosa*	Robert et al. (1989)
Glucose lipids	*Alcanivorax borkumensis*, *Alcaligenes* sp.	(Zinjarde and Pant (2002), and Maneerat (2005))
Sophorolipids	*Torulopsis bombicola*, *Candida bombicola*, *Candida batistae*, *Trichosporon asahii*, *Candida apicola*, *Candida lipolytica*, *Candida bogoriensis*	Konishi et al. (2008), Maneerat and Phetrong (2007). Inoue and Ito (1982), and Cavalero and Cooper (2003)
Ornithine lipids	*Myroides* sp. SM1	
Trehalolipids	*Rhodococcus erythropolis*, *Arthrobacter* sp. *Arthrobacter paraffineus*, *Corynebacterium spp.*, *Mycobacterium* spp.	Maneerat (2005), Shulga et al. (1990), and Suzuki et al. (1974)
Lipopeptides and lipoproteins		
Lipopeptide	*Bacillus licheniformis*, *Arthrobacter* MIS 38, *Bacillus subtilis*, *Acinetobacter calcoaceticus*, *Candida petrophilum*	Mclnerney et al. (1990), Morikawa et al. (1993), and Horowitz et al. (1990)
Surfactin	*Bacillus sp.* AB2	Morikawa et al. (1992)
Subtilisin	*Bacillus subtilis*	Al-Bahry et al. (2013)
Viscosin	*Pseudomonas fluorescens*	Alsohim et al. (2014)
Serrawettin	*Serratia marcescens*	Matsuyama et al. (2011)
Fatty acids, neutral lipids, and phospholipids		
Fatty acid	*Corynebacterium lepus*, *Acinetobacter* sp.	Akit et al. (1981)
Bile acids	*Myroides* sp. SM1	Maneerat (2005)
Neutral lipids	*Nocardia erythropolis*, *Corynebacterium hydrocarbolastus*, *Corynebacterium salvonicum*	Cooper et al. (1979) and Lesik et al. (1989)
Phospholipids	*Thiobacillus thiooxidans*, *Corynebacterium lepus*, *Corynebacterium insidiosum*	Zajic et al. (1977)
Polymeric surfactants		
Emulsan	*Acinetobacter calcoaceticus*	Goldman et al. (1982)
Lipid-carbohydrate-protein	*Yarrowia lipolytica*, *Pseudomonas nautica*	Desai and Banat (1997)
Biodispersan	*Acinetobacter calcoaceticus*	Rosenberg et al. (1988a)
Liposan	*Candida lipolytica*	Cirigliano and Carman (1985)
Carbohydrate-lipid-protein	*Pseudomonas fluorescens*, *Candida tropicalis*	Persson et al. (1988)

liquid hydrocarbons and water, thus increasing the surface area available for bacterial biodegradation. Barkay et al. (1999) reported the significant increase in the mineralization rate of fluoranthene and phenanthrene by *Acinetobacter radioresistens* and *Sphingomonas paucimobilis* EPA505 when supplemented with alasan, respectively. Low-molecular-weight biosurfactant rhamnolipids have been shown to accelerate the biodegradation of hydrocarbons when added to contaminated soils at a concentration above the critical micelle concentration (CMC). Surfactin has been reported for the biodegradation of pesticide (Awashti et al. 1999).

On the basis of the chemical composition, biosurfactant may be classified as glycolipids;

lipopeptides and lipoprotein; phospholipids, fatty acids, and neutral lipid; and polymeric and particulate biosurfactant (Lang and Wullbrandt 1999).

1. *Glycolipids*: Glycolipids are the most studied microbial surfactants. They compose of carbohydrates linked to long-chain aliphatic acids or hydroxyl aliphatic acids by an ester group. Rhamnolipids, trehalolipids, and sophorolipids are the most widely used glycolipids.
   (a) Rhamnolipids: Rhamnolipids are the most studied glycolipids till date. These are made up of L-rhamnose molecules bonded to β-hydroxy fatty acids. The production of rhamnolipids was first reported and extensively studied in *Pseudomonas aeruginosa*. Rhamnolipid 1 (L-rhamnosyl-L-rhamnosyl-β-hydroxydecanoyl-β-hydroxydecanoate) and rhamnolipid 2 (L-rhamnosyl-β-hydroxydecanoyl-β--hydroxydecanoate) are the main glycolipids produced by *P. aeruginosa* (Edward and Hayashi 1965). *Bacillus* and other *Pseudomonas* sp. like *P. chlororaphis* and *P. putida* have also been reported for the production of rhamnolipids. Rhamnolipids are fully commercialized for use in manufacture of cosmetics and pharmaceuticals and for the bioremediation of heavy metals and hydrocarbons.
   (b) Trehalolipids: Trehalolipids are glycolipids composed of disaccharide trehalose hydrophilic part. There exists a wide diversity of trehalolipids depending on the hydrophobic part; it can be a short fatty acid as found in *Rhodococcus* sp. or a long-chain α-branched-β- hydroxy fatty acids as found in *Mycobacterium sp.* Trehalolipids are produced by a number of different genera of microorganisms, such as Mycobacterium, Arthrobacter, Nocardia, Micrococcus, and Corynebacterium. However, trehalose dimycolates produced by *Rhodococcus erythropolis* are the most extensively studied class of trehalolipids (Asselineau and Asselineau 1978; Kretschmer et al. 1982).
   (c) Sophorolipids: These glycolipids are produced mainly by yeast such as *Torulopsis bombicola*, *T. petrophilum*, *Candida bombicola*, *T. petrophilum*, and *T. apicola*. Sophorolipids consist of a hydrophilic carbohydrate, sophorose with a hydrophobic long-chain hydroxy fatty acid. The terminal carboxyl group can be in the lactonic form or hydrolyzed to give an anionic surfactant (Rosenberg and Ron 1999; Cooper and Paddock 1984).
2. *Lipopeptides and Lipoproteins*: Lipopeptides compose of a lipid attached to a polypeptide chain. Surfactin, subtilisin, viscosin, and serrawettin are produced by *Bacillus* sp. AB2, *Bacillus subtilis*, *Pseudomonas fluorescens*, and *Serratia marcescens*, respectively. They are again grouped to surfactin and lichenysin.
   (a) Surfactin: They are amphiphilic molecule synthesized by *B. subtilis* that not only lowers the interfacial tension but also brings about changes in interfacial rheological properties of liquids. It is composed of a seven-amino-acid ring structure coupled to a fatty-acid chain via lactone linkage. They are used in oil recovery and remediation of nondispersive pollutants (Arima et al. 1968).
   (b) Lichenysin: It is similar to surfactin in terms of its structural complexity and physicochemical properties. It is produced by *Bacillus licheniformis* strains and exhibits excellent stability under wide ranges of temperature, pH, and salinity. They may be used in cleaning of oil spills and enhanced oil recovery and bioremediation of sites contaminated with pollutants like hydrocarbons and heavy metals. *B. licheniformis* lipopeptide biosurfactant has been reported to lower the interfacial tension between water and *n*-hexadecane (Rosenberg and Ron 1999).
3. *Fatty Acids, Phospholipids, and Neutral Lipids*: Several bacteria and yeast utilize hydrocarbon and alkanes as carbon source and produce a large quantity of fatty acid and phos-

pholipid surfactants (Cirigliano and Carman 1985). In *Acinetobacter* sp. strain HO1-N, phosphatidyl ethanolamine-rich vesicles are produced, which are able to produce optically clear microemulsions of alkanes in water. *Thiobacillus thiooxidans*, *Corynebacterium lepus*, and *Corynebacterium insidiosum* are some of the organisms reported for production of phospholipid biosurfactant. *R. erythropolis* when grown on alkane produces phosphatidyl ethanolamine and decreases the interfacial tension between water and hexadecane.

4. *Particulate Biosurfactant*: Microemulsions playing an important role in alkane uptake by microbial cells are formed when extracellular membrane vesicles partition hydrocarbons. Vesicles of *Acinetobacter* sp. strain HO1-N are composed of protein, phospholipids, and lipopolysaccharide (Hommel et al. 1987). In *Sphingomonas* sp., the cell surface gets covered with extracellular vesicles when grown on polyaromatic hydrocarbons.
5. *Polymeric Biosurfactants*: Polymeric biosurfactants like emulsan, biodispersan, liposan, alasan, and lipomanan produced by genera Acinetobacter, Pseudomonas, Candida, etc. are also termed as bioemulsifiers. In 1997, Husain et al. reported the production of polymeric biosurfactant composing of proteins, carbohydrates, and lipid at the ratio of 35:63:2, respectively, by *Pseudomonas nautica*. Zinjarde and Pant (2002) also reported the production of emulsifier in the presence of alkanes or crude oil by marine fungi, *Yarrowia lipolytica*. *Acinetobacter calcoaceticus* RAG-1 also produces an extracellular polyanionic amphipathic heteropolysaccharide bioemulsifier (Lawniczak et al. 2013; Cameron et al. 1988). Emulsan, a polyanionic amphiphilic heteropolysaccharide, is a very effective emulsifier even at low concentrations, ranging from 0.01 to 0.001% (Rosenberg et al. 1979). Biodispersan, produced by *A. calcoaceticus* A2, can bind to $CaCO_3$ and TiO_2 allowing their dispersion in water (Rosenberg et al. 1988b). Bioemulsifiers obtained from yeasts include the protein-polysaccharide complex, liposan from *Candida lipolytica* (Cirigliano and Carman 1985), and mannoprotein obtained from *Saccharomyces cerevisiae* (Cameron et al. 1988).

6.7 Properties of Biosurfactant

Biosurfactants have advantages over chemical surfactants because of various reasons. The most important characteristic of biosurfactant is its eco-friendly nature. They are biodegradable and have low toxicity, high specificity, and stability at extreme temperature, pH, and salinity (Lang and Wullbrandt 1999; Kosaric 1992). These properties increase the potential use of biosurfactant in the remediation of xenobiotics such as dyes, heavy metals, hydrocarbons, and other recalcitrant.

(a) *Surface and Interface Activity*: The ability to emulsify, i.e., to reduce the interfacial tension of oil in water, has favored the application of biosurfactant for oil recovery from water (Mulligan 2005). Sophorolipids from *T. bombicola* reduce surface and interfacial tension but are not efficient emulsifier (Cooper and Cavalero 2003). By contrast, liposan is not efficient in reducing surface tension, but has been used successfully to emulsify edible oils. Surfactin and Rhamnolipids efficiently reduce the surface tension of water to 25 mN/m and interfacial tension of water: hexadecane to <1 mN/m. The sophorolipids from *T. bombicola* have been reported to reduce the surface tension to 33 mN/m and the interfacial tension to 5 mN/m. Thus, biosurfactants are more efficient emulsifiers as their CMC is about 10–40 times lower to that of chemical surfactants which means less surfactant is required to get a maximum decrease in surface tension (Lang and Wullbrandt 1999).

(b) *Tolerance to Temperature and pH*: The most significant property of the microbial surfactant is that these surface activities are not much affected by environmental stress of temperature and pH. Lichenysin from *B. licheniformis* JF-2 could withstand the temperature up to 50 °C, pH of 4.5–9.0, and NaCl and calcium ion concentrations up to 50 and 25 g/l, respectively. A lipopeptide obtained from *B. subtilis* LB5a was found to be stable even after autoclaving at 121 ° C for 20 min and after storage for 6 months at –18 °C; the surface activity was stable at a pH range of 5–11 and NaCl concentrations up to 20 % (Cooper et al. 1981).

(c) *Biodegradability*: Unlike synthetic or chemical surfactants, biosurfactant is degraded by bacteria and other microorganisms easily, owing to their simple chemical structure and low toxicity. These compounds donot persist for longer time rendering them harmless to the environment. As most biosurfactants are degraded faster than synthetic surfactants, they are suitable for environmental applications such as bioremediation and dispersion of oil spills.

(d) *Low Toxicity*: Microbial surfactants are generally considered nontoxic products, making it appropriate for pharmaceutical, cosmetic, food, and environmental applications. Glycolipid from *Rhodococcus* species was found to be 50 % less toxic as compared to Tween 80 (Kanga et al. 1997). A biosurfactant produced by *P. aeruginosa* when compared with Marlon A- 350, a synthetic surfactant widely used in the industry, was found to be nontoxic and non-mutagenic, while the chemical-derived surfactant indicated higher toxicity and mutagenicity (Flasz et al. 1998).

(e) *Availability*: Biosurfactants can be produced using relatively inexpensive raw materials available in abundance. The carbon source ranging from hydrocarbons, oils, to carbohydrates may be used separately or in combination for large-scale production. Biosurfactant can also be produced by growing microorganisms on industrial wastes, starchy substrates, oil mill effluent, industrial or municipal waste/by-products, petroleum hydrocarbons, carbohydrates, oil sludge, olive oil mill effluent, lactic whey and distiller wastes, starchy substrates, renewable resources, and industrial or municipal wastewater.

(f) *Specificity*: The culture conditions and presence of specific functional groups attribute to the specificity in the properties of the biosurfactant molecules. These properties are exploited in detoxification of specific pollutants and emulsification of industrial effluents.

6.8 Bioremediation of Hydrocarbons Using Biosurfactant

Biodegradation of a hydrocarbon depends on its bioavailability and state of dispersion. To maximize the biodegradation rate, the water-insoluble substrate should either be solubilized or emulsified. Synthetic detergents used to clean up oil spillages have often led to more destruction of the environment due to the release of toxic by-products. Biosurfactants due to their emulsifying property, low toxicity, and high biodegradability may play an important role in bioremediation process (Silva et al. 2014). Biosurfactant can be widely employed in petroleum industry where it is used to the remediate oil spills and removal of oil residue from storage tanks and in microbial-enhanced oil recovery (Sobrinho et al. 2013). In microbes, the biosurfactant both enhances the bioavailability of inaccessible substrates and helps in survival without desiccation in low moisture content. Biosurfactant production generally requires both a hydrophobic and hydrophilic carbon source in the culture medium. Hence, agro-based or industrial waste products can be used to achieve economically and environmentally attractive way of production (Makkar et al. 2011; Dziegielewska and Adamczak 2013).

Petroleum effluents mainly contain polycyclic PHAs (polynuclear aromatic hydrocarbons), saturated hydrocarbons and nitrogen-sulfur-oxygen compounds. Benzene, ethylbenzene, toluene, styrene, and xylenes are extensively used as fuels and industrial solvents. Moreover, PHAs serve as the primary substrate for the production of pharmaceuticals, polymers, agrochemicals, and many more. The low solubility of these high-molecular-weight hydrophobic compounds leads to their prolonged persistence as these compounds are difficult to access by the biodegrading microorganisms. Biosurfactants can enhance microbial growth on bound substrates by desorbing them from surfaces or by increasing their solubility (Marcoux 2000). The most common role of biosurfactant in hydrocarbon bioremediation includes (a) increasing the substrate bioavailability to microorganisms and (b) interacting with the cell surface which increases the hydrophobicity. Low-molecular-weight biosurfactant that has low critical micelle concentrations (CMCs) increases the solubility and mobility of hydrocarbons by incorporating them into the hydrophobic cavities of micelles (Miller and Zhang 1997).

PAHs are one of the many products obtained in petroleum refining process (Park et al. 1990). Some of the petroleum hydrocarbons are alkanes, cycloalkanes, aromatics, polycyclic aromatic hydrocarbons, asphaltenes, resins, etc. Saturated hydrocarbons with straight chain (n-alkanes) are susceptible to microbial degradation, but the branched alkane and aromatics are difficult to degrade. Many PAHs are suspected to be carcinogenic.

Microorganisms with the ability to degrade hydrocarbon compounds are particularly isolated from petroleum-contaminated sites, and the biosurfactant plays a major role in providing the ability to the microorganisms to grow in such extreme habitat and utilize hydrocarbon as carbon and energy source. According to Itoh and Suzuki (1974), the hydrocarbon utilization ability of a rhamnolipid-negative mutant strain of *P. aeruginosa* could only be restored when it is grown in medium supplemented with rhamnolipid. Since then, a large number of biosurfactant-producing microorganisms with the ability to degrade hydrocarbons have been isolated (Bordoloi and Konwar 2009). *E. fergusonii* KLU01 was isolated and identified to be a hydrocarbon-degrading and heavy metal-tolerant bacterium which also had the ability to produce lipopeptide biosurfactant (Sriram et al. 2011).

Sobrinho et al. (2008) found that biosurfactant produced by *Candida sphaerica* exhibited an oil-spreading efficiency of 75 % when tested on motor oil containing seawater. Alasan was recently shown to significantly enhance the rate of biodegradation of several PAHs (Rosenberg and Ron 1999). Noordman et al. (2002) studied hexadecane degradation by biosurfactant from *P. aeruginosa*, and Rahman et al. (2003) examined the bioremediation of n-alkanes present in petroleum sludge. Research with biosurfactants from *Pseudomonas marginalis* indicated that biosurfactants could solubilize polycyclic aromatic hydrocarbons (PAHs) such as phenanthrene and enhance biodegradation (Burd and Ward 1996).

Naphthalene and phenanthrene are degraded by *Pseudomonas alcaligenes* PA-10. The strain uses these PAHs as carbon and energy sources and also co-metabolizes fluoranthene (Gordon and Dobson 2001). According to Nnamchi et al. (2006), *Pseudomonas aeruginosa* and *Burkholderia cepacia* could degrade anthracene and carbazole. According to Garcia-Junco et al. (2001), *Pseudomonas aeruginosa* 19SJ degraded the phenanthrene by producing rhamnolipid which increases the bioavailability of phenanthrene. According to Arulazhagan and coworkers (2010), *Ochrobactrum* sp., *Enterobacter cloacae*, and *Stenotrophomonas maltophilia* have showed their potential in degradation of fluorene and phenanthrene. Anand S. Nayak et al. (2009) reported that the biosurfactant isolated from *Pseudoxanthomonas* sp. PNK-04 aids in uptake and solubilization of 2-chlorobenzoic acid, 3-chlorobenzoic acid, and 1-methyl naphthalene. Obayori et al. (2009) reported 92.34 % degradation of crude oil and 95.29 % removal of diesel oil by using biosurfactant produced by *Pseudomonas* sp. LP1. Reddy et al. (2010) investigated the biodegradative properties of biosurfactant producing *Brevibacterium* sp. PDM-3 strain. The strain could degrade 93.92 % of the phenanthrene and also had the ability to degrade other polyaromatic hydrocarbons, anthracene, and fluorene. Barkay et al. (1999) tested the solubilization of polyaromatic hydrocarbons (PAHs), phenanthrene (PHE), and fluoranthene (FLA) using bioemulsifier, alasan produced by *Acinetobacter radioresistens* KA53.

6.9 Bioremediation of Heavy Metals Using Biosurfactant

Heavy metals are utilized in various industrial activities. These metal residues get mixed up with the industrial effluent and subsequently released into the environment, contaminating drinking water and food. These toxic heavy metals being nonbiodegradable accumulate in animals and humans undergoing a series of biomagnifications.

Biosurfactants have the ability to form complexes with metals, i.e., the anionic biosurfactant creates complexes with metals in a nonionic form, while cationic biosurfactant can replace the same charged metal ions by ion exchange. Das et al. (2009) investigated the ability of anionic biosurfactant produced by marine bacterium in removing heavy metal from aqueous solutions. It was found that the percentage removal of lead and cadmium was influenced by the concentrations of metals and biosurfactant.

Biosurfactants produced by microorganisms were found to enhance biodegradation when complexes with metals. Sandrin et al. (2000) reported that metal-complexed rhamnolipids enhance metal biodegradation by *Burkholderia* sp. by reducing metal toxicity to microbial consortia. Mata-Sandoval et al. (2001) reported the biodegradation of the pesticides by rhamnolipid and Triton X-100, in liquid cultures. Neilson et al. (2003) reported the removal of lead using rhamnolipids. Surfactin constitutes two negative charges, one on the aspartate and the other on the glutamate residues which allow the binding of metals, magnesium, manganese, calcium, barium, lithium, and rubidium onto the biosurfactant (Thimon et al. 1992). Mulligan et al. (1999b) studied the removal of metals from water by surfactin, using 50,000 Da molecular weight cutoff ultrafiltration membrane. It was indicated that metals, cadmium and zinc, became associated with the surfactin micelles. In 2012, K. Ramani et al. reported a lipoprotein biosurfactant isolated from *Pseudomonas gessardii* grown on slaughter house lipid waste and its ability to remove metal ions in aqueous solution. The lipoprotein biosurfactant was found to have a high efficiency for the removal of heavy metal and ions in aqueous solution followed by moderate absorption efficiency for calcium, copper, and iron. Thus the ability of biosurfactant of marine origin to chelate toxic heavy metals and form an insoluble precipitate could be exploited in treatment of heavy metal-contaminated water.

6.10 Bioremediation of Insecticide and Pesticides Using Biosurfactant

Microbial remediation of pesticides has attracted increasing attention as a safe, effective, and cheap biotechnological approach. Biosurfactants unlike synthetic surfactants have attracted much interest in environmental remediation due to their excellent biodegradability, low toxicity, and diverse functions (Van Dyke et al. 1991; Yamane et al. 2008). Biosurfactants have the ability to disperse low-water-soluble hydrophobic organic pollutants such as beta-cypermethrin, making it bioavailable to the bacteria which break it down to nontoxic fragments. Zhang et al. (2011) reported *Pseudomonas aeruginosa* CH7 which produces rhamnolipid which enhances the isomerization and biodegradation of beta-cypermethrin by promoting adsorption, absorption, and dissolution of the hydrophobic compound. Two strains JC1 and JCN13 of *Serratia* spp. have also been reported for their ability to degrade 92 % beta-CP within 10 days and 89 % within 4 days, respectively. Mata-Sandoval et al. (2000) compared the ability of biosurfactant, rhamnolipid and synthetic surfactant, Triton X-100 in solubilizing pesticides:

trifluralin, coumaphos, and atrazine. It was found that synthetic surfactant was less effective when compared to the rhamnolipid. Awashti et al. (1999) evaluated the ability of surfactin produced by *Bacillus subtilis* in bioremediation of pesticide, endosulfan, which is reported to cause congenital anomalies, mental retardation, physical deformities, cerebral palsy, epilepsy, hydrocephalus, etc. (Kerala et al. 2011).

6.11 Bioremediation of Dyes Using Biosurfactant

In textiles industries, the fixation rate of the dyes used is as low as 50 %, and a large quantity of unabsorbed toxic dyes are released through wastewater discharge at an unprecedented scale. Many works have been done on microbial decolorization and degradation of dyes; however, a few researches have been done and reported on the use of microbial biosurfactant for dye degradation/removal. According to Jadhav et al. (2011), rhamnolipid produced by *Pseudomonas desmolyticum* and *Bacillus* sp. VUS could enhance the ability of bacterial cells and the enzymes involved in the degradation of Brown 3REL dye. It was found that there was a decrease in half the time required for the degradation of dye, Brown 3REL, on permeabilization of *Bacillus* sp. VUS with 1 mg/ml rhamnolipid. It might be due to increase in permeability of cell membrane which in turn facilitates the release of enzyme. It is thought that surfactants stimulate the effects of enzymes like α-amylase, cellulases, xylanases, and phytase. Liu et al. 2008 reported that in the presence of rhamnolipid, there was 161.98 % increase in the activity of Lip by *Phanerochaete chrysosporium*. Liang et al. 2010 found that with increase in activity of enzymes involved in dye degradation, Lip by 86 %, there was simultaneous increase in the biodegradation rate of Brown 3REL by 50 %. Lakshmipathy et al. 2010 reported biosurfactant producing marine actinomycetes, *Streptomyces* spp. VITDDK3, with the ability to decolorize nearly 98 % of the sulfonated azo dye, RR5B, in static condition.

Another chemical that is widely used and poses a great environmental threat is tributyltin (TBT). It is an active agent in the antifouling paints applied to ships, boats, and nets. TBT which is used to check the growth and attachment of marine micro- and macroorganisms onto the submerged objects when leached into the water bodies could pose a serious threat to human health. TBT is a biocide which has been banned in many countries due to its toxic effects on the targeted as well as nontargeted marine organisms. Sampath et al. (2012) reported biosurfactant rhamnolipid producing *Pseudomonas* sp. with the ability to degrade TBT. The emulsifying activity of surfactant produced by the microorganisms resulted in the dissolution of the hydrophobic compound in the aqueous phase, thus increasing their bioavailability for degradation (Table 6.2).

6.12 Conclusions and Future Perspectives

The implementation of biosurfactant and biosurfactant-producing marine bacteria in remediating the environment, polluted by industrial effluents and xenobiotics, has been studied. According to Burd and Ward (1996), a strain of *Pseudomonas marginalis* produces high-molecular-weight biosurfactant composed of protein and lipopolysaccharide which aids in the growth of the strain on PAHs. *Cladosporium resinae* produces fatty acids and phospholipids extracellularly which enhance the alkane degradation. Pyrene hydrocarbons were found to be successfully degraded by different species of Mycobacterium, Corynebacterium, Nocardia, Pseudomonas, Rhodococcus, and *Micrococcus*. *Mycobacterium* sp. and *Corynebacterium* sp. showed a good degradation rate of pyrene making them potential agent to remediate pyrene-polluted area (Choi et al. 1996). Garcia-Junco et al. (2001) reported *Pseudomonas aeruginosa* 19SJ with the ability to degrade the phenanthrene by the aid of rhamnolipid which increases the bioavailability of phenanthrene. Both organic

Table 6.2 Biosurfactant and their applications

Biosurfactant	Microorganism(s)	Action	Reference(s)
Glycolipids			
Rhamnolipids	*Pseudomonas cepacia, Pseudomonas sp., Candida tropicalis, Vibrio fischeri, P. fluorescens, P. aeruginosa, Escherichia coli, Bacillus subtilis, P. aeruginosa* and *P. desmolyticum*	Bioremediation of marine and soil environments Phenanthrene Anthracene Hexadecane Phenol Cadmium Pyrene Dye, Brown3REL	Jadhav et al. (2011), Pei et al. (2010), Cassia et al. (2014), Zeng et al. (2011), Liu et al. (2010), Bondarenko et al. (2010), and Das and Mukherjee (2007)
Sophorolipids	*Torulopsis bombicola, Torulopsis petrophilum, Torulopsis apicola, C. lipolytica*	Emulsification of oil; alkane dissimilation	Whang et al. (2008), Pesce (2002), and Baviere et al. (1994)
Trehalose lipids	*Rhodococcus* spp. *Arthrobacter* sp., *Nocardia erythropolis, Corynebacterium* sp., *Mycobacterium* sp	Enhance bioavailability of hydrocarbons	Muthusamy et al. (2008) and Franzetti et al. (2010)
Other glycolipid	*Pseudozyma hubeiensis*	Bioremediation of marine oil pollution	Pei et al. (2010)
Fatty acids, phospholipids and neutral lipids			
Spiculisporic acid	*Penicillium spiculisporum*	Dispersion of hydrophilic pigments; heavy metal sequestration	Ishigami et al. (1983), (2000), and Hong et al. (1998)
Mannosylerythritol lipid	*Calyptogena soyoae*	Bioremediation of marine environment	Pei et al. (2010)
	Nocardiopsis lucentensis MSA04	Bioremediation of marine environment	Pei et al. (2010)
Lipopeptides			
Surfactin	*Bacillus subtilis*	Enhance solubility of hydrocarbons; removal of heavy metals and chlorinated pesticides from water	Awashti et al. (1999), Arima et al. (1968), and Jennema et al. (1983)
Lichenysin	*B. licheniformis, B. subtilis*	MEOR enhancement of oil recovery	Yakimov et al. (1997) and Thomas et al. (1993)
Serrawettin	*Serratia marcescens*	Enhance bioavailability of hydrocarbons	Lai et al. (2009)
Phospholipids	*Acinetobacter* sp.	Bioremediation	Pei et al. (2010)
Lipopeptide	*Rhodococcus* sp. TW53	remediation of oil pollution in marine environment	Pei et al. (2010)
Polymeric biosurfactants			
Emulsan	*Acinetobacter calcoaceticus* RAG 1	Hydrocarbon emulsifier	Choi et al. (1996)
Alasan	*Acinetobacter radioresistens*	Degradation of polyaromatic compounds	Barkay et al. (1999)
Liposan	*Candida lipolytica*	Bioemulsification hydrocarbon in water	Cirigliano and Carman (1985)

and inorganic contaminants can be treated using biosurfactants through different processes. Lu et al. (2003) isolated biosurfactant producing two bacterial species which can effectively degrade petroleum hydrocarbons, from the oil-contaminated site of Dawu water source area in Zibo City, China.

The property of biodegradability and low toxicity of biosurfactants makes them very promising agents for use in bioremediation of marine environment. However, the high cost associated with production and downstream processing hinders the commercial success of biosurfactants. The use of cheap agro-based and industrial wastes as raw materials for production of biosurfactant and the efficient isolation and purification process and optimization of growth conditions for higher yield are the key steps in making their production more economically feasible. A better understanding on the biosurfactant production in microorganisms and the mechanism involved in the remediation process may pave the way for enhanced clean up of the toxic wastewater pollutants and better remediation technology. These interesting surface-active molecules may serve as an efficient and environmental friendly way to rejuvenate environment.

References

Abdelkader E, Nadjia L, Ahmed B (2011) Degradation study of phenazin neutral red from aqueous suspension by paper sludge. J Chem Eng Process Technol 2:109

Abraham WR (1998) Novel glycine containing glycolipids from the alkane using bacterium *Alcanivorax borkumensis*. Biochim Biophys Acta 1393:57–62

Akit J, Cooper DI, Manninen KI, Zajic JE (1981) Investigation of potential biosurfactant production among phytopathogenic *Corynebacteria* and related soil microbes. Curr Microbiol 6:145–150

Al-Bahry SN, Al-Wahaibi YM, Elshafie AE, Al-Bemani AS, Joshi SJ, Al-akhmari HS, Al-Sulaimani HS (2013) Biosurfactant production by *Bacillus subtilis* B20 using date molasses and its possible application in enhanced oil recovery. Int Biodeterior Biodegrad 81:141–146

Alsohim AS, Taylor TB, Barrett GA, Gallie J, Zhang XX, Altamirano-Junqueira AE, Johnson LJ, Rainey PB, Jackson RW (2014) The biosurfactant viscosin produced by *Pseudomonas fluorescens* SBW25 aids spreading motility and plant growth promotion. Environ Microbiol 16(7):2267–2281

Arima K, Kaknuma A, Tamura G (1968) Surfactin, a crystalline peptide lipid surfactant produced by *Bacillus subtilis*: isolation, characterization and its inhibition of fibrin clot formation. Biochem Biophys Res Commun 31:488–494

Arulazhagan P, Vasudevan N, Yeom IT (2010) Biodegradation of polycyclic aromatic hydrocarbon by a halotolerant bacterial consortium isolated from marine environment. Int J Environ Sci Tech 7(4):639–652

Asimiea OA, Sam-Wobo SO (2011) The impact of hydrocarbon waste from brass oil terminal on the Phytoplankton and Periphyton communities of lower Brass River, Niger Delta, Nigeria. J Emerg Trends Eng Appl Sci 2:729–733

Asselineau C, Asselineau J (1978) Trehalose containing glycolipids. Prog Chem Fat Lipids 16:59–99

Awashti N, Kumar A, Makkar R, Cameotra S (1999) Enhanced biodegradation of endosulfan, a chlorinated pesticide in presence of a biosurfactant. J Environ Sci Health B 34:793–803

Barkay T, Navon-Venezia S, Ron EZ, Rosenberg E (1999) Enhancement of solubilization and biodegradation of polyaromatic hydrocarbons by the bioemulsifier alasan. Appl Environ Microbiol 65:2697–2702

Baviere M, Degouy D, Lecourtier J (1994) Process for washing solid particles comprising a sophoroside solution. US Patent 5,326,407

Bondarenko O, Rahman PKSM, Rahman TJ, Kahru A, Ivask A (2010) Effects of rhamnolipids from Pseudomonas aeruginosa DS10- 129 on luminescent bacteria: toxicity and modulation of cadmium bioavailability. Microb Ecol 59:588–600

Bordoloi NK, Konwar BK (2009) Bacterial biosurfactants in enhancing solubility and metabolism of petroleum hydrocarbons. J Hazard Mater 170:495–505

British Petroleum (2010) Deepwater horizon accident investigation report. Golfo Do Mexico 192

Burd G, Ward OP (1996) Bacterial degradation of polycyclic aromatic hydrocarbons on agar plates: the role of biosurfactants. Biotechnol Tech 10:371–374

Cameron DR, Cooper DG, Neufeld RJ (1988) The mannoprotein of Saccharomyces cerevisiae is an effective bioemulsifier. Appl Environ Microbiol 54:1420

Casas JA, García-Ochoa F (1999) Sophorolipid production by Candida bombicola: medium composition and culture methods. J Biosci Bioeng 88:488–494

Casas JA, Garcia de Lara S, Garcia-Ochoa F (1997) Optimization of a synthetic medium for Candida bombicola growth using factorial design of experiments. Enzym Microbiol Tech 21:221–229

Cassia R, Silva FS, Almeida DG, Rufino RD, Luna JM, Santos VA, Sarubbo LA (2014) Applications of biosurfactants in the petroleum industry and the remediation of oil spills. Int J Mol Sci 15:12523–12542

Cavalero DA, Cooper DG (2003) The effect of medium composition on the structure and physical state of sophorolipids produced by *Candida bombicola* ATCC 22214. J Biotechnol 103:31–41

Choi WJ, Choi H-G, Lee W-H (1996) Effects of ethanol and phosphate on emulsan production by *Acinetobacter calcoaceticus* RAG-1. J Biotechnol 45(3):217–225

Cirigliano MC, Carman GM (1985) Purification and characterization of liposan, a bioemulsifier from *Candida lipolytica*. Appl Environ Microbiol 50(4):846–850

Cooper DG, Cavalero DA (2003) The effect of medium composition on the structure and physical state of sophorolipids produced by *Candida bombicola* ATCC22214. J Biotechnol 103:31–41

Cooper DG, Paddock DA (1984) Production of biosurfactants from *Torulopsis bombicola*. Appl Environ Microbiol 47:173–176

Cooper DG, Zajic JE, Gerson DE (1979) Production of surface-active lipids by *Corynebacterium lepus*. Appl Environ Microbiol 37:4–10

Cooper DG, MacDonald CR, Duff SJB, Kosaric N (1981) Enhanced production of surfactin from *B subtilis* by continuous product removal and metal cation addition. Appl Environ Microbiol 42:408–412

Das K, Mukherjee AK (2007) Comparison of lipopeptide biosurfactants production by *Bacillus subtilis* strains in submerged and solid state fermentation systems using a cheap carbon source: some industrial applications of biosurfactants. Process Biochem 42:1191–1199

Das P, Mukherjee S, Sen R (2009) Biosurfactant of marine origin exhibiting heavy metal remediation properties. Bioresour Technol 100:4887–4890

Desai JD, Banat IM (1997) Microbial production of surfactants and their commercial potential. Microbiol Mol Biol Rev 61:47–64

Dziegielewska E, Adamczak M (2013) Evaluation of waste products in the synthesis of surfactants by yeasts. Chem Pap 67:1113–1122

Edward JR, Hayashi JA (1965) Structure of rhamnolipid from *Pseudomonas aeruginosa*. Arch Biochem Biophys 111:415–421

Elaziouti A, Laouedj N, Ahmed B (2011) Effect of pH solution on the optical properties of cationic dyes in dye/maghnia montmorillonite suspensions. J Chem Eng Process Technol 2:113

Flasz A, Rocha CA, Mosquera B, Sajo C (1998) A comparative study of the toxicity of a synthetic surfactant and one produced by *Pseudomonas aeruginosa* ATCC55925. Med Sci Res 26:181–185

Franzetti A, Bestetti G, Caredda P, La Colla P, Tamburini E (2008) Surface-active compounds and their role in the access to hydrocarbons in Gordonia strains. FEMS Microbiol Ecol 63:238–248

Franzetti A, Gandolfi I, Bestetti G, Smyth TJ, Banat IM (2010) Production and applications of trehalose lipid biosurfactants. Eur J Lipid Sci Tech 112:617–627

Garcia-Junco M, De Olmedo E, Ortega-Calvo JJ (2001) Bioavailability of solid and non-aqueous phase liquid (NAPL)-dissolved phenanthrene to the biosurfactant producing bacterium *Pseudomonas aeruginosa* 19SJ. Environ Microbiol 3:561–569

Gautam KK, Tyagi VK (2006) Microbial surfactants: a review. J Oleo Sci 55:155–166

Goldman SS, Rubinovitz YC, Rosenberg E, Gutnick DL (1982) Emulsan in *Acinetobacter calcoaceticus* RAG-1: distribution of cell-free and cell-associated cross-reacting material. Appl Environ Microbiol 44(1):165–170

Gordon L, Dobson ADW (2001) Fluoranthene degradation in *Pseudomonas alcaligenes* PA-10. Biodegradation 12:393–400

Hassanshahian M (2014) Isolation and characterization of biosurfactant producing bacteria from Persian Gulf (Bushehr provenance). Mar Pollut Bull 86:361–366

Hommel RK, Stuwer O, Stubrerd W, Kleber HP (1987) Production of water soluble surface active exolipids by *Torulopsis apicola*. Appl Microbiol Biotechnol 26:199–205

Hong JJ, Yang SM, Lee CH, Choi YK, Kajiuchi T (1998) Ultrafiltration of divalent metal cations from aqueous solution using polycarboxylic acid type biosurfactants. J Colloid Interf Sci 202:63–73

Horowitz S, Gilbert JN, Griffin WM (1990) Isolation and characterization of a surfactant produced by *Bacillus licheniformis* 86. J Ind Microbiol 6:243–248

Husain DR, Goutx M, Acquaviva M, Gilewicz M, Bertrand JC (1997) The effect of temperature on eicosane substrate uptake modes by a marine bacterium *Pseudomonas nautica* strain 617: relationship with the biochemical content of cells and supernatants. World J Microbiol Biotechnol 13:587–590

Inoue S, Ito S (1982) Sophorolipids from *Torulopsis bombicola* as microbial surfactants in alkane fermentations. Biotechnol Lett 4:3–8

Ishigami Y, Yamazaki S, Gama Y (1983) Surface active properties of biosoap from spiculisporic acid. J Colloid Interf Sci 94:131–139

Ishigami Y, Zhang Y, Ji F (2000) Spiculisporic acid functional development of biosurfactants. Chim Oggi 18:32–34

Itoh S, Suzuki T (1974) Fructose-lipids of *Arthrobacter*, *Corynebacteria*, *Nocardia* and *Mycobacteria* grown on fructose. Agric Biol Chem 38:1443–1449

Jadhav M, Kalme S, Tamboli D, Govindwar S (2011) Rhamnolipid from *Pseudomonas desmolyticum* NCIM-2112 and its role in the degradation of Brown 3REL. J Basic Microbiol 51:1–12

Jennema GE, McInerney MJ, Knapp RM, Clark JB, Feero JM, Revus DE, Menzie DE (1983) A halotolerant, biosurfactants-producing *Bacillus species* potentially useful for enhanced oil recovery. Dev Ind Microbiol 24:485–492

Kanga S, Bonner J, Page C, Mills M, Autenrieth A (1997) Solubilization of naphthalene and methyl-substituted naphthalenes from crude oil using biosurfactants. Environ Sci Technol 31:556–561

Kim SY, Oh DK, Kim JH (1997) Biological modification of hydrophobic group in *Acinetobacter calcoaceticus* RAG-1 Emulsan. J Ferment Bioeng 84:162–164

Konishi M, Fukuoka T, Morita T, Imura T, Kitamoto D (2008) Production of new types of sophorolipids by *Candida batistae*. J Oleo Sci 57:359–369
Kosaric N (1992) Biosurfactants in industry. Pure Appl Chem 64:1731–1737
Kretschmer A, Bock H, Wagner F (1982) Chemical and physical characterization of interfacial active lipids from *Rhodococcus erythropolis* grown on –alkane. Appl Environ Microbiol 44:864–870
Lai CC, Huang YC, Wei YH, Chang JS (2009) Biosurfactant-enhanced removal of total petroleum hydrocarbons from contaminated soil. J Hazard Mater 167:609–614
Lakshmipathy TD, Prasad ASA, Kannabiran K (2010) Production of biosurfactant and heavy metal resistance activity of *Streptomyces* Sp VITDDK3-a novel halo tolerant actinomycetes isolated from saltpan soil. Adv Biol Res 4(2):108–115
Lalithakumari D (2011) Microbes: "A Tribute" to clean environment
Lang S, Wullbrandt D (1999) Rhamnose lipids—biosynthesis, microbial production and application potential. Appl Microbiol Biotechnol 51:22–32
Lawniczak L, Marecik R, Chrzanowski L (2013) Contributions of biosurfactants to natural or induced bioremediation. Appl Microbiol Biotechnol 97:2327–2339
Lesik OY, Karpenko EV, Elysseev SA, Turovsky AA (1989) The surface-active and emulsifying properties of *Candida lipolytica* Y-917 grown on n-hexadecane. Microbiol J 51:56–59
Liang YS, Yuan XZ, Zeng GM, Hu CL et al (2010) Biodelignification of rice straw by *Phanerochaete chrysosporium* in the presence of dirhamnolipid. Biodegradation 21(4):615–624
Lima AS, Alegre RM (2009) Evaluation of emulsifier stability of biosurfactant produced by Saccharomyces lipolytica CCT-0913. Braz Arch Biol Technol 52:285–290
Liu XL, Zeng GM, Tang L, Zhong H et al (2008) Effects of dirhamnolipid and SDS on enzyme production by *Phanerochaete chrysosporium* in submerged fermentation. Process Biochem 43:1300–1303
Liu ZF, Zeng GM, Wang J, Zhong H, Ding Y, Yuan XZ (2010) Effects of monorhamnolipid and Tween 80 on the degradation of phenol by *Candida tropicalis*. Process Biochem 45:805–809
Lu XX, Zhang X, Li GH, Zhang WH (2003) Production of biosurfactant and its role in the biodegradation of oil hydrocarbons. J Environ Sci Health A Tox Hazard Subst Environ Eng 38(3):483–492
Makkar RS, Cameotra SS, Banat IM (2011) Advances in utilization of renewable substrates for biosurfactant production. Appl Microbiol Biotechnol 1:1–5
Maneerat S (2005) Biosurfactants from marine microorganisms. Songklanakarin J Sci Technol 27(6):1263–1272
Maneerat S, Phetrong K (2007) Isolation of biosurfactants producing marine bacteria and characteristics of selected biosurfactant. Songklanakarin J Sci Technol 29:781–791
Marcoux J (2000) Optimization of high-molecular-weight polycyclic aromatic hydrocarbons' degradation in a two-liquid-phase bioreactor. J Appl Microbiol 88:655–662
Mata-Sandoval JC, Karns J, Torrents A (2000) Effects of rhamnolipids produced by *Pseudomonas aeruginosa* UG2 on the solubilization of pesticides. Environ Sci Tech 34:4923–4930
Mata-Sandoval JC, Karns J, Torrents A (2001) Influence of rhamnolipids and Triton X-100 on the biodegradation of three pesticides in aqueous and soil slurries. J Agric Food Chem 49:3296–3303
Matsuyama T, Tanikawa T, Nakagawa Y (2011) Serrawettins and other surfactants produced by *Serratia*. Biosurfactants Microbiol Monogr 20:2093–2120
Mclnerney MJ, Javaheri M, Nagle DP (1990) Properties of the biosurfactant produced by *Bacillus licheniformis* strain JF-2. J Ind Microbiol 5:95–102
Miller RM, Zhang Y (1997) Measurement of biosurfactant-enhanced solubilization and biodegradation of hydrocarbons. Methods Biotechnol 2:59–66
Morikawa M, Ito M, Imanaka T (1992) Isolation of a new surfactin producer *Bacillus pumilus* A-l, and cloning and nucleotide sequence of the regulator gene, psf-1. J Ferment Bioeng 74:255–261
Morikawa M, Daido H, Takao T, Murata S, Shimonishi Y, Imanaka T (1993) A new lipopeptide biosurfactant produced by *Arthrobacter sp*. strain MIS38. J Bacteriol 175:6459–6466
Moussa TAA, Ahmed GM, Abdel-hamid SMS (2006) Optimization of cultural conditions for biosurfactants production from *Nocardia amarae*. J Appl Sci Res 2:844–850
Mulligan CN (2005) Environmental applications for biosurfactants. Environ Pollut 133:183–198
Mulligan CN, Gibbs BF (1990) Recovery of biosurfactants by ultrafiltration. J Chem Technol Biotechnol 47:23–29
Mulligan CN, Yong RN, Gibbs BF (1999) Metal removal from contaminated soil and sediments by the biosurfactant surfactin. Environ Sci Tech 33:3812–3820
Mulligan CN, Yong RN, Gibbs BF (2001) Remediation technologies for metal contaminated soils and groundwater: an evaluation. Eng Geol 60(1–4):193–207
Muthusamy K, Gopalakrishnan S, Ravi TK, Sivachidambaram P (2008) Biosurfactants: properties, commercial production and application. Curr Sci 94:736–774
Navon-Venezia S, Zosim Z, Gottlieb A, Legmann R, Carmeli S, Ron EZ, Rosenberg E (1995) Alasan, a new bioemulsifier from *Acinetobacter radioresistens*. Appl Environ Microbiol 61:3240–3244
Nayak AS, Vijaykumar MH, Karegoudar TB (2009) Characterization of biosurfactant produced by *Pseudoxanthomonas sp* PNK-04 and its application in bioremediation. Int Biodeterior Biodegrad 63(1):73–79

Neilson JW, Artiola JF, Maier RM (2003) Characterization of lead removal from contaminated soils by non-toxic soil-washing agents. J Environ Qual 32:899–908

Nnamchi CI, Obeta JAN, Ezeogu LI (2006) Isolation and characterization of some polycyclic aromatic hydrocarbon degrading bacteria from Nsukka soils in Nigeria. Int J Environ Sci Tech 3:181–190

Noordman WH, Wachter JJJ, de Boer GJ, Janssen DB (2002) The enhancement by biosurfactants of hexadecane degradation by *Pseudomonas aeruginosa* varies with substrate availability. J Biotechnol 94:195–212

Obayori OS, Ilori MO, Adebusoye SA, Oyetibo GO, Omotayo AE, Amund OO (2009) Degradation of hydrocarbons and biosurfactant production by *Pseudomonas sp* strain LP1. World J Microbiol Biotechnol 25:1615–1623

Park KS, Sims RC, Dupont RR (1990) Transformation of PAHs in soil systems. J Environ Eng 116:632–640

Pei XH, Zhan XH, Wang SM, Lin YS, Zhou LX (2010) Effects of a biosurfactant and a synthetic surfactant on phenanthrene degradation by a *Sphingomonas* strain. Pedosphere 20:771–779

Persson A, Osterberg E, Dostalek M (1988) Biosurfactant production by *Pseudomonas fluorescens* 378: growth and product characteristics. Appl Microbiol Biotechnol 29:1–4

Pesce L (2002) A biotechnological method for the regeneration of hydrocarbons from dregs and muds, on the base of biosurfactants. World Patent 02/062: 495

Rahman KSM, Rahman TJ, Kourkoutoas Y, Petsas I, Marchant R, Banat IM (2003) Enhanced bioremediation of n-alkane in petroleum sludge using bacterial consortium amended with rhamnolipid and micronutrients. Bioresour Technol 90:159–168

Ramani K, Chandan Jain S, Mandal AB, Sekaran G (2012) Microbial induced lipoprotein biosurfactant from slaughterhouse lipid waste and its application to the removal of metal ions from aqueous solution. Colloids Surf B: Biointerfaces 97:254–263

Reddy MS, Naresh B, Leela T, Prashanthi M, Madhusudhan NC, Dhanasri G, Devi P (2010) Biodegradation of phenanthrene with biosurfactant production by a new strain of *Brevibacillus sp.* Bioresource Technol 101:7980–7983

Report on health effects of endosulfan and progress of rehabilitation activities in Kerala, Department of Health and family welfare, Government of Kerala. 20th Apr 2011

Robert M, Mercade ME, Bosch MP, Parra JL, Espuny MJ, Manresa MA, Guinea J (1989) Effect of the carbon source on biosurfactant production by *Pseudomonas aeruginosa* 44T 1. Biotechnol Lett 11:871–874

Ron EZ, Rosenberg E (2001) Mini review: natural roles of biosurfactants. Environ Microbiol 3:229–236

Rosenberg E, Ron EZ (1999) High- and low-molecular-mass microbial surfactants. Appl Microbiol Biotechnol 52:154–162

Rosenberg E, Zuckerberg A, Rubinovitz C, Gutnick DL (1979) Emulsifier of *Arthrobacter* RAG-1: chemical and physical properties. Appl Environ Microbiol 37:402

Rosenberg E, Rubinovitz C, Gottlieb A, Rosenhak S, Ron EZ (1988a) Production of biodispersan by *Acinetobacter calcoaceticus* A2. Appl Environ Microbiol 54(2):317–322

Rosenberg E, Rubinovitz C, Legmann R, Ron EZ (1988b) Purification and chemical properties of *Acinetobacter calcoaceticus* A2 biodispersan. Appl Environ Microbiol 54:323

Rufino RD, Sarubbo LA, Campos-Takaki GM (2007) Enhancement of stability of biosurfactant produced by *Candida lipolytica* using industrial residue as substrate. World J Microbiol Biotechnol 23:729–734

Sampath R, Venkatakrishnan H, Ravichandran V, Chaudhury RR (2012) Biochemistry of TBT-degrading marine pseudomonads isolated from Indian coastal waters. Water Air Soil Pollut 223:99–106

Sandrin TR, Chech AM, Maier RM (2000) A rhamnolipid biosurfactant reduces cadmium toxicity during napthalene biodegradation. Appl Environ Microbiol 66:4585–4588

Sarubbo LA, Farias CB, Campos-Takaki GM (2007) Co-utilization of canola oil and glucose on the production of a surfactant by *Candida lipolytica*. Curr Microbiol 54:68–73

Shulga AN, Karpenko EV, Eliseev SA, Turovsky AA, Koronelli TV (1990) Extracellular lipids and surface-active properties of the bacterium *Rhodococcus erythropolis* depending on the source of carbon nutrition. Mikrobiologya 59:443–447

Silva EJ, Rocha e Silva NMP, Rufino RD, Luna JM, Silva RO, Sarubbo LA (2014) Characterization of a biosurfactant produced by *Pseudomonas cepacia* CCT6659 in the presence of industrial wastes and its application in the biodegradation of hydrophobic compounds in soil. Colloids Surf B: Biointerfaces 117:36–41

Sobrinho HBS, Rufino RD, Luna JM, Salgueiro AA, CamposTakaki GM, Leite LFC, Sarubbo LA (2008) Utilization of two agroindustrial by-products for the production of a surfactant by *Candida sphaerica* UCP0995. Process Biochem 43:912–917

Sobrinho HB, Luna JM, Rufino RD, Porto ALF, Sarubbo LA (2013) Biosurfactants: classification, properties and environmental applications. In Recent Dev Biotechnol 11:1–29

Sriram MI, Gayathiri S, Gnanaselvi U, Jenifer PS, Mohan Raj S, Gurunathan S (2011) Novel lipopeptide biosurfactant produced by hydrocarbon degrading and heavy metal tolerant bacterium *Escherichia fergusonii* KLU01 as a potential tool for bioremediation. Bioresour Technol 102:9291–9295

Suzuki T, Tanaka H, Itoh I (1974) Sucrose lipids of *Arthrobacteria*, *Corynebacteria* and *Nocardia* grown on sucrose. Agric Biol Chem 33:190–195

Thanomsub B, Watcharachaipong T, Chotelersak K, Arunrattiyakorn P, Nitoda T et al (2004) Monoacylglycerols: glycolipid biosurfactants produced by a thermotolerant yeast, *Candida ishiwadae*. J Appl Microbiol 96:588–592

Thavasi R, Jayalakshmi S, Banat IM (2011) Effect of biosurfactant and fertilizer on biodegradation of crude oil by marine isolates of *Bacillus megaterium*, *Corynebacterium kutscheri* and *Pseudomonas aeruginosa*. Bioresour Technol 102:772–778

Thimon L, Peypoux F, Michel G (1992) Interactions of surfactin, a biosurfactant from *Bacillus subtilis* with inorganic cations. Biotechnol Lett 14:713–718

Thomas CP, Duvall ML, Robertson EP, Barrett KB, Bala GA (1993) Surfactant-based EOR mediated by naturally occurring microorganisms. Soc Pet Eng 11:285–291

Van Dyke MI, Lee H, Trevors JT (1991) Applications of microbial surfactants. Biotechnol Adv 9:141–152

Varsha YM, Naga Deepthi CH, Chenna S (2011) An emphasis on xenobiotic degradation in environmental clean up. J Bioremed Biodegrad S11:001

Whang LM, Liu PWG, Ma CC, Cheng SS (2008) Application of biosurfactant, rhamnolipid, and surfactin, for enhanced biodegradation of diesel-contaminated water and soil. J Hazard Mater 151:155–163

Williams K (2009) Biosurfactants for cosmetic applications: overcoming production challenges. Basic Biotechnol 5:78–83

Yakimov M, Amro M, Bock M et al (1997) The potential of *Bacillus licheniformis* strains for in situ enhanced oil recovery. J Petrol Sci Eng 18:147–160

Yamane M, Toyo T, Inoue K, Sakai T, Kaneko Y, Nishiyama N (2008) Aquatic toxicity and biodegradability of advanced cationic surfactant APA-22 compatible with the aquatic environment. J Oleo Sci 57:529–538

Yeung CW, Law BA, Milligan TG, Lee K, Whyte LG, Greer CW (2011) Analysis of bacterial diversity and metals in produced water, seawater and sediments from an offshore oil and gas production platform. Mar Pollut Bull 62:2095–2105

Zajic JE, Guignard H, Gerson DE (1977) Emulsifying and surface active agents from *Corynebacterium hydrocarboclastus*. Biotechnol Bioeng 19:1285–1301

Zeng G, Liu Z, Zhong H, Li J, Yuan X, Fu H, Ding Y, Wang J, Zhou M (2011) Effect of monorhamnolipid on the degradation of n-hexadecane by *Candida tropicalis* and the association with cell surface properties. Appl Microbiol Biotechnol 90:1155–1161

Zhang C, Wang S, Yan Y (2011) Isomerization and biodegradation of beta-cypermethrin by Pseudomonas aeruginosa CH7 with biosurfactant production. Bioresour Technol 102:7139–7146

Zinjarde SS, Pant A (2002) Emulsifier from a tropical marine yeast, *Yarrowia lipolytica* NCIM 3589. J Basic Microbiol 42(1):67–73

7

Optimization of Cultural Conditions for Marine Microbial Biosurfactant Production: Future Prospects from Untapped Marine Resources

K.V. Deepika, Ganji Purnachandra Nagaraju, and P.V. Bramhachari

Abstract

Marine microbial biosurfactants are of great importance attributable to their structural and functional diversity and industrial applications. Despite huge number of reports describing biosurfactant (BS)/bioemulsifier (BE) applications and advantages, commercialization of these compounds remains complicated and costly. This is mainly due to the usage of chemically synthesized media for growth of microorganism resulting in huge monetary difference between the investment and achievable productivity from the commercial point of view. Although numerous developments have been taken place in the industrial biosurfactant sectors, unfortunately, large-scale production remains economically challenging for many of such molecules. Therefore, the knowledge on the genetic regulatory mechanisms and statistical and genetic algorithms would help to develop an optimized biosurfactant production method with better product characteristics and acquired capability of utilizing cheap agro-industrial wastes as substrates. This review provides an overview of traditional biosurfactant production methods, statistical optimization techniques, recent biotechnological advances, the role and importance of molecular genetics, metagenomics, and gene regulation mechanisms involved in biosynthesis of various marine microbial surfactants of commercial importance.

K.V. Deepika • P.V. Bramhachari (✉)
Department of Biotechnology and Botany, Krishna University, Machilipatnam 521001, Andhra Pradesh, India
e-mail: veerabramha@gmail.com

G.P. Nagaraju
Department of Hematology and Medical Oncology, Winship Cancer Institute, Emory University, Atlanta, GA 30322, USA

M.M. Naik, S.K. Dubey (eds.), *Marine Pollution and Microbial Remediation*,
DOI 10.1007/978-981-10-1044-6_7

7.1 Introduction

Marine microbes are well known for their numerous biomolecules of paramount significance such as antibiotics, enzymes, biopolymers, pigments and toxins. It has been reported that more than 10,000 metabolites with broad-spectrum biological activities and incredible medicinal properties were isolated from marine microbes until now (Kelecom 2002). Nevertheless, due to the vastness of the marine biosphere, most of the marine microbial worlds remain unexplored. Perhaps only <0.1 % of marine microbial world has been explored so far (Ramaiah 2005). In the midst of diverse marine bioactive compounds, microbial biosurfactants/bioemulsifiers (BSs/BEs) are of immense importance due to their structural and functional diversity and wide range of commercial and industrial applications (Banat 1995a, b; Rodrigues et al. 2006a). Microbial BSs are mainly amphiphilic surface-active agents synthesized by bacteria, fungi, yeast and *Actinomycetes*. They are categorized into glycolipids, glycolipoproteins, glycopeptides, lipopeptides, lipoproteins, fatty acids, phospholipids, neutral lipids, lipopolysaccharides (Banat et al. 2010) and glycoglycerolipids (Wicke et al. 2000). The applications of BSs include detergency, emulsification, foaming, dispersion, wetting, penetrating, thickening, microbial growth enhancement, antimicrobial agents, metal sequestering and enhanced oil recovering. All these remarkable aspects make BSs potential candidates to substitute most of the chemical surfactants that are currently in use. Besides, BSs offer several advantages over chemically synthesized surfactants, such as *in situ* production using economical substrates, lower toxicity, biodegradability and ecological compatibility (Marchant and Banat 2012a, b). Strikingly, marine microbes produce BSs with certain remarkable properties including tolerance and stability to pH (3–12), temperature (20–100 °C) and salinity (0.5–2.0 %) (Nerurkar et al. 2009). The biosurfactant-producing microorganisms reported till date are principally isolated from terrestrial sources. Nevertheless, studies on BEs from marine sources are very scanty.

The marine environment which occupies nearly about three-fourth of the earth's surface is a robust reservoir of diverse microbes including biosurfactant producers. Previous reports on BSs of marine origin mainly focused on applications of biosurfactants in environmental remediation isolated from oil-degrading microbes (Rosenberg et al. 1979; Schulz et al. 1991; Yakimov et al. 1998; Thavasi et al. 2011; Peng et al. 2008) and other possible applications such as medical and industrial sectors have not been studied comprehensively (Rodrigues et al. 2006b; Marchant and Banat 2012a, b). Recently, potential application/properties of marine BSs such as antimicrobial (Mukherjee et al. 2009), biofilm disruption (Kiran et al. 2010d) and nanoparticle synthesis (Kiran et al. 2010a) have gained much attention. Microbial surface-active molecules have prospective applications in environmental, food, pharmaceutical, petroleum recovery and health-care industries (Rodrigues et al. 2006b; Mukherjee et al. 2006; Banat et al. 2010). Although biosurfactants exhibit such important advantages, they have not been yet employed extensively in industry because of relatively high production costs. One promising approach for reducing costs is the utilization of alternative substrates such as agro-industrial wastes (Mercade and Manresa 1994). The main problem related to the use of alternative substrates as culture medium is to find a waste with the proper balance of nutrients that allow cell growth and product accumulation (Makkar and Cameotra 1999). The establishment of waste-based medium for biosurfactant production also encounters another difficulty, with regard to the properties of final product which are reliant on the composition of the culture medium (Besson and Michel 1992).

In recent years, the increasing demand for natural polymers for pharmaceutical, food and other industrial applications has led to a remarkable concern in extracellular polymeric substances produced by marine bacteria (Nichols et al. 2005; Poli et al. 2010). Bacteria-producing polymers with novel structures and innovative properties have been isolated in atypical environments, including extreme environments (Chi and Fang 2005; Nichols et al. 2005). Therefore, marine biosurfactants are one of the prospective targets in sustainable production processes and have been considerably increased in the last years. The global market for industrial surfactants is estimated to be more than US$27 billion (Geys et al. 2014). Taking into consideration of the growing demands of microbial surfactants for numerous industrial applications, it is essential to improve the yield and productivity of biosurfactant by fermentation (Das and Mukherjee 2007). So far, the application of biotechnologically produced surfactants is mainly constrained to certain specialized areas, as production processes, in general, yet cannot compete with synthetic and chemical surfactants from an economical viewpoint. This is mainly due to relatively low product yields, high-priced raw materials, and expensive downstream processing. Moreover, the choice of selected carbon source (sugar or nonsugar sources) in the growth medium represents the first step for the optimization of biosurfactant production (Poli et al. 2010). Therefore, the optimized culture conditions can modify the biosurfactant yield and the structure of polymers by Quesada et al. (2004).

Among several approaches, statistical model-based optimization of media components that crucially affect the product formation has been the most indispensable approach toward promoting the production of biosurfactants. Some of the widely used effective statistical techniques include response surface methodology and factorial design (Sen and Swaminathan 1997; Sivapathasekaran et al. 2010b). Alternatively, mathematical tools such as artificial neural network modeling and genetic algorithm optimization (ANN–GA) are widely used to optimize complex nonlinear problems in the field of bioprocess technology (Pal et al. 2009). The combination of ANN and GA is currently used as a powerful tool for product optimization, which is much more advanced to response surface methodologies (Pal et al. 2009; Sivapathasekaran et al. 2010a), particularly for complex processes, such as biological phenomenon. There are several algorithms that are used for network architecture in ANN models for training the target data set. Among them, the resilient back propagation (RBP) first-order optimization algorithm exhibits improved performance characteristics by recovering the data over-fitting problems during the training, which significantly reduces the time period taken (Mastorocostas 2004; Sivapathasekaran et al. 2010b). Although various statistical methods have been beneficially employed for process optimization in a bioreactor, there are hardly any reports on the use of sophisticated optimization techniques like ANN–GA for enhancing biosurfactant production in a laboratory-scale fermenter. This review summarizes fundamental insights related to the factors influencing the product yields and statistical optimization techniques along with the molecular engineering aspects. Also it emphasizes the modern approaches for optimized biosurfactants and future prospects of both naturally occurring and genetically modified bacterial strains in bioremediation and industrial and therapeutic applications.

7.2 Factors Affecting Biosurfactant Production

Microbial productions of primary or secondary metabolites are regulated by the genetic makeup of the producer strain; however, nutritional and environmental factors play a crucial role on the overall output of the metabolite. In addition, the knowledge about the regulation mechanisms of the chosen microorganism is highly imperative for the production of microbial metabolites on a large scale. Generally, biosurfactant production can be induced by hydrocarbons or water-insoluble substrates. Several factors influence the production of microbial metabolites like the nature of carbon, nitrogen, and phosphate sources; metal ions and other additives used in the media formulation; culture conditions like pH, temperature, agitation rate and oxygen availability; the nature of the selected microorganism; and finally the adopted fermentation strategies for large-scale production.

7.3 Nutritional Factors

7.3.1 The Effect of Carbon Source

Several studies revealed production of biosurfactants by various microorganisms on carbohydrates, water-immiscible substrates and also hydrocarbons (Table 7.1). A significant observation by Banat (1995b) revealed that modest amount of biosurfactant yield was obtained when cells were grown on readily available carbon sources; nevertheless, the biosurfactant production was triggered when all the soluble carbon was consumed and water-immiscible hydrocarbon was available in the medium. Robert et al. (1989) reported production of rhamnolipid from *Pseudomonas aeruginosa* using a variety of carbon sources such as C_{11} and C_{12} alkanes, succinate, pyruvate, citrate, fructose, glycerol, olive oil, glucose and mannitol. Some studies have evidenced the use of water-soluble compounds such as glucose, sucrose, and glycerol as source of carbon substrates for biosurfactant production (Desai and Banat 1997). However, cost-effective substrates like molasses (Makkar and Cameotra 1999), peat hydrolysate (Sheppard and Mulligan 1987) and potato processing effluents (Fox and Bala 2000) were suggested for biosurfactant production. An interesting approach was put forth from the study by Mata-Sandoval et al. (2001); hydrophobic substrates like corn oil, lard (rich in unsaturated and saturated fat) and long-chain alcohols significantly maximized the biosurfactant production (100–165 mg/g substrate), while hydrophilic substrates like glucose and succinate acid were shown with poor yields (12–36 mg/g substrate). Experimental results showed that the presence of either residual substrate or hydrophobic by-products in the rhamnolipid crude mixtures can influence their capabilities to solubilize hydrophobic compounds (Mata-Sandoval et al. 2001). Pagilla et al. (2002) used soluble acetate and sparingly soluble hexadecane as carbon substrate for *Gordonia amarae* growth and biosurfactant production in large-scale batch reactors.

In another study, Ilori et al. (2005) identified diesel and crude oil as best carbon sources for biosurfactant production. Pepi et al. (2005) reported an Antarctic psychrotrophic *Halomonas* sp. ANT-3b producing emulsifying compounds when grown on n-hexadecane, but not on mineral medium supplemented with D-fructose. On contrary, Kokare et al. (2007) isolated a marine *Streptomyces* sp. S1 showing significant growth on maltose yeast extract medium. Despite the various research reports on carbon substrates for biosurfactant production, Raza et al. (2007) stated that the type, quality and quantity of biosurfactant are affected and influenced by the nature of the carbon substrate. Biosurfactant yield was limited with galactose or starch as carbon sources because of inhibition due to the decrease in pH is probably caused by the production of secondary acid such as uronic acid (Abouseoud et al. 2008). A marine *Marinococcus* sp. isolated from Alang coast, Gujarat, was able to utilize crude oil along with glucose to produce lipopeptide biosurfactants (Sakalle and Rajkumar 2009). Das et al. (2009b) investigated that glycerol was the best substrate for biosurfactant production by marine *B. circulans*. Malavenda et al. (2010) isolated an Antarctic marine biosurfactant-producing bacteria *Acinetobacter* sp. 11/4, capable of utilizing soybean oil as carbon source. Interestingly, Al-Nahas et al. (2011) exhibited highest biosurfactant yield of 9.8 g/L in *Pseudoalteromonas* sp. using glucose as a carbon source. Ibacache-Quiroga et al. (2013) isolated a marine biosurfactant-producing bacterial strain *Cobetia* sp. MM1IDA2H-1, growing in the presence of sulfur-containing heterocyclic aromatic hydrocarbon dibenzothiophene (DBT). However, in a recent report Singh et al. (2013) demonstrated that for the production of biosurfactant, the preference carbon source mainly depends on the behavior of strain. More recently, Antoniou et al. (2015) reported few marine hydrocarbon-degrading bacterial strains *Pseudomonas pachastrellae* strain KMM330, *Marinomonas vaga* strain 40, *Thalassospira lucentensis* strain QMT2, *Roseovarius crassostreae*, *Alcanivorax borkumensis* SK2, *Paracoccus marcusii* and *Sulfitobacter pontiacus* ChLG-10, capable of producing biosurfactant using crude oil as a carbon source.

Table 7.1 Various carbon sources used for the production of biosurfactants by marine microorganisms

Microorganisms	Biosurfactant	Carbon source	Yield	Reference
Pseudomonas aeruginosa	Glycolipid	Glucose	1.5 g/L	Guerra-Santos et al. (1984)
Pseudomonas aeruginosa	Glycolipid	Corn oil	46 g/L	Linhardt et al. (1989)
Bacillus cereus	lipopeptide	Sucrose	1.6 g/L	Cooper and Goldenberg (1987)
Alcaligenes sp. PHY 9 L.86	Extracellular polysaccharide–lipid	Tetradecane	1.1 mg/L	Goutx et al. (1987)
Arthrobacter sp. EK 1	Trehalose tetraester	n-alkanes	4.8 g/L	Passeri et al. (1991)
Hahella chejuensis	Exopolysaccharide	Sucrose	9.23 g/L	Ko et al. (2000)
Microbacterium sp. DSM 12583	Glycoglycerolipids	Glucose	200 mg/L	Wicke et al. (2000)
Yarrowia lipolytica, NCIM 3589	lipid–carbohydrate–protein	Alkanes or crude oil		Zinjarde and Pant (2002)
Bacillus licheniformis strain (B3-15)	Lipopeptide	Glucose	165 mg/L	Maugeri et al. (2002)
Geobacillus sp. 4004	Exopolysaccharide	Sucrose	90 mg/L	Moriello et al. (2003)
Halomonas alkaliantarctica strain CRSS	Exocellular polysaccharides	Acetate	2.9 g/g	Poli et al. (2004)
Halomonas sp. ANT-3b	Glycolipid	n-Hexadecane	–	Pepi et al. (2005)
Myroides sp. SM1	Ornithine lipids	Weathered crude oil	–	Maneerat et al. (2006)
Pseudomonas aeruginosa A41	Glycolipid	Olive oil	6.58 g/L	Thaniyavarn et al. (2006)
Yarrowia lipolytica IMUFRJ50682	Glycolipopeptide	Glucose	–	Amaral et al. (2006)
Corynebacterium kutscheri	Glycolipopeptide	Peanut oil cake	6·4 mg/mL	Thavasi et al. (2007)
Rhodococcus sp. TW53	Lipopeptide	n-Hexadecane	3.4 g/L	Peng et al. (2008)
Pseudomonas aeruginosa LBM10	Glycolipid	Soybean oil	1.42 g/L	Prieto et al. (2008)
Azotobacter chroococcum	Glycolipid	Peanut oil cake	4.6 mg/mL	Thavasi et al. (2009)
Bacillus circulans	Lipopeptide	Glycerol	2.9±0.11 g/L	Das et al. (2009b)
Bacillus velezensis strain H3	Lipopeptide	Starch	–	Liu et al. (2010)
Alcanivorax dieselolei strain B-5^{T}	Proline lipid	Diesel oil	–	Qiao and Shao, (2010)
Brevibacterium aureum MSA13	Lipopeptide	Molasses	–	Kiran et al. (2010b)
Nocardiopsis lucentensis MSA04	Glycolipid	Kerosene	–	Kiran et al. (2010c)
Bacillus licheniformis TR7	Lipopeptide	Molasses	3.30 g/L	Saimmai et al. (2011)
Bacillus subtilis SA9	Lipopeptide	Molasses	3.78 g/L	Saimmai et al. (2011)

(continued)

Table 7.1 (continued)

Microorganisms	Biosurfactant	Carbon source	Yield	Reference
Pseudoalteromonas sp. AM	Exopolysaccharide	Glucose	10.51 g/L	Al-Nahas et al. (2011)
Zunongwangia profunda SM-A87	Exopolysaccharide	Lactose	8.90 g/L	Liu et al. (2011)
Lactobacillus delbrueckii	Glycolipid	Peanut oil cake	5.3 g/L	Thavasi et al. (2011)
Pseudomonas nitroreducens TSB.MJ10	Lipopeptide	Sodium benzoate	2.9 g/L	de Sousa and Bhosle (2012)
Leucobacter komagatae 183	Lipopeptide	Commercial sugar	3.03 g/L	Saimmai et al. (2012b)
Oleomonas sagaranensis AT18	Glycolipid	Molasses	5.30 g/L	Saimmai et al. (2012a)
Nocardiopsis sp. B4	Glycolipid	Olive oil	–	Khopade et al. (2012b)
Streptomyces VITSSB2	Exopolysaccharide decanal	Mannitol	–	Shubhrasekhar et al. (2013)
Klebsiella sp. *strain* RJ-03	Glycopeptide	Starch	10.1 ± 0.11 g/L	Jain et al. (2013a)
Klebsiella sp. strain RJ-03	Glycopeptide	Corn powder	15.40 ± 0.21 g/L	Jain et al. (2013b)
Pseudomonas aeruginosa	Glycolipid	Sucrose	1.35 mg/mL	Gomathy and Senthilkumar (2013)
Inquilinus limosus KB3	Lipopeptide	Palm oil decanter cake	5.13 g/L	Saimmai et al. (2013b)
Selenomonas ruminantium CT2	Lipopeptide	Molasses	5.02 g/L	Saimmai et al. (2013a)
Aeribacillus pallidus 418	Exopolysaccharide	Maltose	0.13 g/L	Radchenkova et al. (2013)
Pantoea strain BM39	Exopolysaccharide	Glucose	21.30 g/L	Silvi et al. (2013)
Streptomyces sp. MAB36	Glycolipid	Fructose	–	Manivasagan et al. (2014)

7.3.2 The Effect of Nitrogen Source

Earlier reports suggest that the nature and concentration of the nitrogen source also affected the biosurfactant production (Table 7.2). Nitrogen limitation conditions enhanced the production of rhamnolipid and sophorose lipid biosurfactants. According to Mulligan and Gibbs (1989), nitrates should be reduced to nitrite and then to ammonia for utilizing them as nitrogen sources. On the other hand, lipid formation is the rate-determining factor during the biosynthesis of rhamnolipid. Nitrogen limitation may promote lipid accumulation. Consequently, in comparison with ammonia, the assimilation of nitrate as a nitrogen source is apparently slower, simulating a nitrogen-limiting condition that is favorable to rhamnolipid production. Many strains of *P. aeruginosa* use nitrates, ammonia and amino acids as nitrogen sources (Mulligan and Gibbs 1989). Some sources report that limiting concentrations of nitrogen up to 2–3 g/L maximize the production of rhamnolipid (Linhardt et al. 1989). Generally, the presence of an organic nitrogen source promotes both the specific growth rate and biosurfactant production (Farres et al. 1997). Conversely, biosurfactant production was shown to be higher at lower nitrogen concentration (Gorret et al. 2001). The observations of Mata-Sandoval et al. (2001) have shown that limiting

Table 7.2 Nitrogen sources used for the production of biosurfactants by marine microorganisms

Microorganisms	Biosurfactant	Nitrogen source	Yield	Reference
Alteromonas sp. *strain 1644*	Exopolysaccharide	Ammonium chloride	–	Samain et al. (1997)
Pseudomonas aeruginosa LBM10	Glycolipid	Sodium nitrate	1.42 g/L	Prieto et al. (2008)
Aspergillus ustus MSF3	Glycolipoprotein	Yeast extract	–	Kiran et al. (2009)
Bacillus velezensis strain H3	Lipopeptide	Starch	–	Liu et al. (2010)
Brevibacterium casei MSA19	Glycolipid	Peptone	–	Kiran et al. (2010a)
Brevibacterium aureum MSA13	Lipopeptide	Acrylamide	–	Kiran et al. (2010b)
Nocardiopsis lucentensis MSA04	Glycolipid	Beef extract	–	Kiran et al. (2010c)
Pseudoalteromonas sp. AM	Exopolysaccharide	Meat extract	10.51 g/L	Al-Nahas et al. (2011)
Nocardiopsis sp. B4	Glycolipid	Phenylalanine	–	Khopade et al. (2012b)
Oleomonas sagaranensis AT18	Glycolipid	$NaNO_3$	5.30 g/L	Saimmai et al. (2012a)
Leucobacter komagatae 183	Lipopeptide	Monosodium glutamate	3.03 g/L	Saimmai et al. (2012b)
Selenomonas ruminantium CT2	Lipopeptide	Monosodium glutamate	5.02 g/L	Saimmai et al. (2013a)
Bacillus amyloliquefaciens MB-101	Lipopeptide	peptone	6.76 g/L	Dhasayan et al. (2014)
Nocardiopsis MSA13A	Glycolipid	Yeast extract	–	Kiran et al. (2014)
Streptomyces sp. MAB36	Glycolipid	Yeast extract	–	Manivasagan et al. (2014)

concentrations of nutrients are not only important to maximize the rhamnolipids production but also the supplementation of nitrogen sources through time intervals. It has been reported that organic nitrogen source can promote enhanced cell growth, but proved to be an unfavorable nutritional factor for the production of glycolipid biosurfactant (Kim et al. 2006). Similarly Chen et al. (2007) found that nitrate-based compounds and inorganic nitrogen sources were recognized as an excellent source of nitrogen for the rhamnolipid production by *Pseudomonas aeruginosa* S2, giving a maximum rhamnolipid concentration of 2300 mg/mL. Few investigators reported nitrogen-limiting conditions (C/N ratio of 100) of soybean oil and sodium nitrate as favorable to biosurfactant production by *Pseudomonas aeruginosa*. The same group established the fact that the use of ammonium salts in the form of ammonium chloride favored the cell growth but did not facilitate the enhanced biosurfactant production and likewise caused a significant decrease in pH (4.03) (Prieto et al. 2008).

Kiran and colleagues (2009) investigated that the carbon/nitrogen ratio is one of the most fundamental factors which induced the production of secondary metabolites and demonstrated that C/N ratio of 3:2 (glucose/yeast extract) facilitated the highest biosurfactant production in marine *Aspergillus ustus* (MSF3). A similar study by Gandhimathi et al. (2009) showed that peptone supplementation enhanced the lipopeptide biosurfactant yield in marine *Actinomycetes Nocardiopsis alba* MSA10. However, Kiran et al. (2010c) observed that the maximum production of biosurfactant by *Nocardiopsis lucen-*

tensis MSA04 occurred at a C/N ratio of 0.5 (kerosene/beef extract) predicting that a higher amount of nitrogen source was required by the strain compared to that of carbon source. However, Al-Nahas et al. (2011) reported that inorganic nitrogen sources weakly supported growth and polysaccharide production and resulted in a substantial decrease in biomass and product yield; on the other hand, organic nitrogen sources favor both growth and polysaccharide production. Khopade et al. (2012a) reported yeast extract as best nitrogen source for biosurfactant production by marine *Streptomyces* species B3. The maximum production of the biosurfactant by marine *Nocardiopsis* occurred at a C/N ratio of 2:1 using olive oil and phenylalanine as carbon source and nitrogen sources (Khopade et al. 2012b). These reports evidently signify that C/N ratio is certainly an essential parameter for the production of biosurfactant.

7.3.3 The Effect of Other Factors/ Sources

Production, composition and final yield of biosurfactants also depend on the fermenter design and nutrient composition (Mulligan and Gibbs 1993). Besides trace elements, amino acids, supplementation of nanoparticles, aeration, agitation and rheology of growth media also influence biosurfactant production as evidenced in several studies. Earlier studies by Zobell (1941) reported the use of $SrCl_2$ to facilitate the growth of marine strains which act as a growth stimulant. Morin (1998) reported that fermentation broth may develop non-Newtonian characteristics acting as a pseudoplastic fluid due to the presence of extracellular substances and their metabolic products, as well as the lack of homogeneity in terms of mixing, mass and oxygen. Thus, the rheological shifts during the fermentation process could be used as a parameter to monitor the constancy and quality of extracellular substances and their production. Moreover, the utilization of detergents such as Tween 40 (polyoxyethylene sorbitan monopalmitate), Tween 80 (polyoxyethylene sorbitan monooleate), CHAPS (3-[(3-cholamidopropyl) dimethyl ammonio]-1-hydroxypropane-sulfonate), and Triton X 100 (nonaethylene glycol octylphenol ether) may ameliorate oxygen concentration in the growth media increasing the production of exopolysaccharides (Morin 1998). Furthermore, Mata-Sandoval et al. (2001) revealed that best results of rhamnolipid production were obtained under minimal nutrient conditions which help in directing the cellular metabolism to the production of rhamnolipids rather than to increase the cell population. According to Lee et al. (2001) high aeration rates generally enhanced production of extracellular substances and increased the viscosity of the culture broth of *Hahella chejuensis*. Yeh et al. (2006) showed that the supply of oxygen acts as a limiting factor for growth and lipopeptide biosurfactant production. Biosurfactants production from the *Bacillus* sp. was found to be an aerobic process, requiring aeration for their growth and metabolite formation (Lee and Kim 2004).

Agitation is another key factor for biosurfactant production, which helps to distribute the oxygen evenly in the medium and maintain the dissolved oxygen concentration at the desired level. Previous studies reveal that *Bacillus* sp. grown in a highly agitated reactor environment for better production of metabolites (Joshi et al. 2008). The use of Fe metal nanoparticles as nutrient supplements came into light with the studies of He et al. (2006) to enhance biosurfactant production by marine *Actinobacterium.* This approach would reduce the impact of nonmetallic ions of the metal salts in a fermentation process. In optimizing the medium, besides the carbon and nitrogen sources, the role of trace metals has also been reported to be critical in enhancing lipopeptide biosurfactant production (Wei et al. 2007). Mukherjee et al. (2008) during statistical screening of nutritional parameters by a marine bacterium found that optimal concentration of $MgSO_4$ at 0.3 g/L could yield more biosurfactant compared with when $MgSO_4$ was used at high concentration. Kiran et al. (2010b) reported a marine *Brevibacterium casei* MSA19 producing maximum biosurfactant in the presence of $FeSO_4$ and asparagine in culture medium. Enhanced biosurfactant production by marine *Nocardiopsis* MSA13A through the supplementation of $FeCl_3$ followed by $CuSO_4$ was evidenced by the reports of Kiran et al. (2014).

7.4 Environmental Factors

There have been a number of scientific reports describing individual effects of temperature, pH, salinity, aeration and agitation on biosurfactant production. Desai and Banat (1997) also stated that environmental factors such as pH, temperature, agitation and oxygen availability and growth conditions also affect biosurfactant production by influencing their cellular growth or activity. Studies by Gugliandolo and Maugeri (1998) examined the highest production of exopolysaccharides by a thermophilic isolate *Bacillus thermodenitrificans* strain B3-72 at an optimal temperature of 65 °C and pH 7.0 in aerobic conditions. Yakimov et al. (1998) isolated six heterotrophic, biosurfactant-producing, n-alkane-degrading marine bacterial strains grew at an optimal temperature between 25 and 30 °C, at NaCl concentrations ranging from 1–0% to 12.5%. A study by Moriello et al. (2003) reported a marine *Geobacillus* sp. 4004 produced 90 mg/L of extracellular substances at a temperature of 60 °C at pH 7.0. In addition, Mancuso Nichols et al. (2004) depicted the yield of extracellular substances by *Pseudoalteromonas* CAM025 grown at −2 and at −10 °C was 100 mg exopolysaccharides per gram dry weight of cells. This is attributed to the fact that production at low temperatures may be a mechanism for adaptation to cold temperature for this strain. These results further concluded that, the production of extracellular substances may be a successful evolutionary strategy for bacteria inhabiting extreme sea ice environments. Yeh et al. (2006) reported an optimum temperature of 30 °C for surfactin production from *Bacillus subtilis* ATCC 21 332 in a bioreactor.

A deep-sea psychrotolerant bacterium *Pseudoalteromonas* sp. SM9913 produced high amount of extracellular substances (5.25 g L^{-1}) in the temperature range 30–10 °C (Qin et al. 2007). Previously, Chi et al. (2007) detected the production of 22.34 g/L of exopolysaccharide by marine *Cyanobacterium Cyanothece* sp. 113 at 29 °C, aeration rate of 7.0 l/min and continuous illumination with 4300 lx. Additionally Su and co-workers (2007) achieved 18.4 g/L of exopolysaccharide produced by the same strain *Cyanothece* sp. 113 at 29 °C in an optimized medium with 70.0 g/L of NaCl, 0.9 g/L of $MgSO_4$ and continuous illumination at 86.0 μE/M^2/S. Some authors reported the maximum biosurfactant production from *Bacillus subtilis* 20B (Joshi et al. 2008) and *Bacillus subtilis* HOB2 (Haddad and Wang Ji 2009) was achieved at a low temperature of 30 °C. Chayabutra and Ju (2001) found that the rate of rhamnolipid production by *Pseudomonas aeruginosa* ATCC 10145 significantly increased when pH was in the range of 6.5–6.7. A tropical marine strain of *Yarrowia lipolytica*, NCIM 3589 require an initial pH of 8.0 for extracellular emulsifier production (Zinjarde and Pant 2002). Previous studies revealed that most lipopeptide biosurfactants are produced better in acidic conditions. For example, the enhanced surfactin production by *Bacillus subtilis* ATCC 21 332 (Wei et al. 2004); maximum surfactin concentration from *Bacillus subtilis* ATCC 21 332 at a pH value of 6.25 (Yeh et al. 2006); lipopeptide production by *Bacillus circulans* at a pH of 6.5 (Sivapathasekaran et al. 2010b). The optimized pH value clearly documented that enhanced biosurfactant production by marine strains required acidic conditions. Salinity was found to be one of the critical parameter in the production of biosurfactant. Marine bacteria are generally lysed in hypotonic media and particularly in distilled water (Bertrand and Larsen 1989). Salt concentrations affect biosurfactant production depending on its effect on cellular activity. Some biosurfactants however, were not affected by salt concentrations up to 10 % (w/v), although slight reductions in the critical micelle concentrations (CMCs) were detected (Abu-Ruwaida et al. 1991). NaCl activated biosurfactant activity of many strains, which were isolated from seawater or petroleum reservoirs (Yakimov et al. 1995).

Fernandez-Linares et al. (1996) investigated the effects of NaCl concentration on growth, biodegradation and the different modes of transfer of hydrocarbons to the cell surface. Because cytoplasmic bacterial membrane is permeable to water but not to most other metabolites, hyper- or hypoosmotic shock causes instantaneous efflux or influx of water, which is, respectively, followed by a decrease or an increase of the cytoplasmic volume. A tropical marine strain of *Yarrowia lipolytica*, NCIM 3589, requires sodium chloride at a concentration of 2–3 % for extracellular emulsifier production (Zinjarde and Pant 2002). In another study, *B. subtilis* BBK-1 produces three types of lipopeptides—bacillomycin L, plipastatin and surfactin in the presence of NaCl up to 8 % (Roongsawang et al. 2002). Marine isolate showing an optimum activity in the salt supplemented medium was reported by Kiran et al. (2009). Another marine isolate, *Pseudoalteromonas agarovorans*, was shown to produce high amount of extracellular substances at an increased NaCl concentration of 30 g/L (Choi et al. 2009). Subsequently, Kiran and colleagues (2010b) reported a marine *Brevibacterium casei* MSA19 producing maximum biosurfactant in the presence of 2 % NaCl in culture medium. Al-Nahas et al. (2011) reported that a marine bacterium *Pseudoalteromonas* sp. AM grew only when NaCl concentration was increased from 10 to 30 g/L showing an increase in cell growth and production of extracellular substances. However, in a recent report, NaCl concentration up to 8 % and pH 6–8 worked most excellently in case of halophilic bacteria *Halomonas* sp. BS4 (Donio et al. 2013).

7.5 Statistical Optimization Techniques for the Production of Biosurfactants

The Plackett–Burman design is a widely used statistical design technique for the screening of the medium components (Plackett and Burman 1944). RSM was used to optimize the medium composition. Response surface methodology (RSM) is a collection of different statistical techniques, including designing experiments, building models and evaluating the effects of factors to trigger desirable responses (Li et al. 2002). It has the intense ability to interpret the interactive effects among input variables, which are some of the attractive features of RSM (Montgomery 1997; Al-Araji et al. 2007). The statistical approach could overcome the limitations of classical medium optimization (Lotfy et al. 2007). Response surface methodology (RSM) designs, help designers to quantify the relationship between one or more measured responses and the vital input factors, were successfully applied by many researchers to build reliable models and find the optimal medium for a greater biosurfactant yield (Rodrigues et al. 2006c; Kiran et al. 2010b; Sivapathasekaran et al. 2010b). Therefore, the statistical experimental designs are powerful tools for searching the key factors rapidly from a multivariable system and minimizing the errors in determining the effect of parameters (El-Sersy 2012).

Mukherjee et al. (2008) determined critical factors affecting the growth and biosurfactant production by a marine bacterium using a Plackett–Burman-based statistical screening procedure. Optimized medium demonstrated 84.7 % increase in biosurfactant yield over the unoptimized medium. Statistical optimization of biosurfactant production from the sponge-associated marine fungus *Aspergillus ustus* MSF3 using RSM was reported by Kiran et al. (2009). The optimized bioprocess conditions for the maximum production were pH7.0, temperature 20 °C, salt concentration 3 %, glucose and yeast extract as carbon source and nitrogen sources, respectively. Statistical approaches aid in the formulation of production medium of biosurfactant and may be crucial to enhance the quantity of the product (Kiran et al. 2010b). In another study by Kiran et. al (2010b), optimization and characterization of a new lipopeptide biosurfactant produced by marine *Brevibacterium aureum* MSA13 in solid-state culture were reported. Thus, it is most likely that the statistical optimization with pretreated molasses as substrate and olive oil, acrylamide, $FeCl_3$ and inoculum size as critical control factors resulted in threefold increased lipopeptide production over original conditions. An interesting result came from a study by Sivapathasekaran and colleagues (2010b) by achieving an average biosurfactant concentration of 3.01 ± 0.02 g/L by *Bacillus circulans* through statistical modeling-based response surface methodology (RSM) only after identifying the critical components in modified marine medium using the OFAT technique. Kiran et al. (2010c) validated significant interactive influence of secondary control factors such as copper sulfate and inoculum size in glycolipid biosurfactant production by marine *Nocardiopsis lucentensis* MSA04 using RSM experiments. Enhanced biosurfactant production was achieved under solid-state culture (SSC) conditions using kerosene as carbon source, beef extract as nitrogen source and wheat bran as substrate. The most convincing evidence to maximize biosurfactant production by Marine *Bacillus subtilis* N10 was demonstrated by El-Sersy (2012) by using Plackett-Burman experimental design (PBD), evaluated in terms of emulsification index (E24) (80 %), i.e., 1.14-fold increase when compared to its production in basal conditions. These statistical techniques are therefore used for identifying important factors from among many potential factors. In the analysis of these designs, usually only main effects are estimated.

Statistical optimization of biosurfactant production by marine *Nocardiopsis* sp. MSA13A under solid-state culture was carried out by Kiran et al. (2014). It was found that the production was significantly influenced by the variables such as glucose, yeast extract, ferrous nanoparticles (Fe NPs) and inoculum size either interactively or independently. In another study, Mabrouk et al. (2014) statistically optimized the glycolipopeptide production by soft coral-associated marine *Bacillus* sp. E34 using Plackett-Burman design (PBD) and recorded significant increase (1.28-fold) in the emulsification index. Interestingly, Dhasayan and colleagues (2014) attained maximum biosurfactant production of 6.76 g/L by *B. amyloliquefaciens* MB-101 using response surface statistical optimization methods in glycerol-containing medium under submerged fermentation and final yield was 3.48-fold higher than that of non-optimized normal media. Manivasagan et al. (2014) isolated potential glycolipid biosurfactant producer *Streptomyces* sp. MAB36 from marine sediment samples. Medium composition and culture conditions for the glycolipid biosurfactant production by *Streptomyces* sp. MAB36 were optimized, using two statistical methods: Plackett–Burman design was applied to find out the key ingredients and conditions for the best yield of glycolipid biosurfactant production and central composite design was used to optimize the concentration of the four significant variables, starch, casein, crude oil, and incubation time in terms of emulsification index (E_{24}%).

7.6 Metabolic Engineering for the Production of Biosurfactants

The continuous demand to reduce production costs and to generate compounds with enhanced properties for absolute applications has consequently initiated biosurfactant engineering projects based on a variety of biotechnological, molecular, or genetic approaches (Fig. 7.1). Direct and combinatorial strategies were previously attempted to produce arrays of modified products in particular of lipopeptide derivatives. Several studies were focused on the lipopeptide engineering, dynamics and new structural fea-

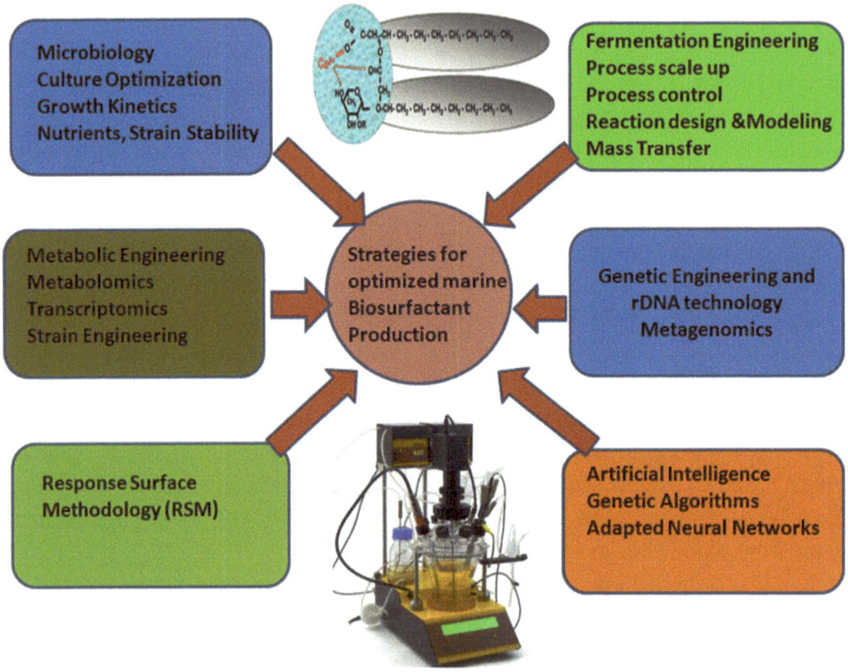

Fig. 7.1 Potential targets for optimization of biosurfactant production

tures of lipopeptide synthetases. Microorganisms produce a set of related lipopeptides (LPs), and the composition of their production profile can be influenced by abiotic and nutritional culture conditions. Biosurfactant production can be modulated by the fermentation conditions and by feeding of specific precursors (Pryor et al. 2007). Further options for the structural tailoring of biosurfactants are to manipulate precursor pathways or modifying enzymes that act posttranslational on the synthesized compounds, e.g., by attachment of specific residues. Mutation of external enzymes involved in selection and activation of the fatty acid can result in the formation of LPs with modified fatty acid chains (Powell et al. 2007). Feeding modified precursors or manipulation of specific precursor pathways can thus result in new products (Amir-Heidari et al. 2008). The N-terminal fatty acid moiety determines to a large extent the biological properties of LPs. Complex combinatorial platforms that comprise a set of different approaches like enzymatic module swapping, complete protein subunit exchanges, modification of accessory tailoring enzymes, or manipulation of precursor structures are often employed (Baltz 2006).

7.7 Concept of Strain Engineering in Biosurfactant Production

Although enormous efforts have been made to investigate the potential candidates to replace *P. aeruginosa*, none of them were successful due to significantly lower yields and final concentrations. Hence, the marine biologists presume to investigate wild-type *P. aeruginosa* strains for economic yields and impending objectives for genetic optimization. Further enhanced yields achievable with these wild-type strains may present a strong basis for further optimization by strain engineering. Several approaches were put forth to optimize the rhamnolipid production in producer strains by genetic modifications (Fig. 7.1). Previous studies suggested few alternatives to be considered in strain engineering: the use of alternative nonpathogenic production strains, broadened substrate spectrum, metabolic spectrum and pathways of producer strains, by-product formation and optimized carbon yield, product formation and genetic regulatory mechanisms and availability of precursor molecules (Henkel et al. 2014). The approach of strain engineering mainly focused on the improvement of product yields which require a complex interactions of genetics (e.g., deletions, insertions, alterations in promoter strength), analytics (e.g., metabolome and transcriptome analysis), and systems biology (e.g., metabolic flux distributions and modeling), to quantify the effects of modification and identify prospective hurdles (Müller and Hausmann 2011). A significant contribution from Koch et al. (1991) described rhamnosyltransferase 1 complex (*rhlAB*) as the key enzyme responsible for rhamnolipid production and genetic modification of rhamnolipid producing *P. aeruginosa* was performed by inserting the *E. coli* lacZY genes into its chromosome so as to make it capable of growing and utilizing lactose from whey for enhanced biosurfactant production. The industrial exploitation of *P. aeruginosa* is limited due to its pathogenic nature and production of wide variety of virulence factors. To surmount this difficulty, the *rhlAB* genes from *P. aeruginosa* were cloned and expressed in heterologous hosts (Ochsner and Reiser 1995). The recombinant strains *P. fluorescens* ATCC 15453 and *P. putida* KT2442 having the rhamnolipid gene cluster demonstrated higher yields of 0.25 and 0.6 g/L, correspondingly, when compared with the parent strain. A similar strategy was adopted by Cha et al. (2008), where the opportunistic pathogen *P. aeruginosa* EMS1 (yielded 5 g/L rhamnolipid) was replaced by *P. putida* by molecular cloning of the *rhlAB* rhamnosyltransferase gene. Apparently, this engineered strain yielded high amount of rhamnolipids by utilizing glucose as a carbon source (Wittgens et al. 2011). However, *P. aeruginosa* PAO1 was not preferred as the appropriate model organism due to its virulence characteristics; it exhibited promising results in rhamnolipid yields. Surprisingly, the yield of 39 g/L after 90 h of cultivation with sunflower oil as substrate and under nitrogen-limiting conditions has been reported by Muller et al. (2011). On the contrary, a yield of 1 g/L of mono-rhamnolipids was obtained by the growth of the nonpathogenic *Pseudomonas chlororaphis* on 2 % glucose (Gunther et al. 2005).

However, traditional engineering by random and targeted genetic alteration, process design and recombinant strategies in rhamnolipid biosynthesis did not succeed. For enhanced process development, there is an urgent need of in-depth information about the rhamnolipid production regulation during bioreactor cultivation to design knowledge-based genetic and process engineering strategies (Müller and Hausmann 2011). No significant differences in the quantity or composition of the rhamnolipid congeners could be produced by manipulating the growth conditions. Sequences for the rhamnolipid genes indicated low levels of strain variation and the majority of polymorphisms did lead to amino acid sequence changes that had no evident phenotypic effect. The expression of the *rhlB* and *rhlC* rhamnosyltransferase genes showed a fixed sequential expression pattern during growth and no significant upregulation could be induced by varying producer strains or growth media. The same study demonstrated that rhamnolipids are highly conserved molecules and their gene expression has a rather stringent control. This leaves little opportunity to manipulate and greatly increase the yield of rhamnolipids from strains of *P. aeruginosa* for biotechnological applications (Perfumo et al. 2013). Increase in yield of natural LPs can be obtained by general engineering approaches of the producer strains. Modifying the gene regulatory regions involved in biosurfactant expression by up-mutations may be considered. Simplified strategy would be to replace weak endogenous promoters with strong promoters that can even be better controlled by stable inducers. The common strategy for molecular engineering approaches is to modify the desired parts of the cloned biosynthetic pathways by standard techniques in *E. coli*, to transfer the modified genetic elements into the producer strain and to obtain stably engineered producer strains by recombination (Koglin et al. 2010). Several authors studied the molecular engineering aspects of production of surfactants (Baltz

2009; Alexander et al. 2010). However, few problems may be associated with the molecular engineering of biosurfactants; in view of that, small change in biosurfactant chemical structure may lead to significant problems for the microbial physiology of producer strains. Even though much knowledge has been gained on biosynthesis of LPs, only few molecules were familiar on their secretion mechanisms and metabolic pathways inside the cell. Compounds with altered bioactivities could turn toxic to the producer or they might impart negative side effects to other cellular processes. Export systems might not efficiently recognize modified LP structures, or they could become overloaded by increased biosurfactant synthesis resulting in intracellular product accumulation. Manipulation of precursor or posttranslational modification pathways could affect other biosynthetic systems with consequences that are difficult to envisage. Co-engineering of associated pathways and enzymes might therefore be indispensable in order to ascertain stable and efficient producer cell lines for modified or newly designed biosurfactants (Koglin et al. 2010).

7.8 Metagenomics for Marine Biosurfactant Discovery

Metagenomics can lead to an invention of novel biosurfactants, tackling issues of low production yields. Metagenomic analyses comprise both sequence-based and function-based strategies. Sequence-based analyses have advanced enormously over the last decade with the progress of next-generation sequencing (NGS) platforms and ensuing scalability, with significant cost reductions. This has enabled the querying of enormous metagenomic sequence datasets to identify protein-coding sequences based on homology to reference sequence data in curated sequence databases (Kanehisa and Goto 2000). Software tools are also now available to query large sequence datasets such as genomes and metagenomes for the presence of gene clusters associated with marine biosurfactant molecules of interest. One such tool, antiSMASH (the antibiotic and secondary metabolite analysis shell), quickly identifies and annotates secondary metabolite gene clusters from genomic sequence data (Medema et al. 2011) and has been used to identify biosurfactant-related genes in the genome of *Serratia marcescens* strain Db11 (Gerc et al.2014). A number of screening methods are available for screening large-insert metagenomic clone libraries for biosurfactant activities. Recent successes include the heterologous production of biosurfactants. Recently, few noteworthy contributions were made in this area with more similar discoveries highlighting the wide scope of these advanced techniques in sustainable production of biosurfactants, cloning of *srwW* gene of *S. marcescens* into *E. coli* using a constitutively expressed promoter (Thies et al. 2014), employing PCR, gene-walking and cloning to express a sophorolipid-related glucosyltransferase gene (*gtf-1*), from *C. bombicola*, in *Saccharomyces cerevisiae* (Solaiman et al. 2014), expression of partial biosurfactant production genes in *S. cerevisiae* (Huang et al. 2014). Nonetheless, the complex regulatory mechanisms governing biosurfactant production may become main difficulty to metagenomic approach of marine biosurfactant synthesis. In view of the fact that production is interlinked to growth through quorum sensing mechanisms and stress or nutrient depletion responses, strategies like addition of precursor molecules to growth media and optimization of bioprocess engineering techniques are highly warranted in the upcoming years. The use of metagenomic DNA from suitable marine environments, proper functional screening and suitable heterologous host systems will collectively aid in the discovery of novel molecules with remarkable surface-active and emulsification properties.

7.9 Modern Approaches for Optimization of Biosurfactant Production

As the process economy plays an imperative role in determining the absolute use of the product, it has become very crucial for researchers to enhance the yield and biosurfactant productivity. The utilization of cheaper substrates has emerged as one of the attractive options, which further justifies the demand of biosurfactant for low-end applications such as enhanced oil recovery, waste water treatment and bioremediation (Das and Mukherjee 2007; Mukherjee et al. 2006; Das et al. 2009a). Further, the adoption of various process intensification strategies (Table 7.3) has resulted in the improved biosurfactant production. Thus, several attempts were made to enhance the product yield and process productivity by modern techniques (Sivapathasekaran and Sen 2013a). Interestingly, Abbasi and Ahmadian (2012) developed a Sugeno-type adaptive neural fuzzy inference (ANFI) modeling technique in order to select the critical medium components for marine biosurfactant production. This model revealed that production process optimization was maximized by involving an artificial neural network and genetic algorithm (ANN-GA)-aided experimental modeling yielding 4.61 ± 0.07 g/L after 22 h of fermentation. In one of the research report, free Fe^{2+} was reported to be sequestered by the marine *Bacillus megaterium* lipopeptide as it was produced. Therefore, Rangarajan et al. (2012) adopted a time-dependent multi-dosing Fe^{2+} feeding strategy for the enhancement in the maximum lipopeptide product yield (YP/S) when compared with glucose mineral salt medium (GMSM) without Fe^{2+} supplementation.

In addition to that, a comparative performance evaluation of batch and semicontinuous reactor operations was demonstrated for the optimized production of a marine microbial surfactant using a Luedeking–Piret model. Interestingly, Sivapathasekaran and Sen (2013a) demonstrated that unsteady state fed-batch (USFB-I) operation showed improvement in the biosurfactant

Table 7.3 Modern approaches for optimization of biosurfactant production in marine microorganisms

Microorganisms	Biosurfactant	Method	Yield	Reference
Bacillus circulans (DMS-2)MTCC 8281	Lipopeptide	CCD, ANN-GA	4.61 ± 0.07 g/L	Sivapathasekaran and Sen (2013b)
Bacillus circulans (DMS-2)MTCC 8281	Lipopeptide	Unsteady state fed-batch strategy	6.2 g/L	Sivapathasekaran and Sen (2013a)
Bacillus megaterium MTCC 8280	Lipopeptide	Bioprocess design	5.34 ± 0.1 g/L	Rangarajan et al. (2015)
Bacillus megaterium	Lipopeptide	Intermittent Fe^{2+} dosing	4.2 ± 0.15 g/L	(Rangarajan et al. 2012)
Bacillus subtilis RB14	Lipopeptide	Heat-induced germination	4 g/L	Rahman et al. (2006)
Bacillus subtilis ATCC 21332	Lipopeptide	Cell immobilization and lipopeptide recovery by pertraction	320 mg/L	Chtioui et al. (2010)
Bacillus subtilis ATCC 21332	Lipopeptide	Bubbleless membrane bioreactor	230 mg/L	Coutte et al. (2010)
Bacillus subtilis GB16	Lipopeptide	Novel oxygenation	–	Lee and Kim (2004)
Bacillus subtilis DSM 21393	Lipopeptide	*In situ* removal by automated surface enrichment	–	Glazyrina et al. (2008)
Bacillus subtilis ATCC 21332	Lipopeptide	Non-foaming modified rotating discs reactor	787 mg/L	Chtioui et al. (2014)

production by about 35 % over batch mode compared to USFB-II. Lee and Kim (2004) developed a novel oxygenation method to achieve significant enhancement in the lipopeptide production by increasing the dissolved oxygen concentration during the microbial production of lipopeptides by *Bacillus* sp. GB16. In another study, Rahman et al. (2006) investigated the production of lipopeptides using induced germination of the spores by heat activation and nutrient supplementation. A new method to remove and separate biosurfactants from complex mixtures by compressing and harvesting the liquid surface layer was developed by Glazyrina et al. (2008). The exploitation of bubbleless bioreactors with aeration by a hollow fiber membrane air–liquid contactor was developed by Coutte et al. (2010) for the production of lipopeptides by *B. subtilis* ATCC 21332. Chtioui et al. (2010) reported the enhanced synthesis of both lipopeptides and especially of the fengycin by the immobilized aerobic cells of *Bacillus subtilis* ATCC 21332. Chtioui et al. (2014) illustrated the production of lipopeptides fengycin and surfactin in rotating disk bioreactor which could be very useful for the production of other molecules using bioprocesses requiring bubbleless oxygen supply. More recently, Rangarajan et al. (2015) elucidated the effect of limiting oxygen and nitrogen sources on lipopeptide selectivity through a bioprocess design for selective enhancement of fengycin production by marine *Bacillus megaterium* resulting in the enhanced lipopeptide production.

The bioremediation of coastal and marine environments is significantly benefited with the use of biosurfactants/bioemulsifiers. Apparently, marine microbes present higher percentage of surfactant production compared to terrestrial species (Maneerat 2000). Since marine source is the critical contaminated site, the microbial population of marine sources may have the inherent capacity to remediate the contaminants at the fastest rate and have robustness in solubilizing and degrading the PAHs (Plante et al. 2008). Spills of oil and hydrocarbons in marine environments stimulate the indigenous community of obligate hydrocarbonoclastic bacteria (OHCB) to thrive, becoming the majority of the total microbial population (Yakimov et al. 2007). The surface-active compounds produced by these organisms mediate the dissolution of hydrocarbons by interacting with the insoluble compounds, reducing the interfacial tension, and making them available to the microbes (Gnanamani et al. 2010). The diverse assortment of hydrocarbons present in crude oil involves resource partitioning by microbial populations, and microbial modification of oil components and the surrounding environment will lead to temporal succession.

Several reports put forth the utilization of marine microbes as potential hydrocarbon and toxic organic pollutant degraders. *Alcanivorax borkumensis* is known for its ability to grow metabolizing only alkanes and producing glycolipid biosurfactants (Yakimov et al. 1998). Melcher et al. (2002) isolated phenanthrene or chrysene strains from San Diego Bay sediments representing the genera of *Vibrio*, *Marinobacter*, *Cycloclasticus*, *Pseudoalteromonas*, *Marinomonas* and *Halomonas*. Coelho et al. (2003) investigated quinoline-degrading marine bacterium *Pseudomonas* sp. strain GU 104 isolated from coastal waters of Goa, India. Another interesting study by Thavasi et al. (2006) described the hydrocarbon-degrading potential of a marine bacterium *Azotobacter chroococcum* isolated from Tuticorin harbor where degradation of 58 % crude oil was evidenced in experimental studies. In cold marine environments, the obligate alkane-degrading psychrophile, *Oleispira*, is commonly associated with oil spills (Coulon et al. 2007). Edlund and Jansson (2008) revealed a remarkable diversity of putative PAH degraders belonging to the genera *Exiguobacterium*, *Shewanella*, *Methylomonas*, *Pseudomonas*, *Bacteroides*, as well as *Deltaproteobacteria* and *Gammaproteobacteria* from marine sediments. The dominant benzo[a]pyrene-degrading bacteria from a marine enrichment were isolated and faster degradation was seen when the three strains (*Ochrobactrum*, *Stenotrophomonas* and *Pseudomonas* spp.) were combined than when tested individually (Luo et al. 2009). New genera of obligate alkane degraders, *Oleibacter* sp., were discovered by Teramoto et al. (2009). Similarly, Kostka et al. (2011) isolated potential oil degraders that belong to *Gammaproteobacteria*, including representatives of genera with known oil degraders (*Alcanivorax*, *Marinobacter*, *Pseudomonas* and Acinetobacter), from Gulf of Mexico beach sands impacted by the deepwater horizon oil spill. Interestingly, Naether et al. (2013) explained that the strategy of *A. borkumensis* to get access to hydrocarbon involves both the solubilization of the substrate within micelles and the subsequent direct uptake of the finely dispersed droplets and also stated cis–trans isomerization of unsaturated fatty acids was proven as one adaptive mechanism of *Alcanivorax borkumensis* SK2 to the toxic organic solvents like chlorophenols and alkanols. Microbial surfactant-mediated degradation of anthracene by marine *Bacillus licheniformis* MTCC 5514 was reported by Swaathy et al. (2014). Furthermore, a particular strain isolated from sediments, *Paracoccus marcusii*, was proposed as an optimal choice for bioremediation purposes as its biomass remains trapped in the hydrocarbon phase and thus remains undisturbed from potential dilution effects by sea currents (Antoniou et al. 2015).

7.10 Future Prospects in Exploitation of Marine Biosurfactants

Large numbers of microbial strains producing biosurfactants have already been isolated from diverse environments, including marine environments (Kiran et al. 2010a, b, c, d). A range of major international companies is actively investigating potential applications for the biosurfactants with interests in a wide range of market niches and potential applications. The following are priorities for further investigation of microbial biosurfactants, principally glycolipids, from marine environments:

1. Limited searches have been made for new producer organisms, since many strains are already known but not fully investigated with different functionalities applicable at low-temperature environments.
2. Investigating microorganisms capable of producing biosurfactants as tools for bioremediation of marine environments.
3. Detailed investigations are warranted on the structure and functionality of marine microbial biosurfactants to identify those with unique characteristics.
4. Studies on novel methodologies are essential in biosurfactant production conditions for optimized yield, downstream processing and raw material consumption.
5. Investigations on assessment of the genetic and metabolic control of biosurfactant production by target marine microorganisms are highly indispensable.
6. Laboratory and field investigations are pertinent for the potential use of marine microbial biosurfactant producers as an augmentation tool for bioremediation.

7.11 Concluding Remarks

Fundamental studies on marine microbial biosurfactants are relatively few in number compared to other polysaccharides. There are indeed many gaps in our knowledge on basic questions such as optimized production, strain engineering, metagenomics, regulatory and metabolic network of many biosurfactants biosynthesis. In-depth information about the biosurfactant production regulation during bioreactor cultivation to design knowledge-based genetic and process engineering strategies is also warranted. The data currently available indicate that marine microbial biosurfactants have considerable biotechnological importance, but further studies are necessary to evaluate the feasibility of their practical application. This may limit the practical application of marine microbial biosurfactants in several prospective applications. Even though diverse microbial biosurfactants have been studied in the last years, only little information is available regarding their hyperproduction and profitable yields. Improve the fermentation processes of marine microbial biosurfactants, it is possible to operate at different levels, by searching for the ideal nutritional conditions resulting in the enhanced yield of biosurfactants with more suitable biotechnological proprieties, as well as to operate at the genomic level by inserting the genes that are directly involved in carbohydrate metabolism. Metagenomic approaches were one of the fast-emerging trends in biosurfactant production. The above strategies, possibly when applied together, can lead to an interdisciplinary approach based on the analysis of complex system biological interactions and supported by statistical and genetic algorithm tools and further predictive studies for the sustainable production of marine biosurfactants.

Acknowledgments The authors gratefully acknowledge University Grants Commission, New Delhi, Govt. of India for financial support as SRF under the Grant No: F.16-1723(SC)/2010(SA-III).

References

Abbasi AA, Ahmadian MT (2012) Adaptive Neural Fuzzy Inference (ANFI) modelling technique for production of marine biosurfactant. In: ASME 2012 International design engineering technical conferences and computers and information in engineering conference (pp 47–52). American Society of Mechanical Engineers

Abouseoud M, Maachi R, Amrane A et al (2008) Evaluation of different carbon and nitrogen sources in production of biosurfactant by *Pseudomonas fluorescens*. Desalination 223:143–151

Abu-Ruwaida AS, Banat IM, Hadirto S et al (1991) Isolation of biosurfactant producing bacteria-product characterisation and evaluation. Acta Biotechnol 11:315–324

Al-Araji L, Rahman R, Basri M et al (2007) Optimisation of rhamnolipids produced by *Pseudomonas aeruginosa* 181 using response surface modelling. Ann Microbiol 57(4):571–575

Alexander DC, Rock J, He X et al (2010) Development of a genetic system for combinatorial biosynthesis of lipopeptides in *Streptomyces fradiae* and heterologous expression of the A54145 biosynthesis gene cluster. Appl Environ Microbiol 76(20):6877–6887

Al-Nahas MO, Darwish MM, Ali AE et al (2011) Characterization of an exopolysaccharide-producing marine bacterium, isolate *Pseudoalteromonas* sp. AM. Afr J Microbiol Res 5:3823–3831

Amaral PFF, Da Silva JM, Lehocky M et al (2006) Production and characterization of a bioemulsifier from *Yarrowia lipolytica*. Process Biochem 41(8):1894–1898

Amir-Heidari B, Thirlway J, Micklefield J (2008) Auxotrophic-precursor directed biosynthesis of nonribosomal lipopeptides with modified tryptophan residues. Org Biomol Chem 6:975–978

Antoniou E, Fodelianakis S, Korkakaki E et al (2015) Biosurfactant production from marine hydrocarbon-degrading consortia and pure bacterial strains using crude oil as carbon source. Front Microbiol 6:274

Baltz RH (2006) Molecular engineering approaches to peptide, polyketide and other antibiotics. Nat Biotechnol 24:1533–1540

Baltz RH (2009) Biosynthesis and genetic engineering of lipopeptides in *Streptomyces roseosporus*. Methods Enzymol 458:511–531

Banat IM (1995a) Biosurfactants characterization and use in pollution removal: state of the art. A review. Acta Biotechnol 15:251–267

Banat IM (1995b) Biosurfactants production and use in microbial enhanced oil recovery and pollution remediation: a review. Bioresour Technol 51:1–12

Banat IM, Franzetti A, Gandolfi I et al (2010) Microbial biosurfactants production, applications and future potential. Appl Microbiol Biotechnol 87:427–444

Bertrand f-c, Larsen H (1989) In: Bianchi M et al (eds) Micro- organisms dans les ecosystems oceaniques. La "bacterie marine": mythe ou realite. Masson, Paris, pp 3–25

Besson F, Michel G (1992) Biosynthesis of iturin and surfactin by *Bacillus subtilis*. Evidence for amino acid activating enzymes. Biotechnol Lett 14:1013–1018

Cha M, Lee N, Kim M et al (2008) Heterologous production of *Pseudomonas aeruginosa* EMS1 biosurfactant in *Pseudomonas putida*. Bioresour Technol 99(7):2192–2199

Chayabutra C, Ju LK (2001) Polyhydroxyalkanoic acids and rhamnolipids are synthesized sequentially in hexadecane fermentation by *Pseudomonas aeruginosa* ATCC 10145. Biotechnol Prog 17:419–423

Chen SY, Lu WB, Wei YH et al (2007) Improved production of biosurfactant with newly isolated *Pseudomonas aeruginosa* S2. Biotechnol Prog 23:661–666

Chi Z, Fang Y (2005) Exopolysaccharides from marine bacteria. J Ocean Univ China 4:67–74

Chi Z, Su CD, Lu WD (2007) A new exopolysaccharide produced by marine *Cyanothece* sp. 113. Bioresour Technol 98(6):1329–1332

Choi D, Piao YL, Shin W et al (2009) Production of oligosaccharide from alginate using *Pseudoalteromonas agarovorans*. Appl Biochem Biotechnol 159:438–452

Chtioui O, Dimitrov K, Gancel F et al (2010) Biosurfactants production by immobilized cells of

Bacillus subtilis ATCC 21332 and their recovery by pertraction. Process Biochem 45(11):1795–1799

Chtioui O, Dimitrov K, Gancel F et al (2014) Selective fengycin production in a modified rotating discs bioreactor. Bioprocess Biosyst Eng 37(2):107–114

Coelho J, Rivonkar CU, Bhavesh NS et al (2003) Biosurfactants production by the quinoline degrading marine bacterium *Pseudomonas* sp. strain GU 104, and its effect on the metabolism of green mussel *Perna viridis* L. Indian J Mar Sci 32(3):202–207

Cooper DG, Goldenberg BG (1987) Surface-active agents from two *Bacillus* species. Appl Environ Microbiol 53(2):224–229

Coulon F, McKew BA, Osborn AM et al (2007) Effects of temperature and biostimulation on oil-degrading microbial communities in temperate estuarine waters. Environ Microbiol 9:177–186

Coutte F, Lecouturier D, Yahia S et al (2010) Production of surfactin and fengycin by *Bacillus subtilis* in a bubbleless membrane bioreactor. Appl Microbiol Biotechnol 87(2):499–507

Das K, Mukherjee A (2007) Comparison of lipopeptide biosurfactants production by *Bacillus subtilis* strains in submerged and solid state fermentation systems using a cheap carbon source: some industrial applications of biosurfactants. Process Biochem 42:1191–1199

Das P, Mukherjee S, Sen R (2009a) Biosurfactant of marine origin exhibiting heavy metal remediation properties. Bioresour Technol 100(20):4887–4890

Das P, Mukherjee S, Sen R (2009b) Substrate dependent production of extracellular biosurfactant by a marine bacterium. Bioresour Technol 100(2):1015–1019

De Sousa T, Bhosle S (2012) Isolation and characterization of a lipopeptide bioemulsifier produced by *Pseudomonas nitroreducens* TSB. MJ10 isolated from a mangrove ecosystem. Bioresour Technol 123:256–262

Desai JD, Banat IM (1997) Microbial production of surfactants and their commercial potential. Microbiol Mol Biol Rev 61:47–64

Dhasayan A, Selvin J, Kiran S (2014) Biosurfactant production from marine bacteria associated with sponge *Callyspongia diffusa*. Biotech 3:1–12

Donio MBS, Ronica FA, Viji VT et al (2013) *Halomonas* sp. BS4, a biosurfactant producing halophilic bacterium isolated from solar salt works in India and their biomedical importance. Springer plus 2(1):149

Edlund A, Jansson JK (2008) Use of bromodeoxyuridine immunocapture to identify psychrotolerant phenanthrene-degrading bacteria in phenanthrene-enriched polluted Baltic Sea sediments. FEMS Microbiol Ecol 65:513–525

El-Sersy NA (2012) Plackett-Burman design to optimize biosurfactant production by marine *Bacillus subtilis* N10. Roman Biotechnol Lett 17(2):7049–7064

Farres J, Caminal G, Lopez-Santin J (1997) Influence of phosphate on rhamnose-containing exopolysaccharide rheology and production by *Klebsiella* I-174. Appl Microbiol Biotechnol 48:522–527

Fernandez-Linares L, Acquaviva M, Bertrand JC et al (1996) Effect of sodium chloride concentration on growth and degradation of eicosane by the marine halotolerant bacterium *Marinobacter hydrocarbonoclasticus*. Syst Appl Microbiol 19(1):113–121

Fox SL, Bala GA (2000) Production of surfactant from *Bacillus subtilis* ATCC 21332 using potato substrates. Bioresour Technol 75:235–240

Gandhimathi R, Kiran GS, Hema TA et al (2009) Production and characterization of lipopeptide biosurfactant by a sponge associated marine actinomycetes *Nocardiopsis alba* MSA10. Bioproc Biosyst Eng 6:825–835

Gerc AJ, Stanley-Wall NR, Coulthurst SJ (2014) Role of the phosphopantetheinyl transferase enzyme PswP, in the biosynthesis of antimicrobial secondary metabolites by *Serratia marcescens* Db10. Microbiology 160:1609–1617

Geys R, Soetaert W, Bogaert V (2014) Biotechnological opportunities in biosurfactant production. Curr Opin Biotechnol 30:66–72

Glazyrina J, Junne S, Thiesen P et al (2008) *In situ* removal and purification of biosurfactants by automated surface enrichment. Appl Microbiol Biotechnol 81(1):23–31

Gnanamani A, Kavitha V, Radhakrishnan N et al (2010) Microbial products (biosurfactant and extracellular chromate reductase of marine microorganisms are the potential agents reduce the oxidative stress induced by toxic heavy metals. Coll Surf B: Biointer 79:334–339

Gomathy C, Senthilkumar R (2013) Production of Rhamnolipid Biosurfactant from a Marine *Pseudomonas aeruginosa*. Int J Res Environ Sci Technol 3(3):86–91

Gorret N, Maubois JL, Engasser JM et al (2001) Study of the effects of temperature, pH and yeast extract on growth and exopolysaccharide production by *Propionibacterium acidi-propionici* on milk microfiltrate using a response surface methodology. J Appl Microbiol 90:788–796

Goutx M, Mutaftshiev S, Bertrand JC (1987) Lipid and exopolysaccharide production during hydrocarbon growth of a marine bacterium from the sea surface. Mar Ecol Prog Ser Oldendorf 40(3):259–265

Guerra-Santos L, Käppeli O, Fiechter A (1984) *Pseudomonas aeruginosa* biosurfactant production in continuous culture with glucose as carbon source. Appl Environ Microbiol 48(2):301–305

Gugliandolo C, Maugeri TL (1998) Temporal variations in heterotrophic mesophilic bacteria from a marine shallow hydrothermal vent off the Island of Vulcano (Aeolian Islands, Italy). Microb Ecol 36(1):13–22

Gunther NW, Nunez A, Fett W et al (2005) Production of rhamnolipids by *Pseudomonas chlororaphis*, a non-pathogenic bacterium. Appl Environ Microbiol 71(5):2288–2293

Haddad NIA, Wang Ji MB (2009) Identification of a biosurfactant producing strain: *Bacillus subtilis* HOB2. Prot Pept Lett 16:7–13

He F, Zhao D, Liu J et al (2006) Stabilization of Fe-Pd nanoparticles with sodium carboxymethyl cellulose for enhanced transport and dechlorination of trichloroethylene in soil and groundwater. Ind Eng Chem Res 46:29–34

Henkel M, Sydatk C, Hausmann R (2014) The prospects for the production of Rhamnolipids on renewable resources. Biosurfactants: production and utilization-processes technologies, and economics, 1st edn. CRC Press, Florida, pp 83–99

Huang FC, Peter A, Schwab W (2014) Expression and characterization of CYP52 genes involved in the biosynthesis of sophorolipid and alkane metabolism from *Starmerella bombicola*. Appl Environ Microbiol 80:766–776

Ibacache-Quiroga C, Ojeda J, Espinoza-Vergara G, Olivero P, Cuellar M, Dinamarca MA (2013) The hydrocarbon-degrading marine bacterium Cobetia sp. strain MM1IDA2H-1 produces a biosurfactant that interferes with quorum sensing of fish pathogens by signal hijacking. Microbial Biotechnol 6(4):394–405

Ilori MO, Amobi CJ, Odocha AC (2005) Factors affecting the production of oil degrading *Aeromonas* sp isolated from a typical environment. Chemosphere 61:985–992

Jain RM, Mody K, Joshi N et al (2013a) Production and structural characterization of biosurfactant produced by an alkaliphilic bacterium, *Klebsiella* sp.: evaluation of different carbon sources. Coll Surf B: Biointer 108:199–204

Jain RM, Mody K, Joshi N et al (2013b) Effect of unconventional carbon sources on biosurfactant production and its application in bioremediation. Int J Biol Macromol 62:52–58

Joshi S, Bharucha C, Desai AJ (2008) Production of biosurfactant and antifungal compound by fermented food isolate *Bacillus subtilis* 20B. Bioresour Technol 99:4603–4608

Kanehisa M, Goto S (2000) KEGG: kyoto encyclopedia of genes and genomes. Nucleic Acids Res 28:27–30

Kelecom A (2002) Secondary metabolites from marine microorganisms. Ann Braz Acad Sci 74:151–170

Khopade A, Ren B, Liu XY et al (2012a) Production and characterization of biosurfactant from marine *Streptomyces* species B3. J Coll Inter Sci 367(1):311–318

Khopade A, Biao R, Liu X et al (2012b) Production and stability studies of the biosurfactant isolated from marine *Nocardiopsis* sp. B4. Desalination 285:198–204

Kim HS, Jeon JW, Kim BH et al (2006) Extracellular production of a glycolipid biosurfactant, mannosylerythritol lipid, by *Candida* sp. SY16 using fed-batch fermentation. Appl Microbiol Biotechnol 70:391–396

Kiran GS, Hema TA, Gandhimathi R et al (2009) Optimization and production of a biosurfactant from the sponge-associated marine fungus *Aspergillus ustus* MSF3. Coll Surf B: Biointer 73(2):250–256

Kiran GS, Sabu A, Selvin J (2010a) Synthesis of silver nanoparticles by glycolipids biosurfactant produced from marine *Brevibacterium casei* MSA19. J Biotechnol 148:221–225

Kiran GS, Anto Thomas T, Selvin J et al (2010b) Optimization and characterization of a new lipopeptide biosurfactant produced by marine *Brevibacterium aureum* MSA13 in solid state culture. Bioresour Technol 101:2389–2396

Kiran GS, Thomas TA, Selvin J (2010c) Production of a new glycolipid biosurfactant from marine *Nocardiopsis lucentensis* MSA04 in solid-state cultivation. Coll Surf B: Biointer 78(1):8–16

Kiran GS, Sabarathnam B, Selvin J (2010d) Biofilm disruption potential of a glycolipid biosurfactant from marine *Brevibacterium casei*. FEMS Immun Med Microbiol 59(3):432–438

Kiran GS, Nishanth LA, Priyadharshini S et al (2014) Effect of Fe nanoparticle on growth and glycolipid biosurfactant production under solid state culture by marine *Nocardiopsis* sp. MSA13A. BMC Biotechnol 14(1):48

Ko SH, Lee HS, Park SH et al (2000) Optimal conditions for the production of exopolysaccharide by marine microorganism *Hahella chejuensis*. Biotechnol Bioprocess Eng 5(3):181–185

Koch AK, Kappeli O, Fiechter A et al (1991) Hydrocarbon assimilation and biosurfactant production in *Pseudomonas aeruginosa* mutants. J Bacteriol 173:4212–4219

Koglin A, Doetsch V, Bernhard F (2010) Molecular engineering aspects for the production of new and modified biosurfactants. In: Biosurfactants. Springer, New York, pp 158–169

Kokare CR, Kadam SS, Mahadik KR et al (2007) Studies on bioemulsifier production from marine *Streptomyces* sp. S1. Indian. J Biotechnol 6(1):78

Kostka JE, Prakash O, Overholt WA et al (2011) Hydrocarbon-degrading bacteria and the bacterial community response in Gulf of Mexico beach sands impacted by the Deepwater Horizon oil spill. Appl Environ Microbiol 77(22):7962–7974

Lee BS, Kim EK (2004) Lipopeptide production from *Bacillus* sp. GB16 using a novel oxygenation method. Enzymol Microb Technol 35(6):639–647

Lee HK, Chun J, Moon EJ et al (2001) *Hahella chejuensis* gen. nov., sp. nov., an extracellular-polysaccharide-producing marine bacterium. Int J Syst Evol Microbiol 51:661–666

Li C, Bai J, Cai Z et al (2002) Optimization of a cultural medium for bacteriocin production by *Lactococcus lactis* using response surface methodology. J Biotechnol 93:27–34

Linhardt RJ, Bakhit R, Daniels L (1989) Microbially produced rhamnolipid as a source of rhamnose. Biotech Bioeng 33:365–368

Liu X, Ren B, Chen M et al (2010) Production and characterization of a group of bioemulsifiers from the marine *Bacillus velezensis* strain H3. Appl Microbiol Biotechnol 87(5):1881–1893

Liu SB, Qiao LP, He HL et al (2011) Optimization of fermentation conditions and rheological properties of exopolysaccharide produced by deep-sea bacterium *Zunongwangia profunda* SM-A87. Plos One 6(11):e26825

Lotfy WA, Ghanem KM, El-Helow ER (2007) Citric acid production by a novel *Aspergillus niger* isolate: I. Mutagenesis and cost reduction studies. Bioresour Technol 98:3464–3469

Luo YR, Tian Y, Huang X et al (2009) Analysis of community structure of a microbial consortium capable of degrading benzo[a]pyrene by DGGE. Mar Pollut Bull 58:1159–1163

Mabrouk ME, Youssif EM, Sabry SA (2014) Biosurfactant production by a newly isolated soft coral-associated marine *Bacillus* sp. E34: statistical optimization and characterization. Life Sci J 11:756–768

Makkar RS, Cameotra SS (1999) Biosurfactant production by microorganisms on unconventional carbon sources-a review. J Surfactant Deterg 2:237–241

Malavenda R, Rizzo C, Lo Giudice A, Michaud L, De Domenico M, Bruni V (2010) Screening for biosurfactant production by Antarctic Marine bacteria. Biol Mar Mediterr 17(1):388–389

Mancuso Nichols CA, Garon S, Bowman JP, Raguénès G, Guezennec J (2004) Production of exopolysaccharides by Antarctic marine bacterial isolates. J Appl Microbiol 96(5):1057–1066

Mancuso-Nichols CA, Guezennec J, Bowman JP (2005) Bacterial exopolysaccharides from extreme marine environments with special consideration of the southern ocean, sea ice, and deep-sea hydrothermal vents: a review. Mar Biotechnol 7:253–271

Maneerat S (2000) Biosurfactants from marine microorganisms. J Sci Technol 27:1263–1272

Maneerat S, Bamba T, Harada K et al (2006) A novel crude oil emulsifier excreted in the culture supernatant of a marine bacterium, *Myroides* sp. strain SM1. Appl Microbiol Biotechnol 70(2):254–259

Manivasagan P, Sivasankar P, Venkatesan J et al (2014) Optimization, production and characterization of glycolipid biosurfactant from the marine *Actinobacterium*, *Streptomyces* sp. MAB36. Bioprocess Biosyst Eng 37(5):783–797

Marchant R, Banat IM (2012a) Biosurfactants: a sustainable replacement for chemical surfactants? Biotechnol Lett 34:1597–1605

Marchant R, Banat IM (2012b) Microbial biosurfactants: challenges and opportunities for future exploitation. Trends Biotechnol 30:558–565

Mastorocostas PA (2004) Resilient back propagation learning algorithm for recurrent fuzzy neural networks. Electron Lett 40:1

Mata-Sandoval JC, Karns J, Torrents A (2001) Effect of nutritional and environmental conditions on the production and composition of rhamnolipids by *P. aeruginosa* UG2. Microbiol Res 155(4):249–256

Maugeri TL, Gugliandolo C, Caccamo D, Panico A, Lama L, Gambacorta A, Nicolaus B (2002) A halophilic thermotolerant Bacillus isolated from a marine hot spring able to produce a new exopolysaccharide. Biotechnol Lett 24(7):515–519

Medema MH, Blin K, Cimermancic P (2011) antiSMASH: rapid identification, annotation and analysis of secondary metabolite biosynthesis gene clusters in bacterial and fungal genome sequences. Nucl Acids Res 39(suppl 2):W339–W346

Melcher RJ, Apitz SE, Hemmingsen BB (2002) Impact of irradiation and polycyclic aromatic hydrocarbon spiking on microbial populations in marine sediment for future aging and biodegradability studies. Appl Environ Microbiol 68:2858–2868

Mercade ME, Manresa MA (1994) The use of agroindustrial by products for biosurfactant production. J Am Oil Chem Soc 71:61–64

Montgomery DC (1997) Response surface methods and other approaches to process optimization. In: Montgomery DC (ed) Design and analysis of experiments. Wiley, New York, pp 427–510

Moriello VS, Lama L, Poli A et al (2003) Production of exopolysaccharides from a thermophilic microorganism isolated from a marine hot spring in flegrean areas. J Ind Microbiol Biotechnol 30(2):95–101

Morin A (1998) Screening of polysaccharide-producing microorganisms, factors influencing the production and recovery of microbial polysaccharides. In: Dumitriu S (ed) Polysaccharides-structural diversity and functional versatility. Marcel Dekker Inc. Publication, New York, pp 275–296

Mukherjee S, Das P, Sen R (2006) Towards commercial production of microbial surfactants. Trends Biotechnol 24:509–515

Mukherjee S, Das P, Sivapathasekaran C et al (2008) Enhanced production of biosurfactant by a marine bacterium on statistical screening of nutritional parameters. Biochem Eng J 42(3):254–260

Mukherjee S, Das P, Sivapathasekaran C et al (2009) Antimicrobial biosurfactants from marine *Bacillus circulans*: extracellular synthesis and purification. Lett Appl Microbiol 48:281–288

Müller MM, Hausmann R (2011) Regulatory and metabolic network of rhamnolipid biosynthesis: traditional and advanced engineering towards biotechnological production. Appl Microbiol Biotechnol 91(2):251–264

Müller MM, Hörmann B, Kugel M et al (2011) Evaluation of rhamnolipid production capacity of *Pseudomonas aeruginosa* PAO1 in comparison to the rhamnolipid over-producer strains DSM 7108 and DSM 2874. Appl Microbiol Biotechnol 89(3):585–592

Mulligan CN, Gibbs BF (1989) Correlation of nitrogen metabolism with biosurfactant production by *Pseudomonas aeruginosa*. Appl Environ Microbiol 55(11):3016–3019

Mulligan CN, Gibbs BF (1993) Factors influencing the economics of biosurfactants. In: Kosaric N (ed) Biosurfactants, production, properties, applications. Marcel Dekker, New York, pp 329–337

Naether DJ, Slawtschew S, Stasik S et al (2013) Adaptation of the hydrocarbonoclastic bacterium *Alcanivorax borkumensis* SK2 to alkanes and toxic organic compounds: a physiological and transcriptomic approach. Appl Environ Microbiol 79(14):4282–4293

Nerurkar AS, Hingurao KS, Suthar HG (2009) Bioemulsifiers from marine microorganisms. J Sci Ind Res 68(4):273

Nichols CM, Garon lardière S, Bowman JP et al (2005) Chemical characterization of exopolysaccharides from Antarctic marine bacteria. Microb Ecol 49:578–589

Ochsner UA, Reiser J (1995) Autoinducer-mediated regulation of rhamnolipid biosurfactant synthesis in *Pseudomonas aeruginosa*. Proc Natl Acad Sci 92(14):6424–6428

Pagilla KR, Sood A, Kim H (2002) *Gordonia* (*norcardia*) *amarae* foaming due to biosurfactant production. Water Sci Tech 46:519–524

Pal MP, Vaidya BK, Desai KM et al (2009) Medium optimization for biosurfactant production by *Rhodococcus erythropolis* MTCC 2794: artificial intelligence verses a statistical approach. J Ind Microbiol Biotechnol 36:747–756

Passeri A, Lang S, Wagner F (1991) Marine biosurfactants, II. Production and characterization of an anionic trehalose tetraester from the marine bacterium *Arthrobacter* sp. EK1. Z Naturforsch C J Biosci 46:204–209

Peng F, Wang Y, Sun F et al (2008) A novel lipopeptide produced by a Pacific Ocean deep-sea bacterium, *Rhodococcus* sp. TW53. J Appl Microbiol 105(3):698–705

Pepi M, Cesàro A, Luit G et al (2005) An Antarctic psychrotrophic bacterium *Halomonas* sp. ANT-3b, growing on n-hexadecane, produces a new emulsifying glycolipid. FEMS Microbiol Ecol 53:157–166

Perfumo A, Rudden M, Smyth TJ et al (2013) Rhamnolipids are conserved biosurfactants molecules: implications for their biotechnological potential. Appl Microbiol Biotechnol 97(16):7297–7306

Plackett RL, Burman JP (1944) The design of optimum multifactorial experiments. Biometrika 33(4):305–325

Plante CJ, Coe KM, Plante RG (2008) Isolation of surfactant-resistant bacteria from natural, surfactant-rich marine habitats. Appl Environ Microbiol 74:5093–5099

Poli A, Moriello VS, Esposito E et al (2004) Exopolysaccharide production by a new *Halomonas* strain CRSS isolated from saline lake Cape Russell in Antarctica growing on complex and defined media. Biotechnol Lett 26(21):1635–1638

Poli A, Anzelmo G, Nicolaus B (2010) Bacterial exopolysaccharides from extreme marine habitats: production, characterization and biological activities. Mar Drugs 8:1779–1802

Powell A, Borg M, Amir-Heidari B et al (2007) Engineered biosynthesis of nonribosomal lipopeptides with modified fatty acid side chains. J Am Chem Soc 129:15182–15191

Prieto LM, Michelon M, Burkert JFM et al (2008) The production of rhamnolipid by a *Pseudomonas aeruginosa* strain isolated from a southern coastal zone in Brazil. Chemosphere 71:1781–1785

Pryor SW, Gibson DM, Hay AG et al (2007) Optimization of spore and antifungal lipopeptide production during the solid-state fermentation of *Bacillus subtilis*. Appl Biochem Biotechnol 143:63–79

Qiao N, Shao Z (2010) Isolation and characterization of a novel biosurfactant produced by hydrocarbon-degrading bacterium *Alcanivorax dieselolei* B-5. J Appl Microbiol 108:1207–1216

Qin G, Zhu L, Chen X et al (2007) Structural characterization and ecological roles of a novel exopolysaccharide from the deep-sea psychrotolerant bacterium *Pseudoalteromonas* sp. SM9913. Microbiology 153(5):1566–1572

Quesada E, Béjar V, Del Moral A et al (2004) Moderately halophilic exopolysaccharide-producing bacteria. In: Ventosa A (ed) Halophilic microorganisms. Springer, Heidelberg, pp 297–314

Radchenkova N, Vassilev S, Panchev I et al (2013) Production and properties of two novel exopolysaccharides synthesized by a thermophilic bacterium *Aeribacillus pallidus* 418. Appl Biochem Biotechnol 171(1):31–43

Rahman MS, Ano T, Shoda M (2006) Second stage production of iturin A by induced germination of *Bacillus subtilis* RB14. J Biotechnol 125(4):513–515

Ramaiah N (2005) Facets and opportunities. In: Ramaiah N (ed) Marine microbiology. National Institute of Oceanography, India, pp 1–6

Rangarajan V, Dhanarajan G, Kumar R et al (2012) Time-dependent dosing of Fe^{2+} for improved lipopeptide production by marine *Bacillus megaterium*. J Chem Technol Biotechnol 87(12):1661–1669

Rangarajan V, Dhanarajan G, Sen R (2015) Bioprocess design for selective enhancement of fengycin production by a marine isolate *Bacillus megaterium*. Biochem Eng J 99:147–155

Raza ZA, Rehman A, Khan MS et al (2007) Improved production of biosurfactant by a *Pseudomonas aeruginosa* mutant using vegetable oil refinery wastes. Biodegradation 18:115–121

Robert M, Mercade ME, Bosch MP et al (1989) Effect of the carbon source on biosurfactant production by *Pseudomonas aeruginosa* 44TI. Biotechnol Lett 11:871–874

Rodrigues L, Banat IM, Teixeira G et al (2006a) Biosurfactants: potential applications in medicine. J Antimicrob Chemother 57:609–618

Rodrigues L, van der Mei HC, Banat IM et al (2006b) Inhibition of microbial adhesion to silicone rubber treated with biosurfactant from *Streptococcus thermophilus* A. FEMS Immun Med Microbiol 46:107–112

Rodrigues L, Teixeira J, Oliveira R et al (2006c) Response surface optimization of the medium components for the production of biosurfactants by probiotic bacteria. Process Biochem 41:1–10

Roongsawang N, Thaniyavarn J, Thaniyavarn S et al (2002) Isolation and characterization of a halotolerant *Bacillus subtilis* BBK-1 which produces three kinds of lipopeptides: bacillomycin L, plipastatin, and surfactin. Extremophiles 6(6):499–506

Rosenberg E, Perry A, Gibson DT et al (1979) Emulsifier of *Arthrobacter* RAG-1: specificity of hydrocarbon substrate. Appl Environ Microbiol 37:409–413

Saimmai A, Sobhon V, Maneerat S (2011) Molasses as a whole medium for biosurfactants production by *Bacillus* strains and their application. Appl Biochem Biotechnol 165(1):315–335

Saimmai A, Rukadee O, Onlamool T et al (2012a) Isolation and functional characterization of a biosurfactant produced by a new and promising strain of *Oleomonas sagaranensis* AT18. World J Microbiol Biotechnol 28(10):2973–2986

Saimmai A, Sobhon V, Maneerat S (2012b) Production of biosurfactant from a new and promising strain of *Leucobacter komagatae* 183. Ann Microbiol 62(1):391–402

Saimmai A, Onlamool T, Sobhon V et al (2013a) An efficient biosurfactant-producing bacterium *Selenomonas ruminantium* CT2, isolated from mangrove sediment in south of Thailand. World J Microbiol Biotechnol 29(1):87–102

Saimmai A, Udomsilp S, Maneerat S (2013b) Production and characterization of biosurfactant from marine bacterium *Inquilinus limosus* KB3 grown on low-cost raw materials. Ann Microbiol 63(4):1327–1339

Sakalle K, Rajkumar S (2009) Isolation of crude oil degrading marine bacteria and assessment for biosurfactant production. Int J Microbiol 7:2

Samain E, Miles M, Bozzi L et al (1997) Simultaneous production of two different gel-forming exopolysaccharides by an *Alteromonas* strain originating from deep sea hydrothermal vents. Carbohydr Polym 34(4):235–241

Schulz D, Passeri A, Schmidt M et al (1991) Marine biosurfactants, I. Screening for biosurfactants among crude oil degrading marine microorganisms from the North Sea. Z Naturforsch C J Biosci 46:197–203

Sen R, Swaminathan T (1997) Application of response surface methodology to evaluate the optimum environmental conditions for the enhanced production of surfactin. Appl Microbiol Biotechnol 47:58–363

Sheppard JD, Mulligan CN (1987) The production of surfactin by *Bacillus subtilis* grown on peat hydrolysate. Appl Microbiol Biotechnol 27:110–116

Shubhrasekhar C, Supriya M, Karthik L et al (2013) Isolation, characterization and application of biosurfactant produced by marine Actinobacteria isolated from saltpan soil from costal area of Andhra Pradesh, India. Res J Biotechnol 8:1

Silvi S, Barghini P, Aquilanti A et al (2013) Physiologic and metabolic characterization of a new marine isolate (BM39) of *Pantoea* sp. producing high levels of exopolysaccharide. Microb Cell Factories 12(1):10

Singh SP, Bharali P, Konwar BK (2013) Optimization of nutrient requirements and culture conditions for the production of rhamnolipid from *Pseudomonas aeruginosa* (MTCC 7815) using *Mesua ferrea* seed oil. Indian J Microbiol 53(4):467–476

Sivapathasekaran C, Sen R (2013a) Performance evaluation of batch and unsteady state fed-batch reactor operations for the production of a marine microbial surfactant. J Chem Technol Biotechnol 88(4):719–726

Sivapathasekaran C, Sen R (2013b) Performance evaluation of an ANN–GA aided experimental modeling and optimization procedure for enhanced synthesis of marine biosurfactant in a stirred tank reactor. J Chem Technol Biotechnol 88(5):794–799

Sivapathasekaran C, Mukherjee S, Ray A et al (2010a) Artificial neural network modeling and genetic algorithm based medium optimization for the improved production of marine biosurfactant. Bioresour Technol 101:2884–2887

Sivapathasekaran C, Mukherjee S, Sen R (2010b) Optimization of a marine medium for augmented biosurfactant production. Int J Chem React Eng 8(1). doi:10.2202/1542-6580.2231

Solaiman DKY, Liu Y, Moreau RA et al (2014) Cloning, characterization and heterologous expression of a novel glucosyltransferase gene from a sophorolipid-producing Candida bombicola. Gene 540:45–53

Su C, Chi Z, Lu W (2007) Optimization of medium and cultivation conditions for enhanced exopolysaccharide yield by marine *Cyanothece* sp. 113. Chin J Oceanol Limnol 25:411–417

Swaathy S, Kavitha V, Pravin AS et al (2014) Microbial surfactant mediated degradation of anthracene in aqueous phase by marine *Bacillus licheniformis* MTCC 5514. Biotechnol Rep 4:161–170

Teramoto M, Suzuki M, Okazaki F et al (2009) Oceanobacter related bacteria are important for the degradation of petroleum aliphatic hydrocarbons in the tropical marine environment. Microbiology 155:3362–3370

Thaniyavarn J, Chongchin A, Wanitsuksombut N et al (2006) Biosurfactant production by *Pseudomonas aeruginosa* A41 using palm oil as carbon source. J Gen Appl Microbiol 52:215–222

Thavasi R, Jayalakshmi S, Balasubramanian T et al (2006) Biodegradation of crude oil by nitrogen fixing marine bacteria *Azotobacter chroococcum*. Res J Microbiol 1:401–408

Thavasi R, Jayalakshmi S, Balasubramanian T et al (2007) Biosurfactant production by *Corynebacterium kutscheri* from waste motor lubricant oil and peanut oil cake. Lett Appl Microbiol 45(6):686–691

Thavasi R, Nambaru VS, Jayalakshmi S et al (2009) Biosurfactant production by *Azotobacter chroococcum* isolated from the marine environment. Mar Biotechnol 11(5):551–556

Thavasi R, Jayalakshmi S, Banat IM (2011) Application of biosurfactant produced from peanut oil cake by *Lactobacillus delbrueckii* in biodegradation of crude oil. Bioresour Technol 102:3366–3372

Thies S, Santiago-Schu¨ b, Kovacˇ ic' F et al (2014) Heterologous production of the lipopeptide biosurfactant serrawettin W1 in Escherichia coli. J Biotechnol 181:27–30

Wei YH, Wang LF, Chang JS (2004) Optimizing iron supplement for enhanced surfactin production from *Bacillus subtilis*. Biotechnol Prog 20:979–983

Wei YH, Lai CC, Chang JS (2007) Using Taguchi experimental design methods to optimize trace element composition for enhanced surfactin production by *Bacillus subtilis* ATCC21332. Process Biochem 42:40–45

Wicke C, Hüners M, Wray V et al (2000) Production and structure elucidation of glycoglycerolipids from a marine sponge-associated *Microbacterium* species. J Nat Prod 63:621–626

Wittgens A, Tiso T, Arndt TT et al (2011) Growth independent rhamnolipid production from glucose using the non-pathogenic *Pseudomonas putida* KT2440. Microb Cell Factories 10(1):80

Yakimov MM, Timmis KN, Wray V et al (1995) Characterization of a new lipopeptide surfactant produced by thermotolerant and halotolerant subsurface *Bacillus licheniformis* BAS50. Appl Environ Microbiol 61:1706–1713

Yakimov MM, Golyshin PN, Lang S et al (1998) *Alcanivorax borkumensis* gen. nov., sp. nov., a new, hydrocarbon- degrading and surfactant-producing marine bacterium. Int J Syst Bacteriol 48:339–348

Yakimov MM, Timmis KN, Golyshin PN (2007) Obligate oil-degrading marine bacteria. Curr Opin Biotechnol 18:257–266

Yeh MS, Wei YH, Chang JS (2006) Bioreactor design for enhanced carrier assisted surfactin production with *Bacillus subtilis*. Process Biochem 41:1799–1805

Zinjarde SS, Pant A (2002) Emulsifier from a tropical marine yeast, *Yarrowia lipolytica* NCIM 3589. J Basic Microbiol 42(1):67–73

Zobell CE (1941) Studies on marine bacteria. I. The cultural requirements of heterotrophic aerobes. J Mar Res 4:173–188

Biosurfactant-Producing Denitrifying Bacteria in Marine Petroleum-Contaminated Environmental Sites

8

Trelita de Sousa

Abstract

Denitrifying bacteria are ubiquitously distributed in marine ecosystems and especially widespread in hydrocarbon-contaminated systems. Their unique flexible respiratory mechanism and ability to degrade a broad range of aliphatic, aromatic, and polycyclic aromatic hydrocarbons enable them to withstand varying oxygen fluxes prevalent in oil-contaminated sites. This chapter emphasizes the significance of denitrifying bacteria in hydrocarbon-contaminated environments. It elucidates the various mechanisms adopted by denitrifying bacteria to degrade hydrocarbons including aliphatic, aromatic, and polycyclic aromatic hydrocarbons and adapt effectively to oxygen fluxes prevalent in petroleum-contaminated systems. In order to facilitate hydrocarbon degradation, denitrifying bacteria produce amphiphilic metabolites like biosurfactants and bioemulsifiers. Such surface-active compounds isolated from indigenous bacteria have received increasing interest over the past few decades due to their important applications in bioremediation projects designed to combat oil spills and in handling, transportation, and recovery of oil. This chapter focuses on the significance of biosurfactant-producing denitrifying bacteria in marine petroleum-contaminated sites for effective use in bioremediation studies.

8.1 Introduction

Petroleum comprises natural gas, condensate, and crude oil and is a complex mixture of various aliphatic, aromatic, and polycyclic aromatic hydrocarbons (Hassanshahian and Cappello 2013; De Sousa 2015). There is no doubt that petroleum commands widespread anthropogenic usage and global economic importance

T. de Sousa (✉)
Department of Microbiology, Goa University,
Taleigao Plateau, Goa 403206, India
e-mail: trelita@gmail.com

M.M. Naik, S.K. Dubey (eds.), *Marine Pollution and Microbial Remediation*,
DOI 10.1007/978-981-10-1044-6_8

(Hassanshahian and Cappello 2013). However, when ruthlessly released into the environment, this vital commodity becomes a serious threat to both aquatic and terrestrial ecosystems including deep-sea, coastal, and estuarine systems, due to their deleterious effects to every marine organism from bacteria, planktons, algae, and fungi to corals, fish, crustaceans, birds, plants, and animals (Ingole and Sivadas 2007; De Sousa 2015).

An estimated 1.3 million tones of petroleum enters the marine environment each year (McGenity et al. 2012). Cataclysmically devastating oil spills like the Exxon Valdez oil spill at Prince William Sound, Alaska (1989), the Prestige oil spill in Galicia (2002), and the Deepwater Horizon in the Gulf of Mexico (2010) only add to the already alarming petroleum burden of coastal and ocean ecosystems (Atlas 2011; Jain et al. 2011; McGenity et al. 2012; De Sousa 2015). Bioremediation using indigenous microorganisms formulates an ideal strategy to overcome this urgent global issue (Atlas 2011; Jain et al. 2011). Petroleum compounds are after all organic substrates, and being in constant contact with these hazardous compounds, the autochthonous microorganisms have developed well-advanced pathways to degrade and utilize them (Atlas 1981; Van Hamme et al. 2003; Chikere and Okpokwasil 2011).

Petroleum hydrocarbons especially the higher polycyclic aromatics are either partially or sparingly soluble in water impeding its bioavailability to the bacteria for effective breakdown (Yadav and Hassanizadeh 2011). This may be resolved by the production of biosurfactants which are potent surface-active agents produced on microbial cell surfaces or excreted extracellularly (Nerurkar et al. 2009; Makkar et al. 2011; Reis et al. 2011). They are amphiphilic in nature and contain a combination of hydrophobic (unsaturated or saturated fatty acids) and hydrophilic (amino acids or peptides; mono-, di-, or polysaccharides) moieties that reduce surface tension and interfacial tension between individual molecules at the surface and interface of two immiscible liquids (De Sousa 2015). These surface-active compounds expedite the degradation of hydrocarbons by increasing their bioavailability through dispersion, desorption, surface, and emulsification properties and enhancing the diffusion of the recalcitrant hydrocarbons into the bacterial cell (Ron and Rosenberg 2001; Anyanwu et al. 2011; De Sousa and Bhosle 2012a).

Because of their facultative ability to switch between O_2 and NO_3^- as their terminal electron acceptor, denitrifying bacteria have been favored in bioremediation projects of hydrocarbons under mixed electron acceptor conditions (Wilson and Bower 1997; Zumft 1997; Nestler et al. 2007; Bae et al. 2009), and the addition of nitrate to oil-contaminated sites has been successfully demonstrated to stimulate and enhance indigenous bioremediation mechanisms (Wilson and Bouwer 1997; Atlas 2011; Van der Zaan et al. 2012; De Sousa 2015). Denitrifying bacteria have been reported to utilize alkanes, toluene, xylene, phenols, cresols, phthalate, cyclohexanol, benzenes (including halobenzenes), benzoate and other aromatic acids, alcohols, aldehydes, and polycyclic aromatic hydrocarbons (Zumft 1997; Chayabutra and Ju 2000; Shinoda et al. 2004; Nestler et al. 2007; Lu et al. 2011). Several new isolates of denitrifiers capable of mineralizing a wide spectrum of aliphatic (Chayabutra and Ju 2000), aromatic (Shinoda et al. 2004; Nestler et al. 2007), and polycyclic aromatic (Lu et al. 2011) compounds have been isolated, and their number is steadily increasing. Research on the hydrocarbon-degrading denitrifying bacteria, *Thauera aromatica* and *Azoarcus evansii*, has significantly advanced the understanding of anaerobic degradation of hydrocarbons (Zumft 1997; Song and Ward 2005; Schmeling and Fuchs 2009).

The prevalence of denitrifying bacteria in petroleum-contaminated sites (Wilson and Bower 1997) and their efficient potential to degrade a wide range of aliphatic, aromatic, and polycyclic aromatic hydrocarbons favor their use in bioremediation of petroleum pollutants especially under prevailing oxygen fluxes (Wilson and Bower 1997; Zumft 1997; Nestler et al. 2007; Bae et al. 2009; De Sousa and Bhosle 2012b; De Sousa 2015). Denitrifying bacteria are also known to produce biosurfactants that assist the uptake of hydrocarbons thereby enabling an effective degradation process (Chayabutra and Ju 2000; De Sousa 2015). This chapter will focus on biosurfactant production by denitrifying bacteria in response to hydrocarbons. It will comprehensively discuss the significance of denitrifying bacteria in petroleum-contaminated ecosystems, the mechanisms adopted to degrade hydrocarbons under both aerobic and anaerobic conditions, the production of surface-active agents by denitrifying bacteria as a strategy to facilitate hydrocarbon degradation, and the various methods of screening for biosurfactant-producing denitrifying bacteria for effective use in oil remediation and recovery.

8.2 Significance of Denitrifying Bacteria in Hydrocarbon-Contaminated Environments

Denitrification is the sequential exergonic dissimilatory reduction of NO_3^- and NO_2^- to the gaseous nitrogen oxides, nitric oxide (NO) and N_2O, and ultimately to dinitrogen gas (N_2) brought about by aerobic to facultatively anaerobic heterotrophs (Knowles 1982; Zumft 1997; De Sousa and Bhosle 2012b). It has significant global implications in the environment especially in the biogeochemical cycling of nitrogen in coastal and estuarine systems (De Sousa and Bhosle 2012b). The process acts as a natural sink for nitrate removal in estuaries (Gruber and Galloway 2008), counteracts eutrophication in estuaries and coastal waterways (Schlesinger 2009), and helps maintain the marine productivity in the ocean (Lam et al. 2009) with crucial implications to global warming and climate change (Zehr 2009). This process has also found important applications in efficient municipal/industrial wastewater treatment (Zhao et al. 2009) and groundwater remediation (Hunter and Shaner 2010). But more significantly, the flexible respiration system of denitrifiers enables them to form ideal cleanup systems in the bioremediation of hydrocarbon contamination (Wilson and Bouwer 1997; Cao et al. 2009).

Denitrification differs from assimilatory nitrate reduction in being involved in energy conservation (bioenergetics) as opposed to cell biosynthesis. However, the assimilatory and dissimilatory reduction of nitrate may occur concurrently, in the sense that when a denitrifying bacterium is able to assimilate nitrate, the reaction may proceed simultaneously with dissimilatory nitrate respiration. The dissimilatory branch comprises denitrification and dissimilatory nitrate reduction to ammonium (DNRA) both commonly initiated by respiratory nitrate reduction (De Sousa and Bhosle 2012b). Formerly it was thought that there is no known bacterium capable of both, denitrification and DNRA (Zumft 1997). However, recent studies have provided evidence that some species of *Paenibacillus*, which is a facultative anaerobic, endospore-forming bacterium, originally included within the genus *Bacillus* and only in 1993 reclassified as a separate genus (Ash et al. 1993), was capable of heterotrophic nitrification, DNRA, and denitrification (Rütting et al. 2011). Out of all the pathways of the intricate nitrogen cycle, denitrification seems to be the most conducive to thrive in petroleum-rich systems especially in the marine systems. Elevated levels of hydrocarbon in petroleum-contaminated sites correlate with increased denitrification rates, rapid nitrate consumption, and net accumulation of dinitrogen (Scott et al. 2014).

During denitrification, a nitrogen oxide, instead of oxygen (O_2), serves as the terminal electron acceptor to generate an electrochemical gradient across the cytoplasmic membrane resulting in the sequential transformation of NO_3^- to N_2 through the concerted action of four sequential enzymes: nitrate reductase (NaR), nitrite reductase (NiR), nitric oxide reductase (NOR), and nitrous oxide reductase (N_2OR) to form the ultimate product, dinitrogen (N_2). The denitrifying genes (*nar* for nitrate respiration, *nir* for nitrite respiration, *nor* for nitric oxide respiration, and *nos* for nitrous oxide respiration) are generally assembled in clusters, and their function and expression is regulated through a complex signal transduction system which is generally triggered by anoxia and the presence of a nitrogenous oxide (Zumft 1997; Stouthamer et al. 1997; De Sousa and Bhosle 2012b).

When bacteria shift to denitrification, the initial reactions in the aerobic electron transport chain, involving the pyridine nucleotides, flavins and quinones, remain unchanged (Fig. 8.1, Wilson and Bouwer 1997; Arai 2011). However, from cyt *b* onwards, separate and specific enzymes, regulated by the redox conditions of ubiquinone, function to determine and control the electron transport to O_2 or NO_3^- (Wilson and Bouwer 1997; Zumft 1997; Arai 2011). Formerly, denitrification was thought to be a strictly anaerobic process mostly due to the inhibitory effect of oxygen on NaR (Knowles 1982; Wilson and Bouwer 1997). However, a co-respiration of O_2 and NO_3^- with both O_2 and NO_3^- (or NO_2^-) serving concomitantly as electron acceptors has been well established during, what is also called, aerobic denitrification (Okada et al. 2005; De Sousa and Bhosle 2012b).

Aerobic denitrification is especially prevalent in systems accumulating organic compounds like hydrocarbons. The abundant availability of electron donors causes an excess of electrons, and these can be simultaneously shuttled to both O_2 and NO_3^- (Wilson and Bouwer 1997; De Sousa and Bhosle 2012b). In fact, research on degradation of hydrocarbons by denitrifying bacteria stems from the fact that dispersal of oil and fuel in the environment may lead to the pollution of deep aquifers which are frequently anaerobic (Wilson and Bouwer 1997; Cao et al. 2009). Dynamic alterations in microbial communities in petroleum hydrocarbon spills and the proliferation of heterotrophs in the presence of hydrocarbons as, invariably, the sole electron donor can, inevitably, result in the rapid depletion of oxygen in these ecosystems through aerobic respiration (Scott et al. 2014).

The availability of organic carbon, the ready occurrence of nitrate, and the prevalence of oxygen fluxes in petroleum-contaminated systems are seemingly tailor-made for the proliferation of denitrifying bacteria (De Sousa and Bhosle 2012b; Scott et al. 2014) including

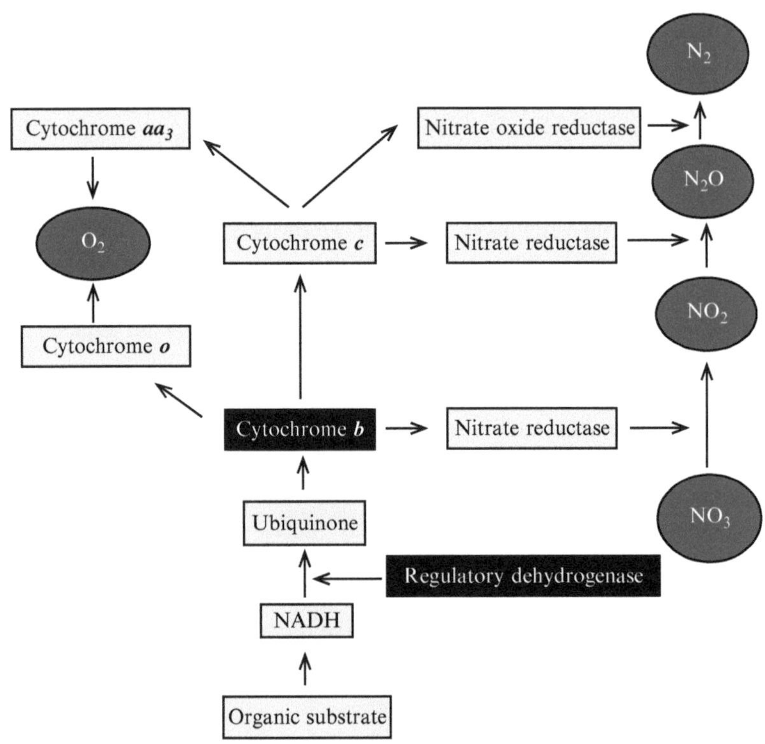

Fig. 8.1 A schematic representation of the electron transport chain in denitrifying bacteria (Wilson and Bouwer 1997; Arai 2011)

Pseudomonas, *Azoarcus*, *Thiosphaera*, *Thauera*, *Rhodopseudomonas*, *Bradyrhizobium*, *Ochrobactrum*, *Paracoccus*, *Mesorhizobium*, *Ensifer*, and *Acidovorax* (Chayabutra and Ju 2000; Shinoda et al. 2004; Song and Ward 2005; Nestler et al. 2007; De Sousa and Bhosle 2012b).

8.3 Mechanism of Hydrocarbon Degradation by Denitrifying Bacteria Under Aerobic and Anaerobic Conditions

O_2 is the most common electron acceptor for microbial respiration (Cao et al. 2009). Therefore, it is not surprising that the breakdown of petroleum compounds, especially the aromatics, is limited by the supply of oxygen (Wilson and Bower 1997). Aerobic degradation of straight-chain aliphatic hydrocarbons (alkanes) requires a membrane-bound monooxygenase in conjunction with soluble rubredoxin and rubredoxin reductase to shunt electrons through NADH to the hydroxylase, in order to convert the alkane to alcohol, which is then oxidized to aldehyde and acid before proceeding to the β-oxidation and tricarboxylic acid cycles (Van Hamme et al. 2003; Wentzel et al. 2007; De Sousa 2015). During aerobic biodegradation of aromatic compounds (benzoate, benzene, toluene), oxygen not only acts as an electron acceptor but is also involved as a highly reactive co-substrate in the initial hydroxylation reactions catalyzed by mono- and dioxygenases. The monooxygenases transform the benzene ring into a few central intermediates such as catechol, protocatechuate, gentisate, and hydroxyl benzoquinols. These are then further cleaved by different dioxygenases to the tricarboxylic acid intermediates via the ortho- or meta-pathways (Cao et al. 2009; Pérez-Pantoja et al. 2010; De Sousa 2015). In the case of polycyclic aromatic hydrocarbons like naphthalene, acenaphthene, anthracene, fluoranthene, pyrene, and chrysene, molecular oxygen is introduced into the aromatic nucleus by multicomponent non-heme iron oxygenase enzyme systems, like the naphthalene dioxygenase, forming a dihydrodiol which then proceeds via the normal aromatic breakdown pathways (Van Hamme et al. 2003; De Sousa 2015).

Although the importance of oxygen in hydrocarbon breakdown is very well apparent, it must be remembered that systems accumulating petroleum hydrocarbons are also inevitably exposed to varying oxygen fluxes (Wilson and Bower 1997). Hydrocarbon-polluted aquifers, aquatic sediments, and submerged soils invariably become anoxic (Cao et al. 2009) mostly because of the quick depletion of oxygen during respiration of easily utilizable substrates, low solubility in water, and low rate of transportation in soils and sediments (Bae et al. 2002). Therefore, microbial processes like denitrification, iron (III) reduction, sulfate reduction, and methanogenesis, capable of utilizing alternative electron acceptors such as nitrate (NO_3^-), iron (Fe^{3+}), sulfate (SO_4^{2-}), and carbon dioxide (CO_2), respectively, are of increasing interest for intrinsic and engineered bioremediation designs (Wilson and Bower 1997; Cao et al. 2009; Pérez-Pantoja et al. 2010; Philipp and Schink 2011). In terms of energy acquisition, NO_3^- is the preferred electron acceptor because its yield is close to that of O_2 in comparison to Fe_3^+, SO_4^{2-}, and CO_2. Furthermore, it is highly water-soluble (660 g/L), does not precipitate oxides, is inexpensive (can be therefore used in in vitro studies), and is nontoxic to aquifer microorganisms at concentrations below 500 mg/L (Wilson and Bower 1997). Van der Zaan et al. (2012) have also reported that the highest rates for anaerobic degradation of benzene were obtained with NO_3^- as the electron acceptor in comparison to SO_4^{2-}, chlorate, and Fe^{3+}.

The anaerobic mechanism for alkane degradation is proposed to transpire either by alkane carboxylation leading to the formation of fatty acids, as seen with sulfate reducers or by addition of alkane to the double bond of fumarate giving succinate substituted with alkane-derived alkyl chains as in the case of denitrifiers (Wentzel et al. 2007). In the anaerobic biodegradation of aromatic compounds, the peripheral pathways catalyzed by the oxygenases converge to benzoyl-CoA (occasionally to resorcinol or phloroglucinol).

Specific multicomponent energy-requiring reductases catalyze the de-aromatizing reactions (Cao et al. 2009; Philipp and Schink 2011). Polycyclic aromatic hydrocarbons are also anaerobically metabolized in a similar way via carboxylation or by the addition to fumarate (Van Hamme et al. 2003).

Denitrifying bacteria can thus utilize hydrocarbons under both aerobic and anaerobic conditions by suitably modifying their pathways in adaptation to the availability of oxygen in the environment. These flexible organisms possess enzymes that enable them to use aerobic degradation pathways to metabolize hydrocarbons in the presence of molecular oxygen, and when oxygen levels diminish, the bacteria switch to nitrate respiration (Wilson and Bouwer 1997; Pérez-Pantoja et al. 2010).

In the case of aliphatic hydrocarbons like hexadecane by *Pseudomonas aeruginosa*, the initial oxygen-requiring transformation of the alkane is taken care of using oxygen as the electron acceptor. The oxygenated metabolites are then further degraded under anaerobic conditions using nitrate as the electron acceptor (Chayabutra and Ju 2000). Such sequential mechanisms are also observed in the degradation of aromatic hydrocarbons like toluene by *Thauera* and *Azoarcus*. The initial degradation of toluene occurred through a dioxygenase-mediated pathway in the presence of oxygen and proceeded to the benzyl-succinate pathway under denitrifying conditions (Shinoda et al. 2004; Song and Ward 2005).

Degradation of especially the aromatic compounds like benzoate and toluene by denitrifying bacteria is enhanced by the presence of readily consumable carbon sources like succinate and acetate (Kariminiaae-Hamedaai et al. 2004; Martínez-Hernández et al. 2009; De Sousa and Bhosle 2012c). Chénier et al. (2003) also reported the enhancement of hexadecane mineralization and denitrification with nutrient amendment through the addition of glucose, ammonium chloride, and phosphorous (K_2HPO_4).

8.4 Biosurfactants Produced by Denitrifying Bacteria in Response to Hydrocarbons

Biosurfactant production is widespread in hydrocarbon-utilizing bacteria. These natural products are a desirable alternative to synthetic surfactants because of their selectivity, biodegradability, low toxicity, and stability (Nerurkar et al. 2009; Vignesh et al. 2011; De Sousa and Bhosle 2012a) and are increasingly in huge demand in hydrocarbon bioremediation, microbial-enhanced oil recovery, agriculture, cosmetic, pharmaceutical, detergent, food, textile, paper, and paint industries (Janek et al. 2010; Lima et al. 2011; Zheng et al. 2012).

Surface-active compounds are mainly categorized on the basis of their chemical composition, function, and point of origin (Vignesh et al. 2011). Biosurfactants reduce surface tension at the air-water interface and include low-molecular-weight molecules including glycolipids, lipopeptides, lipopolysaccharides, polysaccharide-protein complexes, lipoprotein, phospholipids, fatty acids, lipids, polymeric biosurfactant, and particulate biosurfactants (Makkar et al. 2011; Vignesh et al. 2011; De Sousa 2015). Bioemulsifiers, although sometimes also referred to as biosurfactants (Franzetti et al. 2012), on the other hand reduce the interfacial tension between immiscible liquids or at solid-liquid interface and also include mostly low-molecular-weight compounds like lipopeptides and glycolipids (Vignesh et al. 2011; De Sousa and Bhosle 2012a). Extracellular polymeric substances or exopolysaccharides (EPS) or bioemulsans are generally high-molecular-weight emulsifiers which include polymers of polysaccharides, lipopolysaccharides, proteins, or lipoproteins (Franzetti et al. 2012).

Extracellular ionic surfactants like rhamnolipids or sophorolipids (produced by *Pseudomonas* and *Torulopsis* spp., respectively), lipopeptides/

lipoproteins, namely, surfactin or subtilisin (produced by *Bacillus subtilis*), and rhodofactin (produced by *Rhodococcus* sp.) act by emulsification. Extracellular or cell-bound nonionic surfactants, capable of altering the structure of the cell wall, include lipopolysaccharides or liposan (produced by *Candida lipolytica*, *C. tropicalis*, *Rhodococcus erythropolis*, *Mycobacterium* sp., and *Arthrobacter* sp.) and emulsan (synthesized by *Acinetobacter* sp.). Several other effective biosurfactants have also been reported which include xylolipid (produced by *Lactococcus lactis*), mycolates and corynomycolates (produced by *Corynebacteria*, *Mycobacteria*, and *Nocardia* spp.), and ornithinlipids (produced by *Pseudomonas rubescens*, *Gluconobacter cerinus*, and *Thiobacillus ferrooxidans*) (Peng et al. 2008; Nerurkar et al. 2009; Saravanakumari and Mani 2010; Vignesh et al. 2011).

Despite their structural and functional differences, all three types of surface-active agents (biosurfactants, bioemulsifiers, and EPS) have found potential use in bioremediation designs and have garnered an increasing interest over the past several decades in petroleum abatement (Chayabutra and Ju 2000; Martínez-Cánovas et al. 2004; De Sousa and Bhosle 2012a). Although numerous surface-active compounds have been isolated from various marine systems especially those contaminated with petroleum hydrocarbons (Batista et al. 2006; Maneerat and Phetrong 2007; Peng et al. 2008; Nerurkar et al. 2009; Janek et al. 2010; Anyanwu et al. 2011; Franzetti et al. 2012), only a few have been isolated from denitrifying bacteria in response to hydrocarbons such as the rhamnolipid biosurfactant from *Pseudomonas aeruginosa* (Chayabutra and Ju 2000), the lipopeptide bioemulsifier from *Pseudomonas nitroreducens* (De Sousa and Bhosle 2012a; De Sousa 2015), and the exopolysaccharide produced by *Halomonas ventosae* (Martínez-Cánovas et al. 2004). A list of previously isolated biosurfactant/bioemulsifier/EPS-producing marine/estuarine hydrocarbon-degrading denitrifying bacteria is given in Table 8.1.

Besides enhancing growth on hydrophobic substrates like hydrocarbons, biosurfactants/bioemulsifiers assume different functions due to the complexities in chemical structure and surface activity thereby providing varied advantages in different ecological niches. They play an important role in regulating the attachment-detachment of microorganisms to and from surfaces and are therefore also involved in bacterial pathogenesis, quorum sensing, biofilm formation, and conferring antibacterial/antifungal properties to the producing microorganisms (Ron and Rosenberg 2001).

The production of surfactants/emulsifiers by bacteria can be suitably optimized for potential application in bioremediation of hydrocarbons and heavy metals and in enhanced oil recovery

Table 8.1 Biosurfactant/bioemulsifier/exopolysaccharide-producing hydrocarbon-degrading denitrifying bacteria isolated from marine petroleum-contaminated sites

Denitrifying isolate	Type of surface-active compound	Reference
Pseudomonas aeruginosa ATCC 10145	Rhamnolipid biosurfactant	Chayabutra and Ju (2000)
Pseudomonas nitroreducens TSB.MJ10	Lipopeptide bioemulsifier	De Sousa and Bhosle (2012a)
Pseudomonas sp.	Biosurfactant	Grishchenkov et al. (2000)
Brevibacillus sp.	Biosurfactant	Grishchenkov et al. (2000)
Halomonas ventosae	Exopolysaccharide	Martínez-Cánovas et al. (2004)
Dietzia maris	Wax ester-like biosurfactant	Nakano et al. (2011)
Pseudomonas nitroreducens	Rhamnolipid biosurfactant	Onwosi and Odibo (2011)

processes (Anyanwu et al. 2011; Lima et al. 2011; Franzetti et al. 2012; Zheng et al. 2012). The amount of microbial surfactants produced depend primarily on the producer organism and on factors like carbon and nitrogen sources, trace elements, temperature, pH, and aeration conditions (Santos et al. 2002; Onwosi and Odibo 2011; Vignesh et al. 2011). It has been suggested that the production of these surface-active agents is concurrent with the onset of the stationary phase induced by molecular signals involved in quorum sensing (Ron and Rosenberg 2001; Reis et al. 2011). However, certain studies have also reported their synthesis during the exponential growth phase (Batista et al. 2006). Their potential is largely dependent on their chemical nature and their activity may be enhanced by simple media modification (Mutalik et al. 2008; Nerurkar et al. 2009).

8.5 Screening Methods of Denitrifying Bacteria for Biosurfactants

Because of their wide range of applications, various methods and techniques are designed to maximize the isolation of biosurfactants. Biosurfactant production generally depends on the fermentation conditions, environmental factors, and nutrient availability (Santos et al. 2002; Gakpe et al. 2007; Onwosi and Odibo 2011; Vignesh et al. 2011). The denitrifying isolates are cultivated in a suitable medium with the hydrocarbon of choice prior to the screening methods. Care must always be taken to incorporate a nitrate source in the cultivation medium when culturing denitrifying bacteria (De Sousa and Bhosle 2012b). KNO_3 has been observed to be the best source for their cultivation (De Sousa et al. 2013). Succinate also enhances denitrification activity (De Sousa and Bhosle 2012c). Denitrifying bacteria can be screened for biosurfactants by the following methods routinely used for biosurfactants.

8.5.1 Cell Hydrophobicity by the BATH (Bacterial Adherence to Hydrocarbons) Assay

This is one of the most popular and simplest methods used to screen for cell hydrophobicity, and not many changes have been made to the original method proposed by Rosenberg (1984). The optical density of the cells suspended in phosphate buffer before and after mixing with a small volume of hexadecane at high speed is measured. A decrease in absorbance of the aqueous layer after an incubation time of 30 min on mixing with hexadecane reflects the hydrophobicity of the cells.

8.5.2 Emulsification Activity

The cell-free culture broth is used for screening for emulsification activity. In the method described by Maneerat and Phetrong (2007), equal quantities of cell-free supernatant are mixed with hexadecane or xylene at high speed, and the resulting emulsion (De Sousa 2015) is allowed to stabilize for 24 h and the emulsification index (E_{24}), defined as the per cent height of the emulsion layer divided by the total height of the mixture, is calculated.

8.5.3 Quantitative Oil Displacement Assay

In this simple method, cell-free culture broth is carefully placed on the center of an oil film formed by pouring a mixture of paraffin oil and an oil-soluble dye (oil red O) over water in a clean petri plate. An appearance of a clear halo (De Sousa 2015) within 30 s is positive for surface activity (Morikawa et al. 1993).

8.5.4 Qualitative Drop-Collapsing Activity

A small amount of cell-free culture broth is placed onto the surface of an oil droplet on a clean grease-free slide, and its shape is observed for a minute. Flat drops (De Sousa 2015) indicate a positive result, while round drops are negative for surface activity (Youssef et al. 2004).

8.5.5 Surface Tension Measurement and Estimation of Critical Micelle Concentration (CMC)

The surface tension of the cell-free culture broth collected at different time intervals is determined by the ring method using a du Nouy tensiometer (Onwosi and Odibo 2011). The concentration at which micelles begin to form is represented as the CMC which is determined by plotting the surface tension as a function of the biosurfactant concentration. The surface tension can also be measured by the drop-weight and drop-count methods using a Traube's stalagmometer (Fig. 8.2) (De Sousa and Bhosle 2012a).

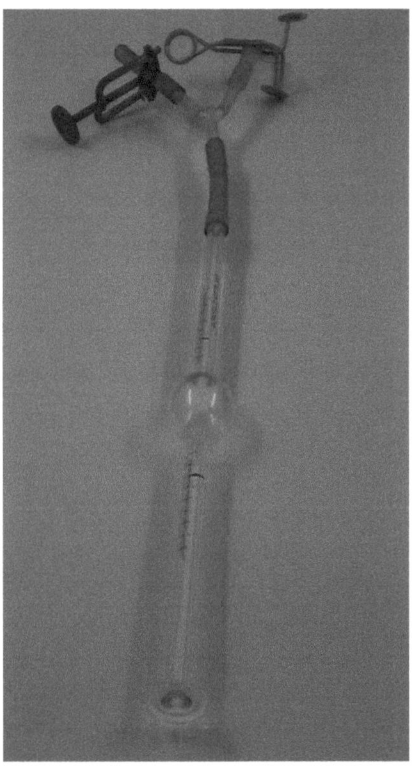

Fig. 8.2 Traube's stalagmometer used to measure surface tension by the drop-weight and drop-count methods

8.5.6 Type of Emulsion

Biosurfactants and bioemulsifiers are characterized on their ability to produce stable emulsions which allow the microorganisms to adhere strongly to the hydrocarbons thus facilitating the degradation process (Nerurkar et al. 2009). They are distinguished on their hydrophile-lipophile balance (HLB): those that possess a low HLB are strongly lipophilic and therefore generally used as water-in-oil emulsifiers, while those that have a high HLB are strongly hydrophilic and appropriately used as oil-in-water emulsifiers (Nerurkar et al. 2009). A drop of oil red O in paraffin oil and/or crystal violet solution is added to the cell-free culture broth. These solutions are then mixed with equal amounts of hexadecane at high speed for 2 min and allowed to settle thereafter, for an hour. If the emulsion is formed in the aqueous phase, then it is an oil-in-water type of emulsion, whereas if the emulsion is formed in the organic phase, it is a water-in-oil type of an emulsion (Fig. 8.3).

8.6 Characteristics of Biosurfactants for Effective Use in Petroleum Bioremediation

Because of their hydrophobicity, bacteria generally adhere to hydrocarbons. However, only the hydrocarbon molecules that are dissolved in the aqueous phase are accessible to microorganisms for biotransformation. Thus, the rate of dissolution of such hydrocarbons is critical for its bioavailability. The amphipathic nature of biosurfactants aids in the dispersion of these particles via dissolution in the aqueous phase thereby rendering them available for uptake by the cell (Ron and Rosenberg 2001; Anyanwu et al. 2011; De Sousa and Bhosle 2012a).

Biosurfactants possess several advantages over chemical surfactants including biodegradability, low toxicity, biocompatibility, digestibility, acceptable economics (cheaper raw materials and lower production costs), better environmental control, and specificity (Gakpe et al. 2007).

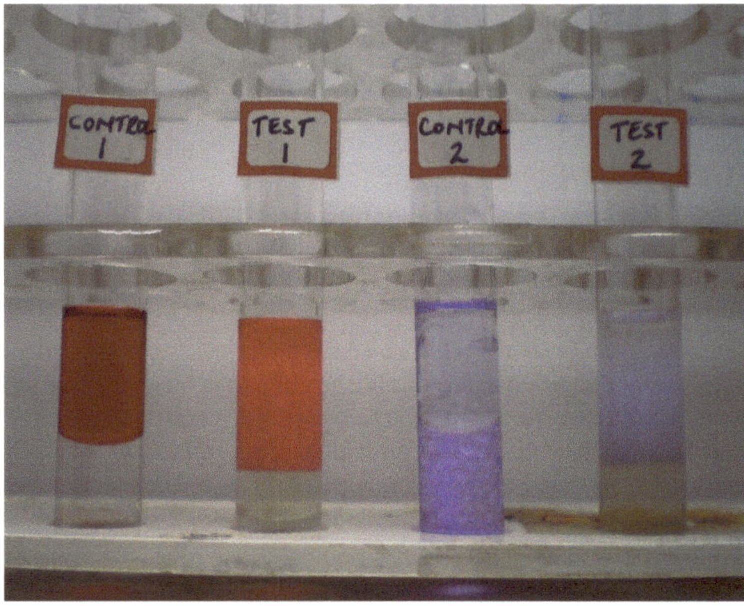

Fig. 8.3 Water-in-oil type of emulsion exhibited by bioemulsifiers produced by a denitrifying strain *Pseudomonas nitroreducens* TSB.MJ10 (De Sousa and Bhosle 2012a; De Sousa 2015). In the absence of the biosurfactant (control tubes), oil-soluble *oil red* O goes into the organic phase, while the water-soluble *crystal violet* goes into the aqueous phase. On mixing the hydrocarbon with the biosurfactant, both *oil red* O and *crystal violet* go into the emulsion layer indicating a water-in-oil type of emulsion as opposed to bringing the dyes into the aqueous layer seen in oil-in-water emulsions

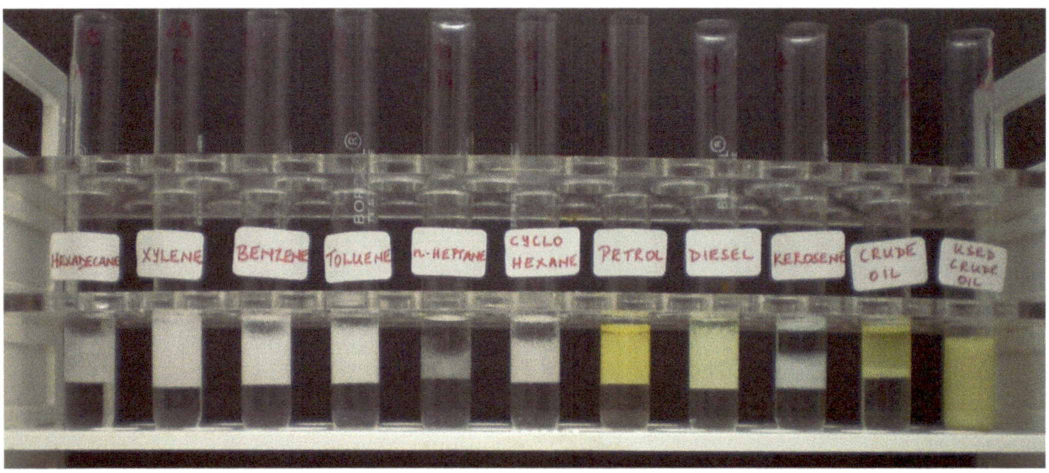

Fig. 8.4 Emulsification of various hydrocarbons by a lipopeptide bioemulsifier produced by a denitrifying strain *Pseudomonas nitroreducens* TSB.MJ10 (De Sousa and Bhosle 2012a; De Sousa 2015)

However, the most important factor governing the suitability of biosurfactants in bioremediation designs is their stability (Janek et al. 2010; Lima et al. 2011; Vignesh et al. 2011; De Sousa and Bhosle 2012a). The stability of biosurfactants is tested against a wide range of pH, temperature, and salt concentrations. It is desirable that the biosurfactants demonstrate a stable E_{24} under extreme pH, temperature, and salt concentrations for effective use under extreme environmental conditions (De Sousa and Bhosle 2012a).

Ideal biosurfactants form stable emulsions with a broad range of aliphatic, aromatic, polycyclic aromatic, and petroleum hydrocarbons (Fig. 8.4). Stable emulsification of petroleum and crude oil facilitates their transportation and management (Lima et al. 2011). Viscosity reductions of petroleum fluids by biosurfactants also help overcome the difficulties encountered during handling, recovery, and transportation of oil, which significantly contribute to their cost-effectiveness (Anyanwu et al. 2011; Lima et al. 2011; De Sousa and Bhosle 2012a; Franzetti et al. 2012).

Reduction of surface tension by biosurfactants defines their efficiency for use in bioremediation studies (Batista et al. 2006; Gakpe et al. 2007; Lima et al. 2011). Lipopeptides show potent surface activity and are therefore generally attributed to the most efficient biosurfactants/bioemulsifiers (Ron and Rosenberg 2001; Janek et al. 2010). Surfactin, a cyclic lipopeptide produced by *Bacillus subtilis*, is regarded as the most active biosurfactant ever discovered and possesses the ability to reduce the surface tension of water from 71.3 to 27.5 mN/m with interesting biological properties such as the formation of ion-conducting pores in membranes (Fox and Bala 2000; Ron and Rosenberg 2001). It is composed of a cyclic heptapeptide linked to a fatty acid (Ron and Rosenberg 2001). However, it must be remembered that not all surface-active agents reduce surface or interfacial tension (Batista et al. 2006; De Sousa and Bhosle 2012a; Franzetti et al. 2012). It is thus apparent that biosurfactants and bioemulsifiers have unique characteristics which determine their exploitation in petroleum bioremediation and abatement at different levels (Batista et al. 2006; Gakpe et al. 2007; Nerurkar et al. 2009; Lima et al. 2011).

8.7 Concluding Remarks

The marine system is vast and there lies an enormous realm of information and knowledge which is still unknown to us. This chapter discusses the significance of denitrification in petroleum-contaminated ecosystems and provides a comprehensive understanding of mechanisms adopted by denitrifying bacteria to degrade and utilize hydrocarbons under both aerobic and anaerobic conditions. It is seen that although the importance of denitrification and application of biosurfactants in petroleum bioremediation is massive, not many biosurfactants have been isolated from denitrifying bacteria. This chapter provides simple methods for effective screening of biosurfactant-producing denitrifying bacteria and encourages the isolation of more surface-active agents including biosurfactants, bioemulsifiers, and exopolysaccharides for effective use in petroleum bioremediation.

References

Anyanwu CU, Obi SKC, Okolo BN (2011) Lipopeptide biosurfactant production by *Serratia marcescens* NSK-1 strain isolated from petroleum-contaminated soil. J Appl Sci Res 7:79–87

Arai H (2011) Regulation and function of versatile aerobic and anaerobic respiratory metabolism in *Pseudomonas aeruginosa*. Front Microbiol 2:1–13. doi:10.3389/fmicb.2011.00103

Ash C, Priest FG, Collins MD (1993) Molecular identification of rRNA group 3 bacilli using a PCR probe test. Proposal for the creation of a new genus *Paenibacillus*. Antonie Van Leeuwenhoek 64:253–260

Atlas, R.M. (1981) Microbial degradation of petroleum hydrocarbons: an environmental perspective. Microbiol Rev 45:180–209

Atlas RM (2011) Oil biodegradation and bioremediation: a tale of the two worst spills in U.S. history. Environ Sci Technol 45:6709–6715

Bae HS, Yamagishi T, Suwa Y (2002) Evidence for degradation of 2-chlorophenol by enrichment cultures under denitrifying conditions. Microbiology 148:221–227

Bae H, Im W, Suwa Y, Lee JM, Lee S, Chang Y (2009) Characterization of diverse heterocyclic amine degrading denitrifying bacteria from various environments. Arch Microbiol 191:329–340

Batista SB, Mounteer AH, Amorim FR, Tótola MR (2006) Isolation and characterization of biosurfactant/bioemulsifier-producing bacteria from petroleum contaminated sites. Bioresour Technol 97:868–875

Cao B, Nagarajan K, Loh K (2009) Biodegradation of aromatic compounds: current status and opportunities for biomolecular approaches. Appl Microbiol Biotechnol 85:207–228

Chayabutra C, Ju L (2000) Degradation of *n*-hexadecane and its metabolites by *Pseudomonas aeruginosa* under microaerobic and anaerobic denitrifying conditions. Appl Environ Microbiol 66:493–498

Chénier MR, Beaumier D, Roy R, Driscoll BT, Lawrence JR, Greer CW (2003) Impact of seasonal variations and nutrient inputs on nitrogen cycling and degradation of hexadecane by replicated river biofilms. Appl Environ Microbiol 69:5170–5177

Chikere CB, Okpokwasil GC (2011) Monitoring of microbial hydrocarbon remediation in the soil. 3 Biotech 1:117–138

De Sousa T (2015) Denitrifying bacteria: physiological response to hydrocarbons. In: Borkar S (ed) Bioprospects of coastal eubacteria. Springer International Publishing, Switzerland, p 39

De Sousa T, Bhosle S (2012a) Isolation and characterization of a lipopeptide bioemulsifier produced by *Pseudomonas nitroreducens* TSB.MJ10 isolated from a mangrove ecosystem. Bioresour Technol 123:256–262

De Sousa T, Bhosle S (2012b) Microbial denitrification and its ecological implications in the marine system. In Satyanarayana et al. (eds) Microbes in environmental management: microbes & environment. Springer, Netherlands, pp 683–700

De Sousa T, Bhosle S (2012c) Implications of benzoate induced alterations in cell morphology and physiology in *Pseudomonas aeruginosa* TMR2.13 for potential application in bioremediation and monitoring approaches. J Bioremed Biodegrad doi: 10.4172/2155-6199.S1-008

De Sousa T, Ingole B, De Sousa S, Bhosle S (2013) Seasonal variations of nitrate reducing and denitrifying bacteria utilizing hexadecane in Mandovi estuary, Goa, West Coast of India. Indian J Geo Mar Sci 42:587–592

Fox SL, Bala GA (2000) Production of surfactant from *Bacillus subtilis* ATCC 21332 using potato substrates. Bioresour Technol 75:235–240

Franzetti A, Gandolfi I, Raimondi C, Bestetti G, Banat IM, Smyth TJ, Papacchini M, Cavallo M, Fracchia L (2012) Environmental fate, toxicity, characteristics and potential applications of novel bioemulsifiers produced by *Variovorax paradoxus* 7bCT5. Bioresour Technol 108:245–251

Gakpe E, Rahman PKSM, Hatha AAM (2007) Microbial biosurfactants – review. J Mar Atmos Res 3:1–17

Grishchenkov VG, Townsend RT, McDonald TJ, Autenrieth RL, Bonner JS, Boronin AM (2000) Degradation of petroleum hydrocarbons by facultative anaerobic bacteria under aerobic and anaerobic conditions. Process Biochem 35:889–896

Gruber N, Galloway JN (2008) An earth-system perspective of the global nitrogen cycle. Nature 451:293–296

Hassanshahian M, Cappello S (2013) Crude oil biodegradation in the marine environments. INTECH 101–135. doi.org/10.5772/55554

Hunter WJ, Shaner DL (2010) Biological remediation of groundwater containing both nitrate and atrazine. Curr Microbiol 60:42–46

Ingole BS, Sivadas S (2007) The slippery coastline. Mar Ecol 1:32–37

Jain PK, Gupta VK, Gaur RK, Lowry M, Jaroli DP, Chauhan UK (2011) Bioremediation of petroleum oil contaminated soil and water. Res J Environ Toxicol 5:1–26

Janek T, Łukaszewicz M, Rezanka T, Krasowka A (2010) Isolation and characterization of two lipopeptide biosurfactants produced by *Pseudomonas fluorescens* BD5 isolated from water from the Arctic Archipelago of Svalbard. Bioresour Technol 101:6118–6123

Kariminiaae-Hamedaani H, Kanda K, Kato F (2004) Denitrification activity of the bacterium *Pseudomonas* sp. ASM-2-3 isolated from the Ariake Sea Tideland. J Biosci Bioeng 97:39–44

Knowles R (1982) Denitrification. Microbiol Rev 46:43–70

Lam P, Lavik G, Jensen MM, Van de Vossenberg J, Schmid M, Woebken D, Gutiérrez D, Amann R, Jetten MSM, Kuypers MMM (2009) Revising the nitrogen cycle in the Peruvian oxygen minimum zone. Proc Natl Acad Sci 106:4752–4757

Lima TMS, Fonseca AF, Leão BA, Mounteer AH, Tótola MR, Borges AC (2011) Oil recovery from fuel oil storage tank sludge using biosurfactants. J Bioremed Biodegrad 2:1–5. doi:10.4172/2155-6199.1000125

Lu X, Zhang T, Fand HH, Leung KMY, Zhang G (2011) Biodegradation of naphthalene by enriched marine denitrifying bacteria. Int Biodeterior Biodegrad 65:204–211

Makkar RS, Cameotra SS, Banat IM (2011) Advances in utilization of renewable substrates for biosurfactant production. AMB Express 1:5

Maneerat S, Phetrong K (2007) Isolation of biosurfactant-producing marine bacteria and characteristics of selected biosurfactant. Songklanakarin J Sci Technol 29:781–791

Martínez-Cánovas MJ, Quesada E, Llamas I, Béjar V (2004) *Halomonas ventosae* sp. nov., a moderately halophilic, denitrifying, exopolysaccharide-producing bacterium. Int J Syst Evol Microbiol 54:733–737

Martínez-Hernández S, Olguín EJ, Gómez J, Cuervo-López FM (2009) Acetate enhances the specific consumption rate of toluene under denitrifying conditions. Arch Environ Contam Toxicol 57:679–687

McGenity TJ, Folwell BD, McKew BA, Sanni GO (2012) Marine crude-oil biodegradation: a central role for interspecies interactions. Aquat Biosyst 8:10

Morikawa M, Daido H, Takao T, Murata S, Shimonishi Y, Imanaka T (1993) A new lipopeptide biosurfactant produced by *Arthrobacter* sp. strain MIS38. J Bacteriol 175:6459–6466

Mutalik SR, Vaidya BK, Joshi RM, Desai KM, Nene SN (2008) Use of response surface optimization for the production of biosurfactant from *Rhodococcus* sp. MTCC 2574. Bioresour Technol 99:7875–7880

Nakano M, Kihara M, Iehata S, Tanaka R, Maeda H, Yoshikawa T (2011) Wax ester-like compounds as biosurfactants produced by Dietzia maris from n-alkane as a sole carbon source. J Basic Microbiol 51:490–498

Nerurkar AS, Hingurao KS, Suthar HG (2009) Bioemulsifiers from marine microorganisms. J Sci Ind Res 68:273–277

Nestler H, Kiesel B, Kaschabek SR, Man M, Schlömann M, Balcke GU (2007) Biodegradation of chlorobenzene under hypoxic and mixed hypoxic denitrifying conditions. Biodegradation 18:755–767

Okada N, Nomura N, Nakajuna-Kambe T, Uchiyama H (2005) Characterization of the aerobic denitrification in *Mesorhizobium* sp. strain NH-14 in comparison with that in related rhizobia. Microbes Environ 20:208–215

Onwosi CO, Odibo FJC (2011) Effects of carbon and nitrogen sources on rhamnolipid biosurfactant production by *Pseudomonas nitroreducens* isolated from soil. World J Microbiol Biotechnol. doi:10.1007/s1127-011-0891-3

Peng F, Wang Y, Sun F, Liu Z, Lai Q, Shao Z (2008) A novel lipopeptide produced by a Pacific Ocean deep-sea bacterium, *Rhodococcus* sp. TW53. J Appl Microbiol. doi:10.1111/j.1365-2672.2008.03816.x

Pérez-Pantoja D, González B, Pieper DH (2010) Aerobic degradation of aromatic hydrocarbons. In: Timmis KN (ed) Handbook of hydrocarbon and lipid microbiology. Springer-Verlag, Berlin, pp 799–837

Philipp B, Schink B (2011) Different strategies in anaerobic biodegradation of aromatic compounds: nitrate reducers versus strict anaerobes. Environ Microbiol. doi:10.1111/j.1758-2229.2011.00304.x

Reis RS, Pereira AG, Neves BC, Freire DMG (2011) Gene regulation of rhamnolipid production in *Pseudomonas aeruginosa* – a review. Bioresour Technol 101:6377–6384

Ron EZ, Rosenberg E (2001) Natural roles of biosurfactants. Environ Microbiol 3:229–236

Rosenberg M (1984) Bacterial adherence to hydrocarbons: a useful technique for studying cell surface hydrophobicity. FEMS Microbiol Lett 22:289–295

Rütting T, Boeckx P, Müller C, Klemedtsson L (2011) Assessment of the importance of dissimilatory nitrate reduction to ammonium for the terrestrial nitrogen cycle. Biogeosciences 8:1779–1791

Santos AS, Sampaio APW, Vasques GS, Santa Anna LM, Pereira N, Freire DMG (2002) Evaluation of different carbon and nitrogen sources in production of rhamnolipids by a strain of *Pseudomonas aeruginosa*. Appl Biochem Biotechnol 98:1025–1035

Saravanakumari P, Mani K (2010) Structural characterization of a novel xylolipid biosurfactant from *Lactococcus lactis* and analysis of antibacterial activity against multi-drug resistant pathogens. Bioresour Technol 101:8851–8854

Schlesinger WH (2009) On the fate of anthropogenic nitrogen. Proc Natl Acad Sci 106:203–208

Schmeling S, Fuchs G (2009) Anaerobic metabolism of phenol in proteobacteria and further studies of phenylphosphate carboxylase. Arch Microbiol 191:869–878

Scott NM, Hess M, Bouskill NJ, Mason OU, Jansson JK, Gilbert JA (2014) The microbial nitrogen cycling potential is impacted by polyaromatic hydrocarbon pollution of marine sediments. Front Microbiol 5:1–8. doi:10.3389/fmicb.2014.00108

Shinoda Y, Sakai Y, Uenishi H, Uchihashi Y, Hiraishi A, Yukawa H, Yurimoto H, Kato N (2004) Aerobic and anaerobic toluene degradation of a newly isolated denitrifying bacterium, *Thauera sp.* strain DNT-1. Appl Environ Microbiol 70:1385–1392

Song B, Ward BB (2005) Genetic diversity of benzoyl coenzyme A reductase genes detected in denitrifying isolates and estuarine sediment communities. Appl Environ Microbiol 71:2036–2045

Stouthamer AH, De Boer APN, Van der Oost J, Van Spanning RJM (1997) Emerging principles of inorganic nitrogen metabolism in *Paracoccus denitrificans* and related bacteria. Antonie Van Leeuwenhoek 71:33–41

Van der Zaan BM, Saia FT, Stams AJM, Plugge CM, De Vos WM, Smidt H, Langenhoff AAM, Gerritse J (2012) Anaerobic benzene degradation under denitrifying conditions: *Peptococcaceae* as dominant benzene degraders and evidence for a syntrophic process. Environ Microbiol 14:1171–1181

Van Hamme JD, Singh A, Ward OP (2003) Recent advances in petroleum microbiology. Microbiol Mol Biol Rev 67:503–549

Vignesh R, Haq BMA, Srinivasan M (2011) Biodegradation prospective of microbes. Int J Environ Sci 2:741–754

Wentzel A, Ellingsen TE, Kotlar H, Zotchev SB, Throne-Holst M (2007) Bacterial metabolism of long-chain *n*-alkanes. Appl Microbiol Biotechnol 76:1209–1221

Wilson LP, Bouwer EJ (1997) Biodegradation of aromatic compounds under mixed oxygen/denitrifying conditions: a review. J Ind Microbiol Biotechnol 18:116–130

Yadav BK, Hassanizadeh M (2011) An overview of biodegradation of LNAPLs in coastal (semi)-arid environment. Water Air Soil Pollut 220:225–239

Youssef NH, Duncan KE, Nagle DP, Savage KN, Knapp RM, McInerney MJ (2004) Comparison of methods to detect biosurfactant production by diverse microorganisms. J Microbiol Meth 56:339–347

Zehr JP (2009) New twist on nitrogen cycling in oceanic oxygen minimum zones. Proc Natl Acad Sci 106:4575–4576

Zhao S, Hu N, Chen Z, Zhao B, Liang Y (2009) Bioremediation of reclaimed wastewater used as landscape water by using the denitrifying bacterium *Bacillus cereus*. Bull Environ Contam Toxicol 83:337–340

Zheng C, Wang M, Wang Y, Huang Z (2012) Optimization of biosurfactant-mediated oil extraction from oil sludge. Bioresour Technol. doi:10.1016/j.biortech.2012.01.073

Zumft WG (1997) Cell biology and molecular basis of denitrification. Microbiol Mol Biol Rev 61:533–616

Interaction of Haloarchaea with Metals

9

Sanika Shivdas Naik and Irene Furtado

Abstract

Haloarchaea are predominant microflora of hypersaline econiches such as solar saltern, salt lakes, and salt deposits and so on. Urbanization and industrialization including mining, agriculture, and waste disposal in coastal countries result in the discharge of effluents containing toxic metal ions into rivers, estuaries, and marine econiches. Saltpans located along the estuary often serve as a sink of these metal toxicants. Moreover, solar salterns are sites where microorganisms thrive and where haloarchaea are predominant indicating their survival in metal-toxicated environment to be the result of resistance mechanism specialized to overcome the stress. This chapter reviews and focuses on the various resistance strategies adopted by Archaea especially haloarchaea to survive the metal-contaminated econiche.

9.1 Introduction

Metal pollution is ceaselessly on rise as a consequence of anthropogenic activities, industrialization, urbanization, and natural sources (Hu et al. 2014). Metal pollutants ultimately find its way into soils and water bodies including freshwater as well as marine seas (Zhuang and Gao 2014) from industrial operations such as mining, smelting of ore, manufacturing of iron and steel products, manufacturing of alkaline storage batteries, and application of agrochemicals in agricultural fields which add a considerable quantity of heavy metals in the environment (Kumar et al. 2011). The concentration of very common heavy metals such as Cd, Cr, Cu, Ni, Pb, and Zn in marine sediments is reported to range from 0.09 to 88.6 μg g^{-1} (Zhuang and Gao 2014). These metal pollutants are accumulated in living organisms including microorganisms, animals, and humans having its toxic effect on the whole food chain and are thus responsible for causing metabolic and physiological disorder (Matyar et al. 2010).

Metal binding capabilities of microorganisms have been explored by various researchers across the world which is proven to be an economical as well as eco-friendly solution to remediate the

S.S. Naik (✉) • I. Furtado
Microbiology Department, Goa University,
Taleigao Plateau, Goa 403206, India
e-mail: sanika2k4@gmail.com

M.M. Naik, S.K. Dubey (eds.), *Marine Pollution and Microbial Remediation*,
DOI 10.1007/978-981-10-1044-6_9

metal-contaminated sites (Ahmed and Malik 2012).

Archaea belong to one of the three domains of life and are inhabited by diverse environments such as shallow or deep-sea anaerobic sediments including free-living and endosymbiotic methanogens, hot springs, or deep-sea hydrothermal vents and highly saline landlocked seas comprising methanogens, sulfate reducers, and extreme halophiles (Silveira et al. 2013). However, Archaea are also reported and explored in non-extreme environments, including marine, freshwater, and mangrove ecosystem belonging to Group I Archaea and Groups II and III *Euryarchaeota* (Silveira et al. 2013; Delong 2007). Also, the discovery of numerous archaeal species from metal-rich environment that contributes to the highly oxidative environment such as mining sites, salterns, and metal-contaminated soils has enhanced the interest in studying metal resistance in these microbes (Maezato and Blum 2012).

9.2 Transformations of Metals by Archaea

Archaebacteria are reported to impact the mobility of metals in the environment by mode of oxidation or reduction of elements with the aim of energy conservation or detoxification or indirectly by altering the pH or redox conditions of their environment, which in turn affects metal precipitation or solubilization (Bini 2010). Archaeal species are also able to transform metals into their insoluble forms in biomineralization processes that lead to the formation of mineral deposits of the corresponding metal ores. In hyperthermophilic habitats, the presence of magnetite and uraninite has been proposed to be due to the activity of *P. islandicum.* Cell cultures of *P. islandicum* are capable of Fe^{3+} oxide and U^{4+} mineral reduction, leading to the formation of ultrafine magnetite and UO_2, respectively (Kashefi et al. 2008a, b).

Other metals, including Tc^{7+}, Cr^{6+}, Co^{3+}, Mn^{4+} (Kashefi and Lovley 2000), and Au^{3+} (Kashefi et al. 2001), are also reduced by *P. islandicum* in the presence of hydrogen as the electron donor. *Pyrococcus furiosus* and the archaeal strain 234 are also able to reduce gold, as well as Fe^{3+}, in the presence of H_2 (Kashefi et al. 2001). Metal transformations leading to the formation of insoluble precipitates may also result from metabolic processes. For example, anaerobic oxidation of methane (AOM) can be carried out using different electron acceptors, including the metals manganese or iron (Beal et al. 2009). However, under such conditions, AOM proceeds at significantly slower rates than those observed for sulfate-dependent AOM (Beal et al. 2009).

Haloarchaea belonging to the domain Archaea comprise 36 genera and 129 species identified (Oren 2012). The members of this family are predominant microflora of hypersaline environments such as solar salterns, soda lakes, and salt deposits (Tabak et al. 2005). These environments serve as avenues for transformation of native microflora into potential metal-resistant strains.

9.3 Metal Resistance in Haloarchaea

9.3.1 Intracellular Proteins

Haloarchaea are reported to have γ-glutamylcysteine (γ-GC) that is analogous to glutathione (GSH) and involved in maintaining a reducing environment within the cell that aids in overcoming oxidative and disulfide stress. The resistance to toxic metal ions is attributed to the thiol group of cysteine in γ-GC that can chelate the metal ions. Further, in case of *Halobacterium salinarum*, a multimeric nonheme ferritin DpsA-like protein is found to sequester iron in response to the oxidative stress exerted by excess iron (Reindel et al. 2005). This protein was downregulated under iron-deficient conditions. It exhibits the features of nonheme bacterial ferritins that are expressed to sequester the excess iron. Their expression is repressed under conditions of iron starvation. Kaur et al. (2006) have also proposed that the chelation of Fe^{2+} by the ferritin DpsA is a mechanism for detoxifying Fe^{2+} in *Halobacterium* sp. strain NRC-1. Transcription of DpsA is upregulated by Fe^{2+}, Co^{2+}, and Zn^{2+} and downregulated by $Mn^{2+}$and Fe deficiency conditions; the transcription regulators for this mechanism have not been identified. Furthermore, among haloarchaea, *Halococcus saccharolyticus*, *Halogeometricum sp.*, *Halorubrum saccharovorum*, and *Haloterrigena turkmenica* are reported to produce carboxylate-type siderophores (Dave et al. 2006). Overexpression of siderophores in haloarchaea increases chelation in case of iron deficiency. On the other hand, repression of the siderophores in the presence of excess iron may avoid uptake (Malki et al. 2009; Dave et al. 2006).

9.3.2 Metabolites

Haloarchaea are able to synthesize exopolysaccharide (EPS) as a protective mechanism for survival under stressed conditions such as nutrient depletion, variation in temperature, and the presence of toxic chemical compounds (Poli et al. 2011). Further, Kawakami et al. (2007) have reported a Ca^{2+}-dependent aggregation system in *Halobacterium salinarum* CCM 2090 wherein the divalent ion binds to certain aggregation factors present on the cell surface which stimulates ionic cross-linking between the EPS molecules resulting in conglomeration of the haloarchaeal cells. Moreover, certain receptor proteins are also demonstrated to be present on the cell surface that interact with Ca^{2+} to form cell aggregates. As explained by Kawakami et al. (2007), Ca^{2+} being the twentieth element found in the fourth row of the periodic table could be replaced by other transition metal ions such as Mn^{2+}, Cr^{2+}, Fe^{2+}, Co^{2+}, Ni^{2+}, Cu^{2+}, and Zn^{2+}, belonging to the same row. It is proposed that during aggregate formation, the distinctive electronic configuration of these metals may be responsible for substituting with Ca^{2+}. This implicates that the tolerance to these metal ions may be mediated in the course of binding with EPS. Furthermore, in the presence of certain other metals lacking this electronic configuration, such as Mg^{2+} and Sr^{2+} (alkali earth metals), Mo^{2+}, $Cd^{2+}$and Sn^{2+}, and $Hg^{2+}$and Pb^{2+} belonging to fifth and sixth periods, respectively, no conglomeration was observed. Bolhuis et al. (2006) have proposed a *cbp*, encoding the cell surface calcium-binding acidic-repeat protein that is involved in Ca^{2+}-dependent aggregation in *Haloquadratum walsbyi*, *Haloarcula marismortui*, *Haloterrigena turkmenica*, and *Halobacterium* sp. strain NRC-1, although its role in this process remains to be demonstrated. Iron uptake studied in *Halobacterium salinarum* by Hubmacher et al. (2007) revealed the role of xenosiderophores (triacetylfusarinin) for iron uptake that is an energy-dependent process, either dependent on the respiratory chain or on the existence of the membrane potential. Iron transport across the cell membrane was proposed to include a reductive step that is located at the cell surface or existence of a ligand exchange mechanism at the cell wall or at the membrane.

9.3.3 Efflux Pumps

As reviewed by Srivastava and Kowshik (2013), efflux pumps (Fig. 9.1) are one of the most common mechanisms of resistance to inorganic ions in microbes including Archaea.

9.3.4 P_{1B}-Type ATPases

The P_{1B}-type ATPases are a large family of integral membrane proteins driven by ATP hydrolysis (Fagan and Saier 1994). Metal ions such as Mg^{2+}, Ca^{2+}, Cu^{2+}, Ag^{2+}, Zn^{2+}, and Cd^{2+} act as substrates to these ATPases (Saier 1994). These transporters serve the purpose of uptake of essential elements and efflux of toxic elements, thus conferring resistance to the expelled metal ion (Snavely et al. 1989; Nies 2003). In a system

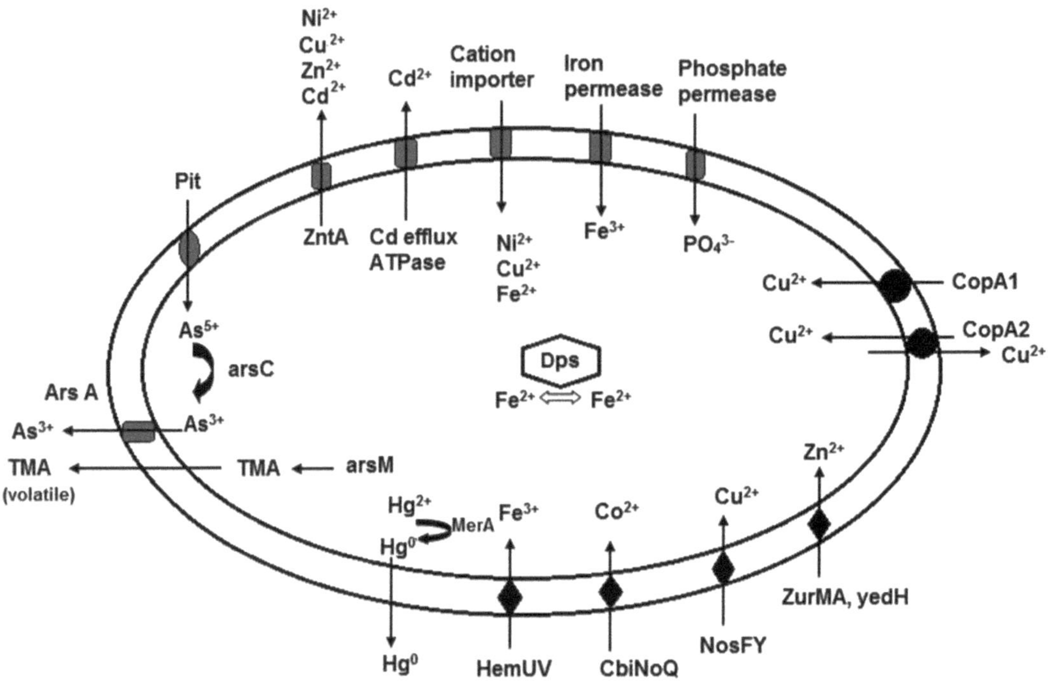

Fig. 9.1 Various metal transporters reported in *Halobacterium* sp. strain NRC-1

level analysis of *Halobacterium* sp. strain *NRC-1* demonstrated by Kaur et al. (2006), the functionality and role of such transporters in metal resistance exhibited upregulation of *yvgX*, a P1B-typeATPase, in response to $Cu^{2+}$and Zn^{2+}metal stress. In bacteria, the *yvgX* family is known to encode two kinds of CopA proteins, CopA1 and CopA2 (Klein and Lewinson 2011). CopA1 is essential for copper influx and tolerance, while CopA2 is involved in the influx/efflux of Cu and its transport to Cu-containing enzyme cytochrome oxidase c (Klein and Lewinson 2011; Raimunda et al. 2011). Archaea are reported to prefer CopA2 over CopA1 (Coombs and Barkay 2005). The *yvgX* of *Halobacterium* sp. strain *NRC-1* was found to be more specific for Cu^{2+} efflux family as the Δ*yvgX* strain was susceptible to Cu^{2+} and not to Zn^{2+}or $Co^{2+}$and therefore belongs to the CopA2 family of proteins and is found in *Haloarcula marismortui*, *Haloarcula hispanica*, and *Haloquadratum walsbyi* (Orell et al. 2012). In uptake of copper, Cu-CPx-type ATPases have also been shown to be involved to satisfy the cellular demands (Tottey et al. 2001; Solioz and Stoyanov 2003). In *Halobacterium* sp. strain *NRC-1*, the *cpx* gene that encodes CPx P1B-type ATPases was found to be downregulated by Fe^{2+}, Cu^{2+}, and Ni^{2+} to prevent the upsurge of toxic metal within the cell (Kaur et al. 2006).

9.3.5 Cation Diffusion Facilitators (CDF) Metal Transporters

The CDF family of transport proteins is ubiquitously present in all three domains of life (Paulsen and Saier, Jr. 1997). CDF functions as homo-dimeric antiporters, transporting metal ions against concentration gradients using H^+/K^+ to create electrochemical gradient. Based upon their substrate specificity, CDFs have been classified as Zn^{2+}-CDF, Fe/Zn-CDF, and Mn-CDF (Montanini et al. 2007). They usually possess six transmembrane domains (TMDs) with a cytoplasmic N- and C-terminal and a histidine loop of variable length between TMD IV and V (Haney et al. 2005; Anton et al. 1999). The amphipathic domains TMD I, II, V, and VI are involved in metal transfer and are the most conserved, while the hydrophobic TMD III and IV are critical for zinc specificity and mutations wherein these domains alter substrate specificity (Montanini et al. 2007). The genome of *Halobacterium* sp. strain *NRC-1* has disclosed a putative CDF Cd $^{2+}$transporter *ZntX*, which confers resistance against Ni^{2+}, Cu^{2+}, and Zn^{2+} besides Cd^{2+} (Kaur et al. 2006). The role of this transporter in metal resistance has been confirmed by upregulation of *ZntA* in response to heavy metals (Cu and/or Zn) and poor growth of Δ*zntA* strain in the presence of Ni^{2+}, Cu^{2+}, Zn^{2+}, and Cd^{2+} (Kaur et al. 2006). In *Haloarcula hispanica* and *Haloarcula marismortui*, *Znt*A transporter for Zn^{2+}transport has also been elucidated. For inorganic metal ion transport, a putative CDF family protein has also been found on the chromosome of *Natrialba magadii* (Orell et al. 2012).

9.3.6 ATP-Binding Cassette (ABC) Transporters

ABC transporters are ubiquitously present in all three domains of life from bacteria to eukaryotes and Archaea. They play a crucial role in substrate uptake, export, and osmoregulation (Albers et al. 2004). ABC transporters for sugar and polypeptide have been found in *Haloferax volcanii* (Hartman et al. 2010), *Haloarcula marismortui* (Baliga et al. 2004), *Halobacterium* sp. *NRC- 1* (Kaur et al. 2006), *Natronomonas pharaonis* (Falb et al. 2005), and *Haloquadratum walsbyi* (Bolhuis et al. 2006). Many of the ABC transporters are metal ion transporters in *Halobacterium* sp. such as *cbiNOQ* for Co^{2+}transport (Roth et al. 1993), *hemUV* for iron uptake (Faraldo-Gomez and Sansom 2003; Schneider and Paoli 2005), *nosFY* for copper (Zumft et al. 1990), and *zurMA* for zinc transport.

9.3.7 Metal Resistance Operons

Haloarchaea are reported to possess large plasmids known as minichromosomes or megaplasmids. These minichromosomes possess genes for antibiotic resistance or metal resistance that may be indispensable for survival of haloarchaea (DasSarma et al. 2009). The model organism *Halobacterium* sp. strain *NRC*-1 harbors pNRC100 (191 kb), one of the two megaplasmids that possess *arsADRC* gene cluster, which confers resistance to arsenate (As^{5+}) and arsenite (As^{3+})/antimonite (Sb^{3+}) (Kaur et al. 2006). As^{5+}can be taken up by the cells through phosphate transporters (pit/pst) and As^{3+} by aquaglyceroporins (glycerophorin membrane transport proteins) (Mukhopadhyay et al. 2002) or hexose transporters (Suzuki and Matsushita 1968) followed by its conversion to $As^{3+}$by arsenate reductase encoded by *arsC*. *arsA* codes for P_{1B}-type ATPase transporters that help in extrusion of As^{3+}/Sb^{3+} from the cell. *arsR* and *arsD* encode trans-acting repressors of the operon. ArsR and ArsD bind to As^{3+}/Sb^{3+} resulting in expression of the *arsA* and *arsC*. Arsenate reductase encoded by *arsC* is expressed weakly in *Halobacterium* sp. strain *NRC-1*, and therefore deletion of *arsC* and *arsADRC* was found to be ineffective in conferring arsenate sensitivity. The operon *arsADRC* was found to be inducible by arsenite and antimonite. *Halobacterium* sp. strain NRC-1 also harbors both arsA in ars operon on the megaplasmid pNRC100 and arsB on the main chromosome. However, arsB was found to play no significant role in arsenic resistance in this organism. Thus, it has been proposed that *Halobacterium* sp. strain *NRC*-1 harbors a novel transporter unrelated to arsB although with a similar function (Wang et al. 2004). Another type of arsenite resistance operon is found in *Halobacterium* sp. strain *NRC-1*, *arsR2M*, which is present upstream of *arsADRC* on pNRC100, where *arsR2* is constitutively expressed while

As^{3+}/Sb^{3+} induces the expression of *arsM* (Wang et al. 2004). The *arsR2* is analogous to *arsR* and *arsM* encodes a putative As^{3+}-methyltransferase that is reported to be analogous to mammalian methyltransferases and S-adenosyl methionine-dependent methyltransferases of *Magnetospirillum magnetotacticum.*

ArsM is involved in converting As^{3+} to methylated species like dimethylarsinate (DMA), trimethylarsine oxide (TMAO), or trimethylarsine (TMA) gas (Cullen and Bentley 2005). *arsM* shows an increased sensitivity to arsenite but fails to do so toward arsenate or antimonite (Wang et al. 2004). Thus, *arsM* confers two possible mechanisms of As^{3+}resistance. First, a concentration gradient is generated that results in the movement of methylated arsenite (negatively charged/uncharged) out of the cell. Second, As^{3+} is eliminated by the formation of volatile trimethylarsine out of the cell (Yuan et al. 2008). In *Halobacterium* sp. strain *NRC-1*, the arsenite resistance is brought about by *arsM* gene present as a part of the *arsR2M* operon (Wang et al. 2004). The *mer* operon confers resistance to mercury in Archaea and bacteria that is involved in detection, regulation, transport, and reduction of Hg^{2+} (Osborn et al. 1997). The *merRHAI* operon of thermoacidophilic archaeon *Sulfolobus solfataricus* is the best studied mercury resistance operon in Archaea. The MerR regulator maintains the operon under control, which then represses the operon in the absence of Hg^{2+} and thus in its presence augments the transcription. $Hg^{2+}$is bound by a TRASH (trafficking, resistance, and sensing of heavy metals) domain MerH, a metallochaperone, while the reduction and detoxification to volatile Hg (0) are carried out by MerA, a mercuric reductase (Schelert et al. 2006). Some *mer* operons carry additional *mer* genes, notably *merB*, an organomercurial lyase that cleaves the C-Hg bonds of organomercurials, and the released Hg^{2+} is reduced to Hg (0) by MerA. *merA* and *merB* genes are also found in *Halobacterium* sp. strain *NRC-1* and *Haloterrigena*, respectively, (Osborn et al. 1997).

9.3.8 Scope of Haloarchaea: Metal Interaction Studies

Although haloarchaea is reported to interact with metal ions also as a bioadsorbent of Mn^{2+}, Pb^{2+}, Cr^{2+}, Ni^{2+}, Zn^{2+}, As^{2+}, Cd^{2+}, Al^{2+}, and Hg^{2+} (Naik and Furtado 2014; Popescu and Dumitru 2009; Williams et al. 2012), it is also explored for synthesis of metal nanoparticles extracellularly (Patil et al. 2014) and intracellularly (Srivastava et al. 2013). Haloarchaeal members are bestowed with unique cell envelope structure consisting of S-layers comprising protein or often glycoprotein subunits that form a self-assembled two-dimensional array enclosing the whole cell surface and glycerol ether core lipids in their membranes (Asker et al. 2002). Three haloarchaeal S-layer glycoproteins have been studied in detail, i.e., the S-layer glycoproteins of *Halobacterium salinarum* (Lechner and Sumper 1987), *Haloferax volcanii* (Sumper et al. 1990), and *Haloarcula japonica* (Wakai et al. 1997). Recently, S-layers of *Bacillus sphaericus* JG-A12 were used as a template to fabricate well-separated Pd, Pt, and Au nanoparticles that are ordered in regular arrays (Pollman and Matys 2007; Merroun et al. 2007). Consequently, haloarchaeal S-layers can be explored as a prospective cell surface structure for nanoparticle studies of nanotechnological significance.

9.3.9 Future Aspects

Haloarchaea being predominant in solar salterns located along the estuary which often serve as a sink of metal toxicants (Chapman and Wang 2001; Litchfield and Gillevet 2002) are sparsely explored for their metal resistance mechanisms. Development of genetic tools and understanding the molecular mechanism underlying the metal resistance pattern in haloarchaea would assist in enlightening the knowledge of metal physiology in haloarchaea that would make haloarchaea a potential candidate for remediation of metal-polluted hypersaline environment (Table 9.1).

Table 9.1 Heavy metal resistance of haloarchaea isolated from various econiche

Haloarchaeal sp.	Isolation site	Metal ion(s)	MIC (mM)	References
Halobacterium sp.	Solar salterns, Spain	Zn^{2+}, Cu^{2+}, Ag^{+}, Hg^{2+}, As^{3+}, Cd^{2+}	0.5, 1–2.5, 0.5, 0.01, 20, ≤2.5	Nieto et al. (1987)
Halocula sp.	Solar salterns, Spain	Zn^{2+}, Cu^{2+}, Ag^{+}, Hg^{2+}, As^{3+}, Cd^{2+}	0.05, 2.5, 0.05, 0.01, 10, 0.05	Nieto et al. (1987)
Halobacterium sp.	Estuarine salterns, Goa, India	Li^{+}, As^{5+}, As^{3+}, Mn^{2+}, Ni^{2+}, Cu^{2+}, $Zn^{2+}/Cd^{2+}/Fe^{2+}/Fe^{3+}$	200, 60, 3, 50, 5, 2.5, 2	Khandavilli et al. (1999)
Halobacterium sp. NRC-1	Salt lake	Fe^{2+}, Mn^{2+}, Co^{2+}, Zn^{2+}, Ni^{2+}, Cu^{2+}	7.5, 2, 0.6, 0.05, 2, 1.2	Kaur et al. (2006)
Haloferax mediterranei	Solar saltern	Zn^{2+}, Ni^{2+}, Pb^{2+}, Cr^{6+}	1, 2.5, 2.5, 5.0	Popescu and Dumitru (2009)
Haloferax sp. TL 5	Telega salt lake, Romania	Zn^{2+}, Ni^{2+}, Pb^{2+}, Cr^{6+}	0.5, 2.5, 2.5, 5.0	Popescu and Dumitru (2009)
Halobacterium saccharovorum, *H. salinarum*, *Natronobacterium gregoryi*	Solar salterns Tamil Nadu, India	Zn^{2+}, As^{3+}, Cd^{2+}	0.01, 0.001, 0.001	Williams et al. (2012)
Haloarcula sp. IRU1	Hypersaline Urmia lake, Iran	As^{5+}	0.65	Taran et al. (2013)
Haloferax volcanii BBK2, *Halorubrum* strain BS17, *Haloarcula japonica* BS2, *Halococcus* strain BK6	Estuarine salterns, Goa, India	Zn^{2+}	1, 0.5, 0.5, 1, respectively.	Salgaonkar et al. (2015)

References

Ahmed M, Malik A (2012) Bioaccumulation of heavy metals by Zn resistant bacteria isolated from agricultural soils irrigated with waste water. Bacteriol J 2(1):12–21

Albers SV, Koning SM, Konings WN, Driessen AJ (2004) Insights into ABC transport in archaea. J Bioenerg Biomembr 36(1):5–15

Anton A, Grobe C, Reibmann J, Pribyl T, Nies DH (1999) CzcD is a heavy metal ion transporter involved in regulation of heavy metal resistance in Ralstonia sp. strain CH34. J Bacteriol 181(22):6876–6881

Asker D, Awad T, Ohta Y (2002) Lipids of Haloferax alexandrinus strain TM(T): an extremely halophilic canthaxanthin producing archaeon. J Biosci Bioeng 93(1):37–43

Baliga NS, Bonneau R, Facciotti MT et al (2004) Genome sequence of *Haloarcula marismortui*: a halophilic archaeon from the dead sea. Gen Res 14(11):2221–2234

Beal EJ, House CH, Orphan VJ (2009) Manganese- and iron dependent marine methane oxidation. Science 325:184–187

Bini E (2010) Archaeal transformation of metals in the environment. FEMS Microbiol Ecol 73(1):1–16

Bolhuis H, Palm P, Wende A et al (2006) The genome of the square archaeon *Haloquadratum walsbyi*: life at the limits of water activity. BMC Genomics 7:169

Chapman PM, Wang F (2001) Assessing sediment contamination in estuaries. Environ Toxicol Chem 20(1):3–22

Coombs JM, Barkay T (2005) New findings on evolution of metal homeostasis genes: evidence from comparative genome analysis of bacteria and archaea. Appl Environ Microbiol 71:7083–7091

Cullen WR, Bentley R (2005) The toxicity of trimethylarsine: an urban myth. J Environ Monit 7(1):11–15

DasSarma S, Capes M, DasSarma P (2009) Haloarchaeal megaplasmids. In: Schwartz E (ed) Microbial megaplasmids. Springer, Berlin, pp 3–30

Dave BP, Anshuman K, Hajela P (2006) Siderophores of halophilic archaea and their chemical characterization. Indian J Exp Biol 44(4):340–344

Delong E (2007) Microbial domains in the ocean: a lesson from the archaea. Oceanography 20(2):124–129

Fagan MJ, Saier MH Jr (1994) P-type ATPases of eukaryotes and bacteria: sequence analyses and construction of phylogenetic trees. J Mol Evol 38(1):57–99

Falb M, Pfeiffer F, Palm P et al (2005) Living with two extremes: conclusions from the genome sequence of *Natronomonas pharaonis*. Gen Res 15(10):1336–1343

Faraldo-Gomez JD, Sansom MSP (2003) Acquisition of siderophores in gram-negative bacteria. Nat Rev Mol Cell Biol 4:105–116

Haney CJ, Grass G, Franke S, Rensing C (2005) New developments in the understanding of the cation diffusion facilitator family. J Ind Microbiol Biotechnol 32(6):215–226

Hartman AL, Norais C, Badger JH et al (2010) The complete genome sequence of *Haloferax volcanii* DS2, a model archaeon. PLoS One 5(3):9605

Hu H, Jin Q, Kavan P (2014) A study of heavy metal pollution in China: current statues, pollution control policies and counter measures. Sustainability 6:5820–5838

Hubmacher D, Matzanke B, Anemuller FS (2007) Iron uptake in the Euryarchaeon *Halobacterium salinarum*. Biometals 20:539–547

Kashefi K, Lovley DR (2000) Reduction of Fe(III) and Mn(IV) and toxic metals at 100 °C by *Pyrobaculum islandicum*. J Bacteriol 66:1050–1056

Kashefi K, Tor JM, Nevin KP, Lovley DR (2001) Reductive precipitation of gold by dissimilatory Fe(III)-reducing bacteria and archaea. Appl Environ Microbiol 67:3275–3279

Kashefi K, Moskowitz BM, Lovley DR (2008a) Characterization of extracellular minerals produced during dissimilatory Fe (III) and U (VI) reduction at 100 degrees C by *Pyrobaculum islandicum*. Geobiology 6:147–154

Kashefi K, Shelobolina ES, Elliott WC, Lovley DR (2008b) Growth of thermophilic and hyperthermophilic Fe(III)- reducing microorganisms on a ferruginous smectite as the sole electron acceptor. Appl Environ Microb 74:251–258

Kaur A, Pan M, Meislin M, Facciotti MT, El-Gewely R, Baliga NS (2006) A systems view of haloarchaeal strategies to withstand stress from transition metals. Genome Res 16(7):841–854

Kawakami Y, Hayashi N, Ema M, Nakayama M (2007) Effects of divalent cations on *Halobacterium salinarum* cell aggregation. J Biosci Bioeng 104(1):42–46

Khandavilli S, Sequeira F, Furtado I (1999) Metal tolerance of extremely halophilic bacteria isolated from estuaries of Goa, India. Ecol Environ Conserv 5(2):149–152

Klein JS, Lewinson O (2011) Bacterial ATP-driven transporters of transition metals: physiological roles, mechanisms of action, and roles in bacterial virulence. Metallomics 3(11):1098–1108

Kumar A, Bisht BS, Joshi VD, Dhewa T (2011) Review on bioremediation of polluted environment: a management tool. Int J Environ Sci 1:1079–1093

Lechner J, Sumper M (1987) The primary structure of a prokaryotic glycoprotein. Cloning and sequencing of the cell surface glycoprotein gene of halobacteria. J Biol Chem 262:9724–9729

Litchfield CD, Gillevet PM (2002) Microbial diversity and complexity in hypersaline environments: a preliminary assessment. J Ind Microbiol Biotechnol 28(1):48–55

Maezato Y, Blum P (2012) Survival of the fittest: overcoming oxidative stress at the extremes of acid, heat and metal. Life 2:229–242

Malki L, Yanku M, Borovok I, Cohen G, Mevarech Aharonowitz MY (2009) Identification and characterization of gshA, a gene encoding the glutamate-cysteine ligase in the halophilic archaeon *Haloferax volcanii*. J Bacteriol 191(16):5196–5204

Matyar F, Akkan T, Ucak Y, Eraslan B (2010) *Aeromonas* and *Pseudomonas*: antibiotic and heavy metal resistance species from Iskenderun Bay, Turkey (northeast Mediterranean Sea). Environ Monit Assess 73:263–273

Merroun M, Rossberg A, Hennig C, Scheinost A, Selenska-Pobell S (2007) Spectroscopic characterization of gold nanoparticles formed by cells and S-layer-protein of *Bacillus sphaericus* JGA12. Mater Sci Eng C 27:288–292

Montanini B, Blaudez D, Jeandroz S et al (2007) Phylogenetic and functional analysis of the Cation Diffusion Facilitator (CDF) family: improved signature and prediction of substrate specificity. BMC Genomics 8:107

Mukhopadhyay R, Rosen BP, Phung LT et al (2002) Microbial arsenic: from geocycles to genes and enzymes. FEMS Microbiol Rev 26(3):311–325

Naik S, Furtado I (2014) Equilibrium and kinetics of adsorption of Mn^{2+} by Haloarchaeon GUSF (MTCC3265). Geomicrobiol J 31(8):708–715

Nies DH (2003) Efflux-mediated heavy metal resistance in prokaryotes. FEMS Microbiol Rev 27(2–3):313–339

Nieto JJ, Ventosa A, Ruiz-Berraquero F (1987) Susceptibility of halobacteria to heavy metals. Appl Environ Microbiol 53(5):1199–1202

Orell A, Navarro CA, Rivero M, Aguilar JS, Jerez CA (2012) Inorganic polyphosphates in extremophiles and their possible functions. Extremophiles. doi:10.1007/s00792-012-0457-9

Oren A (2012) Taxonomy of the family *Halobacteriaceae*: a paradigm for changing concepts in prokaryote systematic. Int J Syst Evol Microbiol 62(2):263–271

Osborn AM, Bruce KD, Strike P, Ritchie DA (1997) Distribution, diversity and evolution of the bacterial mercury resistance (mer) operon. FEMS Microbiol Rev 19(4):239–262

Patil S, Fernandes J, Tangasali R, Furtado I (2014) Exploitation of *Haloferax alexandrinus* for biogenic synthesis of silver nanoparticles antagonistic to human and lower mammalian pathogens. J Clust Sci. doi:10.1007/s10876-013-0621-0

Paulsen IT, Saier MH Jr (1997) A novel family of ubiquitous heavy metal ion transport proteins. J Mem Biol 156(2):99–103

Poli A, Donato P Di, Abbamondi GR, Nicolaus B (2011) Synthesis, production and biotechnological applications of exopolysaccharides and polyhydroxyalkanoates by archaea. Archaea article id 693253:13

Pollman K, Matys S (2007) Construction of S-layer protein exhibiting modified self assembling properties and enhanced metal binding properties. Appl Microbiol Biotechnol 75:1079–1085

Popescu G, Dumitru L (2009) Biosorption of some heavy metals from media with high salt concentrations by halophilic archaea. Biotechnol Biotechnol Equip 23(sup1):791–795

Raimunda D, Gonzalez-Guerrero M, Leeber BW III, Arguello JM (2011) The transport mechanism of bacterial Cu^+-ATPases: distinct efflux rates adapted to different function. Biometals 24(3):467–475

Reindel S, Schmidt CL, Anemuller S, Matzanke BF (2005) Expression and regulation pattern of ferritin-like DpsA in the Archaeon *Halobacterium salinarum*. Bio Met 18(4):387–397

Roth JR, Lawrence JG, Rubenfield M et al (1993) Characterization of the cobalamin (vitamin B12) biosynthetic genes of *Salmonella typhimurium*. J Bacteriol 175(11):3303–3316

Saier MH Jr (1994) Computer-aided analyses of transport protein sequences: gleaning evidence concerning function, structure, biogenesis, and evolution. Microbiol Rev 58(1):71–93

Salgaonkar BB, Das D, Branganca JM (2015) Resistance of extremely halophilic archaea to zinc and zinc oxide nanoparticle. Appl Nanosci 6(2):251–258

Schelert J, Drozda M, Dixit V, Dillman A, Blum P (2006) Regulation of mercury resistance in the crenarchaeote *Sulfolobus solfataricus*. J Bacteriol 188(20):7141–7150

Schneider S, Paoli M (2005) Crystallization and preliminary Xray diffraction analysis of the haem-binding protein HemS from *Yersinia enterocolitica*. Acta Crystallogr 61(8):802–805

Silveira CB, Cardoso AM, Coutinho FH et al (2013) Tropical aquatic archaea show environment-specific community composition. PLoS One 8(9):e76321

Snavely MD, Florer JB, Miller CG, Maguire ME (1989) Magnesium transport in *Salmonella typhimurium*: 28 Mg^{2+} transport by the CorA, MgtA, and MgtB systems. J Bacteriol 171(9):4761–4766

Solioz M, Stoyanov JV (2003) Copper homeostasis in *Enterococcus hirae*. FEMS Microbiol Rev 27(2–3):183–195

Srivastava P, Kowshik M (2013) Mechanisms of metal resistance and homeostasis in haloarchaea. Archaea. doi:10.1155/2013/732864

Srivastava P, Braganca J, Ramanan SR, Kowshik M (2013) Synthesis of silver nanoparticles using haloarchaeal isolate *Halococcus salifodinae* BK3. Extremophiles 17(5):821–831

Sumper M, Berg E, Mengele R, Strobel I (1990) Primary structure and glycosylation of the S-layer protein of *Haloferax volcanii*. J Bacteriol 172:7111–7118

Suzuki Y, Matsushita H (1968) Interaction of metal ions and phospholipids monolayers as a biological membrane model. Ind Health 6(3):128–133

Tabak HH, Lens P, Hullebusch VED, Dejonghe W (2005) Developments in bioremediation of soils and sediments polluted with metals and radionuclides-1. Microbial processes and mechanisms affecting bioremediation of metal contamination and influencing metal toxicity and transport. Rev Environ Sci Biotechnol 4(3):115–156

Taran M, Safari M, Monaza A, Reza JZ, Bakhtiyari S (2013) Optimal condition for the biological removal of arsenic by novel halophilic archaea in different condition and its process optimization. Pol J Chem Technol 15(2):7–9

Tottey S, Rich PR, Rondet SAM, Robinson NJ (2001) Two menkes-type ATPases supply copper for photosynthesis in synechocystis PCC6803. J Biol Chem 276(23):19999–20004

Wakai H, Nakamura S, Kawasaki H, Takada K, Mizutani S, Aono R, Horikoshi K (1997) Cloning and sequencing of the gene encoding the cell surface glycoprotein of *Haloarcula japonica* strain TR-1. Extremophiles 1:29–35

Wang G, Kennedy SP, Fasiludeen Rensing SC, DasSarma S (2004) Arsenic resistance in *Halobacterium* sp. strain NRC-1 examined by using an improved gene knockout system. J Bacteriol 186(10):3187–3194

Williams GP, Gnanadesigan M, Ravikumar S (2012) Biosorption and bio-kinetic properties of solar saltern halobacterial strains for managing Zn^{2+}, As^{2+} and Cd^{2+} metals. Geomicrobiol J. doi:10.1080/01490451.2012.732663

Yuan C, Lu X, Jie Q, Rosen BP, Le XC (2008) Volatile arsenic species released from *Escherichia coli* expressing the asIII S-adenosylmethionine methyltransferase gene. Environ Sci Technol 42(9):3201–3206

Zhuang W, Gao X (2014) Integrated assessment of heavy metal pollution in the surface sediments of the Laizhou Bay and the coastal waters of the Zhangzi Island, China: comparison among typical marine sediment quality indices. PLoS ONE 9(4):e94145

Zumft WG, Viebrock-Sambale A, Braun C (1990) Nitrous oxide reductase from denitrifying *Pseudomonas stutzeri*. Genes for copper-processing and properties of the deduced products, including a new member of the family of ATP/GTP-binding proteins. Eur J Biochem 192(3):591–599

Manganese-Tolerant Bacteria from the Estuarine Environment and Their Importance in Bioremediation of Contaminated Estuarine Sites

10

Flory Pereira

Abstract

Eco-sensitive environs such as coastal and marine ecosystems of the world are constantly facing the risk of destruction. Estuaries, mangroves and solar salterns being fragile econiches are very susceptible to disturbances, both natural and anthropogenic. Metal mining is currently one of the most polluting anthropogenic activities, projected to have a major impact on the ecosystem. Estuaries provide a major pathway for transferring pollutants such as metals into the oceans and back again from the oceans to the rivers which flow into the adjoining areas such as salt marshes, mangroves and salterns. The sediments of these econiches serve as an ecological sink, where the metals concentrate to several orders of magnitude above the normal background levels. Consequently, microorganisms inhabiting these sites are adapted to both the estuarine and their own unique environment. Several groups of multimetal-tolerant bacteria have been isolated from such environments. Manganese (Mn), though an indispensible metal for biological function, could result in toxicity at elevated concentrations. Consumption of solar salt contaminated with metals like manganese could be a major route for human exposure. The mitigation of manganese from such contaminated sites by Mn-tolerant bacteria provides a safe and environment-friendly alternate technology for the future. Manganese-tolerant bacteria are capable of scavenging not only Mn but also many other metal contaminants, viz. Co, Ni, Zn, Cu, Pb, Cd and Hg. Cumulative strategies by which these estuarine bacteria resist high concentrations of manganese include extracellular sequestration, biosorption, precipitation, oxidation and regulation of stress proteins. This article seeks to give an insight into some of the molecular mechanisms adopted by halotolerant

F. Pereira (✉)
Department of Microbiology, PES's Shri Ravi Sitaram Naik College of Arts and Science, Farmagudi, Ponda, Goa, India
e-mail: florycliffy@gmail.com

M.M. Naik, S.K. Dubey (eds.), *Marine Pollution and Microbial Remediation*,
DOI 10.1007/978-981-10-1044-6_10

bacteria from the estuarine environment for tolerating manganese, as they could be a key to effective minimization and mitigation of mining impacts in contaminated estuarine sites.

10.1 Introduction

Due to the toxic and persistent character of metals, contamination with metals in coastal and marine environments is one of the critical issues (Varol and Sen 2012; Zhan et al. 2010) for both the naturally stressed marine ecosystem and humans who are dependent on marine resources for food, recreation and industry (Naser 2013). Estuaries form a zone where a transition from a river environment to a maritime environment takes place (Yang et al. 2014); therefore, they are governed both by riverine and marine influences. Since they receive large volumes of nutrient inputs, both from freshwater runoffs and pollutants entering seawater, they represent some of the most biologically productive ecosystems on the planet (Merrifield et al. 2011; Edgar et al. 2000). They also provide invaluable ecosystem services upon which all organisms depend (Daily et al. 1997). Metals gain entry into coastal and marine environments via varied sources and activities, including anthropogenic disturbances and coastal development. High concentrations of metals are introduced into the environment during metal mining activities and during the processing of metals for industrial use. Other factors that contribute to metal pollution in the estuarine environment are the disposal of wastes and effluents containing metals from various industries such as the agriculture and petroleum industry (Oyetibo et al. 2010). Bonnevie et al. (1994) and Al-Masri et al. (2002) have reported local sources specifically known to be the leading causes of metal pollution in sediments, viz. metal-based industries (e.g. Cd, Cr and Zn from electroplating; Ni, Pb and Cu discharge from smelters; Pb and As from petroleum refineries; paint and dye formulators containing metals such as Cu, Zn, Cd, Cr, Se, Pb and Hg), as well as effluents from plants manufacturing chemicals. On similar lines, Ross (1994) divided metal contamination from anthropogenic sources into five main groups: (1) metals released from natural atmospheric deposits such as copper, chromium, cadmium, uranium, lead, arsenic and mercury; (2) metals from mining and smelting of ores such as iron, manganese, lead, cadmium, mercury and arsenic (Zouboulis et al. 2004); (3) metals used in industries such as cobalt, nickel, zinc, copper, chromium, cadmium, arsenic and mercury; (4) metals in agriculture (zinc, copper, cadmium, selenium, uranium, lead and arsenic); and (5) metals contained in disposed wastes such as copper, zinc, chromium, cadmium, lead, mercury and arsenic. When the threshold levels are exceeded, all metals can be toxic to marine and estuarine organisms, as well as to their consumers at higher trophic levels including man. This critical limit depends on the availability of the metal to the target species, which is further governed by environmental factors.

Until a few years back, not much attention was paid to manganese globally, perhaps because it was commonly perceived to be a trace element that is essential but comparatively non-toxic (Kehres and Maguire 2003). However, manganese, being a constituent of multiple enzymes as well as functioning as an activator of many other enzymes, has a vital role in a number of physiological processes (Nielsen 1999). In vertebrates, it serves as a cofactor for arginase in urea formation (Sekowska et al. 2000); in gluconeogenesis, it serves as a cofactor for pyruvate carboxylase. It also promotes activities of other enzymes such as transferases, for example, glycosyltransferase in cartilage formation (Keen and Zidenberg-Cherr 1996), glutamine synthetase and aminopeptidase P (Yocum and Pecoraro 1999) in protein metabolism (Abell et al. 1995) and farnesyl pyrophosphate synthetase in fat metabolism (Institute of Medicine 2001). Manganese enzymes are particularly essential in detoxification of superoxide radicals by superoxide dismutase (Leach and

Harris 1997). Besides, Mn also has diverse functions within the bacterial cells. Manganese is a cofactor for enzymes such as PEP synthase (Chao et al. 1993), 6-phospho-β-glucosidase (Thompson et al. 1999), 3-phosphoglycerate mutase (Chander et al. 1998) and pyruvate carboxylase (Mukhopadhyay et al. 1998) involved in carbohydrate metabolism.

The importance of manganese in living systems is undeniable. In Cyanobacteria, oxygenic photosynthesis occurs in the reaction centre complex of photosystem II, which requires a tetra-Mn cluster (Morgan et al. 1998). The importance of Mn in bacterial signal transduction has been documented by Missiakas and Raina (1997). In nucleic acid synthesis, endonuclease IV (Hosfield et al. 1999) and ribonuclease HII (Ohtani et al. 2000) are Mn-dependent enzymes. All living cells are protected from reactive oxygen species (ROS) because of its detoxification by the enzymes mangani-catalase (Whittaker et al. 1999) and superoxide dismutase (Fridovich 1995), both of which require Mn^{2+}. Mn^{2+}-containing enzymes such as ortho-phosphatases are known to be involved in regulating vegetative growth and cell density in bacteria during stationary phase, fruiting body development, assimilation of carbon and nitrogen, response to stress, cell segregation (Shi 2004), formation of bacterial spores and cell wall stabilization (Doyle 1989). In addition, non-enzymatic Mn^{2+} is critical for appropriate functioning of diverse products of bacterial origin, such as antibiotics (Archibald 1986) and exopolysaccharides (Appanna 1988), and for oxidation of poly-β-hydroxy butyrate in *Sphaerotilus discophorus* as reported by Stokes and Powers (1967). Kirchner and Grabowski (1972) determined that even in the environment, manganese plays the indirect yet significant role of controlling the availability of nutrients in freshwater, by complexing it with iron.

However, excessive levels of essential elements too can be toxic at elevated levels (Gadd 1992). During the past century, industrialization has been occurring at a very fast pace prompting large-scale production of compounds containing manganese such as metallurgic and chemical products, varnishes, ceramics, fertilizers and fungicide products, alloys of iron and steel as well as livestock feeding supplements (HSDB 2001; ATSDR 2000; IPCS 1999). Manganese, due to its versatility, finds a variety of applications such as in the removal of green colour caused by the presence of iron contaminants in glass, as an additive in various products such as dry cell batteries for the prevention of hydrogen formation and in black paints as a drying agent. The use of manganese containing potassium permanganate as a bleaching agent and as an oxidant for cleaning and disinfection purposes is already well known (ATSDR 2000; HSDB 2001). In some locations, manganese greensands have been used for the treatment of drinking water (ATSDR 2000). Methylcyclopentadienyl manganese tricarbonyl (MMT), an organic manganese compound, finds extensive application in the United States of America (USA) in unleaded petrol as an octane-enhancing agent. Similar use of MMT has also been reported from Canada, South America, Europe and Asia (Lynam et al. 1999). All these products would finally form primary sources of contamination. In addition, sewage sludge and municipal wastewater discharges also contain manganese. Consequently, considering the vast applications of manganese, it is being investigated by scientists as a possible emerging contaminant, especially in the marine environment.

Whether manganese is toxic or not would depend on the permissible/safe levels in our waters and the dietary intake from all sources that has the ability to produce harmful effects. According to estimates by the World Health Organization (WHO), the daily dietary consumption of manganese in adults is between 0.7 and 10.9 mg, with the intake being higher for vegetarians consuming large proportions of manganese-rich nuts, legumes and grains in their diet as compared to the non-vegetarians (WHO 1973). WHO has set the acceptable limit of manganese in drinking water as 0.4 mg/l. Based on a review by Greger (1999), the Food and Nutrition Board of the Institute of Medicine (IOM 2002) sets the tolerable upper intake level of manganese for adults at 11 mg/day and the adequate intake levels at 2.3 mg/day for men and 1.8 mg/day for women. However, based on epidemiological studies on the effects of manganese exposure via drinking water on the central nervous system of humans, it has been estimated that 0.8 mg/kg/day was the lowest observed adverse effect level (LOAEL). More recent epidemiological and experimental studies however confirm that an undeniable link exists between manganese exposure and detrimental changes in the central nervous system. According to Adekunle et al. (2014), manganese is a toxic essential trace element, meaning that though it is necessary for the survival of humans, it is also toxic when present at high concentrations in the human body. Manganese toxicity occurs due to exposure to high levels of manganese, causing 'manganese madness' or 'Parkinson-like' diseases (Gerber et al. 2002; Erikson and Aschner 2003; Perl and Olanow 2007), neurological changes (Santamaria and Sulsky 2010) and abnormalities of the immune (Vartanian et al. 1999) and reproductive (Kim et al. 2012) systems. In mammals, large amounts of manganese have been reported to be toxic to the embryo and foetus and are also known to affect fertility. On the other hand, deficiency of manganese has also been reported to be teratogenic and affect fertility (Gerber et al. 2002). In general, carcinogenicity of compounds containing manganese follows the order $MnCl_2 > KMnO_4 > MnSO_4$. The toxicity though largely depends on the route of administration and the valence state, with Mn(II) being more toxic than Mn(III). Further, manganese toxicity also depends on the associated anion; for example, the chloride ion is less toxic than the citrate ion (Smith 1972). In many cases, metal toxicity is a direct outcome of biomagnification. Although biomagnification of manganese in food chains is insignificant (ATSDR 2000), bacterial mobilization and immobilization of manganese from sediments are noteworthy, especially in waters like that of solar salterns, where consumption of heavy metal-contaminated solar salt could be a major route for human exposure, with potential long-term implications on human health.

10.2 Manganese Biogeochemistry

In order to study manganese as a pollutant, it is imperative to comprehend the behaviour of manganese in the environment. In nature, manganese is present in various oxidation states, from +2 to +7; but the three main oxidation states between which it cycles is Mn^{2+} (which exists as the highly soluble $Mn(H_2O)_6^{2+}$), Mn^{3+} and Mn^{4+} (which precipitate typically as sparingly soluble oxyhydroxides) (Hansel and Francis 2006). Since Mn^{2+} is unstable, it tends to oxidize and precipitate or dissociate to either Mn^{3+} or Mn^{4+} (Tekerlekopoulou and Vayenas 2008). Cycling between these forms is typically due to abiotic or biotic oxidation and reduction reactions (Schamphelaire et al. 2008). The Mn^{2+} form is the only stable form in soil solution, while Mn^{3+} and Mn^{4+} are stable in the solid soil phase (McBride 1994). How manganese is transported and partitioned in water depends upon its solubility, which in turn is governed by other factors such as the oxidation-reduction potential, pH and the associated anions. If neutral conditions exist, then the Eh (redox potential) influences manganese mobility more than the pH. Under conditions where there is a deficiency of oxygen, manganese occurs as the highly soluble Mn^{2+}, but under oxygen-rich conditions, it is found as the oxidized, highly insoluble, stable MnO_2 form (Calvert and Pedersen 1996). The concentration of manganese could thus be much higher in anaerobic groundwaters compared to aerobic surface waters (Marbaniang 2012), because manganese would be released from its minerals and reduced to Mn^{2+} which is its more soluble form, when the conditions are anaerobic. As the depth increases, anaerobic conditions in the soil would increase, thereby increasing the reduction potential. Under such circumstances, oxygen is the first to be used up, and then nitrate and manganese are used as electron acceptors. As the reducing conditions progressively increase, iron gets reduced, followed by sulphate. Therefore, even though manganese comprises a negligible chemical constituent of fresh and marine water, the conversion of soluble Mn^{2+} to insoluble Mn^{3+} and Mn^{4+} oxides and oxyhydroxides is environmentally significant. This is supported by the fact that after oxygen, the oxides and oxyhydroxides of manganese are the most potent oxidizing agents in the environment and therefore may have an important role to play in aquatic geochemical cycles (Sunda and Huntsman 1990; Laha and Luthy 1990). In addition, oxides and oxyhydroxides have a remarkable charge distribution and surface area, which makes them among the most important and strongest sorbents and scavengers of many metals (Tebo et al. 2004). They thus form potentially rich reservoirs of adsorbed metals (Shi et al. 1999; Bratina et al. 1998).

The estuarine environment is known to be a highly dynamic environment, subject to fluctuations in various parameters. Manganese like all other metals is more soluble in freshwater compared to saline water. Pereira et al. (2013) reported that metal cycling and solubility in saltern water are affected by several parameters such as salinity, pH, Eh and dissolved oxygen. According to them, the mean value of manganese recorded in the overlying water of the Ribandar Saltern in Goa, India, was 0.6 ppm during the salt-making season, whereas for the non-salt-making season, it was lower at 0.4±0.3 ppm. Similarly, in the saltern sediment, manganese concentration varied by 34 % with a change in salinity. As the salinity increased, the metal-borne suspended particulate material quickly settled down due to increased coagulation and flocculation, thereby sedimenting the metal. When the salinity or pH of the overlying water decreased, the metals which had sedimented were released back into suspension, thereby making the metals bioavailable to the microorganisms. Saltern sediments have been shown to be dramatically influenced by cyclic changes in salinity and pH of the overlying water, thereby making metals more bioavailable and subsequently resulting in a larger metal-tolerant bacterial population. These metal-tolerant bacteria, originally of estuarine origin, play an important role in maintaining the metal concentration in the saltern well within permissible limits. However, the impact of environmental factors cannot be ignored. Experiments conducted by Spratt and Hodson (1994) showed that manganese oxidation rate in sediments from a salt marsh was linked to salinity and decreased with an increase in hypersaline conditions. However, in a similar study, Atkinson et al. (2007) showed that rather than changes in dissolved oxygen concentration (3–8 mg l^{-1}) or salinity (15–45 %), a change in pH of the overlying water from 5.5 to 8.0, and disturbance of the sediment by physical mixing, resulted in greater metal release and sequestration rates. Balzer (1982) reported that below 16 % saturation, i.e. under hypoxic conditions, the concentration of dissolved manganese increased to levels approaching 1500 μg/l, which is above that normally found in seawater. Huntsman and Sunda (1980) detected an increase in rate of oxidation of manganese when a catalytic surface like manganese dioxide was used or when the pH was increased. Manganese oxides have been reported to be precipitated extremely rapidly following oxidation, in a stream where there was acid mine drainage carrying manganese-rich inflows (Scott et al. 2002). Olaniran et al. (2013) conducted experiments on the solubility of cadmium, wherein they observed that when the pH was increased from 6 to 7 in 1.3 mM phosphate, its solubility decreased 8.8-fold. They demonstrated that metal solubility and bioavailability could decrease by several orders, even with a small pH change. Similar observations were made by Pereira et al. (2013), who found that the level of dissolved manganese in the saltern decreased approximately twofold, when the pH was increased from 7.4 to 8.4. Recent studies by Hallberg and Johnson (2005) concluded that very little oxidation of Mn^{2+} can be observed below pH 8 and that on many occasions it proceeds even more slowly in the presence of iron. Nairn and Hedin (1993) demonstrated that Fe^{2+} impedes biological oxidation of Mn^{2+} at ferrous iron concentrations exceeding 1 mg/l, thus preventing its removal from aerobic wetlands. In a study on the Ribandar Salterns of Goa, India, receiving its input from the adjacent Mandovi Estuary, Pereira et al. (2013) observed that the salterns receive a fair share of contaminants, especially ferrous iron, with concentrations in excess of 1 mg/l. In the presence of such high levels of Fe^{2+}, biological oxidation of Mn^{2+} is expected to be hampered. Thermodynamically, at neutral pH, the spontaneous oxidation of Mn^{2+} to Mn^{4+} should occur when the waters are well aerated. However, because Mn^{2+} requires higher activation energy compared to ferrous iron, the oxidation of Mn^{2+} would be greatly slowed down in non-acidic waters, making it far more stable than ferrous iron. Additionally, iron can form insoluble sulphides, but the same does not happen with manganese; consequently, iron can be removed continuously from the environment, but manganese almost never precipitates as the sulphide forms. Hence, it is generally precipitated as rhodochrosite ($MnCO_3$), and its removal by sulphate-reducing bacteria is uncommon. There have been solitary reports of γ-MnS (rambergite) nanoparticles being produced in the deep anoxic marine basins of the Baltic Sea, by the bacterium *Shewanella oneidensis* MR-1 due to dissimilatory metal reduction. The biogenic manganese sulphide produced was attributed to the simultaneous reduction of MnO_2 and thiosulphate together with the oxidation of H_2 (Lee et al. 2011). However, to date, no such reports of MnS production are available from the estuarine environments. Given the extreme conditions prevalent in salterns, the removal of manganese from saltern waters with concentrations exceeding 1.06 % in the sediment and 0.64 mg L^{-1} in water, respectively, and with a relatively high pollution load index >1 poses quite a challenge. However, the abundance of manganese-tolerant bacteria proves beneficial here (Pereira et al. 2013). Many microorganisms capable of catalysing Mn^{2+} oxidation have been identified, and a few of them are even known to utilize this reaction as a source of energy (Lovley and Phillips 1988; Nealson and Myers 1992; Ehrlich 1996; Tebo et al. 2005). The activation energy barrier for oxidation of manganese being overcome biologically, the adsorption of Mn^{2+} onto MnO_2 and its subsequent precipitation as Mn^{+4} become much more rapid.

10.3 Manganese as a Pollutant in the Marine Environment

The measure of manganese naturally occurring in groundwater is between 1 and 10 ppm, whereas the levels are as high as 100–1000 ppm in river water (Nealson 1983). Barceloux (1999) detected concentrations in freshwater typically ranging from 1 to 200 μg/l, contrary to the large amount (0.2 mg kg^{-1}) found in seawater by Bowen (1979) and Ehrlich (2002). According to ATSDR (2000), US EPA (1984) and Zeri et al. (2000), average manganese concentration in seawater is about 2 μg/l (Barceloux 1999), with the ambient concentrations ranging from 0.4 to 10 μg/l. In anaerobic regions of open seawater, Lewis and Landing (1991, 1992) have reported concentrations as high as 500 μg/l. In the ocean, Glasby (2006) obtained levels of dissolved manganese (Mn^{2+}) up to 0.2–3 nmol kg^{-1}. Smith et al. (1987) sampled 286 rivers and streams from the United States and detected an average dissolved manganese concentration of 24 μg/l. Howe et al. (2004) found manganese ranging from 0.03 to 4.0 μg/l in the north-east Atlantic Ocean, the North Sea, the English Channel and the Indian Ocean. Similar levels of manganese, i.e. 0.2–25.5 μg/l, were reported by Alessio and Lucchini (1996) off the coast of the United Kingdom, from the coastal waters of North Sea and the Irish Sea. In the north-western cities of Madagascar, Ravelonabdra et al. (2010) determined levels of manganese at Nosy Be to be as high as 36.02–119.1 μg/g of sediment. Evaluating the contamination of the Aden Port sediments in Yemen, Nasir et al. (2006) reported levels of manganese ranging from 138.23 to 658.87 μg/g. In acidic waters where the pH is less than 2.5 due to drainage from acid mines, Filipek et al. (1987) recorded dissolved manganese concentrations as high as 4400 μg/l. According to Reimer (1999), dissolved manganese concentrations in natural waters free from anthropogenic pollution range from 10 to 10,000 μg/l, with the concentration generally not exceeding 200 μg/l (Pearson and Greenway 2005). Howe et al. (2004) also documented the levels of manganese from various sites in a report called the Concise International Chemical Assessment Document 63. According to the report, lakes receiving inputs from urban areas, industries as well as windborne dust from old mine dumps contained manganese ranging up to 13,400 mg/kg dry weight; sediments from the river contained concentrations as high as 410–6700 mg/kg dry weight, whereas sediments from the intertidal mudflats had 100–1000 mg/kg. Similarly, concentrations of manganese as high as 3550–8960 mg/kg dry weight have been reported from the Baltic Sea surface sediments and were attributed to the load of ferromanganese in the river.

India has manganese ore mines at Karnataka, Odisha, Madhya Pradesh, Maharashtra and Goa. In Karnataka, in the central estuarine region represented by Gurpur and Swarna Estuary, manganese concentrations as high as 90 ± 5 % and 77 ± 10 %, respectively, have been reported from the mudflat sedimentary environment (Fernandes and Nayak 2015). In the proximity of the mining regions of Jharkhand, Orissa and West Bengal, manganese levels detected were as high as 0.024–0.165 % in the surface sediments of the Dhamra Estuary (Barik and Panda 2014). High levels of manganese (12.59 mg/kg) were obtained in the sediment of Siddheshwar Dam constructed on Purna River in the Hingoli district of Maharashtra state (Parveen and Bhosle 2013). The highest and lowest levels of manganese recorded in the waters of Manjara Dam of Maharashtra were 0.1866 mg/L and 0.057 mg/L, respectively (Mane et al. 2013). In Madhya Pradesh, water samples of Birsinghpur town and its neighbouring areas in the District of Satna were found to be contaminated with manganese ranging from 0.125 to 0.292 mg/l in all surface waters and 0.012–0.248 mg/l in 30 % of groundwater samples (Shrivastava and Mishra 2011), which exceeded the limit of 0.1 mg/l recommended by the WHO. Wasim et al. (2010) detected manganese concentrations as high as 0.022–1.78 mg/l in the surface water of Ganga River around Kolkata in West Bengal, whereas Attri and Kerkar (2011) reported an average manganese concentration of 0.19 ± 0.002 %, in the sediment of the Mandovi Estuary of Goa. This estuary receives a considerable supply of inorganic nitrogenous nutrients from a variety of sources such as mining rejects (De Souza 1999), land runoff (Sardessai and Sundar 2007) and sewage effluents (Ansari et al. 1986). Alagarsamy (2006) reported manganese ranging from below detection limits to 1.61 % from the surface sediments of the same estuary. Interestingly, the Ribandar Saltern, which lies along the Mandovi Estuary and evaporates the water it receives, contained an average manganese concentration of 0.72 ± 0.16 % in the sediment during the non-salt-making season, which showed an increase by 42 % in the salt-making season (Pereira et al. 2013).

Heavy metal contamination in aquatic marine ecosystems is found to be localized near highly urbanized, industrialized areas along the coast. These contaminants being associated with very fine particles often remain in suspension for an infinite period of time. Eventually, they get deposited in the sediments, where they may be concentrated tenfold to 100-fold higher than in solution. Therefore, sediments behave as sinks for all pollutants, from where they are slowly released into the surrounding long after the primary source has disappeared, thereby emerging as an important secondary source of contamination (Temara et al. 1998; Pereira et al. 2013).

Mangroves and estuaries serve as nurseries and breeding grounds for a number of commercially important marine organisms such as shellfish, fish, shrimps and crabs. Exposure of human beings to toxic heavy metals occurs when they consume seafoods contaminated with metal. Metal accumulation by an organism from its surrounding water indirectly reflects the extent of pollution in the estuarine water. The extent of heavy metal bioaccumulated by an organism depends on the rate of uptake of metal by the organism and its physiology. In the case of fish, the major factors that determine the rate of accumulation are age and size of the fish, time taken to complete its life cycle, its feeding habits and ecological needs as well as other environmental parameters such as salinity, pH, hardness and temperature of the water (Canli and Atli 2003). Even at lower trophic levels, aquatic biota can bioconcentrate manganese from water to a large extent. Howe et al. (2004) have estimated bioconcentration factors (BCFs) for various marine species such as fish (35–930), mussels (800–830), phytoplankton (2500–6300), marine and freshwater plants (2000–20,000) and macroalgae (300–5500). Zingde et al. (1976) detected metals such as manganese, zinc, copper and arsenic in the flora and fauna from the coastal waters and estuaries of Goa and attributed it to the ferromanganese ore-bearing landmass and the mining operations. Among seaweeds, they identified *Sargassum tenerrimum* as the most efficient accumulator of trace metals, followed by *Padina tetrastromatica*. Among crustaceans, highest metal values were found in crabs, while prawns, mussels and bivalves showed intermediate levels. Oysters showed least preference for arsenic and manganese. A study by George and Kureishy (1979) on bioaccumulation factor in zooplankton from a limited area of the Bay of Bengal showed that their bioaccumulation factor was highest for Fe (15,638), followed by Co (1955), Zn (1042), Ni (442), Mn (358) and Cu (228), respectively. KesavaRao and Indusekhar (1989) reported the trend in concentration factor of the elements as Mn > Zn > Cu > Co > Ni in seawater and seaweeds (green, brown and red) from Saurashtra coast. A study on trace metals bioaccumulation conducted by Alam et al. (2010) on four soft-bottom polychaetes from the mangrove wetlands of the Indian Sundarban showed Mn enrichment as high as 730 mg/kg body weight. In Uppanar Estuary which lies on the southeast coast of India, metals such as Mn, Pb, Cd and Cr were beyond the permissible limits set by the World Health Organization. As expected therefore, Pravinkumar et al. (2014) found metals accumulated to different levels in the liver, muscle and gill of finfish and tissues and the gill of shellfish collected from both the Uppanar Estuary and the Vellar Estuary from the same vicinity. Tissue samples of finfish and shellfish showed high levels of metal accumulation, i.e. Mg (9.74–31.55), Mn (0.09–11.12), Fe (1.42–9.85), Al (1.26–3.95), Zn (0.31–2.31), Cr (0.25–2.91), B (0.17–0.51), Pb (0.14–0.50), Cd (0.10–0.72), Cu (0.05–0.62), Ni (0.02–0.30) and Co (0–0.18) μg g^{-1} dry weight, respectively. Kasimpur in Aligarh receives wastewater containing pollutants, namely, fly ash and heavy metals, from the Harduaganj Thermal Power Station (HTPS). In a study on fish exposed to effluents from this power plant, Javed and Usmani (2013) noticed bioaccumulation of heavy metals such as Mn in body organs such as the gills, liver and kidney and in muscle tissue and integument of the fish *Mastacembelus armatus*, which exceeded severalfold the maximum permissible limit of Mn (1.0 ppm) set by the FAO (1983). The liver seemed to have the maximum affinity for bioaccumulation of Mn (49.96 mg Kg^{-1}.dry weight), whereas muscle tissue had the least affinity (9.03 mg Kg^{-1}.dry weight). Similar studies for comparison of heavy metal content were conducted by Gawade et al. (2013) on grey mullet fish (*Liza aurata*) and oysters (*Crassostrea madrasensis*) from Gosthani Estuary in Bheemunipatnam which lies on the east coast of India and clam (*Polymesoda erosa*) samples collected from the Mandovi Estuary which lies on the west coast of India. In the Gosthani Estuary, average concentrations of Fe were 754.3 ± 350.2 μg/g in fish gills, 415.1 ± 134.7 μg/g in oyster and 60.5 ± 40.59 μg/g in muscle tissue of fish, respectively, whereas Mn was 42.2 ± 16.7 μg/g in tissue, 380.9 ± 176.1 μg/g in gills and 48.4 ± 20.7 μg/g in oyster, respectively. In the Mandovi Estuary, clam showed average accumulation of 711.7 ± 206.3 μg/g Fe and 53.9 ± 27.6 μg/g Mn. High Fe (79.5 %) and Mn (9.4 %) concentrations in clams collected from Mandovi Estuary have been attributed to Fe and Mn ore mining. The metal accumulated in an organism from its food source thus gets concentrated many times higher at higher tropic levels as biomagnification occurs.

In the estuarine environment, heterotrophic marine bacteria play a significant role in food web dynamics, metal speciation and bioavailability, as well as in the biogeochemical cycling of the metals such as Mn, by accumulating them to varying degrees and releasing them back into the water column in dissolved form, depending on environmental changes, thus either making them more bioavailable than their inorganic forms or by immobilizing them and mitigating toxic levels (Pereira et al. 2013).

10.4 Manganese-Tolerant Bacteria

Extensive studies have been carried out on the use of microorganisms for bioremediation of manganese by sequestration, precipitation or alteration of the oxidation state. Some of the most remarkable manganese-tolerant bacteria are *Bacillus*, *Leptothrix discophora*, *Pseudomonas putida* and *Pedomicrobium* (Geszvain et al. 2013; Ridge et al. 2007; Saratovsky et al. 2006; Villalobos et al. 2003; Brouwers et al. 2000); among fungi *Acremonium*, *Cephalosporium* sp., *Phanerochaete*, *Paraconiothyrium*, *Phoma*, *Periconia* sp., *Coniothyrium* and *Sporothrix* sp. are well known (Sasaki et al. 2006; Saratovsky et al. 2009; Timonin et al. 1972), whereas algae like *Cladophora* and *Gloeothece magna* (Mohamed 2001; Duggan et al. 1992) play an important role in bioremediation of metal in the environment. Because of their ability to catalyse the oxidation or reduction of manganese (Cerrato et al. 2010; Gounot 1994), manganese-oxidizing bacteria which are ubiquitous and found in just about any habitat exposed to high levels of manganese are ideal candidates for manganese cycling, especially in freshwaters, marine waters, sediments and submarine basalt surfaces, hydrothermal vents, ferromanganese nodules and cave deposits (Hansel and Francis 2006). Previous studies conducted by Tebo et al. (2004), on Mn(II)-oxidizing bacteria in freshwater and marine habitats, showed that the majority of these bacteria belonged to either alpha- beta- and gamma-*Proteobacteria*, Gram-positive *Firmicutes* having low GC content or Gram-positive *Actinobacteria* with high GC content. Similar studies by Sullivan and Koppi (1993) on alfisol soil having large Mn deposits yielded *Citrobacter freundii* and *Pseudomonas* sp. capable of Mn(II) oxidation. Fe-Mn nodules and the soils surrounding these nodules were also a rich source of phylogenetically diverse bacteria (He et al. 2008). Hitherto unrecognized, *Roseobacter*-like bacteria were shown to play a role in the oxidation of Mn(II), especially in the coastal environments (Hansel and Francis 2006). Mn(II) oxidizers with activities exceeding 50 mM MnO_2 were isolated by Yang et al. (2013) from among the phyla *Proteobacteria* (*Escherichia*, *Cupriavidus*, *Ralstonia*, *Variovorax*), *Actinobacteria* (*Agromyces*, *Cellulomonas*, *Microbacterium*) and *Firmicutes* (*Bacillus*). Pereira et al. (2012) isolated halotolerant *Chromohalobacter beijerinckii* which was tolerant to manganese at concentrations as high as 10 mM at a salinity of 20 % (w/v), from saltern sediments of Ribandar in Goa, India. This saltern is fed by waters from the Mandovi Estuary which is a waterway for shipping ferromanganese ores. They obtained Mn-tolerant bacteria with counts as high as 10^5–10^6 CFU g^{-1} from the sediment and 10^4 CFU ml^{-1} in the overlying water of the Ribandar Saltern, respectively (Pereira et al. 2012). Of the total viable population tested on nutrient media, they found that 82 % isolates were resistant to manganese, iron and lead at a salinity of 1.5 % (w/v) and 74 % were resistant to manganese at 20 % salinity. At a lower salinity of 1.5 % (w/v), multimetal tolerance to a minimum of five metals was observed in 20 % of Fe and Mn isolates, 18 % of Co isolates and 11 % Ni isolates, with some showing tolerance even to seven metals. Interestingly, at a higher salinity of 20 % (w/v), a higher percentage of bacteria showed multimetal tolerance, i.e. 28 % Mn isolates, 18 % Pb isolates, 14 % Ni isolates and 10 % Fe and Cd isolates.

10.5 Mechanisms of Manganese Tolerance in Bacteria

The importance of metal ions such as manganese as vital cofactors for metalloproteins and enzymes in many biological reactions is well known. Trace metals are required for such metabolic processes such as photosynthesis, carbon fixation, nutrient assimilation and respiration (Morel et al. 1991), in redox processes as a source of electrons in energy production and in anaerobic respiration as electron acceptors (Doelman et al. 1994) and for structural stability (Ji and Silver 1995; Poole and Gadd 1989; Hughes and Poole 1989). The role of

Mn in protecting cells against reactive oxygen species by serving as a cofactor for enzymes that remove harmful free radicals produced during metabolism is well studied (Tseng et al. 2001; Niven et al. 1999). Manganese forms non-proteinaceous manganese antioxidants which can combat oxidative damage without actually being affected by Fenton chemistry. However, to survive in any environment, it is critical for all microorganisms to maintain homeostasis, i.e. they need to keep a balance between the amount of metal actually available to them from their immediate surroundings and that which is taken up, in accordance with physiological needs, because any imbalance would have serious consequences. It is necessary to regulate the metal acquisition system based on the conditions prevailing in the surrounding. During metal starvation, the activity of the transporters and the expression of the relevant genes need to be upregulated, but when there is a surplus of metals, efflux pumps need to be activated (Wakeman and Skaar 2012). Bacteria strictly maintain metal homeostasis by employing numerous transcriptional and biochemical regulators for the uptake and export of metals, thus enabling them to adjust to changing environmental conditions. A combination of two or three fundamental mechanisms is employed for resistance and homeostasis of many metals, which are categorized as (1) use of permeability barriers such as extracellular polysaccharides (EPS) and cell walls; (2) decreased uptake or enhanced efflux; (3) intracellular accumulation; (4) extracellular precipitation as oxalates, phosphates, sulphides, oxides and carbonates; and (5) enzymatic detoxification.

10.5.1 Use of Permeability Barriers such as Cell Walls and Extracellular Polysaccharides (EPS)

Gadd (1990) reported that tolerance to heavy metals could be an intrinsic property of the microorganism, for example, production of extracellular mucilage or polysaccharide or an impermeable cell wall. The positively charged metal ions directly bind to negatively charged functional pyruvyl, carboxyl, phosphoryl, hydroxyl, succinyl and uronyl groups on the exopolymer. Manganese entrapped in the EPS can be oxidized directly by enzymatic reactions involving extracellular polysaccharides (EPS) (Tebo et al. 1997; Beveridge 2005; Ghiorse and Hirsch 1979). The exopolymer matrix which surrounds the cell is the most common site for Mn(II) oxidation in β-*Proteobacteria*, γ-*Proteobacteria* and low-GC Gram-positive bacteria. However, the location may vary. For example, both strains MnB1 and GB1 of *Pseudomonas putida* oxidized Mn(II) enzymatically on the outer membrane glycocalyx (de Vrind et al. 1986; Caspi et al. 1998; Brouwers et al. 1999), whereas strain SG-1 of *Bacillus* sp. was found to deposit it on the exosporium (van Waasbergen et al. 1996; Francis and Tebo 2002). The sheath of the mesophilic bacterium *Leptothrix discophora* SS-1 was shown to be encrusted with manganese oxides due to the sorption of these metal ions. This enzymatic oxidation of Mn(II) was possible due to a protein which is normally present on its extracellular sheath (Adams and Ghiorse 1986, 1987; Brouwers et al. 2000; Corstjens et al. 1997). Deposition of MnO on EPS was also reported by Ghiorse and Hirsch (1979) and Sly et al. (1990) in *Pedomicrobium* spp. In *Leptothrix discophora*, isolated from Pinal Creek, near Globe, Arizona, the outer surface of the cell wall and holdfasts of the bacteria were found to be the main sites for concentration of Mn (Robbins and Corley 2005). Studies by Gutierrez et al. (2012) revealed that EPS of *Halomonas* sp. TG39 showed similar binding capacity for Fe, Mn, Ca, Mg, Si and Al. Interestingly, even though *Halomonas* sp. from Carlsberg Ridge could only oxidize and precipitate Mn^{2+} extracellularly (Fernandes et al. 2005), the same isolate in the absence of Mn^{2+} but in the presence of other metals such as Ni and Co could assimilate metals both intra- and extracellularly (Antony et al. 2010; Sujith et al. 2010). In another study on *R. etli* M4 isolated from an environment contaminated with $MnSO_4$-rich acid mine drainage (AMD), Foster et al. (2000) demonstrated the remarkable ability of the EPS produced by this organism to bind Mn(II) ions in the form of

$MnSO_4$. Pulsawat et al. (2003) showed that the anion bound to Mn can significantly affect both the metal adsorption capacity and binding strength of the biosorbent. They reported a decrease in the specific adsorption capacity (qmax) of manganese to EPS, for example, when sulphate was the corresponding anion, it showed a maximum binding of 62 mg g^{-1}, followed by nitrate (53 mg g^{-1}) and chloride (21 mg g^{-1}). Markus et al. (2002) have reported Mn adsorption of 43 mg g^{-1} by dead biomass of *Lecanora conizaeoides* with Cl^- as the counter ion. In contrast, Mn adsorption by *Sargassaceae* sp. was approximately 100-fold lesser than the EPS from *R. etli* M4 when the anions were sulphate (da Costa et al. 1996). In the case of *Rhizobium radiobacter* F2, Wang et al. (2010) determined maximum biosorption capacities of Mn(II) by extracellular polymeric substance (EPS) to be 28.86 ± 1.28 mg/g under optimum conditions. EPS, bacterial cell surfaces and manganese oxides are all capable of absorbing heavy metals to different degrees and therefore profoundly influence the fate and removal of heavy metals from marine environments. However, reports of manganese-resistant bacteria removing Mn from estuaries, mangroves or salterns by binding to EPS are limited.

10.5.2 Decreased Uptake and Enhanced Efflux

The intracellular levels of Mn^{2+} are tightly controlled by regulating its import and export across the cell envelope, which relies on the ability of manganese-tolerant bacteria to sense Mn^{2+} and maintain homeostasis. Prokaryotes can accumulate Mn in high concentrations in their cytoplasm, implying that the uptake of Mn must be carried out by some energized pumping mechanism (Frausto da Silva and Williams 1991). Such Mn^{2+} uptake systems are seen in many Gram-positive bacteria like *B. subtilis*, *L. plantarum* and *Staphylococcus aureus* (Silver and Lusk 1987). It has been hypothesized that in some bacteria like *Deinococcus radiodurans*, the hyperaccumulation of low molecular weight Mn(II) complexes is a mechanism to protect these organisms from oxidative stress and radiation damage to DNA due to reactive oxygen species.

As per literature available, three kinds of Mn^{2+} uptake systems have been studied (Fig. 10.1). One is a P-type ATPase (*MntA*) with a high specificity for Mn^{2+} as seen in *L. plantarum* (Hao et al. 1999). The second involves members of Nramp family, which is a natural resistance-associated macrophage protein (Nramp). Nramp (*MntH*) is a proton-dependent divalent cation transporter seen in several species of bacteria such as *Escherichia coli*, *Salmonella* and *Brucella abortus*. And the third conforms to the ABC transporter which belongs to the ATP-binding cassette (ABC) superfamily described by Bartsevich and Pakrasi (1995) who discovered the *mntCAB* operon in a cyanobacterium *Synechocystis* sp. PCC 6803, which has a high requirement for Mn^{2+} during photosynthesis. ABC permeases directly utilize the hydrolysis of ATP to pump manganese into the cytoplasm. These transport-

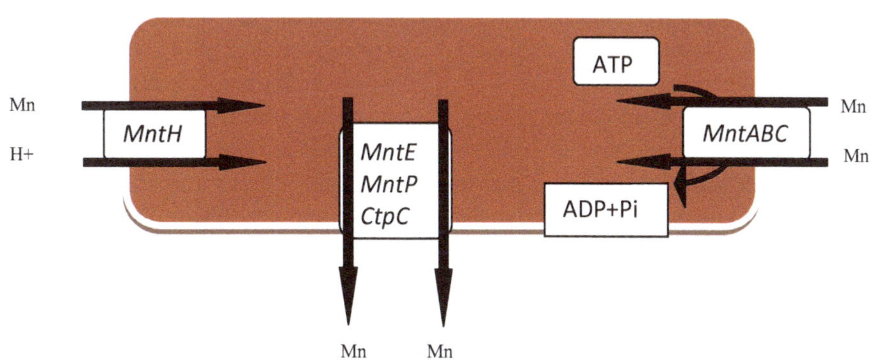

Fig. 10.1 Uptake and efflux pumps for manganese resistance in estuarine bacteria

ers comprise three proteins. One is a cation-binding protein (*MntC*) which is extracellular, the second is *MntA* which is an ATP-binding protein in the cytoplasm, and the third is an intrinsic transmembrane protein (*MntB*) that mediates transport. Substrates such as Mn^{2+} are exclusively transported by inducible uptake systems like ABC transporters in *Streptococcus gordonii* (Kolenbrander et al. 1998). In *B. subtilis*, a bifunctional metallorepressor protein (*MntR*) acts on *mntH* such that the expression of Nramp transporter is repressed when Mn^{2+} abounds in the surrounding, but when Mn^{2+} is low it activates *mntABCD* transcription which encodes for the ABC transporter (Que and Helmann 2000). In contrast to uptake systems, very few bacterial Mn^{2+} efflux systems are known. The only identified Mn^{2+} efflux system is *MntE* in *Streptococcus pneumoniae* (Jakubovics and Valentine 2009) and in *Deinococcus radiodurans* (Sun et al. 2010), which is a new member of the cation diffusion facilitator (CDF) family. In addition to *MntE*, the P-type ATPase *CtpC*, from *M. tuberculosis* and *M. smegmatis*, and *YebN* (renamed *mntP*), a new system for export of Mn^{2+} in *Xanthomonas oryzae pv. oryzae* (Xoo), appear to facilitate manganese efflux. *MntP* was found to be positively regulated by *MntR*, a – transcription regulator dependent on Mn^{2+} (Li et al. 2011). Any deletion in manganese transporting ATPases (*MntP*) and cation-phosphate transporter (*CtpC*) genes resulted in manganese sensitivity due to accumulation of high levels of manganese inside the bacterial cell, implying that there may be a threshold level of tolerance to Mn^{2+} inside bacterial cells. Mn^{2+} functions as an antioxidant below this critical level, but when the threshold level is crossed, it becomes toxic and induces excessive ROS production while decreasing the capacity to scavenge ROS. Manganese therefore plays a significant role in bacterial physiology. In order to survive, the bacterial cells should be able to respond to fluctuations in external Mn^{2+}. Here the coordination between expression and activity of the transporters becomes crucial for the normal functioning of the cells. Many divalent transporters can have broad substrate specificity. Kehres et al. (2000) showed that transport of other ions such as Fe^{2+}, Ni^{2+}, Cu^{2+}, Co^{2+}, Zn^{2+} and Cd^{2+} could also take place via manganese transporters. Transport of other divalent metals by manganese transporters is possible when Mn^{2+} oxidation causes upregulation of the manganese transporters. Such organisms wherein the export systems can be regulated are ideal candidates for bioremediation of contaminated sites. Regulated systems for export of Mn^{2+} may exist in bacteria, but so far none have been reported from marine environments or from mangroves.

10.5.3 Intracellular Accumulation

Intracellular traps can be biosynthesized as a temporary measure for removal of metal ions. Tolerance to heavy metal ions could be due to the existence of metal-binding proteins such as metallothionein that are capable of sequestering metal and detoxifying it or exporting it out of the cells. These metal-binding ligand molecules enable the organisms to survive even in an environment which has toxic levels of metals. Metallothioneins have the potential to sequester Hg, Cd, Pb, Zn, Ag and also Ni but with a lower affinity. There have been reports of metallothioneins in *Cyanobacteria* and several other phyla (Kagi and Nordberg 1979; Hamer 1986; Shaw et al. 1992). The role of metallothioneins in limiting metal availability and thereby conferring protection against the harmful effects of toxic levels of metal is documented in *Anabaena* PCC 7120, *Synechococcus* PCC 7942, *Pseudomonas aeruginosa* and *Pseudomonas putida* (Blindauer et al. 2002), but no reports of Mn sequestering are documented, especially from marine environments. Cyanobacterial metallothionein gene *SmtA* from *Synechococcus* PCC 7942 was found to be hypersensitive to Zn and Cd, but had no affinity for other metals (Turner et al. 1993). However, this was not the case with *E. coli*, where *smtA*, besides having high affinity for Zn and Cd, could also bind Hg and Cu (Shi et al. 1992). Similar displacement of Zn from the protein could also be possible by Mn ions, though no such reports are documented. The presence of 'pseudo-thioneins' has also been reported by Higham et al. (1984) in Cd-adapted *Pseudomonas putida*.

According to Cavat et al. (2003), the existence of metallothionein gene in the bacteria may be an indication that the organism needs to sequester the metals which are present in its environment, in order to prevent toxicity. In fact, high levels of intracellular accumulation of Mn have been shown to enable *Lactobacillus plantarum*, which lacks a superoxide dismutase, to survive under aerobic conditions (Archibald and Fridovich 1981). Similar studies by Daly et al. (2004) showed that *Deinococcus radiodurans* could survive oxidative stress and high levels of ionizing radiation because it maintains low levels of iron (Fe) and high levels of Mn within the cell. Medicis et al. (1986) found that intracellular content of manganese in halophilic bacteria was higher, implying that manganese could play a role in haloadaptation. In many microbes, proteins like phytochelatins (PCs) and metal-binding peptides like the cysteine (Cys)-rich glutathione (GSH) confer resistance to metals by chelating the metals within the cell. In addition to peptides, intracellular phosphates can bind manganese, forming phosphate-Mn complexes which are potent antioxidants that have shown the capacity to react efficiently with superoxide, without displaying any pro-oxidant side effects shown by other redox-active metals (Jensen and Jensen 2014). Hence, in a contaminated environment, metal accumulation, especially by phosphates within the cell, would be particularly significant due to its unique application for remediation. High Mn concentrations may actually be beneficial to the organism because it would result in a shift from a Fe-based metabolism to a Mn-based metabolism. When Fe is replaced by Mn, enzymes having Fe cofactors exposed to the solvent are protected from oxidative stress because now Mn is bound to the cellular protein (Sobota and Imlay 2011). This is particularly significant in an iron- and manganese-rich environment such as an estuary contaminated with ferromanganese ores.

10.5.4 Extracellular Precipitation

The most efficient metal chelators are manganese oxide and organic acids. In manganese-contaminated environments, the most common mechanism employed for its removal is by oxidation of Mn^{2+} to MnO_2. A crust of Mn oxide minerals was observed on the cell surface of *Pseudomonas putida* GB-1 due to extracellular Mn oxidation (Tebo et al. 1997). Here, the Mn oxides not only protect the organism from external reactive oxygen species (ROS) but may serve as a mechanism for reducing the toxicity of Mn(II) and other toxic metals (Tebo et al. 2005). Mn(II) oxidation was visually apparent in *Roseobacter*-like bacteria, as distinct brown oxide particles (<2 mm) after an incubation period of 2–3 days (Hansel and Francis 2006). Besides oxides, oxalates, phosphates, sulphides and carbonates also effectively cause extracellular precipitation of manganese. Organic acids like malonate may be released into the surrounding medium, thereby binding manganese and mitigating toxic levels from the environment.

Precipitation of oxidized Mn was evident in algae and *Cyanobacteria*, when the pH was greater than 8 during photosynthesis (Richardson et al. 1988). *Brachybacterium* sp. strain Mn32A, a manganese-oxidizing bacterium from deep-sea sediments, showed an exceptionally high ability to remove Mn(II) by Mn oxidation, with resistance to Mn(II) as high as 55 mM (Fig. 10.2). Experimentally, the removal of Mn(II) by strain Mn32 occurred in two steps. First, the soluble Mn(II) was oxidized to insoluble biogenic oxides of Mn such as manganite (MnOOH) and birnessite (MnO_2 group); further, these biogenic Mn oxides served as adsorbants for more Mn(II) from the culture media (Wang et al. 2009). It was seen that the Mn oxides generated by strain Mn32 had better capacity to adsorb Zn(II) and Ni(II), i.e. it was two to three times higher than fresh, synthetic commercial MnO_2.

10.5.5 Enzymatic Detoxification

Heterotrophic bacteria can bring about Mn oxidation rapidly using indirect mechanisms such as by modifying the pH or redox conditions of their surrounding or by releasing a chemical such as superoxide which is an oxidant for Mn (Learman

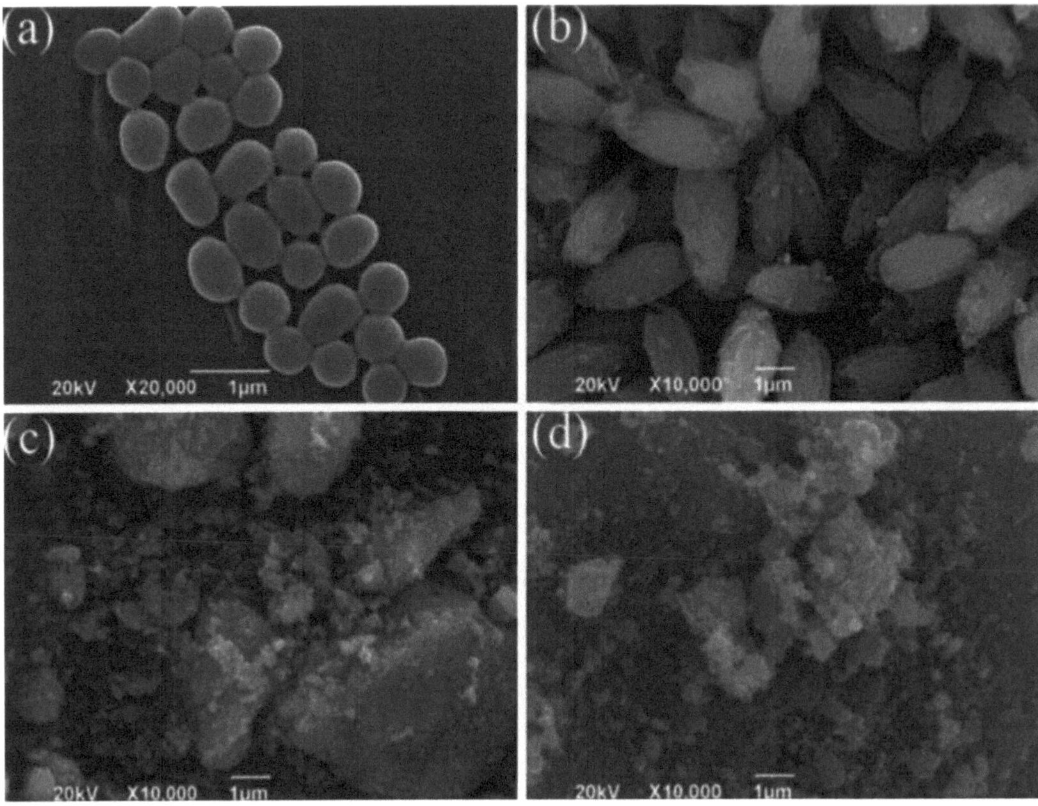

Fig. 10.2 SEM images of (**a**) *Brachybacterium* sp. strain Mn32 cells grown in liquid A medium without $MnCl_2$ for 9 days, (**b**) encrustation of biogenic Mn oxides on the cell surface of strain Mn32 when grown in liquid A medium with 0.2 M $MnCl_2$ for 9 days, (**c**) commercial MnO_2, (**d**) synthetic fresh MnO_2 (Courtesy of Wang et al. 2009)

et al. 2011), as well as by direct mechanisms such as the production of polysaccharides or by enzymatic activity. When the pH of the environment ranges between 6 and 9, the natural indirect oxidation of Mn(II) to insoluble Mn(III, IV) oxides is very slow. However, if the concentration of oxygen is about 3–5 mg L^{-1} and the redox potential is above +200 mV (Schweisfurth et al. 1978), there is an increase in the rate of manganese oxidation by four to five orders, if Mn(II)-oxidizing bacteria are present (Nealson et al. 1988; Tebo 1991; Wehrli et al. 1995; Tebo et al. 1997). Reports show that bacteria can bring about catalysis of Mn to Mn oxides in seawater, by enzymes known as multicopper oxidases (MCOs). MCO genes are essential for Mn oxidation and observed in all Mn(II)-oxidizing bacteria studied to date (Tebo et al. 2005). Corstjens et al. (1997) demonstrated that the *mnxG* gene encoding for MCO was the direct catalyst of Mn(II) oxidation in *Leptothrix discophora* and was also responsible for the oxidation of Mn by spores of the marine *Bacillus* species whose cells became encrusted with amorphous Mn oxides, thereby efficiently removing Mn(II) from solution (Rosson and Nealson 1982; van Waasbergen et al. 1996; Francis and Tebo 2002; Dick et al. 2008). Francis and Tebo (1999) showed that the spores of Mn(II)-oxidizing marine *Bacillus* sp. strain SG-1 could be effectively used for bioremediation of Mn due to their unique capacity to bind metal and precipitate it oxidatively, besides their inherently tough physical nature. Geszvain et al. (2013) identified the *mnxG* gene of *Pseudomonas putida* GB1 by its similarity to the *mnxG* gene of *Bacillus* sp. SG-1. In 1999, Larsen et al. observed a copper-dependent enzyme in *Pedomicrobium* sp. strain ACM 3067 which was similar to the multicopper oxidase which catalysed Mn(II) oxidation. Hansel and Francis (2006) isolated a diverse group of Mn(II)-oxidizing *Roseobacter*-like species able to oxidize soluble Mn(II) rapidly to its insoluble Mn (III, IV) oxides, from the coastal estuary of Elkhorn Slough on the border of Monterey Bay in California. In the coastal waters, these bacteria were found not only to actively participate in manganese (Mn) oxidation and cycling but also to provide a nonconventional method for oxidation of Mn(II) in the photic zone via a photo-oxidation pathway. A protein homologous to multicopper oxidases encoded by *Cum*A, which could actively mineralize Mn oxide from Mn(II) and deposit it on the outer membranes in *Pseudomonas putida*, was reported by Brouwers et al. (1999). Another mechanism for metal resistance in microorganisms exposed to toxic metals is by the upregulation of genes involved in metal homeostasis and the redox control of defence. Analysis by SDS-PAGE to determine which proteins are directly responsible for catalysing the mechanism by which *Chromohalobacter beijerinckii* from saltern sediments of Ribandar, Goa, could oxidize Mn(II) biochemically revealed a protein band of 50 kDa which was upregulated on addition of 10, 100 and 1000 μM of Mn^{2+}, while there was a downregulation of the 53 kDa fraction (Pereira et al. 2012). Mn(II)-oxidizing proteins of varying sizes were reported from diverse *Bacillus* species and were attributed to variations in gene size, or the additional proteins required for activity, or post-translational modifications (e.g. proteolysis or glycosylation) (Francis and Tebo 2002). Boogerd and de Vrind (1987) and Adams and Ghiorse (1987) consistently recovered a 110 kDa protein by SDS-PAGE, which was capable of oxidizing Mn(II) in *L. discophora* SS-1 gels, yet the *mofA* gene for multicopper oxidase is predicted to be a 174 kDa protein (Corstjens et al. 1997). Mn(II)-oxidizing activity in *P. putida* GB-1 could be recognized due to the formation of complexes of high molecular mass of about 250 and 180 kDa in native gels (Nieto et al. 1989). This suggests that several proteins, including the multicopper oxidase *cumA*, are required for activity.

Because minerals of Mn such as Mn oxides are highly charged, they can adsorb and concentrate many other metals. The geochemical cycling of Mn oxides can therefore influence trace metal distribution (Huang 1991). Thus, manganese-tolerant bacteria capable of oxidizing manganese are important in the mitigation of not only manganese but also other metals in contaminated estuarine environments. As observed by Barkay (1987) and Nies (2003), microorganisms from contaminated estuarine sites such as solar salterns may employ various mechanisms, such as a shift towards a more resistant population, or the increased expression of stress-related genes, or the inter- and intraspecies transfer of metal-resistant genes. Haritha et al. (2009) have suggested the there is a strong possibility that transposition of metal resistance traits from plasmid to chromosome may take place during acute metal stress, thereby disseminating the property of metal resistance.

10.6 Ecological Significance of Manganese-Tolerant Estuarine Bacteria as Bioremediators

Induced tolerance in bacteria continuously exposed to a particular pollutant, such as manganese, for a long time provides a mode for indirectly mitigating metal concentrations in the ecosystem (Blanck et al 1988; Hopkin 1993; Chapman et al. 1998) and maintaining it within limits, thus preventing it from adversely affecting living organisms. Due to their exceptional properties such as their capability to grow even at a high anion and cation concentrations, halophilic and halotolerant microorganisms are ideal candidates for bioremediation processes (Ventosa et al. 1998). Hence, not only do they have a requirement for these elements, but they are also naturally tolerant to elements that may be toxic to their non-extremophilic counterparts (Amoozegar et al. 2005). Mondal et al. (2008) used species from the *Ralstonia* family that were phylogenetically related to *Cupriavidus metallidurans* for

Table 10.1 Manganese tolerance mechanisms employed by marine bacteria

Marine sites	Name of bacteria	Type of resistance mechanism	References
Tamar Estuary, Tasmania	*Halomonas* strains	Manganese oxidation	Vojak et al. (1985)
Mandovi and Zuari River, Goa	*V. harveyi*	EPS	Bramhachari and Dubey (2006)
Elkhorn Slough Estuary, California	*Roseobacter*-like species	Mn oxidation	Hansel and Francis (2006)
Carlsberg Ridge	*Halomonas species*	Mn oxidation and extracellular precipitation	Fernandes et al. (2005)
Pinal Creek, Arizona	*Leptothrix discophora*	EPS	Robbins and Corley (2005)

bioremediation studies and recorded removal of 65.2 % Fe, 72.7 % Mn, 98.6 % Cu, 8 % As and 99.3 % Zn, respectively, from water contaminated with metal. Hallberg and Johnson (2005) used a simple fixed bed bioreactor packed with microorganisms capable of oxidation and precipitation of Mn(II) as a highly effective method for the removal of metals from a synthetically prepared metal-rich wastewater. Industries frequently make use of salts and then release effluents containing brine into the environment. Manganese-tolerant bacteria displaying multimetal tolerance and adapted to live in hypersaline habitats would prove useful as biological detoxicants in saline aquatic polluted environments or even as bioindicators of pollution. Table 10.1 gives a summary of manganese tolerance mechanisms reported in bacteria from different marine sites.

10.7 Conclusion

The use of microbes for the detoxification and rehabilitation of metal-contaminated estuarine environments has proved to be one of the most appealing technologies because of its low cost, safety, effectiveness and ease. Microorganisms being metal accumulators or convertors of metals into non-toxic forms have a novel inherent property for remediation of toxic metals. Indigenous metal-tolerant microorganisms, with the ability to naturally remove, attenuate or detoxify toxic substances from estuarine environments such as salterns and mangroves contaminated with ores from mines, pesticides and industrial effluents, need to be explored and harnessed.

Acknowledgement I would like to express my gratitude to my Ph.D. guide and mentor Dr. Savita Kerkar for extending her constant support and guidance. I am also grateful to H. O. D. Department of Biotechnology, Goa University; to the ex-director of the National Centre for Antarctic and Ocean Research (NCAOR), Goa, Shri Rasik Ravindra; and to Dr. K. P. Krishnan, scientist at NCAOR for permitting use of the lab facilities and assistance. I would like to thank Dr. Thamban Meloth (programme director, laboratories at NCAOR) for his constant support. I would also like to place on record the assistance given by Mr. Rupesh Kumar Sinha in the fieldwork.

References

Abell LM, Schineller J, Keck PJ, Villafranca JJ (1995) Effect of metal-ligand mutations on phosphoryl transfer reactions catalyzed by *Escherichia coli* glutamine synthetase. Biochemistry 34:16695–16702

Adams LF, Ghiorse WC (1986) Physiology and ultrastructure of *Leptothrix discophora* SS-1. Arch Microbiol 145:126–135

Adams LF, Ghiorse WC (1987) Characterization of extracellular Mn-oxidizing activity and isolation of a Mn-oxidizing protein from *Leptothrix discophora* SS-1. J Bacteriol 169(3):1279–1285

Adekunle AS, Oyekunle JAO, Baruwa SO, Ogunfowokan AO, Ebenso EE (2014) Speciation study of the heavy metals in commercially available recharge cards coatings in Nigeria and the health implication. Toxicol Rep 1:243–251

Alagarsamy R (2006) Distribution and seasonal variation of trace metals in surface sediments of the Mandovi estuary, West coast of India. Estuar Coast Shelf Sci 67:333–339

Alam MA, Gomes A, Sarkar SK, Shuvaeva OV, Vishnevetskaya NS, Gustaytis MA, Bhattacharya BD, Godhantaraman N (2010) Trace metal bioaccumulation by soft-bottom polychaetes (Annelida) of Sundarban Mangrove Wetland, India and their potential use as contamination indicator. Bull Environ Contam Toxicol 85(5):492–496. doi:10.1007/s00128-010-0110-1

Alessio L, Lucchini R (1996) Manganese and manganese compounds. In: Argentesi F, Roi R, Sevilla Marcos JM (eds) Data profiles for selected chemicals and pharmaceuticals 3, 4, 5. European Commission, Joint Research Centre, Ispra

Al-Masri MS, Aba A, Khalil H, Al-Hares Z (2002) Sedimentation rates and pollution history of a dried lake: Al-Qteibeh Lake. Sci Total Environ 293(1–3):177–189. doi:10.1016/S0048-9697(02)00013-X

Amoozegar MA, Hamedi J, Dadashipour M, Shariatpanahi S (2005) Effect of salinity on the tolerance to toxic metals and oxyanions in native moderately halophilic spore-forming bacilli. World J Microbiol Biotechnol 21:1237–1243

Ansari ZA, Ingole BS, Parulekar AH (1986) Effect of high organic enrichment of benthic polychaete population in an estuary. Mar Pollut Bull 17:361–365

Antony R, Sujith PP, Fernandes SO, Verma P, Khedekar VD, Loka Bharathi PA (2010) Cobalt immobilization by manganese oxidizing bacteria from Indian Ridge System. Curr Microbiol 62:840–849

Appanna VD (1988) Alteration of exopolysaccharide composition in *Rhizobium meliloti* JJ-1 exposed to manganese. FEMS Microbiol Lett 50(2–3):141–144. doi:10.1111/j.1574-6968.1988.tb02927.x

Archibald F (1986) Manganese: its acquisition by and function in the lactic acid bacteria. Crit Rev Microbiol 13:63–109

Archibald FS, Fridovich I (1981) Manganese and defenses against oxygen-toxicity in *Lactobacillus plantarum*. J Bacteriol 145:442–451

Atkinson CA, Jolley DF, Simpson SL (2007) Effect of overlying water pH, dissolved oxygen, salinity and sediment disturbances on metal release and sequestration from metal contaminated marine sediments. Chemosphere 69(9):1428–1437

ATSDR (2000) Toxicological profile for manganese (update). Draft for public comment. US. Department of Health and Human Services, Public Health Service, Agency for Toxic Substances and Disease Registry, Atlanta

Attri K, Kerkar S (2011) Seasonal assessment of heavy metal pollution in tropical mangrove sediments (Goa, India). J Ecobiotechnol 3(8):9–15

Balzer W (1982) On the distribution of iron and manganese at the sediment/water interface: thermodynamic versus kinetic control. Geochim Cosmochim Acta 46:1153–1161

Barceloux DG (1999) Manganese. J Toxicol Clin Toxicol 37:293–307. doi:10.1081/CLT-100102427

Barik SK, Panda CR (2014) Distribution and seasonal variation of trace metal concentration in surface sediment of Dhamra Estuary, East coast of India. Int J Environ Sci 4(5):1015–1022. doi:10.6088/ijes.2014040404540

Barkay T (1987) Adaptation of aquatic microbial communities to Hg stress. Appl Environ Microbiol 53(12):2725–2732

Bartsevich VV, Pakrasi HB (1995) Molecular identification of an ABC transporter complex for manganese: analysis of a cyanobacterial mutant strain impaired in the photosynthetic oxygen evolution process. EMBO J 14(9):1845–1853

Beveridge TJ (2005) Bacterial cell wall structure and implications for interactions with metal ions and minerals. J Nucl Radiochem Sci 6:7–10

Blanck H, Wangberg SA, Molander S (1988) Pollution-induced community tolerance-A new ecotoxicological tool. In: Cairns J, Pratt JR (eds) Functional testing of aquatic biota for estimating hazards of chemicals. ASTM (American Society of Testing and Materials), Philadelphia, pp 219–230

Blindauer CA, Harisson MD, Robinson AK, Parkinson JA, Bowness PW, Sadler PJ, Robinson NJ (2002) Multiple bacteria encode metallothioneins and *SmtA*-like zinc fingers. Mol Microbiol 45(5):1421–1432

Bonnevie NL, Huntley SL, Found BW, Wenning RJ (1994) Trace metal contamination in surface sediments from Newark Bay, New Jersey. Sci Total Environ 144(1):1–16. doi:10.1016/0048-9697(94)90423-5

Boogerd FC, de Vrind JPM (1987) Manganese oxidation by *Leptothrix discophora* SS-1. J Bacteriol 169(2):489–494

Bowen HJM (1979) Environmental chemistry of the elements. Academic Press, London, p 333

Bramhachari PV, Dubey SK (2006) Isolation and characterization of exopolysaccharide produced by *Vibrio harveyi* strain VB23. Lett Appl Microbiol 43:571–577. doi:10.1111/j.1472-765X.2006.01967.x

Bratina BJ, Stevenson BS, Green WJ, Schmidt TM (1998) Manganese reduction by microbes from oxic regions of the Lake Vanda (Antarctica) water column. Appl Environ Microbiol 64(10):3791–3797

Brouwers GJ, de Vrind JPM, Corstjens PLAM, Cornelis P, Baysse C, de Vrind-de Jong EW (1999) CumA, a gene encoding a multicopper oxidase, is involved in Mn^{2+}-oxidation in *Pseudomonas putida* GB-1. Appl Environ Microbiol 65(4):1762–1768

Brouwers GJ, Corstjens PLAM, de Vrind JPM, Verkamman A, de Kuyper M, de Vrind-de Jong EW (2000) Stimulation of Mn^{2+} oxidation in *Leptothrix discophora* SS-1 by Cu^{2+} and sequence analysis of the region flanking the gene encoding putative multicopper oxidase MofA. Geomicrobiol J 17:25–33. doi:10.1080/014904500270468

Calvert SE, Pedersen TF (1996) Sedimentary geochemistry of manganese; implications for the environment of formation of manganiferous black shales. Econ Geol 91(1):36–47

Canli M, Atli G (2003) The relationships between heavy metals (Cd, Cr, Cu, Fe, Pb, Zn) levels and the size of six Mediterranean fish species. Environ Pollut 121:129–136

Caspi R, Tebo BM, Haygood MG (1998) c-type cytochromes and manganese oxidation in *Pseudomonas putida* strain MnB1. Appl Environ Microbiol 64:3549–3555

Cavat JS, Borrelly GP, Robinson NJ (2003) Zn, Cu and Co in cyanobacteria: selective control of metal availability. FEMS Microbiol Ecol 122:165–181

Cerrato JM, Falkinham JO 3rd, Dietrich AM, Knocke WR, McKinney CW, Pruden A (2010) Manganese-oxidizing and -reducing microorganisms isolated from biofilms in chlorinated drinking water systems. Water Res 44(13):3935–3945. doi:10.1016/j.watres.2010.04.037

Chander M, Setlow B, Setlow P (1998) The enzymatic activity of phosphoglycerate mutase from gram-positive endospore-forming bacteria requires Mn and is pH sensitive. Can J Microbiol 44:759–767

Chao YP, Patnaik R, Roof WD, Young RF, Liao JC (1993) Control of gluconeogenic growth by *pps* and *pck* in *Escherichia coli*. J Bacteriol 175:6939–6944

Chapman PM, Wang F, Janssen C, Persoone G, Allen HE (1998) Ecotoxicology of metals in aquatic sediments: binding and release, bioavailability, risk assessment, and remediation. Can J Fish Aquat Sci 55:2221–2243

Corstjens PLAM, de Vrind JPM, Goosen T, de Vrind-de Jong EW (1997) Identification and molecular analysis of the *Leptothrix discophora* SS-1 *mofA* gene, a gene putatively encoding a manganese-oxidizing protein with copper domains. Geomicrobiol J 14:91–108. doi:10.1080/01490459709378037

da Costa ACA, de Mesquita LMS, Tornovsky J (1996) Batch and continuous metals biosorption by a brown seaweed from a zinc producing plant. Miner Eng 9:811–824

Daily GC, Alexander S, Ehrlich PR, Goulder L, Lubchenco J, Matson PA, Mooney HA, Postel S, Schneider SH, Tilman D, Woodwell GM (1997) Ecosystem services: benefits supplied to human societies by natural ecosystems. Issues Ecol 2:1–16

Daly MJ, Gaidamakova EK, Matrosova VY, Vasilenko A, Zhai M, Venkateswaran A, Hess M, Omelchenko MV, Kostandarithes HM, Makarova KS, Wackett LP, Fredrickson JK, Ghosal D (2004) Accumulation of Mn(II) in *Deinococcus radiodurans* facilitates gamma-radiation resistance. Science 306(5698):1025–1028. doi:10.1126/science.1103185

De Souza SN (1999) Effect of mining rejects on the nutrient chemistry of Mandovi estuary, Goa. Indian J Mar Sci 28:198–210

de Vrind JPM, de Vrind-de Jong EW, de Voogt JWH, Westbroek P, Boogerd FC, Rosson RA (1986) Manganese oxidation by spores and spore coats of a marine *Bacillus* species. Appl Environ Microbiol 52:1096–1100

Dick GJ, Torpey JW, Beveridge TJ, Tebo BM (2008) Direct identification of a bacterial manganese(II) oxidase, the multicopper oxidase *MnxG*, from spores of several different marine *Bacillus* species. Appl Environ Microbiol 74:1527–1534, doi:0.1128/AEM.01240-07

Doelman P, Jansen E, Michels M, Van Til M (1994) Effects of heavy metals in soil on microbial diversity and activity as shown by the sensitivity-resistance index, an ecologically relevant parameter. Biol Fertil Soil 17:177–184

Doyle RJ (1989) How cell walls of gram-positive bacteria interact with metal ions. In: Beveridge TJ, Doyle RJ (eds) Metal ions and bacteria. Wiley, New York, pp 275–293

Duggan LA, Wildeman R, Tipping E (1992) The aerobic removal of manganese from mine drainage by an algal mixture containing *Cladophora*. Environ Pollut 57:251–274

Edgar GJ, Barrett NS, Braddon DJ, Last PR (2000) The conservation significance of estuaries: a classification of Tasmanian estuaries using ecological, physical and demographic attributes as a case study. Biol Conserv 92:383–397. doi:10.1016/s0006-3207(99)00111-1

Ehrlich HL (1996) Geomicrobiology, 3rd edn. Marcel Dekker, New York, p 719

Ehrlich HL (2002) Geomicrobiology, 4th edn. Marcel Dekker, New York, p 768

Erikson KM, Aschner M (2003) Manganese neurotoxicity and glutamate GABA interaction. Neurochem Int 43:475–480

FAO (Food and Agriculture Organization) (1983) Compilation of legal limits for hazardous substances in fish and fishery products. FAO Fish Circ 464:5–100

Fernandes MC, Nayak GN (2015) Speciation of metals and their distribution in tropical estuarine mudflat sediments, southwest coast of India. Ecotoxicol Environ Saf 122:68–75. doi:10.1016/j.ecoenv.2015.07.016

Fernandes SO, Krishnan KP, Khedekar VD, Loka Bharathi PA (2005) Manganese oxidation by bacterial isolates from the Indian Ridge System. Biometals 18:483–492

Filipek LH, Nordstrom DK, Ficklin WH (1987) Interaction of acid mine drainage with waters and sediments of West Squaw Creek in the West Shasta Mining District, California. Environ Sci Technol 21:388–396

Foster LJR, Moy YP, Rogers PL (2000) Metal binding capabilities of *Rhizobium etli* and its extracellular polymeric substances. Biotechnol Lett 22:1757–1760

Francis CA, Tebo BM (1999) Marine *Bacillus* spores as catalysts for oxidative precipitation and sorption of metals. J Mol Microbiol Biotechnol 1(1):71–78

Francis CA, Tebo BM (2002) Enzymatic manganese(II) oxidation by metabolically dormant spores of diverse *Bacillus* species. Appl Environ Microbiol 68:874–880

Frausto da Silva JJR, Williams RJP (1991) The biological chemistry of the elements. The inorganic chemistry of life, 2nd edn. Clarendon, Oxford, p 600

Fridovich I (1995) Superoxide radical and superoxide dismutases. Annu Rev Biochem 64:97–112

Gadd GM (1990) Metal tolerance. In: Edwards C (ed) Microbiology of extreme environments. McGraw-Hill, New York, pp 178–210

Gadd GM (1992) Metals and microorganisms: a problem of definition. FEMS Microbiol Lett 100:197–203

Gawade L, Harikrishna CNV, Sarma VV, Ingole BS (2013) Variation in heavy metals concentration in the edible oyster *Crassostrea madrasensis*, clam *Polymesoda erosa* and grey mullet *Liza aurata* from coastline of India. Indian J Sci 2(4):59–63

George MD, Kureishy TW (1979) Trace metals in zooplankton from the Bay of Bengal. Indian J Mar Sci 8:190–192

Gerber GB, Leonard A, Hantson P (2002) Carcinogenicity, mutagenicity and teratogenicity of manganese compounds. Crit Rev Oncol Hematol 42(1):25–34. doi:10.1016/S1040-8428(01)00178-0

Geszvain K, McCarthy JK, Tebo BM (2013) Elimination of manganese(II, III) oxidation in *Pseudomonas putida* GB-1 by a double knockout of two putative multicopper oxidase genes. Appl Environ Microbiol 79:357–366. doi:10.1128/AEM.01850-12

Ghiorse WC, Hirsch P (1979) An ultrastructural study of iron and manganese deposition associated with extracellular polymers of *Pedomicrobium*-like budding bacteria. Arch Microbiol 123:213–226

Glasby GP (2006) Manganese: predominant role of nodules and crusts. In: Schulz HD, Zabel M (eds) Marine geochemistry, 2nd edn. Springer, Berlin/Heidelberg, pp 371–427

Gounot AM (1994) Microbial oxidation and reduction of manganese: consequences in groundwater and applications. FEMS Microbiol Rev 14(4):339–349

Greger JL (1999) Nutrition versus toxicology of manganese in humans: evaluation of potential biomarkers. Neuro Toxicol 20:205–212

Gutierrez T, Biller DV, Shimmield T, Green DH (2012) Metal binding properties of the EPS produced by *Halomonas* sp. TG39 and its potential in enhancing trace element bioavailability to eukaryotic phytoplankton. Biometals 25(6):1185–1194. doi:10.1007/s10534-012-9581-3

Hallberg K, Johnson D (2005) Biological manganese removal from acid mine drainage in constructed wetlands and prototype bioreactors. Sci Total Environ 338:115–124

Hamer DH (1986) Metallothionein. Annu Rev Biochem 55:913–951. doi:10.1146/annurev.bi.55.070186.004405014

Hansel CM, Francis CA (2006) Coupled photochemical and enzymatic Mn(II) oxidation pathways of a planktonic *Roseobacter*-like bacterium. Appl Environ Microbiol 72:3543–3549

Hao Z, Chen S, Wilson DB (1999) Cloning, expression and characterization of cadmium and manganese uptake genes from *Lactobacillus plantarum*. Appl Environ Microbiol 65:4746–4752

Haritha A, Korripally PS, Tiwari A, Kiranmayi P, Rodrigue A, Mohan PM, Singh SS (2009) *MrdH*, a novel metal resistance determinant of *Pseudomonas putida* KT2440, is flanked by metal-inducible mobile genetic elements. J Bacteriol 191(19):5976–5987

He JZ, Zhang LM, Jin SS, Zhu YG (2008) Bacterial communities inside and surrounding soil iron-manganese nodules. Geomicrobiol J 25:14–24

Higham DP, Sadler PJ, Scawen MD (1984) Cadmium-resistant *Pseudomonas putida* synthesises novel cadmium proteins. Science 225:1043–1046

Hopkin SP (1993) In situ biological monitoring of pollutants in ecosystems. In: Calow P (ed) Handbook of ecotoxicology, vol 1. Blackwell, Oxford, UK, pp 397–427

Hosfield DJ, Guan Y, Haas BJ, Cunningham RP, Tainer JA (1999) Structure of the DNA repair enzyme endonuclease IV and its DNA complex: double-nucleotide flipping at a basic sites and three-metal-ion catalysis. Cell 98:397–408

Howe PD, Malcolm HM, Dobson S (2004) Manganese and its compounds: environmental aspects. In: Sheffer M (ed) Concise international chemical assessment document, 63. Ottawa and printed by Wissenchaftliche Verlagsgesellschaft mbH, Stuttgart, p 15

HSDB (2001) Manganese compounds. Bethesda, National Library of Medicine, Hazardous Substances Data Bank. Available at http://toxnet.nlm.nih.gov/cgi-bin/sis/htmlgen?HSDB. Accessed on 29 Oct 2015

Huang PM (1991) Kinetics of redox reactions on manganese oxides and its impact on environmental quality. In: Sparks DL, Suarez DL (eds) Rates of soil chemical processes. Soil Science Society of America, Madison, pp 191–230

Hughes MN, Poole RK (1989) Metal toxicity. In: Hughes MN, Poole RK (eds) Metals and microorganisms. Chapman and Hall, New York, pp 252–302

Huntsman SA, Sunda WG (1980) The role of trace metals in regulating phytoplankton. In: Morris I (ed) The physiological ecology of phytoplankton. University of California, Berkeley, pp 285–328

Institute of Medicine (2001) Dietary reference intakes for vitamin A, vitamin K, boron, chromium, copper, iodine, iron, manganese, molybdenum, nickel, silicon, vanadium, and zinc. National Academy Press, Washington, DC

IOM (2002) Dietary reference intakes for vitamin A, vitamin K, arsenic, boron, chromium, copper, iodine, iron, manganese, molybdenum, nickel, silicon, vanadium and zinc. Institute of Medicine, Food and Nutrition Board. National Academy Press, Washington, DC, pp 10–1–10–22

IPCS (1999) Manganese and its compounds. Geneva, World Health Organization. International programme on chemical safety (Concise International Chemical Assessment Document 12)

Jakubovics NS, Valentine RA (2009) A new direction for manganese homeostasis in bacteria: identification of a novel efflux system in *Streptococcus pneumoniae*. Mol Microbiol 72:1–4. doi:10.1111/j.1365-2958.2009.06637.x

Javed M, Usmani N (2013). Assessment of heavy metal (Cu, Ni, Fe, Co, Mn, Cr, Zn) pollution in effluent dominated rivulet water and their effect on glycogen metabolism and histology of *Mastacembelus armatus*. SpringerPlus. 2:390.doi.org/10.1186/2193-1801-2-390

Jensen AN, Jensen LT (2014) Chapter 1: Manganese transport, trafficking and function in invertebrates. In: Aschner M, Costa LG (eds) Manganese in health and disease. Royal Society of Chemistry, UK. pp 1–27

Ji GY, Silver S (1995) Bacterial resistance mechanisms for heavy metals of environmental concern. J Ind Microbiol 14:61–75

Kagi JHR, Nordberg M (1979) Metallothionein. BirkhauserVerlag, Basel, Stuttgart, pp 1–378

Keen CL, Zidenberg-Cherr S (1996) Manganese. In: Ziegler EE, Filer LJ (eds) Present knowledge in nutrition. ILSI Press, Washington, DC, pp 334–343

Kehres DG, Maguire ME (2003) Emerging themes in manganese transport, biochemistry and pathogenesis in bacteria. FEMS Microbiol Rev 27(2–3):263–290

Kehres DG, Zaharik ML, Finlay BB, Maguire ME (2000) The NRAMP proteins of *Salmonella typhimurium* and *Escherichia coli* are selective manganese transporters involved in the response to reactive oxygen. Mol Microbiol 36:1085–1100

KesavaRao C, Indusekhar VK (1989) Seasonal variation in chemical constituents of certain brown sea weed and sea water from Saurashtra coast: manganese, zinc. copper, nickel, cobalt and molybdenum. Mahasagar 22(2):73–81

Kim SI, Jang YS, Han SH, Choi MJ, Go EH, Cheon YP, Lee JS, Lee SH (2012) Effect of manganese exposure on the reproductive organs in immature female rats. Development and Reproduction 16(4):295–300

Kirchner WB, Grabowski S (1972) Manganese in lacustrine ecosystems: a review. Am Water Resour Assoc 8:1259–1264. doi:10.1111/j.1752-1688.1972.tb05270.x

Kolenbrander PE, Andersen RN, Baker RA, Jenkinson HF (1998) The adhesion-associated *sca* operon in *Streptococcus gordonii* encodes an inducible high-affinity ABC transporter for Mn^{2+} uptake. J Bacteriol 180:290–295

Laha S, Luthy RG (1990) Oxidation of aniline and other primary aromatic amines by manganese dioxide. Environ Sci Technol 24:363–373

Larsen EI, Sly LI, McEwan AG (1999) Manganese(II) adsorption and oxidation by whole cells and a membrane fraction of *Pedomicrobium* sp. ACM 3067. Arch Microbiol 171:257–264

Leach RM, Harris ED (1997) Manganese. In: O'Dell BL, Sunde RA (eds) Handbook of nutritionally essential minerals. Marcel Dekker, New York, pp 335–355

Learman DR, Voelker BM, Vazquez-Rodriguez AI, Hansel CM (2011) Formation of manganese oxides by bacterially generated superoxide. Nat Geosci 4:95–98

Lee JH, Kennedy DW, Dohnalkova A, Moore DA, Nachimuthu P, Reed SB, Fredrickson JK (2011) Manganese sulfide formation via concomitant microbial manganese oxide and thiosulfate reduction. Environ Microbiol Microbiol 13(12):3275–3288. doi:10.1111/j.1462-2920.2011.02587.x

Lewis BL, Landing WM (1991) The biogeochemistry of Mn and Fe in the Black Sea. Deep-Sea Res 38(Suppl 2):773–803

Lewis BL, Landing WM (1992) The investigation of dissolved and suspended particulate trace metal fractionation in the Black Sea. Mar Chem 40(1–2):105–141

Li C, Tao J, Mao D, He C (2011) A novel manganese efflux system, *YebN*, is required for virulence by *Xanthomonas oryzae pv. oryzae*. PLoS One 6(7):10.1371/journal.pone.0021983

Lovley DR, Phillips EJP (1988) Novel mode of microbial energy metabolism: organic carbon oxidation coupled to dissimilatory reduction of iron or manganese. Appl Environ Microbiol 54:1472–1480

Lynam DR, Roos JW, Pfeifer GD, Fort BF, Pullin TG (1999) Environmental effects and exposures to manganese from use of methylcyclopentadienyl manganese tricarbonyl (MMT) in gasoline. Neuro Toxicol 20:145–150

Mane PC, Kadam DD, Chaudhari RD, Bhosle AB (2013) Accumulation of manganese and selenium in water: a case study of Manjara dam, Maharashtra. Int J Curr Res 5(12):4142–4146

Marbaniang DG (2012) Spectrophotometric determination of Manganese in ground water in Shillong City using bismuthate oxidation method. Int J Environ Prot 2(5):22–26

Markus H, Mulack C, Pual A (2002) Manganese adsorption in the epiphytic lichens *Hypogymnia physodes* and *Lecanora conizaeoides*. Environ Exp Bot 48:107–117

McBride MB (1994) Environmental chemistry of soils, 1st edn. Oxford University Press, New York, p 406

Medicis ED, Paquette J, Gauthier JJ, Shapcott D (1986) Magnesium and manganese content of halophilic bacteria. Appl Environ Microbiol 52(3):567–573

Merrifield MS, Hines E, Liu X, Beck MW (2011) Building regional threat-based networks for estuaries in the Western United States. PLoS One 6(2):1–10

Missiakas D, Raina S (1997) Signal transduction pathways in response to protein misfolding in the extracytoplasmic compartments of *E. coli*: role of two new phosphoprotein phosphatases *PrpA* and *PrpB*. EMBO J 16:1670–1685

Mohamed ZA (2001) Removal of cadmium and manganese by a non toxic strain of the fresh water cyanobacterium *Gloeothece Magna*. Wat Res 35:4405–4409
Mondal P, Majumber CB, Mohanty B (2008) Treatment of arsenic contaminated water in a batch reactor by using *Ralstonia eutropha* MTCC 2487 and granular activated carbon. J Hazard Mater 153:588–599
Morel FMM, Hudson RJM, Price NM (1991) Limitation of productivity by trace metals in the sea. Limnol Oceanogr 36:1742–1755
Morgan TR, Shand JA, Clarke SM, Eaton-Rye JJ (1998) Specific requirements for cytochrome c-550 and the manganese stabilizing protein in photoautotrophic strains of *Synechocystis* sp. PCC 6803 with mutations in the domain *Gly*-351 to *Thr*-436 of the chlorophyll-binding protein CP47. Biochemistry 37:14437–14449
Mukhopadhyay B, Stoddard SF, Wolfe RS (1998) Purification, regulation, and molecular and biochemical characterization of pyruvate carboxylase from *Methanobacterium thermoautotrophicum* strain DH. J Biol Chem 273:5155–5166
Nairn B, Hedin RS (1993) Contaminant removal capabilities of wetlands constructed to treat coal mine drainage. In: Moshiri GA (ed) Constructed wetlands for water quality improvement, 7th edn. Lewis Publishers, Boca Raton, pp 187–195
Naser HA (2013) Assessment and management of heavy metal pollution in the marine environment of the Arabian Gulf: a review. Mar Pollut Bull 72(1):6–13. doi:10.1016/j.marpolbul.2013.04.030
Nasir SM, Okbah MA, Kasen SM (2006) Environmental assessment of heavy metal pollution in bottom sediments of Aden Port, Yemen. Int J Oceans Oceanogr 1:99–109
Nealson KH (1983) The microbial manganese cycle. In: Krumbein WE (ed) Microbial geochemistry. Blackwell Scientific Publications, Oxford, pp 191–221
Nealson KH, Myers CR (1992) Microbial reduction of manganese and iron: new approaches to carbon cycling. Appl Environ Microbiol 58:439–443
Nealson KH, Tebo BM, Rosson RA (1988) Occurrence and mechanisms of microbial oxidation of manganese. Adv Appl Microbiol 33:279–318
Nielsen FH (1999) Ultratrace minerals. In: Shils M (ed) Nutrition in health and disease. Williams and Wilkins, Baltimore, pp 283–303
Nies DH (2003) Efflux-mediated heavy metal resistance in prokaryotes. FEMS Microbiol Rev 27:313–339
Nieto JJ, Fernandez-Castillo R, Marquez MC, Ventosa A, Quesada E, Ruiz-Berraquero F (1989) Survey of metal tolerance in moderately halophilic eubacteria. Appl Environ Microbiol 55:2385–2390
Niven DF, Ekins A, al-Samaurai AA (1999) Effects of iron and manganese availability on growth and production of superoxide dismutase by *Streptococcus suis*. Can J Microbiol 45:1027–1032
Ohtani N, Haruki M, Muroya A, Morikawa M, Kanaya S (2000) Characterization of ribonuclease HII from *Escherichia coli* overproduced in a soluble form. J Biochem 127:895–899
Olaniran AO, Balgobind A, Pillay B (2013) Bioavailability of heavy metals in soil: impact on microbial biodegradation of organic compounds and possible improvement strategies. Int J Mol Sci 14(5):10197–10228
Oyetibo GO, Ilori MO, Adebusoye SA, Obayori OS, Amund OO (2010) Bacteria with dual resistance to elevated concentrations of heavy metals and antibiotics in Nigerian contaminated systems. Environ Monit Assess 168:305–314. doi:10.1007/s10661-009-1114-3
Parveen SR, Bhosle AB (2013) Heavy metal contamination in soils near Siddheshwar Dam Maharashtra. India Res J Chem 3(1):6–9
Pearson GF, Greenway GM (2005) Recent developments in manganese speciation. Trends Anal Chem 24:803–809
Pereira F, Krishnan KP, Sinha RK, Kerkar S (2012) Insights on metal-microbe interactions in *Bacillus* sp. and *Chromohalobacter* sp. from a solar saltern. J Ecobiotechnol 4(1):14–24
Pereira F, Kerkar S, Krishnan KP (2013) Bacterial response to dynamic metal concentrations in the surface sediments of a solar saltern (Goa, India). Environ Monit Assess 185(5):3625–3636
Perl DP, Olanow CW (2007) The neuropathology of manganese-induced Parkinsonism. J Neuropathol Exp Neurol 66(8):675–682
Poole RK, Gadd GM (1989) Metals: microbe interactions. IRL Press, Oxford, pp 1–37
Pravinkumar M, Logesh AR, Viswanathan C, Elumalai V, Raffi SM, Kathiresan K (2014) Bioaccumulation of trace elements in fin and shell fishes of Uppanar and Vellar Estuaries at Southeast coast of India. Int J Mar Sci 4(41):1–7. doi:10.5376/ijms.2014.04.0041
Pulsawat W, Leksawasdi N, Rogers PL, Foster LJR (2003) Anions effects on biosorption of Mn(II) by extracellular polymeric substance (EPS) from *Rhizobium etli*. Biotechnol Lett 25:1267–1270
Que Q, Helmann JD (2000) Manganese homeostasis in *Bacillus subtilis* is regulated by *MntR*, a bifunctional regulator related to the diphtheria toxin repressor family of proteins. Mol Microbiol 35:1454–1468
Ravelonabdra PH, Rajoelisoa A, Mong Y, Rajaonarivony M, Ravonizafy C, Niricalaina A (2010) Assessment of heavy metals concentrations in coastal sediments in north-western cities of Madagascar. Afr J Environ Sci Technol 4(2):051–060
Reimer PS (1999) Environmental effects of manganese and proposed freshwater guidelines to protect aquatic life in British Columbia [MSc thesis]. University of British Columbia, Vancouver
Richardson LL, Aguilar C, Nealson KH (1988) Manganese oxidation in pH and O_2 microenvironments produced by Phytoplankton. Limnol Oceanogr 33:352–363
Ridge JP, Lin M, Larsen EI, Fegan M, McEwan AG, Sly LI (2007) A multicopper oxidase is essential for manganese oxidation and laccase-like activity in *Pedomicrobium* sp. ACM 3067. Environ Microbiol 9:944–953
Robbins EI, Corley TL (2005) Microdynamics and seasonal changes in manganese oxide epiprecipita-

tion in Pinal Creek, Arizona. Hydrobiologia 534:165–180
Ross SM (1994) Retention, transformation and mobility of toxic metals in soils. In: Ross SM (ed) Toxic metals in soil-plant systems. Wiley, Chichester, pp 63–152
Rosson RA, Nealson KH (1982) Manganese binding and oxidation by spores of a marine *Bacillus*. J Bacteriol 151:1027–1034
Santamaria AB, Sulsky SI (2010) A risk assessment of an essential element: manganese. J Toxicol Environ Health A 73(2):128–155
Saratovsky I, Wightman PG, Paste PA, Gaillard JF, Poeppelmeier KR (2006) Manganese oxides: parallels between abiotic and biotic structures. J Am Chem Soc 128:11188–11198
Saratovsky I, Gurr SJ, Hayward MA (2009) The structure of manganese oxide formed by the fungus *Acremonium* sp. strain KR21-2. Geochim Cosmochim Acta 73:3291–3300
Sardessai S, Sundar D (2007) Variability of nitrate and phosphate. In: Shetye SR, Dileep Kumar M, Shankar D (eds) The Mandovi and Zuari Estuaries. National Institute of Oceanography, Dona Paula, pp 59–66
Sasaki K, Matsuda M, Hirajima T, Takano K, Konno H (2006) Immobilization of Mn(II) ions by a Mn oxidizing fungus *Paraconiothyrium* sp. like strain at neutral pH. Mater Trans 47:2457–2461
Sekowska A, Danchin A, Risler JL (2000) Phylogeny of related functions: the case of polyamine biosynthetic enzymes. Microbiology 146:1815–1828
Schamphelaire L, Rabaey K, Boeckx P, Boon N, Verstrate W (2008) Outlook for benefits of sediment microbial fuel cells with two bio-electrodes. J Microbial Biotechnol 1(6):446–462
Schweisfurth R, Eleftheriadis D, Gundlach H, Jacobs M, Jung W (1978) Microbiology of the precipitation of manganese. In: Krumbein WE (ed) Environmental biogeochemistry and geomicrobiology. Ann Arbor Science, Ann Arbor, pp 923–928
Scott DT, McKnight DM, Voelker BM, Hrncir DC (2002) Redox processes controlling manganese fate and transport in a mountain stream. Environ Sci Technol 36(3):453–459
Shaw CF III, Stillman MJ, Suzuki KT (1992) Metallothioneins: an overview of metal-thiolate complex formation in metallothioneins. In: Stillman MJ, Shaw CF III, Suzuki KT (eds) Metallothioneins: synthesis, structure and properties of metallothioneins, phytochelatins and metal-thiolate complexes. VCH Publishers, New York
Shi L (2004) Manganese-dependent protein o-phosphatases in prokaryotes and their biological functions. Front Biosci 9:1382–1397
Shi J, Lindsay WP, Huckle JW, Morby AP, Robinson NJ (1992) Cyanobacterial metallothionein gene expressed in *Escherichia coli*. Metal-binding properties of the expressed protein. FEBS Lett 303:159–163
Shi JX, Chen ZY, Hu XG (1999) A study on bacterial abundance and its mineralization in ferromanganese nodule area of the East Pacific Ocean. Donghai Mar Sci 17(3):46–64
Shrivastava KBL, Mishra SP (2011) Studies of various heavy metal in surface and ground water of Birsinghpur Town and its surrounding rural area District Satna (M.P.). Curr World Environ 6(2):271–274
Silver S, Lusk JE (1987) Bacterial magnesium, manganese and zinc transport. In: Rosen BP, Silver S (eds) Ion transport in prokaryotes. Academic, London, pp 165–180
Sly LI, Arunpairojana V, Dixon DR (1990) Binding of colloidal MnO_2 by extracellular polysaccharides of *Pedomicrobium manganicum*. Appl Environ Microbiol 56:2791–2794
Smith RG (1972) Vanadium. In: Lee DHK (ed) Metallic contaminants and human health. Academic Press, New York, pp 153–157
Smith RA, Alexander RB, Wolman MG (1987) Water-quality trends in the nation's rivers. Science 235:1607–1615
Sobota JM, Imlay JA (2011) Iron enzyme ribulose-5-phosphate 3-epimerase in *Escherichia coli* is rapidly damaged by hydrogen peroxide but can be protected by manganese. Proc Natl Acad Sci U S A 108:5402–5407
Spratt HG Jr, Hodson RE (1994) The effect of changing water chemistry on rates of manganese oxidation in surface sediments of a temperate saltmarsh and a tropical mangrove estuary. Estuar Coast Shelf Sci 38:119–135
Stokes JL, Powers MT (1967) Stimulation of polyhydroxybutyrate oxidation in *Sphaerotilus discophorus* by manganese and magnesium. Arch Microbiol 59:295–301
Sujith PP, Khedekar VD, Girish AP, Loka Bharathi PA (2010) Immobilization of nickel by bacterial isolates from the Indian ridge system and the chemical nature of the accumulated metal. Geomicrobiol J 27:424–434
Sullivan L, Koppi A (1993) Manganese oxide accumulations associated with some soil structural pores. II. Composite coatings and translocation. Soil Res 31:227–233
Sun H, Xu G, Zhan H, Chen H, Sun Z, Tian B, Hua Y (2010) Identification and evaluation of the role of the manganese efflux protein in *Deinococcus radiodurans*. BMC Microbiol 10:319. doi:10.1186/1471-2180-10-319
Sunda WG, Huntsman SA (1990) Diel cycles in microbial manganese oxidation and manganese redox speciation in coastal waters of the Bahama Islands. Limnol Oceanogr 35:325–338
Tebo BM (1991) Manganese(II) oxidation in the suboxic zone of the Black Sea. Deep Sea Res 38(Suppl 2):S883–S905
Tebo BM, Ghiorse WC, vanWaasbergen LG, Siering PL, Caspi R (1997) Bacterially mediated mineral formation: insights into manganese(II) oxidation from molecular genetic and biochemical studies. Geomicrobiol Interact Between Microbes Miner 35:225–266
Tebo BM, Bargar JR, Clement BG, Dick GJ, Murray KJ, Parker D, Verity R, Webb SM (2004) Biogenic manganese oxides: properties and mechanisms of formation.

Annu Rev Earth Planet Sci 32:287–328. doi:10.1146/annurev.earth.32.101802.120213

Tebo BM, Johnson HA, McCarthy JK, Templeton AS (2005) Geomicrobiology of manganese(II) oxidation. Trends Microbiol 13:421–428

Tekerlekopoulou AG, Vayenas DV (2008) Simultaneous biological removal of ammonia, iron and manganese from potable water using a trickling filter. Biochem Eng J 39:215–220

Temara A, Skei JM, Gillan D, Warnau M, Jangoux M, Dubois P (1998) Validation of the asteroid *Asterias rubens* (Echinodermata) as a bioindicator of spatial and temporal trends of Pb, Cd, and Zn contamination in the field. Mar Environ Res 45(4–5):341–356

Thompson J, Ruvinov SB, Freedberg DI, Hall BG (1999) Cellobiose-6-phosphate hydrolase (*CelF*) of *Escherichia coli*: characterization and assignment to the unusual family 4 of glycosylhydrolases. J Bacteriol 181:7339–7345

Timonin MI, Illman WI, Hartgerink T (1972) Oxidation of manganous salts of manganese by soil fungi. Can J Microbiol 18:793–799

Tseng HJ, Srikhanta Y, McEwan AG, Jennings MP (2001) Accumulation of manganese in *Neisseria gonorrhoeae* correlates with resistance to oxidative killing by superoxide anion and is independent of superoxide dismutase activity. Mol Microbiol 40(5):1175–1186

Turner JS, Morby AP, Whitton BA, Gupta A, Robinson NJ (1993) Construction of Zn^{2+}/Cd^{2+} hypersensitive cyanobacterial mutants lacking a functional metallothionein locus. J Biol Chem 268:4494–4498

US EPA (1984) Health assessment document for manganese. Final draft. Cincinnati, US Environmental Protection Agency, Office of Research and Development (Report No. EPA/600/8-83/013F). Assessed on 27 Oct 2015

Van Waasbergen LG, Hildebrand M, Tebo BM (1996) Identification and characterization of a gene cluster involved in manganese oxidation by spores of the marine *Bacillus* sp. strain SG-1. J Bacteriol 178:3517–3530

Varol M, Sen B (2012) Assessment of nutrient and heavy metal contamination in surface water and sediments of the Upper Tigris River, Turkey. Catena 92:1–10

Vartanian JP, Sata M, Henry M, Hobson SW, Meyerhans A (1999) Manganese cations increase the mutation rate of human immune deficiency virus type 1 ex vivo. J Gen Virol 80:1983–1986

Ventosa A, Nieto JJ, Oren A (1998) Biology of moderately halophilic aerobic bacteria. Microbiol Mol Biol Rev 62:504–544

Villalobos M, Toner B, Bargar J, Sposito G (2003) Characterization of the manganese oxide produced by *Pseudomonas putida* strain MnB1. Geochim Cosmochim Acta 67:2649–2662

Vojak PWL, Edwards C, Jones MV (1985) Evidence for microbial manganese oxidation in the River Tamar estuary, South West England. Estuar Coast Shelf Sci 20:661–671

Wakeman CA, Skaar EP (2012) Metalloregulation of gram-positive pathogen physiology. Curr Opin Microbiol 15:169–174

Wang W, Shao Z, Liu Y, Wang G (2009) Removal of multi-heavy metals using biogenic manganese oxides generated by a deep-sea sedimentary bacterium – *Brachybacterium* sp. strain Mn32. Microbiology 155:1989–1996. doi:10.1099/mic.0.024141-0

Wang L, Ma F, Feng SS, Sun DZ, Yu B, Xing J (2010) Biosorption characteristics of Extracellular Polymeric Substance (EPS) produced by *Rhizobium radiobacter* for removal of Cu(II) and Mn(II) Ions from aqueous solutions. Bioinforma Biomed Eng (iCBBE): 1–4. doi:10.1109/ICBBE.2010.5515723

Wasim AM, Paramasivam M, Ganguly M, Purkait S, Sengupta D (2010) Assessment and occurrence of various heavy metals in surface water of Ganga river around Kolkata: a study for toxicity and ecological impact. Environ Monit Assess 160(1–4):207–213

Wehrli B, Friedl G, Manceau A (1995) Reaction rates and products of manganese oxidation at the sediment-water interface. In: Huang CP, O'Melia CR, Morgan JJ (eds) Aquatic chemistry: interfacial and interspecies processes. Am Chem Soc, 244:111–134

Whittaker MM, Barynin VV, Antonyuk SV, Whittaker JW (1999) The oxidized (3,3) state of manganese catalase. Comparison of enzymes from *Thermus thermophilus* and *Lactobacillus plantarum*. Biochemistry 38:9126–9136

WHO (1973) Trace elements in human nutrition: manganese. Report of a WHO expert committee, Technical report series no. 532. World Health Organization, Geneva, pp 34–36

Yang W, Zhang Z, Zhang Z, Chen H, Liu J, Ali M, Liu F, Li L (2013) Population structure of manganese-oxidizing bacteria in stratified soils and properties of manganese oxide aggregates under manganese–complex medium enrichment. PLoS One 8(9):e73778

Yang ZF, Sun T, Zhao R (2014) Environmental flow assessments in estuaries related to preference of phytoplankton. Hydrol Earth Syst Sci 18:1785–1791. doi:10.5194/hess-18-1785-2

Yocum CF, Pecoraro VL (1999) Recent advances in the understanding of the biological chemistry of manganese. Curr Opin Chem Biol 3:182–187

Zeri C, Voutsinou-Taliadouri F, Romanov AS, Ovsjany EI, Moriki A (2000) A comparative approach of dissolved trace element exchange in two interconnected basins: Black Sea and Aegean Sea. Mar Pollut Bull 40(8):666–673

Zhan S, Peng S, Liu C, Chang Q, Xu J (2010) Spatial and temporal variations of heavy metals in surface sediments in Bohai Bay, North China. Bull Environ Contam Toxicol 84:482–487

Zingde MD, Singbal SYS, Moraes CF, Reddy CVG (1976) Arsenic, copper, zinc and manganese in the marine flora and fauna of coastal and estuarine waters around Goa. Indian J Mar Sci 5:212–217

Zouboulis AI, Loukidou MX, Matis KA (2004) Biosorption of toxic metals from aqueous solutions by bacteria strains isolated from metal polluted soils. Process Biochem 39:909–916

Applications of Siderophore Producing Marine Bacteria in Bioremediation of Metals and Organic Compounds

11

Teja Gaonkar and Sunita Borkar

Abstract

Siderophores are chelating agents that are produced by bacteria and fungi for iron uptake under its limiting conditions. Low iron content is a peculiar feature of marine ecosystems. Marine microorganisms, like their terrestrial counterparts, successfully overcome iron limitation by production of siderophores. However, marine siderophores structurally differ from their terrestrial counterparts. Microbial siderophores have been known to facilitate heavy metal sequestration and also play a vital role in organic compound degradation. Therefore, such siderophores have potential for bioremediation of metal and organic compound polluted areas. This chapter focuses on the potential use of siderophore-producing marine bacteria in remediation of metal and organic compound.

11.1 Introduction

11.1.1 Bioremediation of Metals

Metals are important in human civilisation. Nevertheless, an unwarranted use of metal ions has contributed to metal pollution of soil, air and water and has adverse effects on human health. However, exposure to heavy metals has continued. They enter into the environment due to natural or anthropogenic activities. Natural sources of heavy metal contamination include seepage from rocks into water, volcanic activity, forest fires and partitioning of polluting elements between sedimentary rocks and their precursor sediments and water. Anthropogenic sources of heavy metal contamination include industrial wastes, mining activity, agricultural practices, automobile emissions and military activity (Banik et al. 2014).

The outcome of exposure of humans to high levels of heavy metals is developmental retardation, numerous types of cancers, kidney damage, disruption of endocrine system and various immunological, neurological and other disorders (Mudgal et al. 2010). Thus, the grave problem of heavy metal pollution requires efficient cleaning of the polluted areas. Many conventional methods have been used over the years for cleaning up metal-polluted areas such as excavation and landfill, thermal treatment, acid leaching and electro

T. Gaonkar (✉) • S. Borkar
P.E.S.'s R.S.N College of Arts and Science,
Farmagudi, Ponda, Goa, India
e-mail: teja22984@gmail.com

M.M. Naik, S.K. Dubey (eds.), *Marine Pollution and Microbial Remediation*,
DOI 10.1007/978-981-10-1044-6_11

reclamation (Banik et al. 2014). However, most of the conventional methods employed for remediation of soil are uneconomical resulting in the deterioration of organic content of soil and its deterioration (Rajkumar et al. 2010).

Heavy metals are not degraded to harmless by-products by any biological, physical or chemical means. However, they are known to be transformed from one oxidation state or organic complex to another (Banik et al. 2014). One of the promising technologies for bioremediation of such soils contaminated with metal is "microbial-assisted phytoremediation". This process utilises plants to bring about absorption, accumulation and detoxification of contaminants in soil through various chemical, biological and physical processes (Prasad et al. 2010) and also involves plant growth-promoting rhizobacteria (PGPR). Rhizosphere bacteria are known to promote plant growth either by making nutrients available to them or by suppressing growth of the plant pathogens. Besides, some of these PGPR exhibit metal resistance and can be used in reclamation of such metal-polluted soils. *Siderophore* production is an important property of PGPR which can be exploited for phytoremediation of soils with metal contamination (Jing et al. 2007).

Iron is required for the growth of living organisms. It essentially participates in fundamental processes such as oxygen metabolism, electron transfer and DNA and RNA synthesis. However, under aerobic conditions, iron is present as ferric hydroxide, a highly insoluble form, which severely limits its availability to microorganisms. The concentration of free Fe^{+3} at pH close to neutrality has been reported to be around 10^{-18} M (Braud et al. 2009). To acquire this limiting iron, microorganisms produce siderophores which are low molecular weight iron-chelating compounds. Siderophores bind to Fe^{+3} and transport it into the cells where it is reduced to Fe^{+2} and released from siderophore (Cornelis and Matthijs 2002).

For such a strategy like phytoremediation or bioremediation to be feasible, it is mandatory to understand the interaction and response of siderophore producers to metals. A study was therefore undertaken to determine the effect of various metals on siderophore production and growth in a selected isolate, *Bacillus amyloliquefaciens* NAR38.1. The isolate is only recently reported for its important role in plant growth promotion (Buensanteai et al. 2008; Mishra and Kumar 2012). Nine different metals were selected for study and were grouped into essential and non-essential ones. The metals used in the study and which are required for the metabolic activities (biotic metals) of the cell were cobalt (Co), zinc (Zn), copper (Cu), manganese (Mn) and molybdenum (Mo). Abiotic metals (Schalk et al. 2011) which are toxic to cells and which were used in the study were aluminium (Al), arsenic (As), cadmium (Cd) and lead (Pb). Literature survey indicates that siderophores can bind and solubilise different metals like Cu, Ni, Zn, Pb and Cd and also the actinides like U(IV), Pu(IV) and Th(IV) (Schalk et al. 2011). The ability of the siderophore to bind a particular metal is defined by ligand functionality which also determines the stability of metal-siderophore complex (Hernlem et al. 1999).

11.1.2 Bioremediation of Organic Compounds

Organic compounds are one of the largest groups of environmental pollutants and are considered toxic to living organisms (Park et al. 2001). Compounds which have benzene ring in their structure rank second in the list of naturally occurring substances. Benzene ring is thermodynamically stable and resists environmental chemical degradation and, therefore, is hazardous to biosphere (Khomenkov et al. 2008). Marine and coastal areas are exposed to aromatic hydrocarbon contamination as a result of anthropogenic shipping activities and urbanisation (Raghavan and Furtado 2000; de Sousa and Bhosle 2012). Accumulation of such aromatics can have adverse effect on indigenous flora affecting membrane transport and other physiologically important activities (de Sousa and Bhosle 2012). *Pseudomonas* and other related bacterial genera have been well documented in bioremediation of such aromatics. *Pseudomonas* can metabolise and degrade a large group of organic compounds. Members of the genus *Pseudomonas* can degrade, transform or co-metabolise numerous natural as well as anthropogenic compounds (Zeyaullah et al. 2009). Metabolism of aromatic compounds proceeds with opening or cleavage of the aromatic ring which is catalysed by the enzyme "oxygenase".

Oxygenases catalyse incorporation of molecular oxygen and require cofactors like copper, flavin and iron. However, the most frequently used cofactor is iron. Amongst the 100 oxygenases reported till date, nearly 50 % have iron incorporated in their structure or require it as their cofactor (Nozaki and Ishimura 1974). Dioxygenases cleave benzene ring to form catechol or phenol derivatives (Song et al. 2000). Catechol is further acted upon by dioxygenases which are categorised into intradiol and extradiol. Intradiol enzymes are red and contain Fe^{+3} in their structure, whereas extradiols are colourless in nature and contain Fe^{+2} (Nozaki and Ishimura 1974). Therefore, degradation or utilisation of aromatic compounds which involves oxygenases will increase iron demand of the cells.

Staijen and Witholt (1998) have studied requirement of trace elements during expression of alkane hydroxylase and have reported an increase in iron requirement. The efficiency of *Pseudomonas* strain to degrade toluene has been reported to decrease with decrease in iron concentration (Dinkla et al. 2001). The study also suggests that microorganisms require high iron concentration to sustain in a medium with benzoate as the source of carbon. Iron requirement increases as it is required as a cofactor by the mono- and dioxygenases for degradation of aromatic hydrocarbons (Dinkla et al. 2001). The literature survey thus indicates that siderophores perform an important role in bioremediation of organic pollutants in iron-limited ecosystems by making it available to the microorganisms. However, not much work has been reported on the role of siderophores in biodegradation of organic compounds. Low iron content being an inherent characteristic of marine environment, siderophore-producing microorganisms have an edge over the non-siderophore-producing bacteria when used in bioremediation of such organic compounds in marine ecosystems. Hence, our study was focussed to reveal the effect of aromatic compound sodium benzoate on siderophore production in a selected isolate, *Pseudomonas aeruginosa* TMR2.13. Benzoate was chosen as the carbon source as it is found as an intermediate in the biodegradation of most of the organic compounds and is the simplest aromatic salt (Cao et al. 2008).

Besides playing a role in degradation of aromatic compounds by satisfying the iron demand of the microorganisms, siderophores also play a direct role in degradation of hydrocarbon and other aromatic compounds. Barbeau et al. (2002) have isolated *Marinobacter hydrocarbonoclasticus*, which degrades oil and also produces a siderophore petrobactin that undergoes a light-mediated decarboxylation when bound to Fe^{+3}. They have proposed that the siderophore may help the bacterium in degradation of petroleum hydrocarbons by facilitating iron uptake. Inoue et al. (2003) have demonstrated the vital role of siderophore pyoverdine in degradation of triphenyltin (TPT). TPT is an organotin used as active component of paints and agrochemicals. The compound enters into the aquatic system by the process of leaching from the antifouling paints and runoff from agricultural fields causing harmful effects in aquatic organisms. Siderophores have also been reported to play role in conferring resistance to toxic levels of tributyltin chloride (TBTCl) in estuarine *Alcaligenes faecalis* strain SD5 (Khanolkar et al. 2015a). Khanolkar et al. (2015b) have reported upregulation of siderophore synthesis in the presence of TBTCl in *Klebsiella pneumoniae* strain SD9.

11.2 Studies on Siderophore-Producing Bacteria from Coastal Ecosystems of Goa

Since indigenous microorganisms are preferred for bioremediation in field interventions, it was envisaged to isolate the microorganisms required for the study. The isolates were obtained from two distinct ecosystems being sand dunes representing nutrient-deficient ecosystem and mangroves representing nutrient-rich ecosystem (Figs. 11.1 and 11.2, respectively). Sand dune ecosystems are described as mounds of sand with vegetation, found along the coastal areas, and are characterised by low nutrients, drought, high salinity and sand erosion (Arun et al. 1999). Plants growing in sand dune habitats help in stabilisation of the same and are controlled by the interaction between the biotic and physicochemical components of the sand matrix (Arun and Sridhar 2004). The interactions between plants and bacteria help plants to settle in ecosystem restoration process (Egamberdiyeva 2005; Glick 1995). Although bacterial diversity plays an important role in establishing vegetation and supporting plant growth, few reports are available on the rhizobacteria associated with plants in sand dune regions. Despite the role of bacterial diversity in sand dune plant communities, very few reports are available on the distribution and abundance of rhizosphere-associated bacteria. Mangrove ecosystems chosen as second sampling site are rich in nutrients in contrast to sand dune ecosystem (Alongi et al. 1993; Holguin et al. 1992; Sengupta and Chaudhuri 1991; Vazquez et al. 2000). Samples collected from both the ecosystems were processed to isolate siderophore-producing bacteria and furthermore to study the effect of metals and sodium benzoate on production of siderophores in selected isolates (Gaonkar 2015; Gaonkar et al. 2012).

Fig. 11.1 Rhizosphere of sand dune creeper *Ipomoea pes-caprae* at Miramar beach

Fig. 11.2 Mangrove ecosystem

11.3 Effect of Metals on Siderophore Production

For isolation of the potential siderophore-producing microorganisms, the mangrove sediment sample was serially diluted, and 100 µl of sample was plated on nutrient agar (NA) and was kept for incubation at room temperature for 48 h. Predominant colonies were selected, and the isolates were purified by repeated streaking on NA. The ability of the isolates to produce siderophores was detected using the chrome azurol sulphonate (CAS) assay described by Schwyn and Neilands (1987). The bacterial isolate giving maximum siderophore production was used for further studies and subjected to routine biochemical tests for its tentative identification according to Bergey's Manual of Systematic Bacteriology (Krieg and Holt 1984; Sneath et al. 1986). The identification was confirmed by 16S rDNA sequencing. The sequence was used for BLAST (Altschul et al. 1990) with the nrdatabase of NCBI. Closely matching sequences were aligned using multiple alignment software programme Clustal X (Thompson et al. 1997). MEGA4.2.1 was used to construct phylogenetic tree. Furthermore, gas chromatography was used for FAME analysis to confirm the identification of the isolate.

By polyphasic taxonomy, the isolate was confirmed to be *Bacillus amyloliquefaciens* NAR38.1 and was selected to study the effect of various metals on production of siderophore by the isolate. It was interesting to note that the isolate showed varied response to all these metals (Zn, Co, Cu, Mo, Mn, As, Pb, Al and Cd) in terms of growth and siderophore production. The effect of metals can be categorised in four groups. One was increase in siderophore production as seen in presence of Zn, Co and Mn. The increase in siderophore production with subsequent sequestration of metal ions is a mechanism to reduce the toxicity of metal ions by reducing the concentration of the same in the medium. Decrease in siderophore production was observed in presence of Mo and As without affecting growth of the isolate. An interesting observation was made with respect to siderophore production in presence of Pb and Al which decreased siderophore production at low concentration, however increased siderophore production at high concentrations. Al^{+3} being similar in size and charge to Fe^{+3} forms complexes with siderophores, thus making it unavailable for iron transport (Garrison and Crumbliss 1987; Hu and Boyer 1996). Therefore, bacteria produce increased amount of siderophore to acquire iron. Naik and Dubey (2011) have also reported lead-enhanced siderophore production in *Pseudomonas aeruginosa* strain 4EA, Cd- and Co-suppressed growth as well as siderophore production. The production of siderophore by *B. amyloliquefaciens* NAR38.1 in the presence of toxic metal signifies that it may play an important role in the uptake and mobilisation of heavy metals or developing metal resistance. Our studies indicated that the organism is a potent isolate for use in bioremediation of metal-contaminated marine or coastal areas (Gaonkar 2015; Gaonkar and Bhosle 2013).

11.4 Isolation of Siderophore-Producing and Sodium Benzoate-Utilizing Bacteria

Our initial attempts to isolate siderophore producers from sand dune (Sect. 11.2) which could also utilise sodium benzoate as the sole source of carbon resulted in the isolation of a very few isolates (Gaonkar et al. 2012). Sand dune samples were therefore enriched in mineral salts medium (MSM) with or without added iron and sodium benzoate as the sole carbon source. Enrichment technique involves providing conditions favourable for growth of bacteria on a particular substrate. A total of 18 and 22 isolates were obtained on enrichment with iron and without iron, respectively. It was observed that all the isolates obtained from the sample enriched in MSM with iron exhibited siderophore production. However, out of 22 isolates obtained from MSM without iron, only 16 showed siderophore production.

11.4.1 Growth of the Selected Isolates in Liquid MSM with Varying Sodium Benzoate Concentrations

Two isolates E6 and E16 which showed maximum siderophore production (Fig. 11.3) as evident from the zone size on CAS plate were selected for identification using biochemical tests (Krieg and Holt 1984). The cultures were tentatively identified as *Pseudomonas* sp. and along with the isolate TMR 2.13 isolated from coastal sand dunes (Gaonkar et al. 2012) were selected to check their growth in MSM with varied concentrations of sodium benzoate as the only carbon source. The two isolates E6 and E16 showed growth in MSM containing up to 1.5 % of sodium benzoate concentration with maximum siderophore production and growth at 0.5 % of sodium benzoate (Figs. 11.4a, 11.4b). Interestingly, TMR2.13 showed growth up to 2 % sodium benzoate (Fig. 11.4c) with maximum siderophore production at 1.5 %. Rothera's test (Norris and Ribbons 1971) showed the ortho mode of ring cleavage for degradation of sodium benzoate in all the three isolates (Fig. 11.5). Significantly, the production of siderophore in presence of benzoate was observed to be higher with TMR2.13 as compared to E6 and E16. Since our objective was to understand the effect of sodium benzoate on siderophore production, TMR2.13 was chosen for further studies.

11.4.2 Effect of Sodium Benzoate on the Morphology of *P. aeruginosa* TMR2.13

To study the effect of sodium benzoate on morphology of the cells, the cells were grown in MSM with the carbon source as 0.4 % glucose or

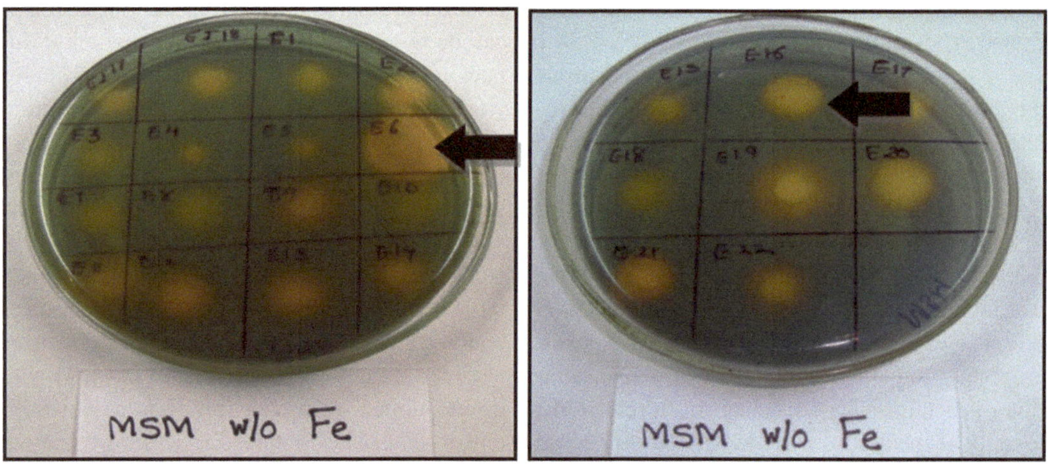

Fig. 11.3 Screening of the enriched isolates on MSM with CAS and 0.2 % sodium benzoate as the carbon source. Isolates E6 and E16 chosen for further studies are marked with *arrows*

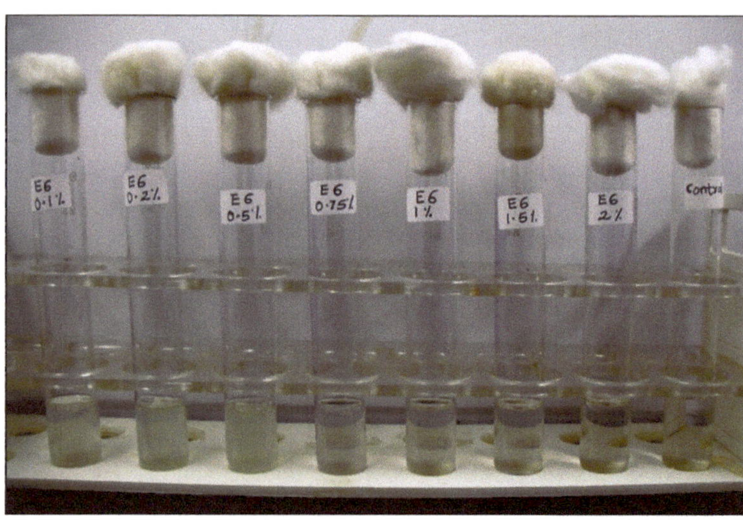

Fig. 11.4a Growth of isolate E6 in different sodium benzoate concentrations

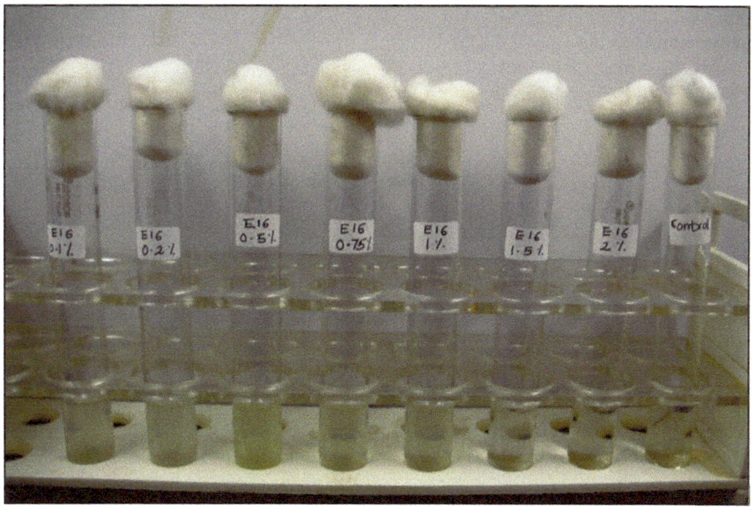

Fig. 11.4b Growth of isolate E16 in different sodium benzoate concentrations

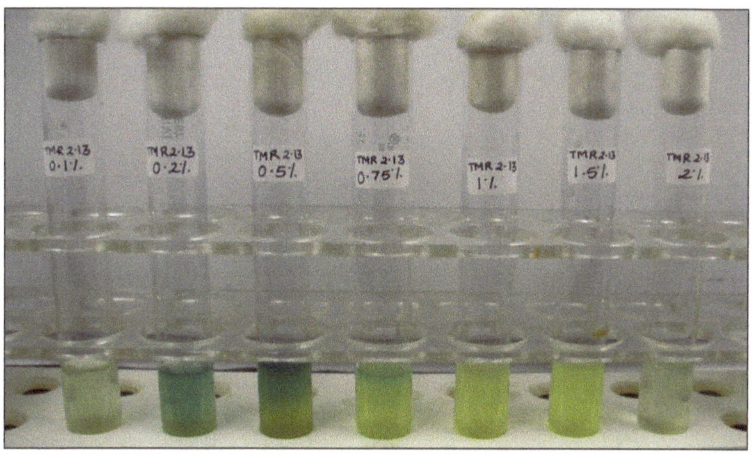

Fig. 11.4c Growth of isolate TMR2.13 in different sodium benzoate concentrations

sodium benzoate. The flasks were incubated at 150 rpm for 24 h. The broth was centrifuged, and the pellet was washed and then suspended in phosphate-buffered saline. The cells were fixed onto a coverslip in 2.5 % glutaraldehyde fixative (pH 7.2–7.4) overnight at 28 °C and then washed using phosphate-buffered saline. The cells were dehydrated with increasing concentrations of acetone (30 %, 50 %, 70 % and 90 %) for 10 min each and finally in 100 % acetone for 30 min. The coverslips were placed onto stubs which were then placed in a sputter coater (JEOL JFC1600). After sputtering the specimens with platinum, the stubs were placed onto the electron microscope sample chamber and observed using JEOL JSM – 6360LV scanning electron microscope.

It was observed that the cells of TMR2.13 showed increase in length by 0.47 μM when grown in presence of sodium benzoate as compared to when grown in presence of glucose (Fig. 11.6). Reports have indicated that one of the mechanisms to overcome the toxicity of aromatic hydrocarbons is alterations in structure and permeability of the cell membrane (de Sousa and Bhosle 2012; Sikkema et al. 1995). Such alterations in cell morphology have also been reported for *Aeromonas caviae* strain KS-1 when grown in

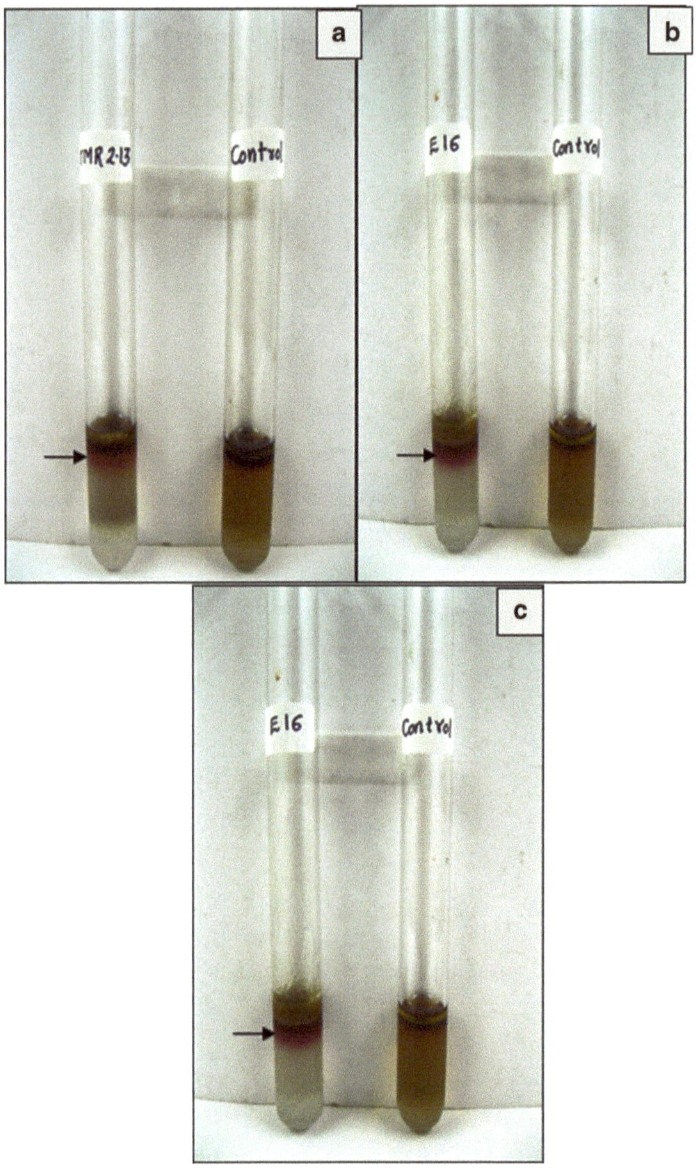

Fig. 11.5 Formation of purple ring by the Rothera's test (marked with an *arrow*) depicting ortho pathway in (**a**) TMR2.13 (**b**) E6 and (**c**) E16

presence of pollutants like 1.4 mM lead nitrate and 1 mM tributyltin chloride (Shamim et al. 2012).

Biosynthesis and secretion of siderophore is known to be related to the requirement of iron for metabolism of specific growth substrates. The presence of aromatic compounds along with easily metabolisable co-substrates supports the production of siderophores. Hence it was envisaged to study siderophore production in presence of various carbon sources.

11.4.3 Effect of Various Carbon Sources and Media on Siderophore Production in *P. aeruginosa* TMR2.13

Isolate TMR 2.13 was grown in presence of different carbon sources as follows:

(I) Succinate basal medium (BM) supplemented with the following carbon sources:
   1. BM + 0.4% succinate

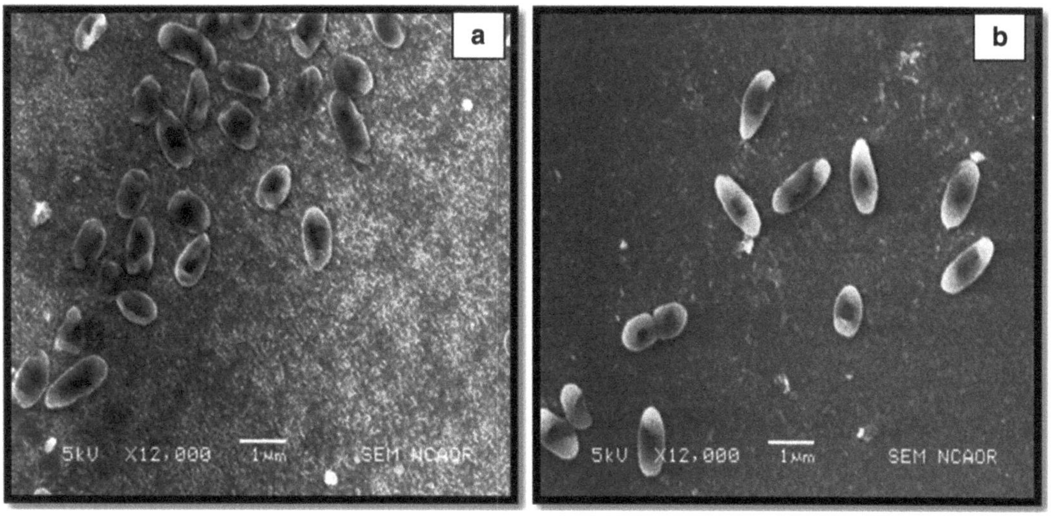

Fig. 11.6 SEM of *P. aeruginosa* TMR2.13 grown in: (**a**) MSM with 0.4 % glucose as the carbon source and without iron. (**b**) MSM with 0.4 % sodium benzoate as the carbon source and without iron

2. BM + 30mg / L $FeSO_4.7H_2O$ + 0.4% succinate
3. BM + 30mg / L $FeCl_3.6H_2O$ + 0.4% succinate
4. BM + 0.4% glucose
5. BM + 0.4% glucose + 30mg / L $FeCl_3.6H_2O$
6. BM + 0.4% sodium benzoate
7. BM + 0.4% sodium benzoate + 30mg / L $FeCl_3.6H_2O$

(II) Nutrient broth (NB)
(III) Typtone yeast extract glucose broth (TYB)
(IV) Glutamate medium
(V) Casamino acid medium (CAA)

It was observed that sodium benzoate upregulated siderophore production when added as the sole carbon source in basal medium as compared to glucose or sodium succinate as the carbon source. Addition of iron in the succinate medium at a concentration of 30 mg/L suppressed siderophore production when succinate or glucose was the carbon source but not with sodium benzoate. Higher concentration of iron (60 mg/L) was required to suppress siderophore production with sodium benzoate as the carbon source (Gaonkar et al. 2012), indicating increased demand of iron for its degradation. Siderophore production was also inhibited in nutrient broth, tryptone yeast extract glucose broth, glutamate and casamino acid medium which could be attributed to contamination of media components with Fe. This study confirmed sodium benzoate as the best carbon source for siderophore production by *Pseudomonas aeruginosa* TMR2.13 amongst the various carbon sources tested.

Results obtained have thus highlighted the impact of aromatic compound metabolism on siderophore production. Presence of sodium benzoate augments siderophore production since catechol 1,2-dioxygenase involved in the degradation of the aromatic compound requires iron as a cofactor (de Sousa and Bhosle 2012). The cells sense iron deficiency in the growth medium and secrete siderophores to meet the iron demand. The ability to produce siderophores is relevant in biodegradation of aromatic pollutants and would prove to be useful in restoration of contaminated ecosystems deficient in iron.

11.5 Conclusion

Iron concentration is one of the many factors influencing efficient cleanup of organic pollutants and metals. In natural environments, the availability of iron is usually low, more so, in marine and coastal ecosystem. This study has helped to understand the response of our isolate to aromatic compound sodium benzoate and also to metal ions. Our studies indicate that it may be feasible to use siderophore-producing bacteria for bioremediation of organic compounds. Such isolates may also play a role in phytoextraction as they exhibit metal resistance and siderophore production in presence of metals. Further investigations and field trials are required to extend the present work to natural ecosystems exposed to influx of such pollutants.

Acknowledgement Authors wish to thank Prof. Saroj Bhosle, Dept of Microbiology, Goa University for her valuable guidance to carry out this work. Fellowship provided by MoES, Delhi, to Dr. Teja Gaonkar to carry out the research work is greatly acknowledged.

References

Alongi DM, Christoffersen P, Tirendi F (1993) The influence of forest type on microbial-nutrient relationships in tropical mangrove sediments. J Exp Mar Biol Ecol 171:201–223

Altschul SF, Gish W, Miller W, Myers EW, Lipman DJ (1990) Basic local alignment search tool. J Mol Biol 215:403–410

Arun AB, Sridhar KR (2004) Symbiotic performance of fast growing rhizobia isolated from the coastal sand dune legumes of west coast of India. Biol Fertil Soils 40:435–439

Arun AB, Beena KR, Raviraja NS, Sridhar KR (1999) Coastal sand dunes—a neglected ecosystem. Curr Sci 77:19–21

Banik S, Das KC, Islam MS, Salimullah M (2014) Recent advancements and challenges in microbial bioremediation of heavy metals contamination. JSM Biotechnol Biomed Eng 2(1):1035

Barbeau K, Zhang G, Live DH, Butler A (2002) Petrobactin, a photoreactive siderophore produced by the oil-degrading marine bacterium *Marinobacter hydrocarbonoclasticus*. J Am Chem Soc 124:378–379

Braud A, Hannauer M, Mislin GLA, Schalk IJ (2009) The pseudomonas aeruginosa pyochelin-iron uptake pathway and its metal specificity. J Bacteriol 191(11):3517–3525

Buensanteai N, Yuen GY, Prathuangwong S (2008) The biocontrol bacterium *Bacillus amyloliquefaciens* KPS46 produces auxin, surfactin and extracellular proteins for enhanced growth of soybean plant. Thai J Agric Sci 41:101–116

Cao B, Geng A, Loh KC (2008) Induction of ortho and meta cleavage pathways in *Pseudomonas* in biodegradation of high benzoate concentration: MS identification of catabolic enzymes. Appl Microbiol Biotechnol 81:99–107

Cornelis P, Matthijs S (2002) Diversity of siderophore-mediated iron uptake systems in fluorescent pseudomonads: not only pyoverdines. Environ Microbiol 4:787–798

De Sousa T, Bhosle S (2012) Implications of benzoate induced alterations in cell morphology and physiology in *Pseudomonas aeruginosa* TMR2.13 for potential application in bioremediation and monitoring approaches. J Bioremed Biodegrad S1:008

Dinkla IJT, Gabor EM, Janssen DB (2001) Effects of iron limitation on the degradation of toluene by pseudomonas strains carrying the TOL (pWWO) plasmid. Appl Environ Microbiol 67:3406–3412

Egamberdiyeva D (2005) Characterization of *Pseudomonas* species isolated from the rhizosphere of plants grown in serozem soil, semi arid region of Uzbekistan. Scientific World Journal 5:501–509

Gaonkar T (2015) Eubacterial siderophores and factors modulating their production. In: Borkar S (ed) Bioprospects of Eubacteria: coastal ecosystems of Goa. Springer International Publishing Switzerland, p 25–38

Gaonkar T, Bhosle S (2013) Effect of metals on a siderophore producing bacterial isolate and its implications on microbial assisted bioremediation of metal contaminated soils. Chemosphere 93:1835–1843. http://dx.doi.org/10.1016/j.chemosphere.2013.06.036

Gaonkar T, Nayak P, Garg S, Bhosle S (2012) Siderophore-producing bacteria from a sand dune ecosystem and the effect of sodium benzoate on siderophore production by a potential isolate. Scientific World Journal 2012:1–8

Garrison JM, Crumbliss AL (1987) Kinetics and mechanism of aluminium (III) siderophore ligand exchange: mono (deferriferrioxamine B) aluminium (III) formation and dissociation in aqueous acid solution. Inorg Chim Acta 138:61–65

Glick BR (1995) The enhancement of plant growth by free-living bacteria. Can J Microbiol 41:109–117

Hernlem BJ, Vane LM, Sayles GD (1999) The application of siderophores for metal recovery and waste remediation: examination of correlations for prediction of metal affinities. Water Res 33:951–960

Holguin G, Guzman MA, Bashan Y (1992) Two new nitrogen-fixing bacteria from the rhizosphere of mangrove trees: their isolation, identification and in vitro interaction with rhizosphere *Staphylococcus* sp. FEMS Microbiol Ecol 101:207–216

Hu X, Boyer GL (1996) Siderophore-mediated aluminum uptake by *Bacillus megaterium* ATCC 19213. Appl Environ Microbiol 62:4044–4048

Inoue H, Takimura O, Kawaguchi K, Nitoda T, Fuse H, Murakami K, Yamaoka Y (2003) Tin-carbon cleavage of organotin compounds by pyoverdine from *Pseudomonas chlororaphis*. Appl Environ Microbiol 69:878–883

Jing Y, He Z, Yang X (2007) Role of soil rhizobacteria in phytoremediation of heavy metal contaminated soils. Zhejiang Univ Sci B 8:192–207

Khanolkar D, Dubey SK, Naik MM (2015a) Tributyltin chloride (TBTCl)- enhanced exopolysaccharide and siderophore production in an estuarine *Alcaligenes faecalis* strain. Arch Environ Contam Toxicol. doi:10.1007/s00244-010-0120-9

Khanolkar D, Dubey SK, Naik MM (2015b) Biotransformation of tributyltin chloride to less toxic dibutyltin dichloride and monobutyltin trichloride by *Klebsiella pneumonia* strain SD9. Int Biodeterior Biodegrad 104:212–218

Khomenkov VG, Shevelev AB, Zhukov VG, Zagustina NA, Bezborodov AM, Popov VO (2008) Organisation of metabolic pathways and molecular genetic mechanisms of xenobiotic degradation in microorganisms: a review. Appl Biochem Microbiol 44:117–135

Krieg RN, Holt GJ (1984) Bergey's manual of systematic bacteriology, vol 1. Williams &Wilkins, Baltimore, USA

Mishra VK, Kumar A (2012) Plant growth promoting and phytostimulatory potential of *Bacillus Subtilis* and *Bacillus Amyloliquefaciens*. ARPN J Agric Biol Sci 7:509–518

Mudgal V, Madaan N, Mudgal A, Singh RB, Mishra S (2010) Effect of toxic metals on human health. Open Nutraceuticals J 3:94–99

Naik MM, Dubey SK (2011) Lead-enhanced siderophore production and alteration in cell morphology in a Pb-resistant *Pseudomonas aeruginosa* strain 4EA. Curr Microbiol 62:409–414

Norris JR, Ribbons DW (1971) Methods in microbiology, vol 6A. Academic Press, London

Nozaki M, Ishimura Y (1974) Oxygenases. In: Neilands JB (ed) Microbial iron metabolism, 1st edn. Academic Press, New York, pp 417–441

Park SH, Ko YJ, Kim CK (2001) Toxic effects of catechol and 4-chlorobenzoate stresses on bacterial cells. J Microbiol 39(3):206–212

Prasad MN, Freitas H, Fraenzle S, Wuenschmann S, Markert B (2010) Knowledge explosion in phytotechnologies for environmental solutions. Environ Pollut 158:18–23

Raghavan TM, Furtado I (2000) Tolerance of an estuarine halophilic archaebacterium to crude oil and constituent hydrocarbons. Bull Environ Contam Toxicol 65:725–731

Rajkumar M, Ae N, Prasad M, Freitas H (2010) Potential of siderophore-producing bacteria for improving heavy metal phytoextraction. Trends Biotechnol 28(3):142–149

Schalk IJ, Hannauer M, Braud A (2011) New roles for bacterial siderophores in metal transport and tolerance. Environ Microbiol 13:2844–2854

Schwyn B, Neiland JB (1987) Universal chemical assay for the detection and determination of siderophores. Anal Biochem 160:47–56

Sengupta A, Chaudhuri S (1991) Ecology of heterotrophic dinitrogen fixation in the rhizosphere of mangrove plant community at the Ganges river estuary in India. Oecologia 87:560–564

Shamim K, Naik MM, Pandey A, Dubey SK (2012) Isolation and identification of Aeromonas caviae strain KS-1 as TBTC –and lead-resistant estuarine bacteria. Environ Monit Assess. doi:10.1007/s10661-012-2940-2

Sikkema J, de Bont JA, Poolman B (1995) Mechanisms of membrane toxicity of hydrocarbons. Microbiol Rev 59:201–222

Sneath AHP, Mair SN, Sharpe EM, Holt JG (1986) Bergey's manual of systematic bacteriology, vol 2, Williams and Wilkins Bacteriology Symposium. Academic Press, London

Song J, Sung J, Kim YM, Zylstra GJ, Kim E (2000) Roles of the *meta*- and the *ortho*-Cleavage pathways for the efficient utilization of aromatic hydrocarbons by *Sphingomonas yanoikuyae* B1. J Microbiol 38:245–249

Staijen IE, Witholt (1998) Synthesis of alkane hydroxylase of *Pseudomonas oleovorans* increases the iron requirement of alk+ bacterial strains. Biotechnol Bioeng 57:228–237

Thompson JD, Gibson TJ, Plewniak F, Jeanmougin F, Higgins DG (1997) The CLUSTAL X windows interface: flexible strategies for multiple sequence alignment aided by quality analysis tools. Nucleic Acids Res 25:4876–4882

Vazquez P, Holguin G, Puente ME, Lopez-Cortes A, Bashan Y (2000) Phosphate-solubilizing microorganisms associated with the rhizosphere of mangroves in a semiarid coastal lagoon. Biol Fertil Soils 30:460–468

Zeyaullah M, Abdelkafe AS, Zabya WB, Ali A (2009) Biodegradation of catechols by micro-organisms – a short review. Afr J Biotechnol 8:2916–2922

Bacterial Degradation of Algal Polysaccharides in Marine Ecosystem

12

Md Imran, Preethi B. Poduval, and Sanjeev C. Ghadi

Abstract

Complex polysaccharides (CPs) such as agar, alginates, carrageenan, etc., are largely present as component of cell wall of seaweeds, promoting structural integrity and shielding the host from pathogens or predators. Marine bacteria-degrading CPs are ubiquitous and have been isolated from diverse sources such as costal water, sediments, and deep sea, as well as from the surface of seaweeds and crustaceans. Polysaccharide-degrading enzymes (carbohydrases) such as agarase, alginate lyase, and carrageenase are produced by CP-degrading bacteria that aid in depolymerizing agar, alginate, and carrageenan, respectively. The resultant oligosaccharides are subsequently degraded to various intermediates that are further channelized to energy conversion pathways. Several of these carbohydrases have been purified, and study of their biochemical properties such as molecular weight, substrate specificity, pH, and temperature range for optimal activity indicates a vast diversity. Recently, several multiple CP-degrading bacteria (epiphytic/free form) have been isolated. *Saccharophagus degradans* and *Microbulbifer* species are the dominant groups of multiple polysaccharide-degrading bacteria and play a significant role in recycling of carbon from CPs. In natural marine ecosystem, the cell wall of seaweeds comprises an array of CPs in mixed proportion. The production of diverse carbohydrases by CP-degrading bacteria enables it to decompose seaweeds into single cell detritus as well as generate reducing sugars from degraded polysaccharides. Thus, CP-degrading bacteria are potential candidates for eco-friendly degradation of algal waste.

12.1 Introduction

Polysaccharides are linear or branched chain polymers consisting of repeating unit of monosaccharide units that are linked by glycosidic bonds. In

M. Imran • P.B. Poduval • S.C. Ghadi (✉)
Department of Biotechnology, Goa University, Taleigao plateau, Goa 403206, India
e-mail: saga@unigoa.ac.in

M.M. Naik, S.K. Dubey (eds.), *Marine Pollution and Microbial Remediation*,
DOI 10.1007/978-981-10-1044-6_12

marine ecosystem, predominant polysaccharides such as agar, alginate, carrageenan, etc., are designated as complex polysaccharides (CPs) and are predominantly present in the cell wall of seaweeds. The repeating units that create these polysaccharides are modified by diverse functional groups making them recalcitrant. Thus, recalcitrant CPs are aptly adapted to confer structural integrity to host marine organisms.

In marine ecosystem, to prevail over oligotrophic conditions, CP-degrading bacteria form complex association with seaweeds containing CPs. Further with the aid of carbohydrases, the algal CPs are degraded by bacteria into biological useable form. Thus, polysaccharide-degrading bacteria play a vital role in recycling of carbon from algal CPs. The present review reports the various predominant algal CPs in marine ecosystem, the CP-degrading bacteria that have been isolated from various sources, the polysaccharide degradation pathway, and biochemical properties of depolymerizing enzymes. Additionally, the recent focus on multiple CP-degrading bacteria and their potential in degradation of algal waste is also discussed.

12.2 Bacterial Degradation of Agar

12.2.1 Sources of Agar

Agar is mainly obtained from seaweeds such as *Gelidium*, *Gracilaria*, and *Porphyra* sp. that belong to Rhodophyceae family (Fu and Kim 2010). The cell wall of red seaweed primarily consists of cellulose and agar (Kang et al. 2003). The presence of an array of heterologous polysaccharides in the cell wall of seaweeds offers a unique niche promoting formation of biofilm comprising CP-degrading bacteria (Lemos et al. 1985). These epiphytic bacteria degrade the algal cell wall polysaccharides and utilize the degraded by-products or alternatively utilize the organic compounds that are leached from partially decomposed seaweeds as carbon source.

12.2.2 Screening for Agarolytic Bacteria

The agarolytic bacteria form a predominant group among the CP-degrading bacteria. They have been isolated from various marine sources such as coastal water and sediments as well as from marine algae, marine mollusks, etc. The agarolytic bacteria form depression/craters or zones of clearance on seawater-based agar medium (Fig. 12.1a). The agarolytic activity of bacterial strain can be further confirmed by a dye-based plate screening method (Hodgson and Chater 1981). A distinct clearance zone around the agarolytic bacterial colonies is observed after spreading Lugol's iodine (Fig. 12.1b).

12.2.3 Diversity of Agarolytic Bacteria

Most agarolytic bacteria belong to *Proteobacteria* and *Bacteroidetes* phyla. Agarolytic bacteria belonging to *Gammaproteobacteria* class of the *Proteobacteria* phylum include genera such as

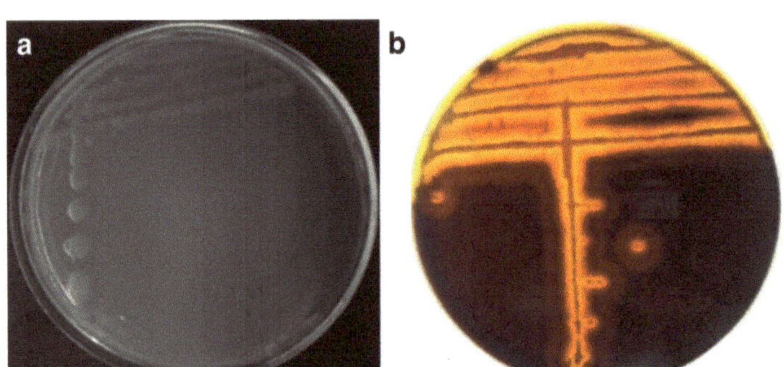

Fig. 12.1 (**a**) *Microbulbifer* strain CMC-5 depicting craters on seawater-based medium containing 1.8 % agar; (**b**) Observation of light brownish clearance zone around agarolytic colonies after adding Lugol's iodine

Pseudomonas (Ha et al. 1997), *Alteromonas* (Leon et al. 1992; Potin et al. 1993), *Pseudoalteromonas* (Vera et al. 1998), *Vibrio* (Aoki et al. 1990; Araki et al. 1998; Sugano et al. 1993), *Alterococcus* (Shieh and Jean 1998), *Microbulbifer* (Ohta et al. 2004b), *Agarivorans* (Ohta et al. 2005b), *Thalassomonas* (Ohta et al. 2005a), and *Saccharophagus* (Ekborg et al. 2005).

Agarolytic bacteria belonging to phylum *Bacteroidetes* have been isolated and constitute the *Bacteroidia* class that includes the genus *Marinilabilia* (Veldkamp 1961), the *Flavobacteriia* class that includes the genera *Flavobacterium* (Van der Meulen et al. 1974), *Cellulophaga* (Lewin 1969), and *Zobellia* (Barbeyron et al. 2001), and the *Sphingobacteriia* class that includes the genera *Cytophaga* (Veldkamp 1961), *Persicobacter* (Stanier 1941), and *Microscilla* (Zhong et al. 2001).

12.2.4 Biochemical Properties of Agarase

Table 12.1 summarizes the biochemical properties of purified agarase from agarolytic bacteria that have been isolated from various sources. The β-agarase has been predominantly reported over α-agarase in bacteria. The optimal pH for most agarase enzymes is around 7.0, with few agarases demonstrating activity in slightly alkaline pH. Although most of the agarases demonstrate optimum temperature in the range of 30–40 °C, agarases from *Thalassomonas* sp. and *Microbulbifer* strain CMC-5 depict optimal activity at 45 and 50 °C, respectively.

12.2.5 Agar Degradation Pathway

Agar extracted from *Gracilaria* and *Gelidium* has agarobiose (comprising of β-D-galactopyranose

Table 12.1 Biochemical properties of purified agarase (*Alteromonas agarlyticus* GJ1B and *Thalassomonas sp. JAMB-A33* produce α-agarase; all other bacterial strains produced β-agarase)

Agarolytic bacteria isolated from various sources	Biochemical characteristic of agarase			
	Mol wt. (KDa)	Optimal temperature (°C)	Optimal pH	References
Marine algae				
Microbulbifer strain CMC-5	59	50	7.0	Jonnadula et al. (2009)
Pseudomonas atlantica	32	–	7.0	Morrice et al. (1983)
Pseudoalteromonas antarctica N-1	33	–	7.0	Vera et al. (1998)
Alteromonas sp. SY 37–12	39.5	35	7.0	Wang et al. (2006a)
Vibrio sp. AP-2	20	–	5.5	Aoki et al. (1990)
Seawater				
Alteromonas sp. C-1	52	30	6.5	Leon et al. (1992)
Alteromonas agarlyticus GJ1B	360	–	7.2	Potin et al. (1993)
Cytophaga sp.	–	40	7.2	Duckworth and Turvey (1969)
Vibrio sp. JT0107	107	30	8.0	Sugano et al. (1993)
Bacillus megaterium	15	40	6.6	Khambhaty et al. (2008)
Marine sediment				
Agarivorans sp. HZ105	58	–	6.0–9.0	Hu et al. (2008)
Vibrio sp. PO303	87.5	38–55	6.5–7.5	Araki et al. (1998)
Thalassomonas sp. JAMB-A33	85	45	8.5	Ohta et al. (2005a)
Marine mollusks				
Agarivorans albus YKW-34	50	40	8.0	Fu et al. (2008)

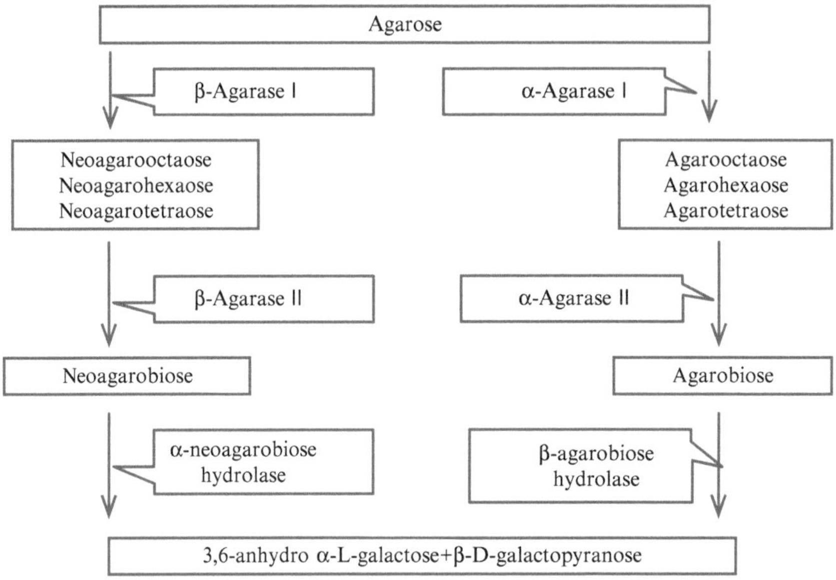

Fig. 12.2 Role of α-agarase and β-agarase in agar degradation (Chi et al. 2012)

and 4-O-linked α-L-galactopyranose) as main repeating moiety. Degradation of agar to β-D-galactopyranose and 3,6-anhydro-α-L-galactose involves several enzymes. On the basis of their mode of action, agarases are classified into two groups: α-agarase (EC 3.2.1.158) and β-agarase (EC 3.2.1.81) (Araki 1959). The β-agarase hydrolyzes the β-(1,4) glycosidic bonds to produce neoagaro-oligosaccharides with β-D-galactopyranose residues at their reducing ends. The α-agarases, on the other hand, hydrolyze the α-(1,3) glycosidic linkages of neoagarose repetition moieties and produce agaro-oligosaccharides with 3,6-anhydro-α-L-galactose residues at their reducing ends (Fig. 12.2).

12.3 Alginate Degradation

12.3.1 Sources of Alginate

Brown seaweeds belonging to Phaeophyceae are the main source of alginate in the marine environment. Alginate is the main structural component of cell wall of brown algae comprising up to 40 % of dry weight of seaweeds. It consist of (1-4)-linked β-D-mannuronate (M) and α-L-glucuronate (G) residues in varying proportions (Matsubara et al. 2000). Some bacteria are also known to synthesize alginate (Clementi 1997; Albrecht and Schiller 2005). Marine macroalgae such as *Laminaria japonica*, *Laminaria hyperborea*, *Sargassum* sp., *Ascophyllum nodosum*, *Macrocystis pyrifera*, *Laminaria digitata*, and others have been exploited for commercial production of alginate.

12.3.2 Screening for Alginolytic Bacteria

The glycosidic bond in alginate is cleaved by alginate lyase by β-elimination reaction (Gacesa 1992). Alginate degradation by bacteria can be detected by plate assay method (Gacesa and Wusteman 1990). Cetylpyridinium chloride is added to plates containing alginolytic culture, and the presence of clearance zone around colonies confirms the alginate degradation by bacterial strain (Fig. 12.3). Based on the substrate specificity, alginate lyases are classified as polymannuronate (M), polyguluronate (G), and polyMG-specific lyases.

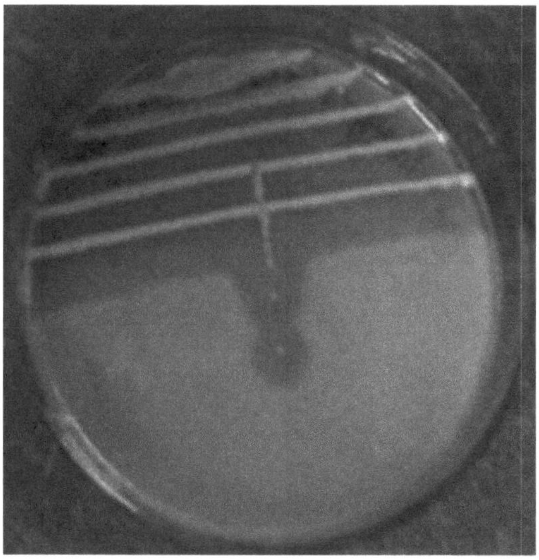

Fig. 12.3 *Microbulbifer* strain CMC-5 grown on seawater-based agar medium containing 1 % sodium alginate showing clearance zone around bacterial colonies after addition of cetylpyridinium chloride (Gacesa and Wusteman, 1990)

12.3.3 Diversity of Alginolytic Bacteria

Several reports on production of alginate lyases from algae, marine animals, marine bacteria, marine fungi, and viruses are available (Wong et al. 2000; Suzuki et al. 2006; Kawamoto et al. 2006; Iwamoto et al. 2001). Among the marine bacteria, alginate lyases have been reported from members of *Pseudomonas* and *Vibrio* (Wong et al. 2000). Alginate lyase is also reported from *Agarivorans* sp. that was isolated from the deep-sea sediments. Other marine bacteria that produce alginate lyase are *Pseudomonas* sp. (Min et al. 1977), *Vibrio haliotioli* (Sugimura et al. 2000), *Klebsiella aerogenes* type 25 (Lange et al. 1989), *Pseudomonas alginovora* XO17 (Boyen et al. 1990), *Vibrio harveyi* AL-128 (Kitamikado et al. 1992), *Corynebacterium* sp. ALY-1 (Matsubara et al. 1998), *Pseudoalteromonas* sp. SM0524, etc. Additionally, Dong et al. (2012) isolated 21 alginolytic bacteria from the *Laminaria* sp. collected from the Arctic Ocean. These 21 bacterial strains belonged to the five genera, viz., *Psychromonas*, *Pseudoalteromonas*, *Polaribacter*, *Psychrobacter*, *and Winogradskyella.*

12.3.4 Biochemical Properties of Alginate Lyase

Table 12.2 indicates the alginate lyases that have been purified from alginolytic bacteria isolated from diverse sources in marine ecosystem. Unlike other CP-degrading bacteria, alginolytic bacteria have also been reported from the gut of turban snails (Table 12.2). Although some of the reported alginate lyase have either polyM or polyG specificity, few of the alginate lyase hydrolyzed polyM as well as polyG substrates.

12.3.5 Alginate Degradation Pathway

The alginate degradation pathway has been characterized in *Sphingomonas* sp. A1 (Fig. 12.4). The three endo-alginate lyases (A1, A1-II, and A1-III) and single exo-alginate lyase (A1-IV) from *Sphingomonas* strain A1 degrade alginate to constituent monosaccharides. All the three endotype alginate lyases are product of a single gene (Yonemoto et al, 1993). A1-I (65 kDa) is autocatalytically processed into A1-II (25 kDa) and A1-III (40 kDa) (Hisano et al. 1994). These three endotype alginate lyases produce unsaturated di-, tri-, and tetrasaccharides from alginate (Hashimoto et al. 1998; Yoon et al. 2000). The unsaturated oligosaccharides are further degraded into constituent monosaccharides by A1-IV (Miyake et al. 2003).

12.4 Carrageenan Degradation

12.4.1 Sources of Carrageenan

Carrageenan is a sulfated polysaccharide commonly present in certain red seaweeds such as *Kappaphycus alvarezii*, *Eucheuma denticulatum*, *Chondrus crispus*, and *Gigartina skottsbergii*. The anionic polymers such as agars and carrageenan amount to 50 % of the dry mass of seaweed. The quantity and location of sulfated ester and the presence of 3,6-anhydro-bridges in the α-linked residues are the decisive factor for classification of carrageenans. The κ (kappa), ι (iota), and λ (lambda) carrageenans are the most industrially

Table 12.2 Biochemical properties of alginate lyase purified from alginolytic bacteria

	Biochemical properties of alginate lyase				
Alginolytic bacteria isolated from various sources	Substrate specificity	Molecular mass (KDa)	Optimum pH	Optimum temperature (°C)	References
Marine algae					
Alginovibrio aquatilis	–	110	8	–	Stevens and Levin (1977)
Alteromonas sp. strain H-4	PolyM, PolyG	32	7.5	30	Sawabe et al. (1992)
Beneckea pelagia	–	–	8	25	Sutherland and Keen (1981)
Halomonas marina	PolyM	39	–	–	Kraiwattanapong et al. (1999)
Photobacterium sp.	PolyM, PolyG	30	–	–	Malissard et al. (1995)
Pseudomonas sp.	PolyG	94	7.5	–	Muramatsu and Sogi (1990)
Pseudomonas alginovora (strain X017)	PolyM	28	7.5	–	Boyen et al. (1990)
Vibrio sp.	PolyM	24	–	–	Chavagnat et al. (1996)
Vibrio alginolyticus	PolyG	47	8.2	–	Kitamikado et al. (1992)
Vibrio harveyi AL-128	–	57	7.8	–	Kitamikado et al. (1990)
Vibrio sp. YWA	PolyM	62.5	7	25	Wang et al. (2006b)
Pseudoalteromonas sp. SM0524	PolyG	32	8.5	50	Li et al. (2011)
Turban shell gut					
Vibrio sp. YKW-34	PolyG, PolyM	60	7.0	40	Fu et al. 2007
Sea mud					
Vibrio sp. QY105	PolyG, PolyM	37	7.0	38	Wang et al. (2013)
Deep-sea sediment					
Agarivorans sp. JAM-A1m	PolyG, PolyM	31	10.0	30	Kobayashi et al. (2009)

exploited carrageenans and are apparent by the respective occurrence of single, dual, or ternary ester-sulfate groups on every recurring disaccharide unit.

12.4.2 Screening for Carrageenan-Degrading Bacteria

In comparison to other polysaccharide-degrading bacteria reported in literature, very few bacterial strains have been reported to hydrolyze carrageenans. Carrageenan-degrading bacteria can also be detected by the plate assay method (Ohta and Hatada 2006). A clearance zone is observed around carrageenan-degrading bacteria after spreading phenol red (Fig. 12.5).

12.4.3 Diversity of Carrageenan-Degrading Bacteria

The carrageenan-degrading bacteria that have been reported from marine ecosystem belong to

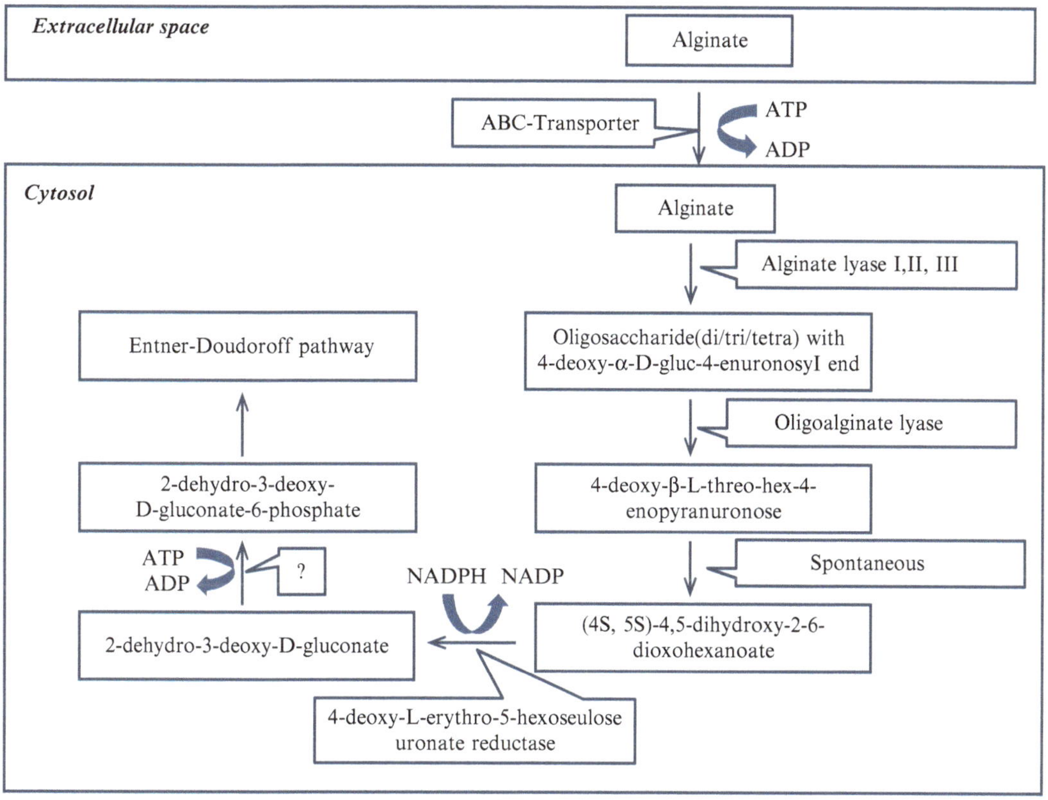

Fig. 12.4 Alginate degradation pathway in *Sphingomonas* sp. (Hashimoto et al. 2005)

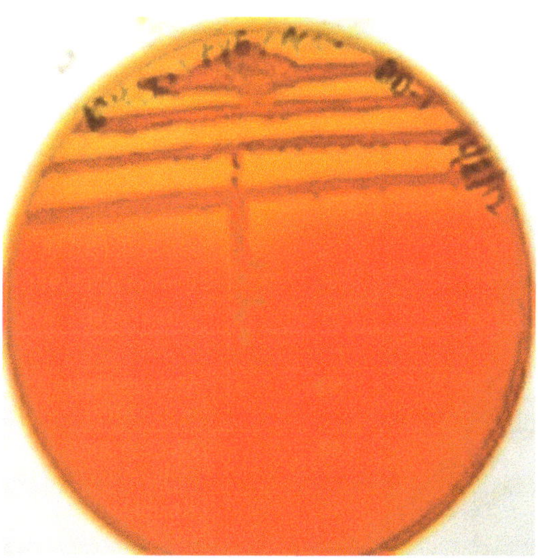

Fig. 12.5 *Microbulbifer* strain CMC-5 grown on seawater-based agar medium containing 1 % carrageenan depicting yellowish clearance zone after addition of phenol red (Ruijssenaar and Hartsman 2001)

γ-*Proteobacteria*, *Flavobacteria*, or *Spingobacteria*. Nine carrageenan-degrading bacteria belonging to the γ-*Proteobacteria* were isolated from North Atlantic Ocean. One of the strain that was identified as *Pseudoalteromonas carrageenovora* (Gauthier et al. 1995) produced two different carrageenases with activity against κ-carrageenan and λ-carrageenan, respectively (Weigl and Yaphe 1966), whereas another strain identified as *Alteromonas fortis* (Potin 1992) produced only one extracellular carrageenase with specificity for ι-carrageenan (Greer and Yaphe 1984). Further, besides the two β-agarases, κ-carrageenase and ι-carrageenase were secreted by *Z. galactanivorans*, an *epiphytic Flavobacterium* isolated from the red alga *Delesseria sanguinea* (Potin et al. 1991; Barbeyron et al. 2000; Jam et al. 2005). Additionally, κ-carrageenase has also been reported from *Cytophaga* strain 1 k-C783 (Sarwar et al. 1987).

12.4.4 Biochemical Properties of Carrageenase

Table 12.3 depicts the biochemical properties of carrageenase that have been characterized. Most of the reported carrageenases were predominantly of k type. The carrageenases have divergent molecular weight ranging from 30 to 128 kDa, whereas the optimal pH for activity was 7–8. The optimal temperature range for carrageenase activity ranged from 25 to 55 °C.

12.4.5 Carrageenan Degradation Pathway

Carrageenases specifically cleaves the β-(1→4) linkage of their respective substrates. Among the carrageenan-degrading bacteria, *P. carrageenovora* has been widely studied. The κ-carrageenase purified from *P. carrageenovora* yields k-neocarratetraose sulfate and k-neocarrabiose sulfate as end products (McLean and Williamson 1979b; Weigl and Yaphe 1966). Subsequently, k-neocarratetraoses are degraded by a κ-neocarratetraose hydrolase (McLean and Williamson 1981), and a specific sulfatase subsequently hydrolyzes the 4-sulfate group of κ-neocarrabioses (Fig. 12.6a) (McLean and Williamson 1979a). *P. carrageenovora* also produced an extracellular enzyme complex that demonstrates λ-carrageenase activity (Johston and McCandless 1973) (Fig. 12.6b). Pathway for the degradation of κ-carrageenan is shown in Fig. 12.6a and was observed in some other species including *Coraliomargarita akajimensis* DSM 45221 (Mavromatis et al. 2010), *Pseudoalteromonas* sp. AJ5 (Ma et al. 2008), *Rhodopirellula baltica* (Glockner et al. 2003), and *Zobellia galactanivorans* (Barbeyron et al. 1998). *Pseudoalteromonas* sp. CL19 degrades

Table 12.3 Biochemical properties of carrageenase purified from carrageenan-degrading bacteria

Carrageenan-degrading bacteria and their source of isolation	Biochemical properties of carrageenase				
	Type of carrageenase	Mw (KDa)	Optimal pH	Optimal temperature (°C)	References
Seaweed					
Cytophaga-like bacterium	κ	40	7.0	40	Potin et al. (1991)
Tamlana sp. HC4	κ	66.4	8.0	30	Sun et al. (2010)
Vibrio sp. CA-1004	κ	35	8.0	40	Araki et al. (1999)
Cytophaga MCA-2	κ	30	–	–	Mou et al. (2004)
Pseudoalteromonas-like bacterium WZUC 10	κ	40	7.2	–	Zhou et al. (2008)
Pseudoalteromonas sp. QY203	κ	34	7.2	45	Li et al. (2013)
Psuedoalteromonas porphyrae	ι	40	8.0	55	Liu et al. (2011)
Zobellia galactanovorans	κ	45	7.5	30	Potin et al. (1991)
Cytophaga sp. 1 κ-C783		100	7.6	25	Sarwar et al. (1987)
Seawater					
Bacillus subtilis	κ	66	8.0	50	Vijayaraghavan et al. (2012)
Pseudomonas carragenovora	κ	35	8.0	40	McLean and Williamson (1979b)
Pseudomonas elongata	κ	128	5.6–7.7	40	Khambhaty et al. (2007)
Deep sea					
Pseudoalteromonas sp. strain CL19	λ	100	7.0	35	Ohta and Hatada (2006)

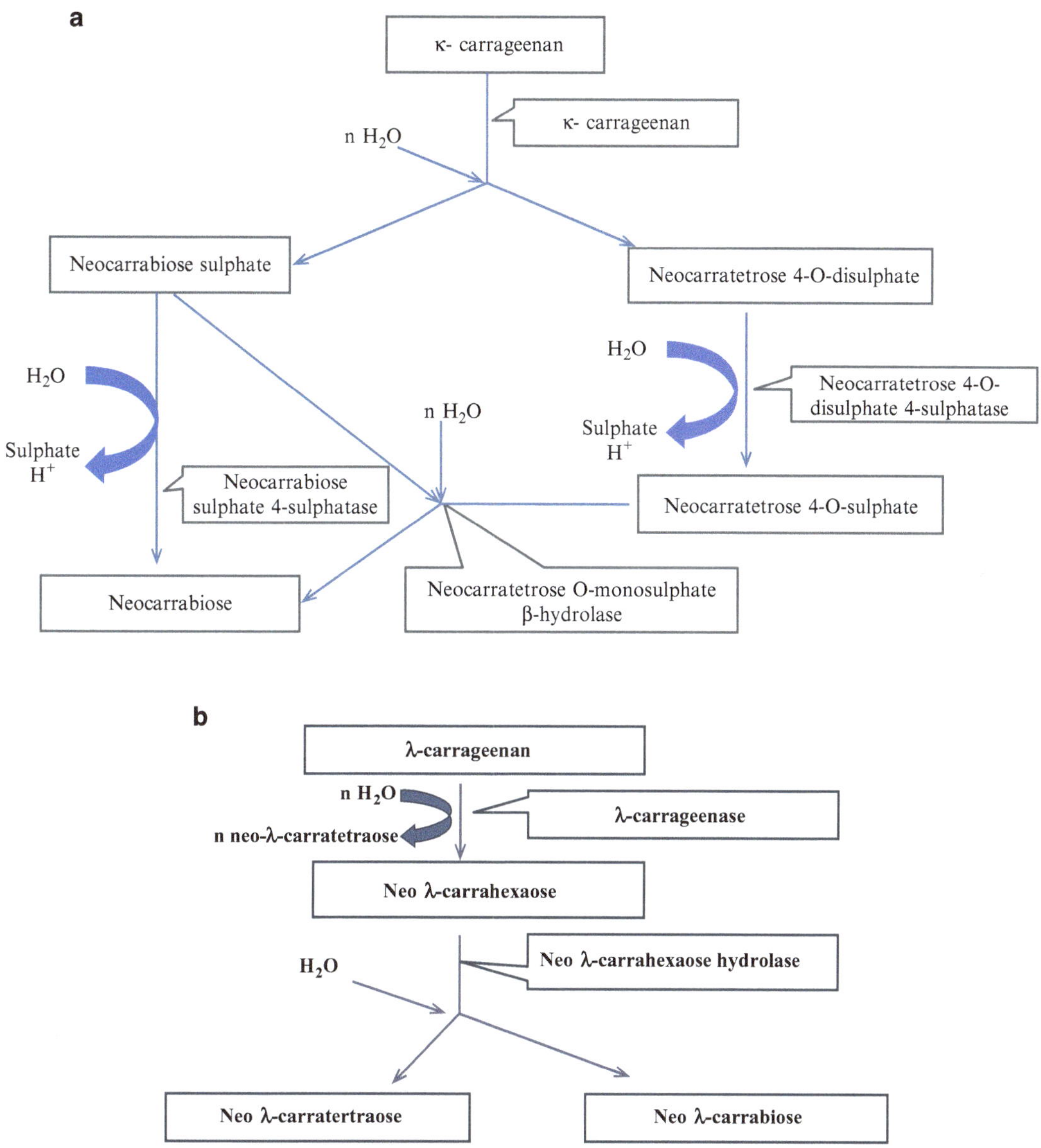

Fig. 12.6 Pathway for degradation of carrageenan in *P. carrageenovora*. (**a**) Degradation of κ-carrageenan, (**b**) degradation of ʎ-carrageenan

λ-carrageenan by pathway as shown in Fig. 12.6b (Ohta and Hatada 2006).

12.5 Multiple Polysaccharide-Degrading Bacteria

Predominantly, most of the earlier studies on CP-degrading bacterial strains report degradation of individual polysaccharides, i.e., the strains were either reported as agarolytic or alginolytic or carrageenolytic. Recently, a great deal of interest has been generated on bacteria that degrade more than one CP and are referred as multiple polysaccharide-degrading bacteria. Among the multiple polysaccharide-degrading bacteria, *Microbulbifer* sp. is more dominant in their ability to degrade more than one polysaccharide as observed in strains of *M. hydrolyticus*, *M. arenaceous*, *M. elongatus*, *and M. salipaludis* (Table 12.4). *Saccharophagus degradans*, earlier classified as *Microbulbifer degradans* 2–40 (Ensor et al. 1999), was observed

Table 12.4 Multiple polysaccharide-degrading bacteria isolated from various sources

Sr. No.	Bacteria	Source of isolation	Polysaccharides degraded by bacteria	References
1	*Microbulbifer mangrovi* DD-13	Mangrove ecosystem	Agar, alginate, chitin, carboxymethyl cellulose, laminarin, pectin, pullulan, starch, carrageenan, β-glucan, and xylan	Vashist et al. (2012)
2	*Microbulbifer hydrolyticus*	Marine pulp mill effluent	Agar, cellulose, carrageenan, chitin, starch, and xylan	Gonzalez et al. (1997)
3	*Microbulbifer elongatus*	–	Agar, cellulose, chitin, xylan, and starch	Yoon et al. (2003b)
4	*Microbulbifer salipaludis*	Salt marsh	Agar, starch, and xylan	Yoon et al. (2003a)
5	*Microbulbifer* sp. CMC-5	Decomposing seaweeds	Agar, alginate, carrageenan, cellulose, xylan, and chitin	Jonnadula et al. (2009)
6	*Microbulbifer* sp. 6532A	Drifting seaweed	Alginate, cellulose, fucoidan, laminarin, and agar	Wakabayshi et al. (2012)
7	*Saccharophagus degradans*	Decaying salt marsh cord grass, *Spartina alterniflora*	Agar, alginate, chitin, cellulose, fucoidan, laminarin, pectin, pullulan, starch, and xylan	Ekborg et al. (2005)
8	*Saccharophagus* Myt-1	Marine sediments	Agar, alginate, chitin, cellulose, fucoidan, laminarin, pectin, pullulan, starch, and xylan	Sakatoku et al. (2011)
9	*Microbacterium oxydans*	Marsh	Alginate and laminarin	Kim et al. (2013)
10	*Bacillus* sp. SYR4	Sandbar	Agar and carrageenan	Kang and Kim (2015)
11	*Simiduia* sp. strain TM-2	Marine sediment	Alginate, cellulose, and agar	Tawara et al. (2015)

to degrade ten polysaccharides, viz., cellulose, agar, chitin, alginate, fucoidan, pectin, laminarin, pullulan, xylan, and starch (Ekborg et al. 2005). Vashist et al. (2012) also reported a novel species of *Microbulbifer mangrovi* DD-13 from the mangroves of Goa, India, and was observed to degrade 11 polysaccharides such as agar, alginic acid, chitin, carboxymethyl cellulose, laminarin, pectin, pullulan, starch, carrageenan, β-glucan, and xylan (Vashist et al. 2012).

12.6 Degradation of Seaweeds by CP-Degrading Bacteria

Seaweeds have diverse applications and can be used in fertilizer, fungicides, and herbicides as well as for isolation of phycocolloids, such as alginate, carrageenan, and agar. Seaweeds are a favorite food item, particularly in Asian countries as it has several health benefits. The worldwide increase in consumption of seaweed has given a boost to seaweed aquaculture industry, culminating in surge in seaweed waste, mostly from industrial processes. The use of seaweeds as a depolluting agent for cleaning inland sea areas and eutrophied seawater has also contributed to the generation of seaweed waste (Tang et al. 2007, 2011). As a result, the disposal as well as recycling of organic carbon from seaweed waste is essential for the preservation of the marine environment (Tang et al. 2009). Many microbes have been successfully used in disposal of seaweed as well as for production of valuable products from seaweed waste. *Microbulbifer* strain CMC-5, a multiple polysaccharide-degrading bacteria, decomposes red seaweed thallus and produces reducing sugar in the growth medium (Jonnadula et al. 2009). Similarly, *Microbacterium oxydans*, a novel alginate and laminarin-degrading bacterium, produces reducing sugars when grown in the culture medium with brown seaweed as sole carbon source (Kim et al. 2013). Additionally, a carrageenan and agar-degrading *Bacillus* sp. SYR4 has also been reported in the saccharification of complex polysaccharides from red seaweeds into reducing sugars (Kang and Kim 2015). *Microbulbifer* strain 6532A is reported to decompose seaweed thalli to single cell detritus particles that can be used as feed material in aquaculture industry (Wakabayashi et al. 2012). *Saccharophagus* strain Myt-1 is also capable of decomposing several types of seaweeds, viz., green, brown, as well as red algae (Sakatoku et al. 2011). Tang et al. (2011) reported improved composting of *Undaria pinnatifida* by using a microbial consortium containing *Halomonas* and *Gracilibacillus* sp. isolated from marine environments.

Marine bacteria participate in carbon recycling by utilizing/degrading CPs. Enzymatic degradation of CPs is the most challenging task for the microorganisms. In order to achieve this task, microorganisms produce various extra- and intracellular polysaccharide-hydrolyzing enzymes. However, in natural system CPs are usually part of complex system and are present in varying proportions as well as in combination with other polysaccharides. Ideally, degradation of mixed CPs would require microbial consortia degrading different polysaccharides. In this scenario, the role of multiple polysaccharide-degrading bacteria would be vital in recycling of carbon from mixed CPs from marine ecosystem.

References

Albrecht MT, Schiller NL (2005) Alginate lyase (AlgL) activity is required for alginate biosynthesis in *Pseudomonas aeruginosa*. J Bacteriol 187:3869–3872

Aoki T, Araki T, Kitamikado M (1990) Purification and characterization of a novel β-agarase from *Vibrio* sp. AP-2. Eur J Biochem 187:461–465

Araki C (1959) Seaweed polysaccharides. In: Wolfrom ML (ed) Carbohydrate chemistry of substances of biological interest. Pergamon, London, pp 15–30p

Araki T, Hayakawa M, Zhang L, Karita S, Morishita T (1998) Purification and characterization of agarases from a marine bacterium, *Vibrio* sp. PO-303. J Mar Biotechnol 6:260–265

Araki T, Hashitomo Y, Morishita T (1999) Purification and characterization of k-carrageenase from a marine bacterium *Vibrio* sp.CA-1004. Fish Sci 65:937–942

Barbeyron T, Gerard A, Potin P, Henrissat B, Kloareg B (1998) The kappa-carrageenase of the marine bacterium *Cytophaga drobachiensis*, structural and phylogenetic relationships within family-16 glycoside hydrolases. Mol Biol Evol 15(5):528–537

Barbeyron T, Michel G, Potin P, Henrissat B, Kloareg B (2000) iota-Carrageenases constitute a novel family of glycoside hydrolases, unrelated to that of kappa-carrageenases. J Biol Chem 275:35499–35505

Barbeyron TL, Haridon S, Corre E, Kloareg B, Potin P (2001) *Zobellia galactanovorans* gen. nov., sp. nov., a marine species of Flavobacteriaceae isolated from a red alga, and classification of [*Cytophaga*] uliginosa (ZoBell and Upham 1944) Reichenbach 1989 as Zobellia uliginosa gen. nov., comb. nov. Int J Syst Evol Microbiol 51:985–997

Boyen C, Bertheau Y, Barbeyron T, Kloareg B (1990) Preparation of guluronate lyase from *Pseudomonas alginovora* for protoplast isolation in *Laminaria*. Enzyme Microb Technol 12:885–890

Chavagnat F, Duez C, Guinand M, Potin P, Barbeyron T (1996) Cloning, sequencing and overexpression in *Escherichia coli* of the alginate-lyase-encoding aly gene of *Pseudomonas alginovora*: identification of three classes of alginate lyases. Biochem J 319:575–583

Chi WJ, Chang YK, Hong SK (2012) Agar degradation by microorganisms and agar-degrading enzymes. Appl Microbiol Biotechnol 94(4):917–930

Clementi F (1997) Alginate production by *Azotobacter vinelandii*. Crit Rev Biotechnol 17:327–361

Dong S, Yang J, Zhang X, Shi M, Song X, Chen X, Zhang Y (2012) Cultivable alginate lyase-excreting bacteria associated with the arctic brown alga *Laminaria*. Mar Drugs 10:2481–2491

Duckworth M, Turvey JR (1969) An extracellular agarase from a *Cytophaga* species. Biochem J 113:139–142

Ekborg N, Gonzalez JM, Howard MB, Taylor LE, Hutcheson SW, Weiner RM (2005) *Saccharophagus degradans* gen. nov., sp. nov., a versatile marine degrader of complex polysaccharides. Int J Syst Evol Microbiol 55(4):1545–1549

Ensor LA, Stosz SK, Weiner RM (1999) Expression of multiple complex polysaccharide degrading enzyme systems by marine bacterium strain 2–40. J Ind Microbiol Biotechnol 23:123–126

Fu XT, Kim SM (2010) Agarase: review of major sources, categories, purification method, enzyme characteristics and applications. Mar Drugs 8(1):200–218

Fu XT, Lin H, Kim SM (2007) Purification and characterization of a Na+/K+ dependent alginate lyase from turban shell gut *Vibrio* sp. YKW-34. Enzyme Microb Technol 41(6–7):828–834

Fu XT, Lin H, Kim SM (2008) Purification and characterization of a novel β-agarase, AgaA34, from *Agarivorans albus* YKW-34. Appl Microbiol Biotechnol 78:265–273

Gacesa P (1992) Enzymic degradation of alginates. Int J Biochem 24:545–552

Gacesa P, Wusteman FS (1990) Plate assay for simultaneous detection of alginate lyases and determination of substrate specificity. Appl Environ Microbiol 56:2265–2267

Gauthier G, Gauthier M, Christen R (1995) Phylogenetic analysis of the genera *Alteromonas, Shewanella*, and *Moritella* using genes coding for small-subunit rRNA sequences and division of the genus *Alteromonas* into two genera, *Alteromonas* (emended) and *Pseudoalteromonas* gen. nov., and proposal of twelve new species combinations. Int J Syst Bacteriol 45:755–761

Glockner FO, Kube M, Bauer M, Teeling H, Lombardot T, Ludwig W, Gade D, Beck A, Borzym K, Heitmann K, Rabus R, Schlesner H, Amann R, Reinhardt R (2003) Complete genome sequence of the marine *planctomycete Pirellula* sp. strain 1. Proc Natl Acad Sci U S A 100(14):8298–8303

Gonzalez JM, Mayer F, Moran MA, Hodson RE, Whitman WB (1997) *Microbulbifer hydrolyticus* gen. nov., sp. nov., and *Marinobacterium georgiense* gen. nov., sp. nov., two marine bacteria from a lignin-rich pulp mill waste enrichment community. Int J Syst Bacteriol 47:369–376

Greer CW, Yaphe W (1984) Purification and properties of iota carrageenase from a marine bacterium. Can J Microbiol 30:1500–1506

Ha JC, Kim JT, Kim SK, Oh TK, Yu JH, Kong IS (1997) Beta agarase from *Pseudomonas* sp. W7: purification of the recombinant enzyme from *Escherichia coli* and the effects of salt on its activity. Biotechnol Appl Biochem 26:1–6

Hashimoto W, Okamoto M, Hisano T, Momma K, Murata K (1998) *Sphingomonas* sp. A1 lyase active on both polyD-mannuronate and heteropolymeric regions in alginate. J Ferment Bioeng 86:236–238

Hashimoto W, Miyake O, Ochiai A, Murata K (2005) Molecular identification of *Sphingomonas* sp. A1 alginate lyase (A1-IV') as a member of novel polysaccharide lyase family 15 and implications in alginate lyase evolution. J Biosci Bioeng 99(1):48–54

Hisano T, Nishimura M, Yamashita T, Sakaguchi K, Murata K (1994) On the self-processing of bacterial alginate lyase. J Ferment Bioeng 78:109–110

Hodgson DA, Chater KF (1981) A chromosomal locus controlling extracellular agarase production by *Streptomyces coelicolor* A3 (2) and inactivation by chromosomal integration of plasmid SCP1. J Gen Microbiol 124:339–348

Hu Z, Lin BK, Xu Y, Zhong MQ, Liu GM (2008) Production and purification of agarase from a marine agarolytic bacterium *Agarivorans* sp. HZ105. J Appl Microbiol 106:181–190

Iwamoto Y, Araki R, Iriyama K, Oda T, Fukuda H, Hayashida S, Muramatsu T (2001) Purification and characterization of bifunctional alginate lyase from *Alteromonas* sp. strain No. 272 and its action on saturated oligomeric substrates. Biosci Biotechnol Biochem 65:133–142

Jam M, Flament D, Allouch J, Potin P, Thion L, Kloareg B, Czjzek M, Helbert W, Michel G, Barbeyron T (2005) The endobeta-agarases AgaA and AgaB from the marine bacterium *Zobellia galactanivorans:* two paralogue enzymes with different molecular organizations and catalytic behaviours. Biochem J 385:703–713

Johston KH, McCandless EL (1973) Enzymic hydrolysis of the potassium chloride soluble fraction of carrageenan: properties of "lambda carrageenases" from *Pseudomonas carrageenovora*. Can J Microbiol 19(7):779–788

Jonnadula RC, Verma P, Shouche YS, Ghadi SC (2009) Characterization of *Microbulbifer* strain CMC-5, a new biochemical variant of *Microbulbifer elongatus* type strain DSM6810T isolated from decomposing seaweeds. Curr Microbiol 59:600–607

Kang S, Kim JK (2015) Reuse of red seaweed waste by a novel bacterium, Bacillus sp. SYR4 isolated from a sandbar. World J Microbiol Biotechnol 31:209–217

Kang NY, Choi YL, Cho YS, Kim BK, Jeon BS, Cha JY, Kim CH (2003) Cloning, expression and characterization of a beta-agarase gene from a marine bacterium, *Pseudomonas* sp, SK38. Biotechnol Lett 25(14):1165–1170

Kawamoto H, Horibe A, Miki Y, Kimura T, Tanaka K, Nakagawa T, Kawamukai M, Matsuda H (2006) Cloning and sequencing analysis of alginate lyase genes from the marine bacterium *Vibrio* sp. O2. Mar Biotechnol 8:481–490

Khambhaty Y, Mody K, Jha B (2007) Purification and characterization of k-carrageenase from a novel γ-Proteobacterium, *Pseudomonas elongatus* (MTCC 5261) syn. *Microbulbifer elongates* comb.nov. Biotechnol Bioprocess Eng 12:668–675

Khambhaty Y, Mody K, Jha B (2008) Purification, characterization and application of a novel extracellular agarase from a marine *Bacillus megaterium*. Biotechnol Bioprocess Eng 13:584–591

Kim EJ, Fathoni A, Jeong GT, Jeong HD, Nam TJ, Kong IS, Kim JK (2013) Microbacterium oxydans, a novel alginate- and laminarin-degrading bacterium for the reutilization of brown-seaweed waste. J Environ Manage 130:153–159

Kitamikado M, Yamaguchi K, Tseng CH, Okabe B (1990) Method designed to detect alginate-degrading bacteria. Appl Environ Microbiol 56:2939–2940

Kitamikado M, Tseng CH, Yamaguchi K, Nakamura T (1992) Two types of bacterial alginate lyases. Appl Environ Microbiol 58:2474–2478

Kobayashi T, Uchimura K, Miyazaki M, Nogi Y, Horikoshi K (2009) A new high-alkaline alginate lyase from a deep-sea bacterium *Agarivorans* sp. Extremophiles 13(1):121–129

Kraiwattanapong J, Tsuruga H, Ooi T, Kinoshita S (1999) Cloning and sequencing of a *Delaya marina* gene encoding for alginate lyase. Biotechnol Lett 21:169–174

Lange B, Wingender J, Winkler UK (1989) Isolation and characterization of an alginate lyase from *Klebsiella aerogenes*. Arch Microbiol 152:302–308

Lemos ML, Toranzo AE, Barja JL (1985) Antibiotic activity of epiphytic bacteria isolated from intertidal seaweeds. Microb Ecol 11:149–163

Leon O, Quintana L, Peruzzo G, Slebe JC (1992) Purification and properties of an extracellular agarase from *Alteromonas* sp. strain C-1. Appl Environ Microbiol 58:4060–4063

Lewin RA (1969) A classification of flexibacteria. J Gen Microbiol 58:189–206

Li JW, Dong S, Song J, Li CB, Chen XL, Xie BB, Zhang YZ (2011) Purification and characterization of a bifunctional alginate lyase from *Pseudoalteromonas* sp. SM0524. Mar Drugs 9(1):109–123

Li S, Jia P, Wang L, Yu W, Han F (2013) Purification and characterization of a new thermostable k-carrageenase from the marine bacterium *Pseudoalteromonas* sp. QY203. J Ocean Univ Chin 12:155–159

Liu GL, Li Y, Chi Z, Chi ZM (2011) Purification and characterization of κ-carrageenase from the marine bacterium *Pseudoalteromonas porphyrae* for hydrolysis of κ-carrageenan. Process Biochem 46:265–271

Ma Y, Dong S, Liu S, Mou H, Jiang X (2008) Optimization of kappa-carrageenase production by *Pseudoalteromonas* sp. AJ5. Wei Sheng Wu Xue Bao 48(6):757–764

Malissard M, Chavagnat F, Duez C, Vacheron MJ, Guinand M (1995) Overproduction and properties of the mannuronate alginate lyase AlxMB. FEMS Microbiol Lett 126:105–112

Matsubara Y, Kawada R, Iwasaki K, Oda T, Muramatsu T (1998) Extracellular poly (a-L-guluronate) lyase from *Corynebacterium* sp.: purification, characterization, and conformational properties. J Protein Chem 17:29–36

Matsubara Y, Kawada R, Iwasaki K, Kimura Y, Oda T, Muramatsu T (2000) Cloning and sequence analysis of a gene (aly PG) encoding poly-(α-L-guluronate) lyase from *Corynebacterium* sp. strain ALY-1. J Biosci Bioeng 89:199–202

Mavromatis K, Abt B, Brambilla E, Lapidus A, Copeland A, Deshpande S, Nolan M, Lucas S, Tice H, Cheng JF, Han C, Detter JC, Woyke T, Goodwin L, Pitluck S, Held B, Brettin T, Tapia R, Ivanova N, Mikhailova N, Pati A, Liolios K, Chen A, Palaniappan K, Land M, Hauser L, Chang YJ, Jeffries CD, Rohde M, Goker M, Bristow J, Eisen JA, Markowitz V, Hugenholtz P, Klenk HP, Kyrpides N (2010) Complete genome sequence of *Coraliomargarita akajimensis* type strain (04OKA010-24). Stand Genomic Sci 2(3):290–299

McLean MW, Williamson FB (1979a) Glycosulphatase from *Pseudomonas carrageenovora*, purification and some properties. Eur J Biochem 101:497–505

McLean MW, Williamson FB (1979b) κ-Carrageenase from *Pseudomonas carrageenovora*. Eur J Biochem 93:553–558

McLean MW, Williamson FB (1981) Neocarratetraose 4-O-monosulphate beta hydrolase from *Pseudomonas carrageenovora*. Eur J Biochem 113:447–456

Min KH, Sasaki SF, Kashiwabara Y, Nishizawa K (1977) Substrate specificity of endo-polyguluronide lyase of *Pseudomonas* sp. J Biochem 81:547–553

Miyake O, Hashimoto W, Murata K (2003) An exotype alginate lyase in *Sphingomonas* sp. A1: overexpression in *Escherichia coli*, purification, and characterization of alginate lyase IV (A1-IV). Protein Expr Purif 29:33–41

Morrice LM, McLean MW, Williamson FB, Long WF (1983) beta-agarases I and II from *Pseudomonas atlantica*. Purifications and some properties. Eur J Biochem 135:553–558

Mou HJ, Jiang XL, Liu ZH, Guan HS (2004) Structural analysis of kappa-carrageenan oligosaccharides released by carrageenase from marine *Cytophaga* MCA-2. J Food Biochem 28:245–260

Muramatsu T, Sogi T (1990) Characterization of alginate lyases from a marine bacterium. Comp Biochem Physiol 97B:103–108

Ohta Y, Hatada Y (2006) A novel enzyme, lambda-carrageenase, isolated from a deep-sea bacterium. J Biochem 140:475–481

Ohta Y, Nogi Y, Miyazaki M, Li Z, Hatada Y, Ito S, Horikoshi K (2004) Enzymatic properties and nucleotide and amino acid sequences of a thermostable β-agarase from the novel marine isolate JAMB-A94. Biosci Biotechnol Biochem 68:1073–1081

Ohta Y, Hatada Y, Miyazaki M, Nogi Y, Ito S, Horikoshi K (2005a) Purification and characterization of a novel α-agarase from a *Thalassomonas* sp. Curr Microbiol 50:212–216

Ohta Y, Hatada Y, Ito S, Horikoshi K (2005b) High-level expression of a neoagarobiose-producing β-agarase gene from *Agarivorans* sp. JAMB-A11 in *Bacillus subtilis* and enzymic properties of the recombinant enzyme. Biotechnol Appl Biochem 41:183–191

Potin P (1992) Recherche, production, purification et caractérisation de galactane-hydrolases pour la préparation d'oligosaccharides des parois d'algues rouges. Thèse de Doctorat, Université de Bretagne Occidentale, Brest, Bretagne, France

Potin P, Sanseau A, Le Gall Y, Rochas C, Kloareg B (1991) Purification and characterization of a new kappa-carrageenase from a marine *Cytophaga*-like bacterium. Eur J Biochem 201:241–247

Potin P, Richard C, Rochas C, Kloareg B (1993) Purification and characterization of the α-agarase from

Alteromonas agarlyticus (Cataldi) comb. nov., strain GJ1B. Eur J Biochem 214:599–607

Ruijssenaar HJ, Hartsman S (2001) Plate screening methods for the detection of polysaccharases-producing microorganisms. Appl Microbiol Biotechnol 55:143–149

Sakatoku A, Wakabayashi M, Tanaka Y, Tanaka D, Nakamura S (2011) Isolation of a novel Saccharophagus species (Myt-1) capable of degrading a variety of seaweeds and polysaccharides. Microbiol Open 1:2–12

Sarwar G, Matayoshi S, Oda H (1987) Purification of a kappa carrageenase from marine *Cytophaga* species. Microbiol Immunol 31:869–877

Sawabe T, Ezura Y, Kimura T (1992) Purification and characterization of an alginate lyase from marine *Alteromonas* sp. Nippon Suisan Gakkaishi 58:521–527

Shieh WY, Jean WD (1998) *Alterococcus agarolyticus*, gen.nov. sp. nov., a halophilic thermophilic bacterium capable of agar degradation. Can J Microbiol 44:637–645

Stanier RY (1941) Studies on marine agar-digesting bacteria. J Bacteriol 42:527–559

Stevens RA, Levin RE (1977) Purification and characteristics of an alginase from *Alginovibrio aquatilis*. Appl Environ Microbiol 33:1156–1161

Sugano Y, Terada I, Arita M, Noma M, Matsumoto T (1993) Purification and characterization of a new agarase from a marine bacterium, *Vibrio* sp. strain JT0107. Appl Environ Microbiol 59:1549–1554

Sugimura I, Sawabe T, Ezura Y (2000) Cloning and sequence analysis of *Vibrio halioticoli* genes encoding three types of polyguluronate lyase. Mar Biotechnol 2:65–73

Sun F, Ma Y, Wang Y, Liu Q (2010) Purification and characterization of novel k-carrageenase from marine *Tamlana* sp. HC4, *Chin J Oceanol*. Limnology 28(6):1139–1145

Sutherland IW, Keen GA (1981) Alginases from *Beneckea pelagia* and *Pseudomonas* spp. J Appl Biochem 3:48–57

Suzuki H, Suzuki K, Inoue A, Ojima T (2006) A novel oligoalginate lyase from abalone, *Haliotis discus hannai*, that releases disaccharide from alginate polymer in an exolytic manner. Carbohydr Res 341:1809–1819

Tang JC, Wei JH, Maeda K, Kawai H, Zhou Q, Hosoi-Tanabe S, Nagata S (2007) Degradation of the seaweed wakame (Undaria pinnatifida) by a composting process with the inoculation of Bacillus sp. HR6. Biocontrol Sci 12:47–54

Tang JC, Taniguchi H, Chu H, Zhou Q, Nagata S (2009) Isolation and characterization of alginate-degrading bacteria for disposal of seaweed wastes. Lett Appl Microbiol 48:38–43

Tang J, Wang M, Zhou Q, Nagata S (2011) Improved composting of Undaria pinnatifida seaweed by inoculation with Halomonas and Gracilibacillus sp. isolated from marine environments. Bioresour Technol 102:2925–2930

Tawara M, Sakatoku A, Tiodjio RE, Tanaka D, Nakamura S (2015) Cloning and characterization of a novel agarase from a newly isolated bacterium Simiduia sp. strain TM-2 able to degrade various seaweeds. Appl Biochem Biotechnol 177:610–623

Van der Meulen HJ, Harder W, Veldkamp H (1974) Isolation and characterization of *Cytophaga flevensis* sp. Nov., a new agarolytic flexibacterium. Antonie Van Leeuwenhoek 40:329–346

Vashist P, Nogi Y, Ghadi SC, Verma P, Shouche YS (2012) *Microbulbifer mangrovi* sp. nov., a polysaccharide degrading bacterium isolated from the mangroves of India. Int J Syst Evol Microbiol. doi:10.1099/ijs.0.042978-0

Veldkamp H (1961) A study of two marine agar-degrading, facultatively anaerobic myxobacteria. J Gen Microbiol 26:331–342

Vera J, Alvarez R, Murano E, Slebe JC, Leon O (1998) Identification of a marine agarolytic *Pseudoalteromonas* isolate and characterization of its extracellular agarase. Appl Environ Microbiol 64:4378–4383

Vijayaraghavan R, Prabha VK, Rajendran S (2012) Isolation and characterization of K-carrageenase from *Bacillus subtilis*. World Sci Technol 2(9):13–18

Wakabayashi M, Sakatoku A, Noda F, Noda M, Tanaka D, Nakamura S (2012) Isolation and characterization of Microbulbifer species 6532A degrading seaweed thalli to single cell detritus particles. Biodegradation 23:93–105

Wang JX, Mou HJ, Jiang XL, Guan HS (2006a) Characterization of a novel β-agarase from marine *Alteromonas* sp. SY37–12 and its degrading products. Appl Microbiol Biotechnol 71:833–839

Wang YH, Yu GL, Wang XM, Lv ZH, Zhao X, Wu ZH, Ji WS (2006b) Purification and characterization of alginate lyase from marine *Vibrio* sp. YWA. Acta Biochim Biophys Sin 38(9):633–663

Wang Y, Guo EW, Yu WG, Han F (2013) Purification and characterization of a new alginate lyase from a marine bacterium *Vibrio* sp. Biotechnol Lett 35(5):703–708

Weigl J, Yaphe W (1966) The enzymic hydrolysis of carrageenan by *Pseudomonas carrageenovora*: purification of a kappa-carrageenase. Can J Microbiol 12:939–947

Wong TY, Preston LA, Schiller NL (2000) Alginate lyase: review of major sources and enzyme characteristics, structure-function analysis, biological roles, and applications. Ann Rev Microbiol 54:289–340

Yonemoto Y, Tanaka H, Hisano T, Sakaguchi K, Abe S, Yamashita T, Murata K (1993) Bacterial alginate lyase gene: nucleotide sequence and molecular route for generation of alginate lyase species. J Ferment Bioeng 75(5):336–342

Yoon HJ, Hashimoto W, Miyake O, Okamoto M, Mikami B, Murata K (2000) Overexpression in *Escherichia coli*, purification, and characterization of

Sphingomonas sp. A1 alginate lyases. Protein Expr Purif 19:84–90

Yoon JH, Kim IG, Shin DY, Kang KH, Park YH (2003a) *Microbulbifer salipaludis* sp. nov., a moderate halophile isolated from a Korean salt marsh. Int J Syst Evol Microbiol 53:53–57

Yoon JH, Kim H, Kang KH, Oh TK, Park YH (2003b) Transfer of *Pseudomonas elongata* Humm 1946 to the genus *Microbulbifer* as *Microbulbifer elongatus* comb. nov. Int J Syst Evol Microbiol 53:1357–1361

Zhong Z, Toukdarian A, Helinski D, Knauf V, Sykes S, Wilkinson JE, Bryne CO, Shea T, DeLoughery C, Caspi R (2001) Sequence analysis of a 101-kilobase plasmid required for agar degradation by a *Microscilla* isolate. Appl Environ Microbiol 67:5771–5779

Zhou MH, Ma JS, Li J, Ye HR, Huang KX, Zhao XW (2008) A k-carrageenase from a newly isolated *Pseudoalteromonas*-like Bacterium, WZUC10. Biotechnol Bioprocess Eng 13:545–551

Impact of Pollution on Phytoplankton and Implications for Marine Econiches

13

Priya Mallika D'Costa, Maria Shamina D'Silva, and Ravidas Krishna Naik

Abstract

Pollution in marine environments has been widely studied because of its impact, both direct and indirect, on human livelihood. Pollution may arise from different sources, the most well known being influx of domestic sewage, industrial waste and mining effluents. Shipping activities also contribute to pollution, either through accidental oil spills, ballasting and deballasting activities or bioinvasion. Pollution due to microplastics, radiation and heat (thermal pollution) is also gaining prominence. The impact of these varying sources of pollution on marine econiches is wide ranging; different trophic levels are affected. A pivotal trophic level impacted by pollution in the marine environment is phytoplankton, known for their role as microscopic primary producers and base of aquatic food webs. Pollution affects phytoplankton communities at different levels – abundance, growth strategies, dominance and succession patterns. Even if no direct changes in phytoplankton communities are visible, pollutants may accumulate in phytoplankton and be passed on to other trophic levels in a cascading manner, resulting in biomagnification of certain pollutants. This article focusses on the effects of different types of pollution (both point sources and non-point sources) on phytoplankton communities. The anthropocentric concept of 'pollution' and the links between pollution, eutrophication and harmful algal blooms (HABs) are also analyzed. An understanding of the synergistic interactions between these aspects and climate change effects will be useful to devise suitable remediation strategies for future use.

P.M. D'Costa (✉)
Department of Microbiology, Goa University, Goa 403206, India
e-mail: priyadcosta@unigoa.ac.in

M.S. D'Silva
Independent Researcher, Delfterstrasse 32, 5004 Aarau, Switzerland
e-mail: shaminanio@gmail.com

R.K. Naik
ESSO-National Centre for Antarctic and Ocean Research, Vasco-da-Gama, Goa 403804, India
e-mail: ravi@ncaor.gov.in

M.M. Naik, S.K. Dubey (eds.), *Marine Pollution and Microbial Remediation*,
DOI 10.1007/978-981-10-1044-6_13

13.1 Introduction

Marine pollution is a well-studied phenomenon. In simple words, it is the entry/discharge of contaminants into a marine environment, resulting in adverse effects on the physical, chemical and biological characteristics of the environment. According to the definition by GESAMP (1991), marine pollution is the introduction by man, directly or indirectly, of substances or energy into the marine environment (including estuaries) resulting in such deleterious effects as harm to living resources and marine life; hazards to human health; hindrance to marine activities, including fishing; impairment of quality for use of seawater; and reduction of amenities.

The term 'pollution' is clearly an anthropocentric concept. According to the Gaia hypothesis, the earth will take care of itself, irrespective of the effect on humans (Lovelock and Margulis 1996). The changes in biotic communities occurring as a result of exposure to different pollutants could all be viewed as factors driving the succession of the biosphere on the whole, with the fittest species surviving under the given set of conditions. However, for mankind, the future of the human race is paramount. This colours our perception of 'pollution'. Therefore, we view 'pollution' in terms of the reduction it brings about in the usability of the environment for humans.

Pollution of marine environments has been the focus of several studies. The types of marine pollution are varied and include pollution due to sewage, agricultural products, industrial effluents, organic compounds, antibiotics, POPs, plastics, heat (thermal pollution) and biological pollution or bioinvasion (Williams 1996). These pollutants originate either from point sources or non-point sources. A point source of pollution is usually a single, identifiable source of pollution, e.g. a sewage effluent discharge outlet. A non-point source of pollution refers to pollution from diffuse sources and is, therefore, comparatively difficult to regulate, e.g. polluted agricultural effluent draining into a waterbody. Non-point source pollutants may have arisen from a point source, but are so classified because of their ability to be transported long distances and accumulate other pollutants in the process.

Studies on marine pollution cover a range of aspects including the types of pollutants in marine environments, their fate in marine systems and, most importantly, their effects on different levels of food webs. The visibility of the problem is enhanced for organisms belonging to higher trophic levels. In contrast to the visible effects of pollution on macroscopic marine flora and fauna, the effects of pollution on phytoplankton are subtle and noticed only after the resultant changes have cascaded through other trophic levels and become noticeable. This is due to the fact that phytoplankton are microscopic, photosynthetic organisms, invisible to the naked eye. They include diatoms, dinoflagellates, cyanobacteria and other algal groups as well, even in the size range of 0.2–2 μm, termed picophytoplankton (e.g. *Prochlorococcus*, *Synechococcus*). These single-celled organisms function as primary producers in the food web and, therefore, form the base of the food web in the open sea. Being ubiquitous in aquatic systems, they contribute significantly to climatic processes (Jeffrey and Vesk 1997). However, this very characteristic of having an 'ubiquitous' distribution also implies that they are susceptible to the various types of pollution affecting aquatic systems. Therefore, this chapter focusses on phytoplankton and attempts to decipher the effects of different types of pollution on phytoplankton.

Diatoms and dinoflagellates, both belonging to Kingdom Protista, were the earliest known dominant phytoplankton groups in marine environments. They differ in several aspects (Table 13.1) and most notably in their life cycle strategies. Diatoms are either centric or pennate forms depending on their symmetry. Centric diatoms have radial symmetry, whereas pennate diatoms have bilateral symmetry. An interesting characteristic of dinoflagellates is the variation in their mode of nutrition, ranging from autotrophy,

Table 13.1 Comparison of diatoms and dinoflagellates, major phytoplankton groups

Characteristics	Diatoms	Dinoflagellates
Cell wall	Silica	Cellulose
Flagella	Absent	Two flagella present
Eyespot	Absent	Present in most species
Pigments	Fucoxanthin, chlorophyll	Peridinin, xanthophyll, chlorophyll
Nutrition	Autotrophy, heterotrophy (assimilation of organic matter)	Autotrophy, mixotrophy, heterotrophy
r- vs k-selection	r-selected	k-selected
Poisoning/syndrome caused	Amnesic shellfish poisoning (ASP)	Diarrhetic shellfish poisoning (DSP)
		Neurotoxic shellfish poisoning (NSP)
		Paralytic shellfish poisoning (PSP)
		Ciguatera fish poisoning (CFP)
		Putative estuary associated syndrome (PEAS)
Toxic species	*Pseudo-nitzschia* spp. (ASP)	*Gambierdiscus toxicus* (CFP)
		Dinophysis acuminata, D. fortii, Prorocentrum lima (DSP)
		Karenia brevis (NSP)
		Alexandrium spp., *Gymnodinium* spp., *Pyrodinium* spp. (PSP)
		Protoperidinium crassipes, Azadinium spinosum (AZP)
		Pfiesteria piscicida (PEAS)
Toxins produced[a]	Domoic acid (ASP)	Ciguatoxin, maitotoxin (CFP)
		Okadaic acid (DSP)
		Brevetoxin (NSP)
		Saxitoxin (PSP)
		Azaspiracid poisoning (AZP)

Tomas (1997), Van Dolah et al. (2001), and James et al. (2010)

[a]Toxin involved in PEAS still not identified

mixotrophy to heterotrophy (Tomas 1997). Mixotrophic dinoflagellates have the ability to switch between autotrophic and heterotrophic mode of nutrition depending on the environmental conditions. Thingstad et al. (1996) have described mixotrophy as being a strategy of 'eating one's competitor'. Mixotrophic dinoflagellates can sustain growth in habitats with low nutrient concentrations by switching to phagotrophy. Additionally, under conditions of reduced availability of particulate food, they survive by switching to photosynthesis (Jost et al. 2004).

Based on the evolutionary characteristics of phytoplankton, diatoms are reported to be r-selected organisms, whereas dinoflagellates are considered k-selected organisms. r-selected species are characterized by rapid growth response in response to nutrients and/or opportunistic appearance in turbulent environments (Margalef 1978; Kilham and Kilham 1980). k-selected species are specialists in environments with limited resources, especially under conditions of intense competition with neighbouring species (Harper 1977). Patil (2003) opined that the 'r- vs k-selection' is a rather simplistic comparison and can be noticed not only between two groups of phytoplankton (diatoms and dinoflagellates) but also within diatoms themselves (centric and pennate diatoms).

13.2 Major Pollutants in Coastal Systems and Their Effect on Phytoplankton

Phytoplankton respond to different types of pollution not only through changes in abundance but also dominance and succession patterns and community structure parameters (species evenness, diversity and richness). Species richness, evenness and diversity are mathematical diversity indices, which function as measures of biodiversity. Species richness refers to the number of different species in a sample, community or taxonomic group. It is often represented by Margalef's species richness (d) that calculates the number of species present, taking into account the number of individuals. Species evenness computes how equivalent the community is on a numerical basis. In other words, it is a measure of how evenly the individuals in a community are distributed among the different species. It is usually represented by Pielou's evenness index (denoted by J'). Its values range between 0 and 1. Species diversity refers to the number of different species that are represented in a data set/community and is inclusive of both species richness and species evenness. It can be expressed in terms of the Shannon-Wiener diversity index (H') (Colwell 2009). These parameters are excellent indicators of the status of phytoplankton communities and, being indices, are comparable across different studies.

13.2.1 Sewage Pollution

Sewage pollution includes pollution due to domestic/municipal waste as well as effluents from industries, slaughterhouses, etc. Large volumes of wastes, produced daily from highly populated urban areas, are carried by drainage systems and finally find their way into rivers or marine environments. In terms of volume, sewage pollution is the major type of pollution affecting waterbodies in coastal environments.

Sewage effluent discharged into coastal waters contains a diverse mixture of harmful substances including bacterial, viral and protozoan pathogens. High concentrations of pathogens have been reported in raw sewage: up to 4×10^9 bacteria per litre and 10^7 viruses per litre (HMSO 1990). These include the bacteria-*Salmonella* spp., *Escherichia coli*, *Streptococcus* sp., *Staphylococcus aureus* and *Pseudomonas aeruginosa*, the fungi-*Candida* and viruses-enterovirus, hepatitis, poliomyelitis, influenza and herpes.

Sewage also contains a variety of organic and inorganic wastes, in addition to toxic chemicals such as organochlorines, organotins and heavy metals (HMSO 1990). Since sewage contains abundant organic substances, it supports a high microbial load, elevated levels of microbial decomposition and, therefore, reduced concentration of dissolved oxygen. These conditions lead to a high biological oxygen demand (BOD), which is the most prominent characteristic of sewage. This can have consequences ranging from decrease in dissolved oxygen concentrations to below critical levels required by aquatic life, to production of toxic gases (ammonia, hydrogen sulphide), smothering of the benthic flora and fauna by solid substances suspended in sewage and increase in heavy metal concentration.

Effect on phytoplankton: The main effects of sewage pollution on phytoplankton are related to its high nutrient concentration, resulting from the abundant microbial activity and decomposition processes. The surge in nutrient concentrations can stimulate phytoplankton to grow to high numbers. This increases the food available for primary consumers and, consequently, for organisms higher up in the food web. This is usually one of the reasons for an increase in the population of fish in an area. However, this increase may be short-lived, especially if there is a shift in the dominant phytoplankton from non-toxic to toxic species. Under such conditions, in spite of the high phytoplankton abundance, fish species feeding on these phytoplankton may accumulate these toxins or undergo mass mortality (Hernandez et al. 1998).

A change in the dominant algal groups may also occur, leading to eutrophication conditions. Changes in phytoplankton community composition resulting from sewage and industrial pollution have been documented in several studies. These changes include reduction in the complexity of the phytoplankton community (Taslakian and Hardy 1976), increase in production and chlorophyll *a* (Thompson and Ho 1981; Pan and Rao 1997; Ramaiah et al. 1998) and increase in potentially harmful diatom species – *Pseudo-nitzschia* sp. (Pan and Rao 1997).

13.2.2 Industrial Pollution

Industrial pollution is caused due to the efflux of heavy metals and trace elements, which are by-products of many industrial processes. Mining effluents also contribute to this type of pollution. Industrial pollution is closely linked with sewage pollution. Islam and Tanaka (2004) attribute this to industrial areas being densely populated or, in other words, well-established cities being chosen as the site for new industries to be set up. Though toxicity of sewage effluent is only partially due to its metallic compounds (Seidl et al. 1998), metals are known to be toxic to biota, especially when they occur as cations. This is because, as cations, they have the ability to bind to short carbon chains and, therefore, bioaccumulate and concentrate in marine organisms over long time periods. The ten most toxic metals, listed by Davies (1978), in order of decreasing toxicity, are mercury, cadmium, silver, nickel, selenium, lead, copper, chromium, arsenic and zinc. Based on the high affinities of the widely distributed heavy metals (e.g. mercury, silver, copper) for sulphur and their tendency to bind to sulphhydryl groups of proteins and enzymes in living organisms, Davis (1978) stated that the effects of metals on organisms are associated with interference in the metabolic processes involving sulphur-containing constituents.

Effect on phytoplankton: Exposure of phytoplankton to metals results in an increase in the cell membrane permeability in phytoplankton and other marine algae, leading to disruption of cell integrity. Additionally, an increase in cell size was seen in phytoplankton grown in the presence of copper and mercury (Davies 1978). Similar heavy metal-induced changes in cell shape of phytoplankton were also reported by Kayser (1976). Probably, the inhibition of cell division in phytoplankton by metals was responsible for the production of very large cells of phytoplankton (Davies 1978).

13.2.3 Agricultural Waste and Organic Compounds

Agricultural waste includes fertilizers (both natural and synthetic) and pesticides and, given their widespread use, is also an important source of pollution in waterbodies. Compared to synthetic fertilizers, natural fertilizers/manure can provide soils with 100 times more hazardous products for the equivalent plant nutrient content due to the biological concentration processes they have undergone (Joly 1993). Pesticides, another widely used class of products in agriculture, also contribute to pollution of waterbodies mainly due to their toxicity and persistence. In fact, Duursma and Marchand (1974) have stated that pesticides and their residues are among the most devastating agents for aquatic ecosystems and organisms affecting all levels of the food chain from the lowest to the top level. Many of the pesticides currently in use are organochlorines, organophosphates, PAHs and organometals, and are grouped as 'persistent organic pollutants' (POPs). POPs have wide-ranging distribution; they have been reported from the Arctic to Antarctic and from intertidal to abyssal depths (Islam and Tanaka 2004).

Islam and Tanaka (2004) have suggested two mechanisms for agricultural waste products as pollutants – bioconcentration and biomagnification. Bioconcentration refers to the accumulation of chemical from the surrounding medium into an organism by virtue of the lipophilicity of many chemicals, whereas biomagnification refers to the increasing concentration of a chemical as food energy is transformed within the food chain. Due to biomagnification, the concentration of the chemical will keep on increasing (get magnified) as it passes up the food web and will be maximum in the top-level predators, in most instances, man. The same holds true for POPs. The effects of accumulation of POPs seen in higher organisms include lesions, tumours, cancers (especially in fish and animals), immunosuppression, hormonal disturbances, reproductive failure, imposex/sex change in gastropods, poor fish health, excessive slime on fish scales and gills, egg shell thinning and death (Islam and Tanaka 2004 and references therein).

Effect on phytoplankton: POPs cause a decrease in the abundance and viability of NE Atlantic Ocean phytoplankton even at trace concentrations (Echeveste et al. 2010). Endosulfan, a widely used agricultural pesticide, is taken up from the water column by phytoplankton and zooplankton (DeLorenzo et al. 2002). Pesticide accumulation in phytoplankton has also been reported by Rose and McIntire (1970), Stange and Swackhamer (1994) and DeLorenzo et al. (2001). Islam and Tanaka (2004) state that these changes in phytoplankton are chronic in nature and go unnoticed. Interestingly, the changes that are observed in phytoplankton are not due to pesticide exposure alone but due to a combination of environmental stresses (eutrophication and pathogens) (Islam and Tanaka 2004). It is pertinent to note that such accumulation in the base of the food web will result in biomagnification of pesticide residues at higher trophic levels.

13.2.4 Antibiotic Pollution

Another kind of pollution affecting aquatic systems is that caused by antibiotics. The widespread use of antibiotics ensures their discharge into aquatic systems through domestic and hospital-sourced effluents. Though antibiotics are susceptible to biodegradation by natural bacterial communities, some synthetic antibiotics may be refractive to biodegradation. Additionally, antibiotics are more stable when adsorbed onto sediment (Halling-Sørensen et al. 1998) and, thus, may severely impact benthic microalgal communities.

Antibiotics have been widely studied for their effect on bacteria and the development of antibiotic resistance in non-target organisms (Martinez 2009). Such studies were confined to only bacteria, with other trophic levels being neglected. Since then, the scope of non-target organisms has been widened and now includes diatoms (Campa-Cordova et al. 2006; D'Costa and Anil 2012; Hagenbuch and Pinckney 2012; Pinckney et al. 2013), crustaceans (Wollenberger et al. 2000), duckweeds (Brain et al. 2004) and green algae (Halling-Sørensen 2000; Eguchi et al. 2004; Lai et al. 2009).

Effect on phytoplankton: Antibiotics can affect phytoplankton either directly or through effects on the bacteria associated with phytoplankton. This depends on the mode of action of the antibiotic. Antibiotics such as ß-lactams (penicillin) act on cell wall of bacteria and, therefore, do not affect phytoplankton directly. On the contrary, aminoglycoside antibiotics (e.g. streptomycin), which affect protein synthesis, can inhibit bacteria as well as phytoplankton. This is because the ribosomes in phytoplankton organelles (the site of protein synthesis) are 70S, similar to prokaryotic ribosomes. Few antibiotics have been studied for their effect on phytoplankton. Tylosin has been reported to bring about a reduction in growth, biomass and primary productivity of benthic microalgae (Pinckney et al. 2013). Similarly, penicillin, streptomycin and chloramphenicol modify the community structure of benthic diatom communities, especially when tested in combination (D'Costa and Anil 2012). Exposure to 0.4 mgml^{-1} penicillin, 0.2 mgml^{-1} streptomycin and 0.04 mgml^{-1} chloramphenicol reduced benthic diatom abundance by 99–100 %, probably due to the high concentrations used and synergistic effects. Interestingly, enhanced diatom emergence (133 % compared to control) was observed in 0.2 mgml^{-1} streptomycin treatment (D'Costa and Anil 2012). Separate experiments carried out with natural biofilms indicated that even low concentrations of penicillin (50 µgml^{-1}) inhibited diatoms, whereas 50 µgml^{-1} streptomycin did not (unpublished data). Comparable antibiotic-specific and species-specific trends were seen for the effect of different concentrations of penicillin and streptomycin on diatom monocultures – *Coscinodiscus* sp., *Cylindrotheca* sp., *Navicula* sp. and *Thalassionema* sp. (unpublished data). Antibiotic-specific and species-specific adverse effects have been noted for the effect of ciprofloxacin, lincomycin and tylosin on *Cylindrotheca closterium* and *Navicula ramosissima* (Hagenbuch and Pinckney 2012) and for chloramphenicol, erythromycin and furazolidone on *Isochrysis galbana* (phytoflagellate) and *Chaetoceros gracilis* (diatom, Campa-Cordova et al. 2006).

13.2.5 Oil Pollution

The incidence of oil spills has increased over the years, concurrently with the development of shipping activities. The rise in shipping traffic has contributed to its increase. Most oil pollution events result from accidental oil spills from tankers/cargo ships. Coastal refineries, where crude oil is processed to produce fuels, solvents and lubricants, also pose a significant risk. Accidental spillage as well as release of contaminated water used in these processes is the means by which oil can enter the environment. Other sources of oil include run-off from land and oily wastes discharged by ships in the absence of effective legislation regulating the same.

Oil, consisting of a variety of polycyclic aromatic hydrocarbons, affects a variety of organisms at different trophic levels, ranging from echinoderms, molluscs, sea stars, sea urchins, marine mammals, to, most noticeably, birds. The visible effects are loss of mobility, due to the oil forming a layer over the surface of the organism and affecting buoyancy and heat insulation. Other effects include heavy secretion and erosion of the mucus membrane; severe eye irritation leading to blindness; changes in cell membrane leading to cancer, carcinomas and papillomas in bottom-feeding fish (Russel and Kotin 1956); nervous abnormalities; sensitivity to environmental fluctuations; and loss of egg viability (Islam and Tanaka 2004).

Effect on phytoplankton: Oil may have a blanketing effect on phytoplankton, and especially on benthic microalgal assemblages, under still conditions. The hydrocarbon components of oil may affect phytoplankton in several ways. Hydrocarbon molecules displace lipid compounds in plasma membranes, thus affecting their semipermeability. Dissolution of the hydrocarbons in the lipid phase of the chloroplasts could also inhibit photosynthesis. Hydrocarbons also inhibit the tricarboxylic acid cycle and oxidative phosphorylation in mitochondrial membranes (Islam and Tanaka 2004).

13.2.6 Biological Pollution/Biological Invasions

The increase in shipping activities, in terms of both volume and distance covered, has led to the problem of 'biological invasion' or 'bioinvasion'. Ships transfer organisms/species from one geographical area to another, either through ballast water or through attachment on ship hulls. Bioinvasive species documented so far include the comb jelly (*Mnemiopsis leidyi*) in the Black Sea, the zebra mussel (*Dreissena polymorpha*) in North America and the black striped mussel (*Mytilopsis sallei*) in India (Mumbai and Visakhapatnam). Conditions during transit and in the recipient port will determine the establishment success of transported biota. If an organism encounters plenty of food sources and negligible levels of predators, it will have higher chances of establishing in the 'recipient' area. Of course, environmental conditions of light, turbidity, dissolved oxygen, nutrients, etc. must also be favourable.

Effect on phytoplankton: Bioinvasive species of phytoplankton have not been reported so far. Rather than indicating non-occurrence, this could simply be a case of 'invisibility' due to the small size of phytoplankton and the necessity of microscopic analysis to study them. In case of phytoplankton, an important aspect that needs to be considered is that very often, even a single cell is sufficient to establish a population. This is mainly the case for HAB-causing dinoflagellates, which produce resting stages called cysts, during their life cycle. The cysts settle to the sediment and undergo a mandatory dormancy period before germinating on the onset of favourable environmental conditions. Such accumulations of cysts are often termed 'seed beds' (Joyce et al. 2005) and may serve as inocula for blooms, when transported through ballast water to a favourable environment, especially when coupled with physical processes like eddies and cyclones (Naik et al. 2011). It must also be noted that for a species to be designated as 'bioinvasive', prior information about the baseline diversity should be available. Therefore, keeping this in mind, Anil et al. (2002) have suggested that studies on biodiversity of target coastal areas are imperative.

13.2.7 Plastic Pollution

Owing to the numerous applications and large volumes of plastics used, they constitute the majority of litter deposited in aquatic environments. A study by the five Gyres Institute reveals that more than five trillion plastic pieces weighing over 250,000 t are floating in the world's oceans (Eriksen et al. 2014). They have been reported in a variety of niches – distant gyres; remote beaches; in the bodies of dead fish, birds and whales; and even in Arctic ice (Safina 2013; de Stephanis et al. 2013; Obbard et al. 2014).

The chemical composition of plastics as well as their recalcitrant nature is of concern. In addition to bisphenol A (BPA), a suspected endocrine disruptor, plastics also contain monomers, plasticizers, flame retardants and antimicrobials (Seltenrich 2015). Once in the aquatic environment, plastic particles undergo turbulence, photodegradation and oxidation processes to get converted into smaller particles called microplastics (<1 or 5 mm). Microplastics are often mistaken for food particles by zooplankton, primary consumers in food webs. Microplastics also serve as reservoirs of persistent, bioaccumulative, toxic compounds (Seltenrich 2015) and as vectors for transfer of rafting species leading to bioinvasion (Winston et al. 1997).

The recalcitrant nature of plastics ensures their persistence in the environment for long periods of time. Plastic litter poses problems to wildlife through ingestion, suffocation, entanglement and ghost fishing, apart from reducing the aesthetic quality of the environment (Gregory 1999). Eventually, plastic materials sink to the bottom and settle onto sediment. It inhibits the transfer of gases and nutrients at the sediment-water interface and smothers the benthic flora and fauna.

Effect on phytoplankton: Plastics floating in the water column decrease the amount of light available to phytoplankton lower in the water column and sediment. A recent study (Long et al. 2015) has revealed that microplastics are incorporated and concentrated in phytoplankton aggregates and ultimately sink faster to the sea floor. In addition, the blanketing effect of plastics on sediment is detrimental to not only benthic fauna but also to the resident microalgal assemblages present there.

13.2.8 Thermal Pollution

Discharges from coastal power plants are detrimental to the aquatic biota in two ways: elevated temperature and residual chlorine. Water temperature is one of the main factors affecting aquatic organisms, especially their survival rate, growth and reproduction (Langford 1990; Davison 1991). An increase in seawater temperature leads to a reduction in dissolved oxygen concentrations and, consequently, increases the metabolic rate of organisms (Poornima et al. 2005). Residual chlorine also affects organisms by diffusing through their cell membrane and inhibiting a range of metabolic activities (Strauss and Puckorius 1984).

Effect on phytoplankton: The elevated temperature is stressful for freshwater, benthic microalgal communities. Phytoplankton abundance decreased with an increase in temperature (Devinney 1980) and chlorine concentration (James 1967; Eppley et al. 1976). Chuang et al. (2009) reported that a chlorine content of 0.2 ppm greatly suppressed phytoplankton productivity, regardless of the effluent temperature, whereas periphyton productivity was more influenced by elevated water temperatures. Poornima et al. (2005) noted that phytoplankton in the effluents of coastal power plants experience the combined effect of temperature and chlorine since they are drawn into the cooling circulation systems and subsequently released back into the sea along with the effluents. They reported a decrease in chlorophyll content and productivity in the receiving waters of a thermal power plant at Kalpakkam, India, and attributed it to mainly chlorination rather than elevated temperature.

13.3 Characteristics of Phytoplankton That Influence Their Response to Pollution

13.3.1 Diversity of Econiches

The various econiches that phytoplankton inhabit (water column, sediment, biofilms) are responsible for their differential responses to the same type of pollution. For example, phytoplankton in the water column (plankton) are more susceptible to pollution as the water environment provides a medium for the pollutant to disperse in. The upside of this is that the pollutant tends to get diluted due to dispersion in water. The reverse is true for phytoplankton living in sediments (microphytobenthos). Microphytobenthos are more prone to pollutants that have a tendency to adsorb onto sediment and may be exposed to various concentrations of the pollutant, depending on the distance between the diatoms and the point source of pollution in the sediment. Several diatoms have devised a strategy to combat this problem; they often grow attached to a substrate, in layers one above the other. This is the 'biofilm' mode of growth and, by virtue of the layered nature of growth, confers resistance towards pollutants, including metals and antibiotics. In the bacterium *Acinetobacter haemolyticus*, biofilm formation increases resistance towards antibiotics and metals compared to planktonic cells (Gaidhani et al. 2014). Similar differences in resistance profiles between planktonic and fouling diatoms may also exist and, therefore, influence the effect of pollutants on phytoplankton.

13.3.2 Differences in Mobility: Epipsammic vs Epipelic Diatoms

Benthic diatoms differ with regard to their mobility within the sediment. They are of two types. Epipsammic diatoms grow attached to sand grains and are immobile or have restricted movement (Round 1965). On the contrary, epipelic diatoms move actively in surface sediment (Round 1979) and produce mucopolysaccharides that are responsible for stabilization of the sediment (Paterson 1997). They have a higher chance of escaping unfavourable conditions/gradients and, therefore, tend to be more resistant to pollutants. This could be the reason why *Amphora* and *Navicula*, both epipelic fouling diatoms, are dominant in most coastal environments in India, irrespective of the extent of pollution.

13.3.3 Interactions with Epibiotic and Loosely Associated Microbes

In nature, phytoplankton are always associated with other microorganisms, either epibiotic, loosely associated or within the cells. In fact, Rehnstam-Holm et al. (2010) have termed phytoplankton as bacterial 'incubators'. *Vibrio* species, known as pathogens, are also associated with phytoplankton (Main et al. 2015). Co-occurring bacteria are also apparent in both diatom and dinoflagellate cultures. They not only influence growth rates, exopolymer production, toxin production and interactions of phytoplankton with other organisms but also affect their response to pollutants (Amin et al. 2012). For example, non-axenic *Amphora coffeaeformis* (diatom) has higher tolerance to metals compared to axenic *Amphora coffeaeformis* (Thomas and Robinson 1987). In view of this, studies on the effect of pollutants on phytoplankton must also consider the effects of the phytoplankton-associated microbiota.

13.4 Monsoons in Tropical and Sub-tropical Environments and Effects on Phytoplankton

Monsoon systems are a major climatic factor affecting phytoplankton community structure in tropical and subtropical environments. Two monsoon systems – southwest (SW) monsoon and northeast (NE) monsoon – influence the Indian subcontinent. They occur during different periods: SW monsoon from June to September and NE monsoon from October to December. The monsoons are characterized by pulsed supply of nutrients (due to increased nutrient influx through riverine input), lowered salinity, increased turbulence and cloud cover. Pulsed supply of nutrients is an important factor inducing phytoplankton diversity (Yamamoto and Hatta 2004).

Monsoons influence physico-chemical variables such as nutrients, light (termed 'bottom-up' control), sinking and mortality (loss factors) and grazing populations ('top-down' control). These factors, along with interactions across trophic levels, regulate phytoplankton community structure (Reynolds 1980; Assmy et al. 2007). They play significant but dissimilar roles in modulating phytoplankton communities, especially in monsoon-influenced environments. They can also be grouped under autogenic and allogenic control mechanisms. Autogenic mechanisms include competitive interactions between component species and drive the system towards biotic interaction-driven succession. On the contrary, allogenic mechanisms include the physico-chemical variables, which function as external drivers that shape community succession (Reynolds 1980).

Monsoon patterns influence diatom and dinoflagellate communities and bring about changes in their dominance and succession strategies. For example, diatoms, the dominant phytoplankton group in tropical and subtropical monsoon-influenced environments (e.g. the coasts of India), often bloom to large numbers, subsequent to the SW monsoon. Blooms of diatoms, *Fragilariopsis (=Fragilaria) oceanica* and *Skeletonema costatum*, are an annual recurring feature. Dinoflagellates, which are present at comparatively lower concentrations, grow to high levels along the west coast of India during periods when there is a break in the monsoon, accompanied by sunny, calm weather (D'Silva et al. 2012b and references therein).

The relevance of monsoon in influencing the diatom and dinoflagellate communities (through shifts in species dominance patterns) has been highlighted by D'Costa and Anil (2010), D'Costa et al. (2008) and D'Silva et al. (2012a) in studies carried out along the west coast of India. In Mumbai port (west coast of India), the pre-monsoon period is characterized by lower nutrient levels and higher suspended particulate matter, oxygen saturation levels and bacterial abundance during pre-monsoon compared to post-monsoon. This is associated with highly diverse diatom and dinoflagellate communities in pre-monsoon (D'Costa et al. 2008; D'Costa and Anil 2010), whose interactions determine the succession of the phytoplankton community (autogenic control). A seasonal cycling between vegetative and resting stages of autotrophic and heterotrophic dinoflagellates was observed. This was attributed to changes in physico-chemical variables (bottom-up control) and zooplankton grazers (top-down control) (Gaonkar et al. 2010) driven by the SW monsoon. Terrigenous input, influenced by the monsoon, is also an important regulatory factor along the west coast of India (D'Silva et al. 2012a). *Skeletonema costatum*, a chain-forming diatom, grows to high numbers, subsequent to the SW monsoon in response to the increased nutrient concentrations and lowered salinity (D'Costa and Anil 2010), leading to a decrease in species richness, evenness and diversity, resulting in allogenic control being more prevalent during post-monsoon periods (D'Costa 2010).

The intensity and duration of monsoons are also relevant; different dinoflagellates dominated in the water column in Mumbai port subsequent to two dissimilar monsoon events (D'Costa et al. 2008). Therefore, monsoons are an important and sometimes overriding factor, influencing phytoplankton communities.

13.5 Links Between Pollution, Eutrophication and HABs

Eutrophication is the progression of organic enrichment of an ecosystem, generally through nutrient inputs (Nixon 1995), from both point and non-point sources of pollution. Eutrophication can occur naturally due to the influx and accumulation of nutrients to waterbodies resulting in changes in species composition and primary production. Anthropogenic eutrophication, resulting from human activities, is more widely known and is driven by the nutrients – nitrogen and phosphorous (Howarth and Marino 2006). The extreme enrichment of aquatic systems with anthropogenic sources of nutrients via surface water, groundwater and air causes transformation of oligotrophic waterbodies to mesotrophic, eutrophic and finally hypertrophic stage. Mesotrophic and eutrophic phases exhibit intermediate to high nutrient concentrations and show increasing water quality problems, whereas the hypertrophic phase is the excessive enrichment of ecosystems with anthropogenic sources of nutrients.

Eutrophication causes major changes in species composition, structure and function of marine communities over large areas. In many cases, the responding dominant species are not toxic and may be beneficial to coastal productivity until they exceed the assimilative capacity of the system, after which hypoxia and other adverse effects occur. When that threshold is reached, seemingly harmless species can have negative impacts. Phytoplankton communities, under eutrophication conditions, generally exhibit an increase in biomass and productivity (Riegman 1995). Changes in community structure are also observed. For example, in Visakhapatnam harbour, along the east coast of India, the dinoflagellate cyst assemblages in the inner eutrophic stations of the harbour were markedly different compared to the outer stations, which had comparatively lower nutrient concentrations. The inner stations were dominated by *Protoceratium reticulatum*, a harmful dinoflagellate, and also exhibited low species diversity (D'Silva et al. 2013). The phytoplankton community may shift from diatom dominance to flagellate dominance. A shift in phytoplankton-sized classes may also be observed; small-sized nanoplankton (e.g. microflagellates and coccoids) tend to dominate in eutrophic waterbodies (Kimor 1992). Macroalgae and filamentous algae dominate and often become a nuisance, leading to a decrease in dissolved oxygen levels, affecting the benthic fauna, recreational uses and tourism potential (Riegman 1995; Rosenberg and Nilsson 2005).

Eutrophication conditions usually lead to the occurrence of harmful algal blooms (HABs), also called red tides. HABs are proliferations of algae that have the potential to damage food webs when they accumulate in high concentrations. They often produce toxic compounds that are biomagnified across trophic levels and cause several poisoning syndromes (ciguatera fish poisoning, amnesic/paralytic/diarrhetic/neurotoxic shellfish poisoning). Other effects associated with HABs are oxygen depletion in water due to respiration and algal decay processes, suffocation of fish and other benthic organisms and shading of the benthic flora (Hallegraeff 1993; Glibert et al. 2005; Backer and McGillicuddy 2006). It has also been shown that an increase in the production of allelochemicals occurs when some HAB species are subjected to unbalanced nitrogen or phosphorus conditions, which in turn are directly caused by eutrophication (Granéli and Johansson 2003; Fistarol et al. 2005). The production of allelochemicals provides a competitive edge to species and allows them to survive the biotic interactions in environments driven by autogenic control mechanisms.

HABs are a nuisance/global problem in many temperate environments and Southeast Asian countries. Comparatively, they are not as widespread in India. A review of bloom occurrences in Indian waters from 1908 to 2009 points out that a total of 101 cases have been reported (D'Silva et al. 2012b). Majority of the blooms reported along the west coast of India are caused by dinoflagellates, whereas diatom blooms prevail along the east coast. *Noctiluca scintillans* and *Trichodesmium erythraeum* are the most commonly occurring forms (D'Silva et al. 2012b).

Glibert et al. (2005) have reported that HABs have been expanding globally, in spatial extent, in duration of blooms and in intensity. A critical question has been the extent to which this change is associated with eutrophication and/or accelerated by climate or other factors (Paerl and Scott 2010). The incidence of anthropogenic eutrophication has been increasing, especially in coastal environments. The increase in use of urea-based fertilizers over the last five decades has also contributed to coastal eutrophication (Glibert et al. 2006). Aquaculture activities, due to their potential for loading/discharging effluents rich in polluting agents, are another contributory factor. These factors, responsible for eutrophication, intensify the occurrence of HABs globally. Phosphorus loading is often cited as the major cause of HABs in freshwaters (Oliver and Ganf 2000).

13.6 Links Between Pollution and Climate Change and Effects on Phytoplankton

Anthropogenic activities are also responsible for several climate change events such as ocean acidification, global warming, El Niño/Southern Oscillation (ENSO) and the Pacific Decadal Oscillation (PDO) (Boyd and Doney 2003; Hutchins et al. 2009; Fu et al. 2012).

13.6.1 Ocean Acidification

Ocean acidification results from the accumulation of CO_2 in the atmosphere due to anthropogenic activities that have increased concentrations of CO_2 and bicarbonate ($HCO_{3\,-}$) in seawater. Both these are inorganic carbon sources that can be utilized by phytoplankton for photosynthesis and growth. However, this change in seawater carbon chemistry is closely related with a decrease in ocean pH; a drop of roughly 0.1units from pre-industrial levels has already been observed (Stocker et al. 2013). If anthropogenic CO_2 emissions are not controlled, ocean pH could drop by an additional 0.6 units to levels lower than has occurred in the past 300 million years (Caldeira and Wickett 2003; Meehl et al. 2007). Ocean acidification has been widely reported, even in high-latitude marine environments (Fabry et al. 2009). Orr et al. (2005) and Raven et al. (2005) envisage that tropical regions and the Northern and Southern Oceans will be affected the most.

The first organisms to be affected by ocean acidification are the ones with a calcareous shell, corals, molluscs, foraminiferans and coccolithophorids, due to the dissolution of their biogenic calcium carbonate ($CaCO_3$) shells (Riebesell et al. 2000). The future predicted changes in ocean acidification and global warming may synergistically affect the settlement and recruitment of marine invertebrates. The future predicted levels of elevated pCO_2 and temperature will affect the diet of invertebrate larvae, thereby influencing their development and metamorphosis (Baragi and Anil 2015). Not all organisms are negatively affected; a diverse range of macroalgal species exhibits resilience under acidification conditions (Porzio et al. 2011).

Effect on phytoplankton: Elevated pCO_2 affects calcification of phytoplankton and photosynthetic carbon fixation (Sett et al. 2014). Elevated pCO_2 elicited positive, negative or no response in the diatoms – *Chaetoceros brevis*, *Navicula directa* and *Nitzschia lecointei* (Boelen et al. 2011; Torstensson et al. 2012; Johnson et al. 2013). Baragi et al. (2015) studied the effect of pCO_2 and temperature on *Navicula distans*, a fouling diatom; they reported that the future predicted changes in pCO_2 and temperature will result in reduction in size of *N. distans* cells and increase in associated picoperiphytes and heterotrophic bacteria.

13.6.2 Global Warming

Global warming, due to anthropogenic activities, results in pronounced changes in marine ecosystems. These effects are not due to the rise in temperature alone, but the concurrent changes in chemistry and circulation of the oceans which influence population dynamics. Harley et al. (2006) have stated that sweeping community-level changes may occur due to climatic impacts on one/few leverage species. Climate warming may also increase the incidence of disease transmission, pathogen development and survival rates and host susceptibility (Harvell et al. 2002). An increase in temperature may also affect the invasion success of nonindigenous species. Warmer temperatures affect those species strongly who are already near their tolerance limit (Laubie 2001), and especially benthic organisms, resulting in empty niches that are available for colonization by new species. Occhipinti-Ambrogi (2007) argues that the responses of these organisms are subtle and cannot be discerned by exposing different life stages to a range of temperatures. Additionally, the effect of seasonal fluctuations on these responses must also be considered. Daufresne et al. (2009), in a meta-analysis of the effect of climate change on body size of ectothermic aquatic organisms (bacteria, phytoplankton, zooplankton and fish), have concluded that global warming results in a significant increase in the proportion of small-sized species and young age classes and a decrease in size at age.

Effect on phytoplankton: Yvon-Durocher et al. (2015) have carried out a study spanning 5 years using artificially warmed ponds (an increase of 4 °C). At the end of the incubation period, the phytoplankton communities in those ponds exhibited greater biomass, change in species composition, species richness and diversity. The communities were dominated by larger species, which were more resistant to grazing by zooplankton, primary consumers in the food web. These changes were mediated through influences on metabolic rates and species interactions. The subsequent increase in biodiversity and productivity of the phytoplankton community also highlights that a warming environment might not always be detrimental for all ecosystems (Yvon-Durocher et al. 2015).

13.6.3 The El Niño/Southern Oscillation (ENSO) and the Pacific Decadal Oscillation (PDO)

ENSO and PDO are large-scale patterns of climate variability. Both ENSO and PDO have warm and cool phases, lasting for 6–18 months and 20–30 years, respectively (Mantua and Hare 2002). The warm phases are characterized by anomalously warm sea surface temperatures in the eastern and equatorial Pacific Ocean, enhanced stratification and reduced upwelling of nutrient-rich water along the eastern Pacific coast (Rasmusson and Carpenter 1982; Rasmusson and Wallace 1983).

Effect on phytoplankton: ENSO events influence the periodicity of *Pyrodinium bahamense* var. *compressum* blooms in Southeast Asia (Maclean 1989; Usup and Azanza 1998). A series of red tides in Hong Kong waters between mid-March and mid-April 1998, which caused US$32 million damage, has been linked with the high magnitude of El Nino in 1997–1978 (Yin et al. 1999).

13.7 Conclusion and Future Prospects

From the above overview of the impact of different types of pollution on phytoplankton, it is apparent that pollution affects phytoplankton not only through reduction in abundance but also dominance and succession strategies. The complexity of these responses is increased by several factors:

- Firstly, the influence of environmental fluctuations such as monsoons on the life cycle strategies of phytoplankton. Life cycle strategies of phytoplankton have evolved analogous to that observed in land plants (Cheke 2007). These are further influenced by monsoon systems that are especially relevant in tropical and subtropical environments (e.g. India). Yet, the exact mechanisms, by which environmental fluctuations regulate sexuality, formation and germination of resting stages, aggregation and sinking and deployment of defence mechanisms in phytoplankton, leading to selection of phytoplankton species, are still unknown (Smetacek and Cloern 2008).

- Secondly, the influence of climate change events on the behaviour of HAB species. These changes include range expansions/contractions of warm-water HAB species, spread of HAB species into new environments and shifts in the timing or magnitude of the algal blooms (Moore et al. 2009; Hallegraeff 2010; Stocker et al. 2013). Fossil records and long-term phytoplankton monitoring data indicate that climate warming in the future could influence HABs by changing their geographic range and shifts towards increased and earlier blooms (Dale et al. 2006). These will have implications for the *Vibrio* species (pathogens) associated with them, mainly *Vibrio cholerae* and *V. parahaemolyticus*. An increase in the extent and frequency of HABs may also increase the risk for *Vibrio* epidemics in coastal areas (Main et al. 2015).
- Thirdly, an important aspect to be considered is that, in natural environments, pollutants are encountered as complex mixtures, implying that pollutants may act synergistically on phytoplankton. Complex mixtures of organic pollutants have an important toxic and, possibly, synergistic effect on phytoplankton abundance, viability and concentrations of chlorophyll *a* at even trace concentrations (Echeveste et al. 2010). The toxicity of these complex mixtures of organic pollutants exceeds by 10^3 times the toxicity expected for a single pollutant. This explains why the concentrations of single pollutants, such as phenanthrene and pyrene, at which lethality of phytoplankton is apparent, are high compared to field concentrations (Echeveste et al. 2010).

Studies on these aspects are crucial in deciphering the impact of pollution on the base of food webs. Future research must focus on not only the impact of pollution on ecologically dominant and keystone species, but must also take into account community-level changes and the ecological mechanisms underlying these changes. This will provide an understanding of the ability of populations to survive the impact of pollution which may be markedly different from the response of individual phytoplankton species. Such analyses will be helpful in providing more accurate inferences about the cascading effects of pollutants across food webs and ultimately on fisheries, the backbone of many economies.

Acknowledgements The authors are grateful to the Harmful Algal Bloom Study Programme 'SHABASHI' funded by NF-POGO. PD acknowledges the Science and Engineering Board, Department of Science and Technology, India, for the SERB-DST Fast Track Project for Young Scientists (DST No: SB/YS/LS-319/2013). RN greatly acknowledges the support provided by the Director, ESSO-NCAOR, MoES, and Dr. Anilkumar (Group Director). This is NCAOR Contribution No. 14/2016.

References

Amin SA, Parker MS, Armbrust EV (2012) Interactions between diatoms and bacteria. Microbiol Mol Biol Rev 76(3):667–684

Anil AC, Venkat K, Sawant SS et al (2002) Marine bioinvasion: concern for ecology and shipping. Curr Sci 83(3):214–218

Assmy P, Henjes J, Klaas C, Smetacek V (2007) Mechanisms determining species dominance in a phytoplankton bloom induced by the iron fertilization experiment EisenEx in the Southern Ocean. Deep Sea Res I 54:340–362

Backer L, McGillicuddy D (2006) Harmful algal blooms. Oceanography 19(2):94–106

Baragi LV, Anil AC (2015) Interactive effect of elevated pCO_2 and temperature on the larval development of an inter-tidal organism, *Balanus amphitrite* Darwin (Cirripedia: Thoracica). J Exp Mar Biol Ecol 471:48–57

Baragi LV, Khandeparker L, Anil AC (2015) Influence of elevated temperature and pCO_2 on the marine periphytic diatom *Navicula distans* and its associated organisms in culture. Hydrobiologia 762:127–142

Boelen P, Van de Poll WH, Van der Strate HJ et al (2011) Neither elevated nor reduced CO_2 affects the photophysiological performance of the marine Antarctic diatom *Chaetoceros brevis*. J Exp Mar Biol Ecol 406:38–45

Boyd PW, Doney SC (2003) The impact of climate change and feedback processes on the ocean carbon cycle. In: Ocean biogeochemistry. Springer, Berlin, pp 157–193

Brain RA, Johnson DJ, Richards SM et al (2004) Effects of 25 pharmaceutical compounds to *Lemna gibba* using a seven-day static-renewal test. Environ Toxicol Chem 23:371–382

Caldeira K, Wickett ME (2003) Oceanography: anthropogenic carbon and ocean pH. Nature 425(6956):365

Campa-Cordova AI, Luna-Gonzalez A, Ascencio F et al (2006) Effects of chloramphenicol, erythromycin, and furazolidone on growth of *Isochrysis galbana* and *Chaetoceros gracilis*. Aquaculture 260(1):145–150

Cheke RA (2007) Thinking long term. Science 318(5850):577

Chuang YL, Yang HH, Lin HJ (2009) Effects of a thermal discharge from a nuclear power plant on phytoplankton and periphyton in subtropical coastal waters. J Sea Res 61(4):197–205

Colwell RK (2009) Biodiversity: concepts, patterns and measurement. In: Levin SA (ed) The Princeton guide to ecology. Princeton University Press, Princeton, pp 257–263

D'Costa PM (2010) Studies on phytoplankton-bacterial interactions. Ph.D. thesis. Goa University

D'Costa PM, Anil AC (2010) Diatom community dynamics in a tropical, monsoon-influenced environment: west coast of India. Cont Shelf Res 30(12):1324–1337

D'Costa PM, Anil AC (2012) The effect of antibiotics on diatom communities. Curr Sci 102:1552–1559

D'Costa PM, Anil AC, Patil JS et al (2008) Dinoflagellates in a mesotrophic, tropical environment influenced by monsoon. Estuar Coast Shelf Sci 77(1):77–90

D'Silva MS, Anil AC, Borole DV et al (2012a) Tracking the history of dinoflagellate cyst assemblages in sediments from the west coast of India. J Sea Res 73:86–100

D'Silva MS, Anil AC, Naik RK et al (2012b) Algal blooms: a perspective from the coasts of India. Nat Hazards 63(2):1225–1253

D'Silva MS, Anil AC, Sawant SS (2013) Dinoflagellate cyst assemblages in recent sediments of Visakhapatnam harbour, east coast of India: influence of environmental characteristics. Mar Pollut Bull 66(1):59–72

Dale B, Edwards M, Reid PC (2006) Climate change and harmful algal blooms. In: Ecology of harmful algae. Springer, Berlin, pp 367–378

Daufresne M, Lengfellner K, Sommer U (2009) Global warming benefits the small in aquatic ecosystems. Proc Natl Acad Sci U S A 106(31):12788–12793

Davies AG (1978) Pollution studies with marine plankton. II. Heavy metals. Adv Mar Biol 15:381–508

Davison IR (1991) Environmental effects on algal photosynthesis: temperature. J Phycol 27(1):2–8

de Stephanis R, Giménez J, Carpinelli E et al (2013) As main meal for sperm whales: plastics debris. Mar Pollut Bull 69(1):206–214

DeLorenzo ME, Scott GI, Ross PE (2001) Toxicity of pesticides to aquatic microorganisms: a review. Environ Toxicol Chem 20(1):84–98

DeLorenzo ME, Taylor LA, Lund SA et al (2002) Toxicity and bioconcentration potential of the agricultural pesticide endosulfan in phytoplankton and zooplankton. Arch Environ Contam Toxicol 42(2):173–181

Devinney JS (1980) Effects of thermal effluents on communities of benthic marine macroalgae. Environ Manage 11:225–242

Dolah FM, Roelke D, Greene RM (2001) Health and ecological impacts of harmful algal blooms: risk assessment needs. Hum Ecol Risk Assess 7(5):1329–1345

Duursma EK, Marchand M (1974) Aspects of organic marine pollution. Oceanogr Mar Biol 12:315–431

Echeveste P, Dachs J, Berrojalbiz N et al (2010) Decrease in the abundance and viability of oceanic phytoplankton due to trace levels of complex mixtures of organic pollutants. Chemosphere 81(2):161–168

Eguchi K, Nagase H, Ozawa M et al (2004) Evaluation of antimicrobial agents for veterinary use in the ecotoxicity test using microalgae. Chemosphere 57:1733–1738

Eppley RW, Renger EH, Williams PM (1976) Chlorine reactions with seawater constituents and the inhibition of photosynthesis of natural marine phytoplankton. Estuar Coast Mar Sci 4(2):147–161

Eriksen M, Lebreton LC, Carson HS et al (2014) Plastic pollution in the world's oceans: more than 5 trillion plastic pieces weighing over 250,000 tons afloat at sea. PLoS One 9(12):e111913

Fabry VJ, McClintock JB, Mathis JT et al (2009) Ocean acidification at high latitudes: the bellwether. Oceanography 22(4):160–171

Fistarol GO, Legrand C, Graneli E (2005) Allelopathic effect on a nutrient-limited phytoplankton species. Aquat Microb Ecol 41:153–161

Fu FX, Tatters AO, Hutchins DA (2012) Global change and the future of harmful algal blooms in the ocean. Mar Ecol Prog Ser 470:207–233

Gaidhani SV, Raskar AV, Poddar S et al (2014) Time dependent enhanced resistance against antibiotics and metal salts by planktonic and biofilm form of *Acinetobacter haemolyticus* MMC 8 clinical isolate. Indian J Med Res 140(5):665

Gaonkar CA, Krishnamurthy V, Anil AC (2010) Changes in the abundance and composition of zooplankton from the ports of Mumbai, India. Environ Monit Assess 168(1–4):179–194

GESAMP (IMO/FAO/UNESCO/WMO/WHO/IAEA/UN/UNEP Joint Group of Experts on the Scientific Aspects of Marine Pollution) (1991) Reducing environmental impacts of coastal aquaculture. Rep Stud GESAMP 47:35

Glibert PM, Anderson DM, Gentien P et al (2005) The global, complex phenomena of harmful algal blooms. Oceanography 18(2):136–147

Glibert PM, Harrison JO, Heil CY et al (2006) Escalating worldwide use of urea – a global change contributing to coastal eutrophication. Biogeochemistry 77:441–463

Granéli E, Johansson N (2003) Increase in the production of allelopathic substances by *Prymnesium parvum* cells grown under N-or P-deficient conditions. Harmful Algae 2(2):135–145

Gregory MR (1999) Plastics and South Pacific Island shores: environmental implications. Ocean Coast Manag 42(6):603–615
Hagenbuch IM, Pinckney JL (2012) Toxic effect of the combined antibiotics ciprofloxacin, lincomycin, and tylosin on two species of marine diatoms. Water Res 46:5028–5036
Hallegraeff GM (1993) A review of harmful algal blooms and their apparent global increase*. Phys Chem Chem Phys 32(2):79–99
Hallegraeff GM (2010) Ocean climate change, phytoplankton community responses, and harmful algal blooms: a formidable predictive challenge1. J Phycol 46(2):220–235
Halling-Sørensen B (2000) Algal toxicity of antibacterial agents used in intensive farming. Chemosphere 40:731–739
Halling-Sørensen B, Nielsen SN, Lanzky PF et al (1998) Occurrence, fate and effects of pharmaceutical substances in the environment – a review. Chemosphere 36(2):357–393
Harley CD, Randall Hughes A, Hultgren KM et al (2006) The impacts of climate change in coastal marine systems. Ecol Lett 9(2):228–241
Harper JL (1977) Population biology of plants. Academic, London
Harvell CD, Mitchell CE, Ward JR et al (2002) Climate warming and disease risks for terrestrial and marine biota. Science 296(5576):2158–2162
Hernandez M, Robinson I, Aguilar A et al (1998) Did algal toxins cause monk seal mortality? Nature 393(6680):28–29
HMSO (1990) Fourth report: pollution of beaches, vol 1. House of Commons Environment Committee, Her Majesty's Stationary Office, London
Howarth RW, Marino R (2006) Nitrogen as the limiting nutrient for eutrophication in coastal marine ecosystems: evolving views over three decades. Limnol Oceanogr 51:364–376
Hutchins DA, Mulholland MR, Fu FX (2009) Nutrient cycles and marine microbes in a CO_2-enriched ocean. Oceanography 22:128–145
Islam MS, Tanaka M (2004) Impacts of pollution on coastal and marine ecosystems including coastal and marine fisheries and approach for management: a review and synthesis. Mar Pollut Bull 48(7):624–649
James WG (1967) Mussel fouling and use of exomotive chlorination. Chem Ind 24:994
James KJ, Carey B, O'Halloran J et al (2010) Shellfish toxicity: human health implications of marine algal toxins. Epidemiol Infect 138(07):927–940
Jeffrey SW, Vesk M (1997) Introduction to marine phytoplankton and their pigment signatures. In: Jeffrey SW, Mantoura RFC, Wright SW (eds) Phytoplankton pigments in oceanography: guidelines to modern methods, Monographs on oceanographic methodology 10. UNESCO Publishing, Paris, pp 37–84
Johnson VR, Brownlee C, Rickaby RE et al (2013) Responses of marine benthic microalgae to elevated CO_2. Mar Biol 160(8):1813–1824
Joly C (1993) Plant nutrient management and the environment. Water Reports (FAO)
Jost C, Lawrence CA, Campolongo F et al (2004) The effects of mixotrophy on the stability and dynamics of a simple planktonic food web model. Theor Popul Biol 66(1):37–51
Joyce LB, Pitcher GC, Du Randt A et al (2005) Dinoflagellate cysts from surface sediments of Saldanha Bay, South Africa: an indication of the potential risk of harmful algal blooms. Harmful Algae 4(2):309–318
Kayser H (1976) Waste-water assay with continuous algal cultures: the effect of mercuric acetate on the growth of some marine dinoflagellates. Mar Biol 36(1):61–72
Kilham P, Kilham SS (1980) The evolutionary ecology of phytoplankton. In: Morris I (ed) The physiological ecology of phytoplankton. Blackwell, Oxford, pp 571–597
Kimor B (1992) The impact of eutrophication on phytoplankton composition in coastal marine ecosystems. Sci Total Environ 1099:871–878
Lai HT, Hou JH, Su CI et al (2009) Effects of chloramphenicol, florfenicol, and thiamphenicol on growth of algae *Chlorella pyrenoidosa, Isochrysis galbana*, and *Tetraselmis chui*. Ecotoxicol Environ Safe 72(2):329–334
Langford T (1990) Ecological effects of thermal discharges. Springer Science and Business Media, Heidelberg, Germany
Laubie L (2001) Climatic changes and trends in marine invertebrates: a need for relevant observing networks and experimental ecophysiology. Atti Associazione Ital Oceanologia Limnol 14:15–24
Long M, Moriceau B, Gallinari M et al (2015) Interactions between microplastics and phytoplankton aggregates: impact on their respective fates. Mar Chem 175:39–46
Lovelock J, Margulis L (1996) The Gaia Hypothesis. In: Scharper SB (ed) Redeeming the time: a political theology of the environment. A and C Black Publishers, London, UK
Maclean JL (1989) Indo-Pacific red tides, 1985–1988. Mar Pollut Bull 20(7):304–310
Main CR, Salvitti LR, Whereat EB et al (2015) Community-level and species-specific associations between phytoplankton and particle-associated *Vibrio* species in Delaware's inland bays. Appl Environ Microbiol 81(17):5703–5713
Mantua NJ, Hare SR (2002) The pacific decadal oscillation. J Oceanogr 58:35–44
Margalef R (1978) Life-forms of phytoplankton as survival alternatives in an unstable environment. Oceanol Acta 1:493–509
Martinez JL (2009) Environmental pollution by antibiotics and by antibiotic resistance determinants. Environ Pollut 157:2893–2902
Meehl GA, Stocker TF, Collins WD et al (2007) Global climate projections. Clim Chang 3495:747–845

Moore SK, Mantua NJ, Hickey BM, Trainer VL (2009) Recent trends in paralytic shellfish toxins in Puget sound, relationships to climate, and capacity for prediction of toxic events. Harmful Algae 8:463–477

Naik RK, Hegde S, Anil AC (2011) Dinoflagellate community structure from the stratified environment of the Bay of Bengal, with special emphasis on harmful algal bloom species. Environ Monit Assess 182(1–4):15–30

Nixon SW (1995) Coastal marine eutrophication: a definition, social causes, and future concerns. Ophelia 41:199–219

Obbard RW, Sadri S, Wong YQ et al (2014) Global warming releases microplastic legacy frozen in Arctic Sea ice. Earth's Future 2(6):315–320

Occhipinti-Ambrogi A (2007) Global change and marine communities: alien species and climate change. Mar Pollut Bull 55(7):342–352

Oliver RL, Ganf GC (2000) Freshwater blooms. In: Whitton BA, Potts M (eds) The ecology of cyanobacteria. Kluwer Academic Publishers, Dordrecht, pp 149–194

Orr JC, Fabry VJ, Aumont O et al (2005) Anthropogenic ocean acidification over the twenty-first century and its impact on calcifying organisms. Nature 437(7059):681–686

Paerl HW, Scott JT (2010) Throwing fuel on the fire: synergistic effects of excessive nitrogen inputs and global warming on harmful algal blooms. Environ Sci Technol 44(20):7756–7758

Pan Y, Rao DS (1997) Impacts of domestic sewage effluent on phytoplankton from Bedford Basin, eastern Canada. Mar Pollut Bull 34(12):1001–1005

Paterson DM (1997) Biological mediation of sediment erodibility: ecology and physical dynamics. In: Burt N, Parker R, Watts J (eds) Cohesive sediments. Wiley, Chichester, pp 215–229

Patil (2003) Studies on ecology of diatoms. Ph.D. thesis. Goa University

Pinckney JL, Hagenbuch IM, Long RA et al (2013) Sublethal effects of the antibiotic tylosin on estuarine benthic microalgal communities. Mar Pollut Bull 68:8–12

Poornima EH, Rajadurai M, Rao TS et al (2005) Impact of thermal discharge from a tropical coastal power plant on phytoplankton. J Thermal Biol 30(4):307–316

Porzio L, Buia MC, Hall-Spencer JM (2011) Effects of ocean acidification on macroalgal communities. J Exp Mar Biol Ecol 400(1):278–287

Ramaiah N, Ramaiah N, Nair VR (1998) Phytoplankton characteristics in a polluted Bombay harbour-Thana-Bassein creek estuarine complex. Indian J Mar Sci 27:281–285

Rasmusson EM, Carpenter TH (1982) Variations in tropical sea surface temperature and surface wind fields associated with the Southern Oscillation/El Niño. Mon Weather Rev 110(5):354–384

Rasmusson EM, Wallace JM (1983) Meteorological aspects of the El Nino/southern oscillation. Science 222(4629):1195–1202

Raven J, Caldeira K, Elderfield H et al (2005) Ocean acidification due to increasing atmospheric carbon dioxide. The Royal Society, London

Rehnstam-Holm AS, Godhe A, Härnström K et al (2010) Association between phytoplankton and *Vibrio* spp. along the southwest coast of India: a mesocosm experiment. Aquat Microb Ecol 58(2):127–139

Reynolds CS (1980) Phytoplankton assemblages and their periodicity in stratifying lake systems. Holarct Ecol 3:141–159

Riebesell U, Zondervan I, Rost B et al (2000) Reduced calcification of marine plankton in response to increased atmospheric CO_2. Nature 407(6802):364–367

Riegman R (1995) Nutrient-related selection mechanisms in marine phytoplankton communities and the impact of eutrophication on the planktonic food web. Water Sci Technol 32(4):63–75

Rose FL, McIntire CD (1970) Accumulation of dieldrin by benthic algae in laboratory streams. Hydrobiologia 35(3–4):481–493

Rosenberg R, Nilsson HC (2005) Deterioration of soft-bottom benthos along the Swedish Skagerrak coast. J Sea Res 54(3):231–242

Round FE (1965) The epipsammon: a relatively unknown freshwater algal association. Br Phycol Bull 2:456–462

Round FE (1979) Occurrence and rhythmic behaviour of *Tropidoneis lepidoptera* in the epipelon of Barnstable Harbor, Massachusetts, USA. Mar Biol 54:215–217

Russel FE, Kotin P (1956) Squamous papilloma in the white croaker. J Nat Cancer Inst 18:857–861

Safina C (2013) No refuge: tons of trash covers the remote shores of Alaska. Yale Environment 360

Seidl M, Huang V, Mouchel JM (1998) Toxicity of combined sewer overflows on river phytoplankton: the role of heavy metals. Environ Poll 101(1):107–116

Seltenrich N (2015) New link in the food chain? Marine plastic pollution and seafood safety. Environ Health Perspect 123(2):A34

Sett S, Bach LT, Schulz KG et al (2014) Temperature modulates coccolithophorid sensitivity of growth, photosynthesis and calcification to increasing seawater pCO_2. PLoS One 9(2):e88308

Smetacek V, Cloern JE (2008) On phytoplankton trends. Science 319:1346–1348

Stange K, Swackhamer DL (1994) Factors affecting phytoplankton species-specific differences in accumulation of 40 polychlorinated biphenyls (PCBs). Environ Toxicol Chem 13(11):1849–1860

Stocker T, Qin D, Plattner G et al (2013) IPCC, 2013: climate change 2013: the physical science basis. Contribution of working group I to the fifth assessment report of the intergovernmental panel on climate change

Strauss SD, Puckorius PR (1984) Cooling-water treatment for control of scaling, fouling, corrosion. Power 128(6):S1–S24

Taslakian MJ, Hardy JT (1976) Sewage nutrient enrichment and phytoplankton ecology along the central coast of Lebanon. Mar Biol 38(4):315–325

Thingstad TF, Havskum H, Garde K, Riemann B (1996) On the strategy of "eating your competitor": a mathematical analysis of algal mixotrophy. Ecology 77:2108–2118

Thomas TE, Robinson MG (1987) The role of bacteria in the metal tolerance of the fouling diatom *Amphora coffeaeformis* Ag. J Exp Mar Biol Ecol 107:291–297

Thompson GB, Ho J (1981) Some effects of sewage discharge upon phytoplankton in Hong Kong. Mar Pollut Bull 12(5):168–173

Tomas CR (ed) (1997) Identifying marine phytoplankton. Academic, San Diego, pp 1–858

Torstensson A, Chierici M, Wulff A (2012) The influence of increased temperature and carbon dioxide levels on the benthic/sea ice diatom *Navicula directa*. Polar Biol 35:205–214

Usup G, Azanza RV (1998) Physiology and bloom dynamics of the tropical dinoflagellate *Pyrodinium bahamense*. In: Anderson DM, Cembella AD, Hallegraeff GM (eds) Physiological ecology of harmful algal blooms, vol G41. Springer, Berlin, pp 81–94

Williams C (1996) Combating marine pollution from land-based activities: Australian initiatives. Ocean Coast Manag 33(1–3):87–112

Winston JE, Gregory MR, Stevens LM (1997) Encrusters, epibionts, and other biota associated with pelagic plastics: a review of biogeographical, environmental, and conservation issues. In: Marine debris. Springer, New York, pp 81–97

Wollenberger L, Halling-Sørensen B, Kusk KO (2000) Acute and chronic toxicity of veterinary antibiotics to *Daphnia magna*. Chemosphere 40:723–730

Yamamoto T, Hatta G (2004) Pulsed nutrient supply as a factor inducing phytoplankton diversity. Ecol Model 171:247–270

Yin K, Harrison PJ, Chen J et al (1999) Red tides during spring 1998 in Hong Kong: is El Niño responsible? Mar Ecol Prog Ser 187:289–294

Yvon-Durocher G, Dossena M, Trimmer M, Woodward G, Allen AP (2015) Temperature and the biogeography of algal stoichiometry. Glob Ecol Biogeogr 24(5):562–570

Selenium Pollution in the Marine Environment and Marine Bacteria in Selenium Bioremediation

14

Lakshangy S. Charya

Abstract

Selenium (Se), a metalloid, is a micronutrient essential to biological systems at lower concentrations but becomes toxic as the level increases. Among the soluble forms of Se, selenite is more toxic than selenate to most living organisms. Selenium pollution is a global phenomenon and is associated with a wide range of human activities, from basic agricultural practices to the modern industrial processes that increase the threat of widespread impacts to aquatic life. Soluble forms of selenium, being mobile, reach groundwaters, whereas other static forms remain in soils. Aquatic organisms living in waters contaminated with Se or wildlife consuming plants from selenium-rich soils may be harmed as they accumulate a level of selenium higher than required by their bodies. Although selenium may prove very risky, resulting in long-term serious effects on aquatic life and fishery resources, selenium contamination in the aquatic environment often goes unnoticed by environmental biologists. The permissible level of total selenium in the aquatic environment is about 2 μg/l. To avoid adverse effects on marine aquatic life, the drainage water should be treated to minimize selenium content before it flows into rivers. The conventional physicochemical methods employed in selenium removal, although effective, may prove to be quite expensive. Recent studies suggest the use of microbiological resources to detoxify selenium to be the most simple and economical method. Science is advancing with newer approaches to tackle this problem of selenium pollution.

14.1 Introduction

Heavy metals or metalloids are natural constituents of the Earth crust, but indiscriminant human activities have drastically altered their geochemical cycles and biochemical balance (Duruibe

L.S. Charya (✉)
Department of Microbiology, Goa University, Taleigão, India
e-mail: lakshangy@yahoo.com

M.M. Naik, S.K. Dubey (eds.), *Marine Pollution and Microbial Remediation*,
DOI 10.1007/978-981-10-1044-6_14

et al. 2007). Any metal (or metalloid) species is regarded as a "contaminant" if it occurs in a form or concentration that is not required or causes ill effects on humans or the environment. Metals/metalloids that are persistent environmental contaminants include lead (Pb), cadmium (Cd), mercury (Hg), arsenic (As), chromium (Cr), copper (Cu), selenium (Se), nickel (Ni), silver (Ag), tin (Sn), tellurium (Te), and zinc (Zn) (Singh et al. 2011). These metal species are notable contaminants because they persist in the environment for a long time and accumulate easily in living organisms to a toxic level (Zhuang and Gao 2013; Coelho et al. 2015). Metals/metalloids are continuously released into the biosphere both by natural causes such as volcanoes and natural weathering of minerals and rocks, and by anthropogenic activities including mining, combustion of fuels, industrial and urban sewage, and agricultural practices (Singh et al. 2011). These mining activities and the discharge of industrial and urban sewage are the main causes of metal pollution in the marine environment (Duruibe et al. 2007). Heavy metals are emitted in both elemental and compound (organometallic and inorganic) forms. Metals from mining areas are leached out and carried by water downstream or as runoff to the sea. Occurrence of these trace elements, as pollutants or contaminants causing significant hazards to life forms in various ecosystems, has motivated governments all over the globe to initiate different preventive measures to tackle the environmental problems arising from these elements (McIntyre 2003).

Living systems require heavy metals in varying amounts. Some heavy metals, such as calcium (Ca), cobalt (Co), iron (Fe), magnesium (Mg), manganese (Mn), molybdenum (Mo), Ni, Se, Cr, Cu, and Zn, are known to be biologically important to humans, and their allowance in the day-to-day diet and in medicines has been recommended (Lloyd 2003; Singh et al. 2011; Coelho et al. 2015). However, some heavy metals such as As, Cd, Pb, and methylated forms of Hg that do not have any significance in human biochemistry and physiology can exhibit toxicity even when consumed at very low concentrations (Nolan 2003; Young 2005; Duruibe et al. 2007). Even those metals that have biological significance should be taken in the diet at recommended concentrations because excess intake could result in poisoning or toxicity and could be detrimental to the organisms (Singh et al. 2011). An inadequate supply of these micronutrients results in a variety of deficiency diseases or syndromes. In biological systems, the essential metals are significant as components of several key enzymes and take part in redox reactions. Arsenic, chromium, and mercury have been reported to affect cell organelles and components such as the cell membrane, endoplasmic reticulum, lysosome, mitochondria, nuclei, and enzymes involved in metabolism and damage repair. Metal ions also interact with DNA and cellular proteins, which causes damage to these biomolecules (Duruibe et al. 2007; Tchounwou et al. 2012).

Emphasizing the occurrence of metals/metalloids as major marine pollutants and their devastating effects on living forms, the 1956 episode of mercury poisoning (originally called Minamata disease) in Japan is a case in point that provides a glimpse into the effects of bioconcentration of methylmercury in the food web and its eventual impacts on human beings (Harada 1995; Duruibe et al. 2007).

Selenium was highlighted as being an essential element for the first time in 1957 (Mayland 1994). It forms a key component of several functional selenoproteins in living organisms, except for yeasts and higher plants (Duruibe et al. 2007; Hesketh 2008; Janz et al. 2010). The chemical form of Se that exists in the environment is of major concern because it exhibits a dual function of being a biologically important element and a potential toxicant that pollutes the environment (Janz et al. 2010). Immediate understanding about its reactivity in biological systems is primary as it fills has significant biological roles. Extensive use of this element has resulted in its accumulation in the environment, creating a threat to living organisms. Selenium pollution of aquatic systems results from the discharge of waste materials from certain mining, agricultural, petrochemical, and industrial manufacturing operations (Lemly 2004). Various methods are employed to remove these toxicants from the ecosystem because of environmental concerns. Bioremediation using microbes, green plants, and enzymes are eco-friendly, cost-effective, and an efficient alternative to target heavy metals and metalloids. The interest in metal transformations mediated by microorganisms is on the rise as these processes are important in the cycling of both organic and inorganic metal species in a range of environments, thus forming the basis for applications of innovative processes in the field of biotechnology (Singh and Prasad 2015).

14.2 Selenium

Selenium, a metalloid, is an essential trace element for living organisms. However, it can have adverse effects on human health as the margin that distinguishes between its toxicity and its beneficence is narrow. Moreover, its distribution in the environment is not uniform, and therefore it is called a 'double-edged sword' element (Vriens et al. 2014; Fernández-Martínez and Charlet 2009). Selenium has been identified as a toxic element for a long time; for example, selenium toxicity causes selenosis, which includes symptoms (in humans) such as garlic odor on the breath, gastrointestinal disorders, hair loss, sloughing of nails, fatigue, irritability, neurological damage, cirrhosis of the liver, pulmonary edema, and eventually death. The significance of selenium was highlighted in 1957. Selenium is an important structural component of many enzymes such as thioredoxin reductase, glutathione peroxidase, and deiodinases. These enzymes take part in various biological processes such as antioxidation, reproduction, muscle function, and tumor prevention (Frakenberger and Benson 1994; Mehdi et al. 2013).

14.2.1 Chemical Properties of Selenium

Selenium is the 67th most abundant element in the Earth's crust with a concentration of 50 ppb. Its concentration in soil is 5 ppm and in seawater 0.2 ppb. Selenium belongs to group 16 and period 4; its atomic number is 34. The electronic configuration of selenium is [Ar] 4s23d104p4. It is present in nature in various oxidation states: 0, −2, +4, +6 (UNEP and GESAMP 1988; Mehdi et al. 2013). It occurs naturally in six isotopic forms, ^{80}Se being the most abundant (49.82 %) and ^{77}Se the least abundant (7.58 %). The boiling and melting points of selenium are 685 °C and 220.5 °C, respectively. It belongs to the family of chacolegens. Selenium resembles sulfur in terms of atomic size, bond energies, ionization potentials, and main oxidation states; it can combine with other elements such as bromine, chlorine, fluorine, hydrogen, and phosphorus and thus forms compounds similar to those of sulfur. It is stable and does not oxidize at ordinary temperatures. Selenium in the atmosphere usually exists in its methylated forms such as dimethyl selenide (DMSe, CH_3SeCH_3) or dimethyl diselenide (DMDSe, $CH_3SeSeCH_3$) with a concentration of 1 ng/m^3 of air. Selenium dioxide is formed by the combustion of selenium in air. Selenium is present in nature and in organisms in organic or inorganic forms. The major organic forms of selenium are selenomethionine (Semet) and selenocysteine (Secys); the inorganic forms are selenite (SeO_3^{-2}), selenide (Se^{2-}), selenate (SeO_4^{-2}), and the selenium element (Se^0) (Martens 2003; Mehdi et al. 2013).

14.2.2 Uses of Selenium

Selenium is widely used commercially in pigments and glassmaking. Cadmium sulfide selenide is used in an array of pigments in ceramics, paints, and plastic. Because of its photoelectric and semiconducting properties, selenium has been widely used in the electronic and electrical industries. Selenium is used in a few types of DC power surge protectors and a type of fluorescent quantum dots (Li et al. 2011). It is used in the photoreceptor drums of photocopiers (xerograghy). Selenium is used in laser printers, solar photovoltaic cells, and X-ray machines, in vulcanization of natural and synthetic rubber, and is a major by-product of copper refineries (UNEP and GESAMP 1988).

Selenium has wide applications in agriculture and horticulture. Sodium selenite and selenate are used as additives and dietary supplements in animal feed. It is also used by pharmaceutical companies in manufacturing of dietary supplements (Reilly 2006). Selenium and its compounds are used in a range of products: selenium sulfide in antidandruff shampoos and in fungicides, and radioactive selenium as an aid in medicine for diagnosis and visualization of certain malignant tumors.

Selenium salts in trace amounts are required for cellular function in different organisms; however, at higher concentration they could be toxic. Selenium also forms an ingredient in many multivitamins and other dietary supplements, including infant formula (Reilly 2006). Selenium is a component of some biologically important enzymes in plants and animals (Mehdi et al. 2013). Selenium requirements in plants vary from species to species (Fairweather-Tait et al. 2010).

The recent advances in the field of nanotechnology have made possible the synthesis of various metals and metalloids in nano-size particles, which because of their size exhibit unique properties that are exploited in various fields. Selenium nanoparticles exhibit various properties such as catalytic activity toward hydration and oxidation reactions, relatively low melting point, high photoconductivity, with high piezoelectric, thermoelectric, and nonlinear optical responses. It has great importance in commercial applications that range from copying machines to electrical rectifiers (De Jong and Borm 2008). Because of their unique properties, selenium nanoparticles act as photovoltaic and semiconductor, antioxidant, and chemoprotective agents. Selenium has also been shown to inhibit the growth of *Staphylococcus aureus* and therefore could be used in the field of medicine (Tran and Webster 2011).

14.2.3 Role of Selenium in the Body

Selenium is required as a trace element, and its lack causes ill effects in human health. Selenium serves as an important component of selenoproteins that take part in numerous biological functions, such as defense as an antioxidant and the synthesis of DNA and thyroid hormones, and also serve in fertility and reproduction (Rayman 2000; Ujang Tinggi 2003, 2008; Mehdi et al. 2013). Both organic and inorganic selenium compounds are readily metabolized in the body. Some of the important selenoproteins include glutathione peroxidase, deiodinase-I, selenoprotein-P, and thioredoxin reductase. Glutathione peroxidase (antioxidant enzyme) neutralizes the organic hydro- and hydrogen peroxides in the intracellular and extracellular compartments. Deiodinase-I is present in the thyroid, kidneys, liver, and brown fat and is important in thyroid hormone metabolism that converts inactive thyroxine into active 3,3′,5′-triiodothyronine. Selenoprotein-P (glycoprotein) aids in homeostasis and the transport of selenium in tissues and also acts as an extracellular antioxidant that is known to eliminate peroxynitrite (Fairweather-Tait et al. 2010). Thioredoxin reductase acts as an antioxidant and regulates intracellular redox potential; it also helps reduce thioredoxin level, stimulates DNA synthesis, and inhibits apoptosis (Mehdi et al. 2013).

Besides selenoproteins, selenium is also known to have importance in the immune system and reproduction. Selenium is found to be present in large amounts in the spleen, liver, and lymph nodes and is involved in stimulation of antibody formation and activity of helper T cells, cytotoxic T cells, NK cells, and phagocytic cells (Burk 1994; Finch and Turner 1996; Mehdi et al. 2013). Studies by Davis et al. have revealed the involvement of different selenoproteins in cancer prevention, and selenium deficiency may be proven as a cancer-promoting factor (Davis et al. 2012). Many studies have also highlighted the importance of selenium in human and animal reproduction and fertility, placenta retention, embryonic implantation, synthesis of testosterone and sperm, and sperm mobility (Mistry et al. 2012).

14.2.4 Selenium Toxicity

The trace element selenium occurs naturally in water, soils, and living organisms, including foods. It forms an important part of nutrition as several Se-bearing proteins incorporate selenocysteine, recently known as the 21st essential amino acid. However, excess uptake of selenium is toxic to biota. The amount of selenium that could be beneficial or lead to toxicity depends on the species of an individual and the conditions in which it is grown. Cases of selenium toxicity in humans and crop damage are rare: toxicity in aquatic biota including fishes and birds is the major concern (Higashi et al. 2005). A popular example indicating the toxic effect of selenium is the aquatic bird disaster at California's Kesterson Reservoir in the early 1980s (Ohlendorf 2002) where selenium toxicity caused the death and deformities of embryos and chicks. Other adverse effects of selenium pollution on wild aquatic birds that have resulted from the discharge of subsurface agricultural drain water and other sources in the aquatic environment include reduced growth, impaired reproduction with teratogenesis, histopathological lesions, and altered hepatic glutathione metabolism (Spallholz and Hoffman 2002).

The toxic impacts of selenium to 20 species of a fish community inhabiting Belews Lake, North Carolina, that was contaminated with wastewater from a coal-fired power plant, caused chronic selenium poisoning. The symptoms in fish include telangiectasia (swelling) of gill lamellae, reduced hematocrit and hemoglobin (anemia), elevated lymphocytes, teratogenic deformities of the head, mouth, fins, and spine, corneal cataracts, exopthalmus (popeye), reproductive failure (ovarian pathology leading to decreased production of viable eggs, and mortality of eggs in post-hatch period as a result of Se bioaccumulation), and pathological alterations in heart, kidney, liver, and ovary (e.g., vacuolization of parenchymal hepatocytes, intracapillary proliferative glomerulonephritis, severe pericarditis and myocarditis, and necrotic and ruptured mature egg follicles) (Lemly 2002).

In the aquatic habitat, although the wild biota is directly exposed to the water, selenium toxicity chiefly occurs through the diet. Table 14.1 summarizes the safe level of selenium in various aquatic ecosystems and water used for various applications (Nagpal, Lands and Parks 2001). The acceptable level of selenium in food is about 5 mg/kg, and in seleniferous areas feeds with Se concentrations above this value are considered toxic, the uptake of which may lead to chronic intoxication. Selenium in drinking water is limited to 0.01 ppm. Selenium is vital to humans as various enzymes containing selenium are important in maintaining the health of an individual, such as glutathione peroxidase, that is present in most body tissues and catalytically degrades hydrogen peroxide (Parkman and Hultberg 2002).

The various compounds and forms of selenium have different behavior in the environment. Soluble forms can be easily transportable, may seep and reach groundwaters, whereas other forms that are comparatively immobile may persist in soils. The aquatic biota living in waters contaminated with selenium or plants growing in Se-rich soils can accumulate the element to high

Table 14.1 Summary of water quality guidelines for selenium

Water use	Guideline for total selenium
Drinking water	10 μg/l maximum
Aquatic life (freshwater)	2.0 μg/l mean
Aquatic life (marine)	2.0 μg/l mean
Aquatic life (sediments)	2.0 μg/g (dry weight) mean
Aquatic life (tissue)	1.0 μg/g body weight (wet weight mean)
Wildlife	4.0 μg/l mean
Irrigation	10.0 μg/l mean
Livestock watering	30.0 μg/l mean

Source: Nagpal, Lands and Parks (2001)

levels. Wildlife feeding on these plants would exceed the required concentration of selenium and may undergo the effects of severe intoxication.

There are different modes by which selenium can enter the body. Selenium can enter through ingestion of selenium containing food or water, contacting the skin with selenium and its compounds, and inhalation of air holding selenium. The degree of toxicity of selenium in water and food depends on the length of exposure and the amount ingested. Intoxication from inhalation of selenium in air is found mostly in workers because of their job-related activities. Symptoms of selenium intoxication include fatigue, dizziness, and irritation of the digestive tract, and in extreme cases may lead to accumulation of fluid in the lungs and bronchitis. Continuous ingestion of a large amount of selenium for an extended period may cause tooth decay, brittle hair, deformed nails, depression, and loss of feeling in the arms and legs, and in extreme conditions can lead to death. Skin that comes in contact with selenium compounds might become red, followed by development of rash, swelling, and pain. Contact with the eyes can cause irritation and burning.

Humans and animals encountering selenium exposure are reported to develop a number of systemic effects. However, knowledge about the different mechanisms of intoxication exerted by selenium and its compounds is very limited. Selenium is known to replace sulfur in biomolecules, specifically when the selenium concentration exceeds that of sulfur in the organism. The most important indicators of chronic selenium toxicity include hair, nail, and skin damage. The International Agency for Research on Cancer has concluded that selenium and its compounds are not carcinogenic to humans. Selenium exposure up to the permissible level is not known to exhibit any adverse effect on human health. On the other hand, efforts are being made to meet the selenium requirement of living beings to prevent the effects of Se deficiency. In Finland, the selenium intake in cattle and humans has been reported as being very low and, as a corrective measure all over the country, selenite was added to fertilizers so that the intake of selenium in diet is increased (Parkman and Hultberg 2002). In Sweden, lakes with fish accumulating high concentrations of Hg are being treated with selenium. Certain areas where the soil is rich in selenium (e.g., seleniferous soils) are recognized as numerous incidents of selenium poisoning in animals have been reported. Besides diarrhea, excessive salivation, shallow breathing, and garlic odor of the breath, the other clinical signs of severe selenium poisoning in animals include dysponea, vomiting, tetanic spasms, and death from respiratory failure.

Selenium has a very narrow range for its concentration that is considered optimal for intake in healthy individuals as this element exhibits dual properties of being both essential and toxic. Whether selenium is deficient or in excess, it can lead to endemic disease. Plants and animals found in soils with low selenium concentration develop selenium deficiency (and vitamin E deficiency) symptoms such as liver necrosis in swine and rats, exudative diathesis in chickens, and muscular dystrophy in sheep and cattle. Signs specific for selenium deficiency in the absence of vitamin E deficiency include pancreatic degeneration in chicks, and poor growth, reproductive failure, vascular changes, and cataracts in rats.

Worldwide there are very few selenium deposits and thus it considered as a rare element. Selenium is largely obtained as a by-product of the electrolytic refining of copper, and Canada is one of the leading selenium producers in the world. In natural waters selenium concentration varies widely but usually it is low. However, in seleniferous areas, the Se concentration is high and may exceed acceptable limits (Nagpal, Lands and Parks 2001).

14.2.5 Occurrence of Selenium in the Environment

Selenium is present in the environment as selenate/selenite oxyanions, wherein the oxidation states are +6 and +4, respectively: elemental selenium and selenide (Dhanjal and Cameotra 2010). The oxyanions of selenium are toxic at high concentrations, with selenite being the most toxic. In soils, selenium is mostly present in the form of elemental selenium and selenate: selenite oxyanions, such as selenate salts and ferric selenite or in its organic form. Selenate salts in soil are easily taken up by plants (Reilly 2006). Vegetables such as turnips, peas, beans, carrots, tomatoes, beets, potatoes, and cucumbers contain a maximum of 6 mg g^{-1} of selenium (Mehdi et al. 2013). Selenium is also found in water, and it originates from atmospheric deposits or soil drainage and subsoils containing a high level of selenium.

In the aquatic environment Se is adsorbed to particulate and colloidal materials (UNEP and GESAMP 1988), and in marine waters, three main species of dissolved selenium exist as selenite, selenate, and organic selenide. The organic selenide (seleno-amino acids in peptides) exist maximally because of primary productivity, bioluminescence, dissolved free amino acids, and pigments (Cutter and Bruland 1984).

Selenium in water bodies is seen to accumulate in aquatic food webs to such an extent that the top predatory aquatic and aquatic-dependent organisms are mainly exposed to selenium while obtaining their diet (Keith 2002). Some marine animals are seen to bioaccumulate much greater concentrations of Se than reported in freshwater species (Muir et al. 1999). Many taxa from bacteria to brine shrimp (*Artemia* spp.) to marine mammals to seabirds show high concentrations of Se (Brix et al. 2004; Dietz et al. 2000; Oremland et al. 2004). Selenium is known to be important in salt tolerance. Bivalves or marine vertebrates enzymatically fix selenium as components of organic osmolytes. Marine birds differ from freshwater birds in the way in which they partition selenium among different tissues. Some marine birds accumulate relatively high concentrations of selenium in their livers, which might have consequences from toxic effects (Janz et al. 2010).

In the United States (US), two anthropogenic factors that are responsible for releasing high levels of selenium and its mobilization in the aquatic systems include the procurement, processing, and combustion of fossil fuels and the irrigation of seleniferous soils for crop production in arid and semiarid regions of the country. Coal fly ash from power stations becomes distributed on land, and selenium seeps from these ashes into water bodies and eventually accumulates in aquatic systems. Also, deposits of Cretaceous marine shales have been observed to weather, producing high-Se soils in many areas of the western US (Lemly 1993a).

14.3 Microbial Selenium Cycle in Nature

In nature selenium is found commonly in four inorganic oxidation states. Selenate (+6), the most oxidized form of selenium, and selenite (+4) are highly water soluble and are known to be toxic to biological systems at relatively low concentrations (parts per million). Elemental selenium (+0) is highly insoluble in water and therefore has minimum toxicity. The most reduced form of selenium is selenide (−2), which may be a highly toxic gas but is seldom a biological threat because it is readily oxidized to insoluble elemental selenium in the presence of air. Groundwater contamination by selenium occurs primarily a result of its solubilization and migration from soil by chemical, physical, and biological processes. In the mid-1980s, the mass death of wildlife in the former Kesterson Wildlife Refuge was reported to occur from naturally occurring selenium in runoff from irrigated farm soils (Ohlendorf 2002).

The selenium cycle includes different organisms such as bacteria, fungi, and plants that act on selenium and its compounds to complete the cycle. These organisms reduce the most oxidized form of the element, and certain other organisms oxidize the reduced element to the initial state. In the selenium cycle it has been observed that *Astragalus* species metabolize the most oxidized forms of selenium, selenate or selenite, to selenide, and certain microorganisms might be involved in oxidation of elemental selenium to selenite (Maiers et al. 1988).

An array of microorganisms is reported to possess the ability to oxidize or reduce a wide variety of selenium-containing compounds (Levine 1924). Microbial genera observed to have this ability include *Acinetobacter*, *Aeromonas*, *Arthrobacter*, *Bacillus*, *Candida*, *Cephalosporium*, *Citrobacter*, *Corynebacterium*, *Flavobacterium*, *Fusarium*, *Micrococcus*, *Neurospora*, *Penicillium*, *Pseudomonas*, *Salmonella*, *Scopulariopsis*, and *Selenomonas*. Although microbial reduction of selenite to insoluble elemental selenium and selenide has been widely reported, reports of selenate being reduced to selenite, elemental selenium, or selenide are less numerous. The reported extent of reduction is lower for selenate than for selenite. Microbial reduction of elemental selenium to selenide and oxidation of elemental selenium to selenite and selenate have also been reported (Maiers et al. 1988). Parulekar and Naik (2015) has isolated selenite-reducing bacteria from Mandovi estuary in Goa in an attempt to use these isolates for selenium bioremediation (Fig. 14.1) (Parulekar and Naik 2015).

Various enzymatic systems have been proposed to catalyze the reduction of selenite in bacteria. In *Thauera selenatis*, the reaction might be catalyzed by a periplasmic dissimilatory nitrite reductase (DeMoll-Decker and Macy 1993). Also, intracellular selenite reduction can be driven by reduced thiols, such as glutathione, in microorganisms. Selenite reacts with glutathione to form selenodiglutathione (GSSe-SG), which can be further reduced by NADPH to unstable selenopersulfide (GS-Se-) in the presence of glutathione reductase. Then, dismutation of GS-Se- will produce GSH and Se(0) (Li et al. 2014).

In semiarid seleniferous soils, the plants that acuumulate selenium form a considerable part of the selenium cycle. When these plants die, they release various organic selenium compounds that they have synthesized into the soils as they decompose. In an aquatic ecosystem three fates are possible for dissolved selenium: either it can persist in free solution, or it can bind with suspended solids or sediments, or it can be consumed and absorbed by organisms. Over time, the maximum amount of selenium is ingested by organisms or bound to other solids, and gradually the selenium becomes concentrated in the top layer of sediment as the suspended material settles. However, because the flow in an aquatic ecosystem is dynamic, the presence of selenium in the sediments is merely temporary, and it is continuously cycled in the system.

In 1989, an important discovery about the biogeochemical movement of selelenium into the soil was reported. Anaerobic microorganisms were found to reduce selenate and selenite, the two toxic oxyanions of selenium, into insoluble elemental selenium and hydrogen selenide (Fig. 14.2). These microbially mediated reactions effectively removed these toxicants from the solution. This phenomenon was found extensively in the environment, and the rates of "dissimilatory reduction" were rapid as compared to reoxidation. Several new species of anaerobic

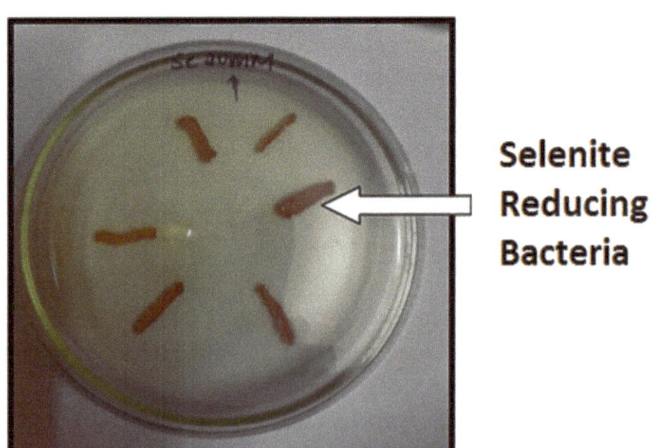

Fig. 14.1 Selenite-reducing bacteria isolated from Mandovi Estuary, Goa (Reproduced from Parulekar and Naik 2015)

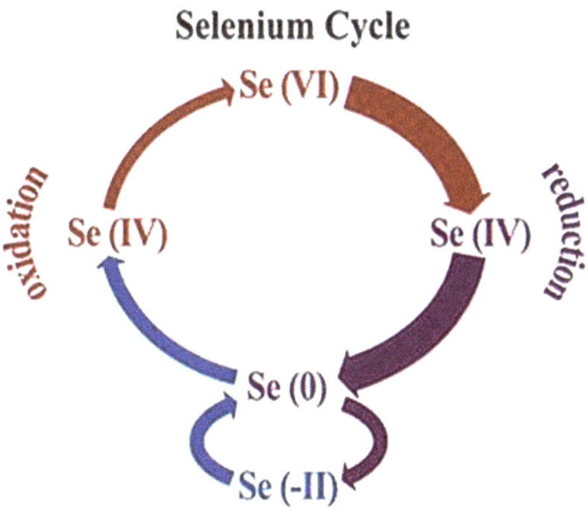

Fig. 14.2 The selenium cycle in nature demonstrating the anaerobic reduction of selenate through selenite to elemental selenium and then to selenide. The *arrows* on the right side were drawn thicker to indicate that the reaction rates of dissimilatory reduction are orders of magnitude more rapid than bacterial oxidation (Reproduced from Dowdle and Oremland 1998; Herbel et al. 2003)

microorganisms were isolated and described that could grow by using either selenate or selenite as their respiratory electron acceptor for the oxidation of organic carbon substrates such as lactate to acetate and carbon dioxide (Blum et al. 1998).

Biogeochemical cycling of selenium in aquatic ecosystems is summarized in Fig. 14.3, which illustrates the movement of selenium from contaminated water and sediment to its way up: it is seen that there are different interrelated paths for selenium. Most of the paths eventually lead up the food web, and during its mobilization the chemical form of selenium changes from an inorganic to different organic forms. Some of the Se (organic forms) is lost from the aquatic system by the process of volatilization (Higashi et al. 2005).

14.4 Bioremediation

Selenium pollution is a global phenomenon and is associated with a wide range of human activities, from basic agricultural practices to the modern industrial processes that increase the threat of widespread impacts to aquatic life. Human activities that increase waterborne concentrations of selenium are increasing. Important sources of selenium contamination in aquatic habitats are often overlooked by environmental biologists and ecological risk assessors because of preoccupation with other higher priority pollutants, yet selenium may pose the most serious long-term risk to aquatic habitats and fishery resources (Lemly 2004).

Interest in pollution of the environment by selenium and the different remediation strategies has escalated during the past three decades. Selenium is known to be an essential micronutrient for animals and humans, although it does not have much function in plant nutrition. However, at elevated concentrations, selenium becomes toxic to animals and humans. A major outbreak of selenium intoxication was recognized in the 1980s in California at the Kesterson National Wildlife Refuge; hence, the entire nation became aware of the potential of selenium as an environmental pollutant. Consequently, the major focus was to develop management strategies and remediation technologies for minimizing the impact of naturally occurring selenium on the biological environment (Banuelos et al. 2002).

Scientific community handling of selenium remediation suggests that organisms resistant to elevated levels of soluble selenium exist and are capable of reducing selenate in concentrations as high as 100 mg/l to less oxidized forms. Although some physicochemical techniques for the removal of selenate are available, they are inefficient at present. Efficient treatment methods currently can remove selenium from water only if the selenium occurs as selenite. The potential exists to use adapted indigenous bacteria in a treatment system that removes soluble and toxic forms of selenium from water: this may be accomplished by allowing microorganisms to reduce selenate to selenite and precipitated selenium, using chemical treatments to precipitate any remaining selenite, and physically filtering the precipitate from the water (Maiers et al. 1988).

The permissible level of total selenium in the aquatic environment is about 2 μg/l. To avoid adverse effects on marine aquatic life, the drainage water should be treated to minimize Se

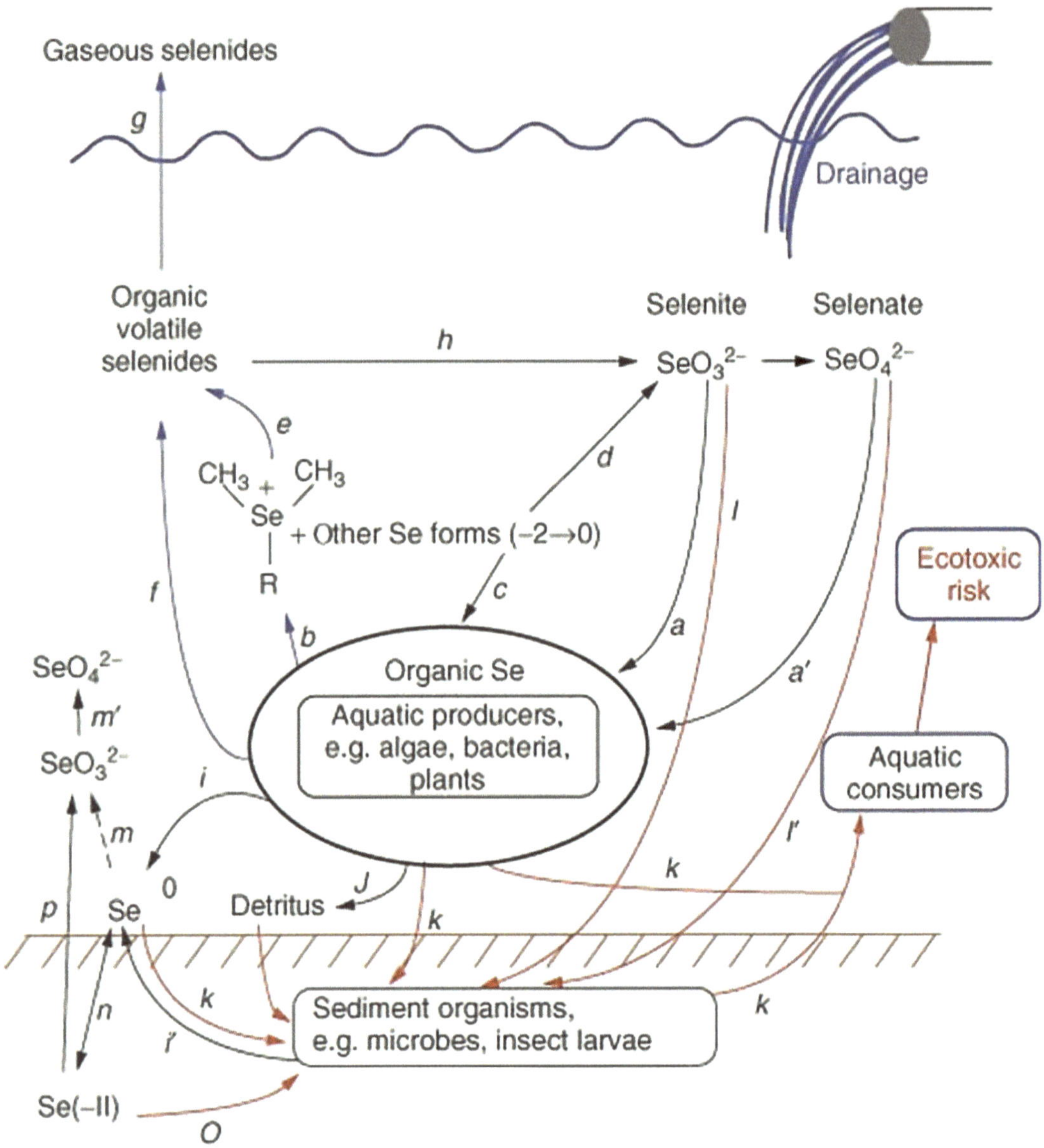

Fig. 14.3 Biogeochemical cycling of Se in aquatic ecosystem. (*a*) Uptake and transformation of Se oxyanions by aquatic primary and secondary producers; much of the biotransformation pathway is yet to be defined. (*b*) Release of selenonium and other organic Se metabolites by aquatic producers. (*c*) Uptake of organic Se compounds by aquatic producers. (*d*) Abiotic oxidation of organic Se compounds to Se oxyanions. (*e*) Release of alkylselenides from selenonium or other alkylated Se precursors through abiotic reaction. (*f*) Release of alkylselenides from selenonium or other alkylated Se precursors through aquatic producers. (*g*) Volatilization of alkylselenides into the atmosphere. (*h*) Oxidation of alkylselenides to Se oxyanions. (*i*) Formation of red amorphous Se element by aquatic and sediment producers. (*j*) Detrital formation from aquatic producers. (*k*) Se bioaccumulation into the food chain with potential ecotoxic consequences; the toxic form(s) are yet to be defined. (*l*) Assimilation of waterborne selenium oxyanions into sediment biota. (*m*) Oxidation of sediment Se (0) to oxyanions. (*n*) Reduction of sediment Se (0) to Se (–II) or vice versa. (*o*) Assimilation of sediment Se (–II) into sediment biota. (*p*) Oxidation of sediment Se (–II) to selenite. (Reproduced from Higashi et al. 2005)

content before it flows into the river. The conventional physicochemical methods employed in selenium removal such as adsorption, reverse osmosis, or chemical precipitation, although effective, may prove to be quite expensive (Lee 1989).

Recent studies suggest that microbial selenium detoxification could be the simplest, most effective, and economic remedial alternative. The microbial reduction of toxic oxyanions of Se (SeO_4^{2-} and SeO_3^{2-}) into insoluble Se(0) or methylation of these species to nontoxic dimethylselenide (DMSe) has proved to be a potential bioremediation strategy for cleanup of Se-contaminated water and sediments (Frankenberger and Arshad 2001).

Cantafio et al. have recognized the potential of a gram-positive motile bacterium, *Thauera selenatis. Thauera selenatis* is a selenate-respiring bacterium that grows both aerobically and anaerobically. *T. selenatis* has a unique metabolism and therefore can be applied in selenium remediation strategy. *T. selenatis* respires anaerobically, by using either selenate or nitrate as the terminal electron acceptor, reducing the selenate to selenite and nitrate to nitrite, when grown with a carbon source such as acetate. The enzymes (reductases) responsible for reduction of selenate and nitrate are not identical. The selenate reductase is found in the periplasm, whereas the nitrate reductase is present in the cytoplasmic membrane. When grown with nitrate (present in drainage water) and selenate, *T. selenatis* reduces both these electron acceptors concomitantly (Rech and Macy 1992). All the other selenite-reducing organisms appear to do so with a nonspecific enzyme system (e.g., nitrate reductase). For example, certain denitrifying bacteria are able to reduce selenate once the nitrate is completely exhausted. In *T. selenatis* the nitrite reductase enzyme is capable of reducing selenite to elemental selenium during denitrification. Hence, active denitrification is, therefore, required for the complete reduction of selenite to elemental selenium by *T. selenatis* (DeMoll-Decker and Macy 1993; Cantafio et al. 1996).

In an earlier experiment, a laboratory-scale biological reactor inoculated with the *T. selenatis* was used to lower the selenium oxyanion level in drainage water below 5 mg/l. The success of this trial paved the way to use *T. selenatis* in a pilot-scale biological reactor system for selenium remediation (Macy et al. 1993). Cantafio et al. (1996) described a simple, effective, and environmentally sound method for the bioremediation of selenium from agricultural drainage water using *Thauera selenatis* on a pilot scale. A medium-packed pilot-scale biological reactor system was constructed at the Panoche Water District, San Joaquin Valley, California and inoculated with *T. selenatis*. The results obtained were very promising, and selenium oxyanion concentrations (selenate plus selenite) in the drainage water were reduced by 98 %. Later, the reactor effluent was analyzed: about 91–96 % of the total selenium recovered was elemental selenium, and most of this elemental selenium could be removed with Nalmet 8072, a commercially available precipitant-coagulant (Cantafio et al. 1996).

Another bacterium, *Enterobacter cloacea*, was found to be very active in reduction of Se oxyanions present in irrigation drainage water into elemental selenium, and the process could be enhanced manyfold by supplementing organic amendments and monitoring different environmental conditions. Another process of biomethylation of selenium in water and soil sediments was found to be active and highly dependent on specific carbon amendments, activators (cofactors), and different physiological growth parameters. Moreover, Se biomethylation was protein/peptide limited rather than limited by nitrogen, amino acids, or carbon. Based on the success and potential results of these studies, a biotechnology prototype was developed that could be applied in cleanup exercises of polluted water and sediments (Frankenberger and Arshad 2001).

Selenium toxicity is encountered in arid and semiarid regions of the world with alkaline, seleniferous soils derived from marine sediments. Once the element is present in waters and soils at elevated levels, the use of conventional physico-chemical techniques for removal of Se leads to different complications. Besides bioremediation with the use of microorganisms, the phytoremediation technique could also prove to be an useful strategy for managing selenium levels in the environment (Bañuelos et al. 2002). The technology involves using plants in combination with microorganisms associated with the plants to extract, accumulate, and volatilize Se. Once the plant roots absorb the selenium, it is translocated to the shoot where it can be easily managed and disposed from the site. Therefore, plant species used for phytoremediation not only reduce the Se load eventually entering agricultural effluent but also the harvested crop can be carefully blended with animal forage and used as a supplement to feed animals suffering from selenium deficiency (Bañuelos 2001).

Although biovolatilization of selenium is an important approach as it is a natural biogeochemical method for removal of selenium, the major disadvantage with this method is mobilizing selenium up the food web. This tendency has been particularly troublesome in attempts to utilize aquatic vascular plants to volatilize selenium. For instance, a small amount of selenium is volatilized by vascular plants whereas most of it is made available in food web materials such as the shoots and roots. Although the shoots could be managed and disposed, most of the selenium that is present in the below-ground portions of the plants (Terry and Zayed 1998) is not practically feasible to manage and difficult to harvest.

Fan and Higashi (1998, 2000) have described selenium volatilization by algal species naturally occurring in the environment as an alternate remediation process in terminal basins. The main concept of this remediation process includes a combination of volatilization of selenium with interrupting its accumulation in the food web (Higashi et al. 2003). The photosynthetic algae are important in volatilizing selenium while serving as food to macroinvertebrates (brine shrimp). The brine shrimp feed on the algae and prevent algal accumulation and participation in the detrital cycle. The brine shrimp, in turn, are harvested as a product to be marketed, and thus interrupt the accumulation of Se in the food web well before it could affect birds and fish. To conclude, both volatilization of selenium by algae and harvesting of brine shrimp results in a net removal of Se from the aquatic system (Higashi et al. 2005).

Among the recent advances in bioremediation of metal-polluted sites are included the use of biosurfactants and nanotechnology. Microbial surface-active metabolites are metal-complexing agents that have been reported to be effective in the remediation of metal-contaminated environments (Mulligan et al. 2001; Singh and Cameotra 2004; Das et al. 2009): these are less toxic, biodegradable, compounds with better environmental compatibility, which could be very promising in future.

The application of nanotechnology for remediation of metal contaminants might prove promising in the near future. Nanotechnology may provide a superior remedy to purify air and water resources by utilizing nanoparticles as a catalyst or sensing system (Fulekar et al. 2014; Singh and Prasad 2015). Multi-walled carbon nanotubes (CNTs) were shown to remove copper (II), lead (II), cadmium (II), and zinc (II) from aqueous solutions and contaminated water with great success (Salam 2013; Yu et al. 2014). Nanoparticles from biological sources such as the plant *Euphorbia macroclada* are suggested for removal and detoxification of heavy metals (especially Pb, Cd, Cu, and Zn), from polluted environments (Mohsenzadeh and Rad 2011). These results have opened the way to use nanoparticles synthesized by microorganisms for removal of metal contaminants from the environment with great success in the near future.

14.5 Conclusion

Heavy metal contamination of ecosystems is a major environmental concern. Human activities that increase waterborne concentrations of selenium are increasing and the threat of widespread impacts to aquatic life is greater than ever before.

Although selenium is considered a toxic poison and the line demarking it as beneficial or harmful is very thin, its toxicity from increased levels in polluted sites has tremendous impacts on wildlife. To reduce the level of metal contamination, several remediation technologies have been implemented. These techniques include immobilization methods with the help of low-cost absorbents, application of some physicochemical methods, and biology-based techniques, that is, bioremediation. Nanotechnology is also used to enhance remediation properties and open up new possibilities for the metal remediation technique. Among all techniques, an eco-friendly remediation technology including bioremediation using microorganisms, whole cells, or their metabolites such as biosurfactants or microbial nanoparticles should provide a promising remedy to clean up metal contaminants from the environment and marine water bodies, providing a vast area to explore for such potential microorganisms to remediate selenium pollution in the marine environment.

References

Banuelos GS (2001) The green technology of selenium phytoremediation. Biofactors 14(1-4):255–260

Banuelos GS, Lin ZQ, Wu L, Terry N (2002) Phytoremediation of selenium-contaminated soils and waters: fundamentals and future prospects. Rev Environ Health 17(4):291–306

Blum SJ, Bindi AB, Buzzelli J, Stolz JF, Oremland RS (1998) *Bacillus arsenicoselenatis* sp. nov., and *Bacillus selenitireducens* sp. nov.: two haloalkaliphiles from Mono Lake, California which respire oxyanions of selenium and arsenic. Arch Microbiol 171:19–30

Brix KV, DeForest DK, Cardwell RD, Adams WJ (2004) Derivation of a chronic site-specific water quality standard for selenium in the Great Salt Lake, Utah, USA. Environ Toxicol Chem 23:606–612

Burk RF (1994) Selenium in biology and human health. Springer, New York, p 221. ISBN 978-1461275978

Cantafio AW, Hagen KD, Lewis GE, Bledsoe TL, Nunan KM, Macy JM (1996) Pilot-scale selenium bioremediation of San Joaquin drainage water with *Thauera selenatis*. Appl Environ Microbiol 62(9):3298–3303

Coelho LM, Rezende HC, Coelho LM, de Sousa PAR, Melo DFO, Coelho NMM (2015) Bioremediation of polluted waters using microorganisms. In: Shiomi N (ed) Advances in bioremediation of wastewater and polluted soil. Sci Technol Med open access doi: 10.5772/60770

Cutter GA, Bruland KW (1984) The marine biogeochemistry of selenium: a re-evaluation. Am Soc Limnol Oceanogr 29(6):1179–1192

Das P, Mukherjee S, Sen R (2009) Biosurfactants of marine origin exhibiting heavy metal remediation properties. Bioresour Technol 100:4887–4890

Davis CD, Tsuji PA, Milner JA (2012) Selenoproteins and cancer prevention. Annu Rev Nutr 32:73–95

De Jong WH, Borm PJA (2008) Drug delivery and nanoparticles: applications and hazards. Int J Nanomed 3(2):133–149

DeMoll-Decker H, Macy JM (1993) The periplasmic nitrite reductase of *Thauera selenatis* may catalyze the reduction of selenite to elemental selenium. Arch Microbiol 160:241–327

Dhanjal S, Cameotra SS (2010) Aerobic biogenesis of selenium nanospheres by *Bacillus cereus* isolated from coalmine soil. Microb Cell Factories 9:52–52

Dietz R, Riget F, Born EW (2000) An assessment of selenium to mercury in Greenland marine animals. Sci Total Environ 245:15–24

Dowdle PR, Oremland RS (1998) Microbial oxidation of elemental selenium in soils and bacterial cultures. Environ Sci Technol 32:3749–3755

Duruibe JO, Ogwuegbu MOC, Egwurugwu JN (2007) Heavy metal pollution and human biotoxic effects. Int J Phys Sci 2(5):112–118

Fairweather-Tait SJ, Collings R, Hurst R (2010) Selenium bioavailability: current knowledge and future research requirements. Am J Clin Nutr 91(5):1484S–1491S

Fan TW-M, Higashi RM (1998) Biochemical fate of selenium in microphytes: natural bioremediation by volatilization and sedimentation in aquatic environments. In: Frankenberger WT, Engberg RA (eds) Environmental chemistry of selenium. Dekker, New York, pp 545–563

Fan TW-M, Higashi RM (2000) Microphyte-mediated selenium biogeochemistry and its role in in situ selenium bioremediation. In: Terry N, Banuelos GS (eds) Phytoremediation of contaminated soil and water. CRC Press, Boca Raton, pp 283–302

Fernández-Martínez A, Charlet L (2009) Selenium environmental cycling and bioavailability: a structural chemist point of view. Rev Environ Sci Biotechnol 8(1):81–110

Finch JM, Turner RJ (1996) Effects of selenium and vitamin E on the immune responses of domestic animals. Res Vet Sci 60:97–106

Frankenberger WTJ, Arshad M (2001) Bioremediation of selenium-contaminated sediments and water. Biofactors 14(1-4):241–254

Frankenberger WTJ, Benson S (1994) Selenium in the environment. Dekker, New York, 472 pp. ISBN 0-8247-8993-8

Fulekar MH, Pathak B, Kale RK (2014) Nanotechnology: perspective for environmental sustainability. In: Fulekar MH, et al (eds) Environment and sustainable development, vol 12. Springer, New Delhi, pp 87–114

Harada M (1995) Minamata disease: methylmercury poisoning in Japan caused by environmental pollution. Crit Rev Toxicol 25(1):1–24

Herbel MJ, Blum JS, Borglin S, Oremland RS (2003) Reduction of elemental selenium to selenide: experiments with anoxic sediments and bacteria that respire Se-oxyanions. Geomicrobiol J 20:587–602

Hesketh J (2008) Nutrigenomics and selenium: gene expression patterns, physiological targets, and genetics. Annu Rev Nutr 28:157–177

Higashi RM, Rejmankova EJ, Gao S, Fan TWM (2003) Mitigating selenium ecotoxic risk by combining foodchain breakage with natural remediation. University of California Salinity/Drainage Program Annual Report

Higashi RM, Cassel TA, Skorupa JP, Fan TWM (2005) Remediation and bioremediation of selenium contaminated waters. In: Lehr JH, Keeley J (eds) Water encyclopedia: water quality and resource development. Wiley, Hoboken, pp 355–360

Janz DM, DeForest DK, Brooks ML et al (2010) Selenium toxicity to aquatic organisms. In: Chapman PM, Adams WJ, Brooks ML (eds) Ecological assessment of selenium in the aquatic environment. Society of Environmental Toxicology and Chemistry (SETAC), Pensacola, pp 139–230

Keith GS (2002) Development of aquatic life criteria for selenium: a regulatory perspective on critical issues and research needs. Aquat Toxicol 57:101–113

Lee E (1989) Current options in treatment of agricultural drainage water. In: Huntley ME (ed) Biotreatment of agricultural wastewater. CRC Press, Boca Raton, pp 33–45

Lemly AD (1993) Guidelines for evaluating selenium data from aquatic monitoring and assessment studies. Environ Monit Assess 28:83–100

Lemly AD (2002) Symptoms and implications of selenium toxicity in fish: the Belews Lake case example. Aquat Toxicol 57:39–49

Lemly AD (2004) Aquatic selenium pollution is a global environmental safety issue. Ecotoxicol Environ Saf 59:44–56

Levine VE (1924) The reducing properties of microorganisms with special reference to selenium compounds. J Bacteriol 10:217–263

Li X, Xu H, Chen Z, Chen G (2011) Biosynthesis of nanoparticles by microorganisms and their applications. J Nanomater Article ID 270974, 16 pages, doi:10.1155/2011/270974

Li DB, Cheng YY, Wu C, Li WW, Li N, Yang ZC, Tong ZH, Yu HQ (2014) Selenite reduction by *Shewanella oneidensis* MR-1 is mediated by fumarate reductase in periplasm. Sci Rep 4:3735. doi:10.1038/srep03735

Lloyd JR (2003) Microbial reduction of metals and radionuclides. FEMS Microbiol Rev 27:411–425

Macy JM, Lawson S, DeMoll-Decker H (1993) Bioremediation of selenium oxyanions in San Joaquin drainage water using *Thauera selenatis* in a biological reactor system. Appl Microbiol Biotechnol 40:588–594

Maiers DT, Wichlacz PL, Thompson DL, Bruhn DF (1988) Selenate reduction by bacteria from a selenium-rich environment. Appl Environ Microbiol 54(1):2591–2593

Martens DA (2003) Selenium. In: Stewart BA, Howel TA (eds) Encyclopedia of water science. Dekker, New York, pp 840–842. ISBN 0-8247-4241-9

Mayland H (1994) Selenium in plant and animal nutrition. In: Frankenberger WT Jr, Benson S Jr (eds) Selenium in the environment. Dekker, New York, pp 29–45

McIntyre T (2003) Phytoremediation of heavy metals from soils. In: Scheper T (ed) Adv Biochem Eng Biotechnol 78:97–123

Mehdi Y, Hornick JL, Istasse L, Dufrasne I (2013) Selenium in the environment, metabolism and involvement in body functions. Molecules 18:3292–3311

Mistry HD, Pipkin FB, Redman CW (2012) Selenium in reproductive health. Am J Obstet Gynecol 206:21–30

Mohsenzadeh F, Rad AC (2011) Application of nanoparticles of *Euphorbia macroclada* for bioremediation of heavy metal polluted environments. Int Conf Nanotechnol Biosens IPCBEE 25:16–24

Muir D, Braune B, DeMarch B, Norstrom R, Wagemann R, Lockhart L, Hargrave B, Bright D, Addison R, Payne J, Reimer K (1999) Spatial and temporal trends and effects of contaminants in the Canadian Arctic marine ecosystem: a review. Sci Total Environ 230:83–144

Mulligan CN, Yong CN, Gibbs BF (2001) Heavy metal removal from sediments by biosurfactants. J Hazard Mater 85:111–125

Nagpal NK, Lands and Parks (2001) Ambient water quality guidelines for selenium: overview. Environment Protection Division, British Columbia Ministry of Environment. ISBN 0-7726-4626-0

Nolan K (2003) Copper toxicity syndrome. J Orthomol Psychiatry 12(4):270–282

Ohlendorf HM (2002) The birds of Kesterson Reservoir: a historical perspective. Aquat Toxicol 57(1-2):1–10

Oremland RS, Herbel MJ, Blum JS, Langley S, Beveridge TJ, Ajayan PM, Sutto T, Ellis AV, Curran S (2004) Structural and spectral features of selenium nanospheres produced by Se-respiring bacteria. Appl Environ Microbiol 70:52–60

Parkman H, Hultberg H (2002) Occurrence and effect of Selenium in the environment–a literature review. IVL-report B1486, IVL Swedish Environmental Research Institute

Parulekar K, Naik M (2015) Isolation of selenite and tellurite reducing bacteria from Mandovi estuary and its potential for synthesis of selenium and tellurium nanoparticles. MSc dissertation, Department of Microbiology, Goa University

Rayman MP (2000) The importance of selenium to human health. Lancet 356:233–241, 55

Rech S, Macy JM (1992) The terminal reductases for selenate and nitrate respiration in *Thauera selenatis* are two different enzymes. J Bacteriol 174:7316–7320

Reilly C (2006) Selenium in food and health, 2nd edn. Springer, New York. ISBN 978-0-387-33243-7

Salam MA (2013) Removal of heavy metal ions from aqueous solutions with multi-walled carbon nanotubes: kinetic and thermodynamic studies. Int J Environ Sci Technol 10(4):677–688

Singh P, Cameotra SS (2004) Enhancement of metal bioremediation by use of microbial surfactants. Biochem Biophys Res Commun 319:291–297

Singh A, Prasad SM (2015) Remediation of heavy metal contaminated ecosystem: an overview on technology advancement. Int J Environ Sci Technol 12:353–366

Singh R, Gautam N, Mishra A, Gupta R (2011) Heavy metals and living systems: an overview. Ind J Pharmacol 43(3):246–253

Spallholz JE, Hoffman DJ (2002) Selenium toxicity: cause and effects in aquatic birds. Aquat Toxicol 57:27–37

Tchounwou PB, Yedjou CG, Patlolla AK, Sutton DJ (2012) Heavy metal toxicity and the environment. EXS 101:133–164. doi:10.1007/978-3-7643-8340-4_6

Terry N, Zayed A (1998) Phytoremediation of selenium. In: Frankenberger WT, Engberg RA (eds) Environmental chemistry of selenium. Dekker, New York, pp 633–655

Tinggi U (2003) Essentiality and toxicity of selenium and its status in Australia: a review. Toxicol Lett 137:103–110

Tinggi U (2008) Selenium: its role as antioxidant in human health. Environ Health Prev Med 13(2):102–108

Tran PA, Webster TJ (2011) Selenium nanoparticles inhibit *Staphylococcus aureus* growth. Int J Nanomedicine 6:1553–1558

UNEP, GESAMP (1988) Arsenic, mercury and selenium in the marine environment. UNEP regional seas reports and studies no. 92, GESAMP report and studies no. 28. www.unep.org/regionalseas/publications/reports/RSRS/pdfs/rsrs092.pdf

Vriens B, Lenny HE, Winkel M, Lenz MB (2014) Assessing global cycling of selenium: Analytical tools to quantify volatile species and their fluxes. In: Banuelos GS, Zhi-Qing L, Xuebin (eds) Selenium in the environment and human health. CRC Press, Roca Baton

Young RA (2005) Toxicity profiles: toxicity summary for cadmium. Risk Assessment Information System (RAIS), University of Tennessee. rais.ornl.gov/tox/profiles/cadmium.shtml

Yu JG, Zhao XH, Yu LY, Jiao FP, Jiang JH, Chen XQ (2014) Removal, recovery and enrichment of metals from aqueous solutions using carbon nanotubes. J Radioanal Nucl Chem 299:1155–1163. doi:10.1007/s10967-013-2818y

Zhuang W, Gao X (2013) Acid-volatile sulfide and simultaneously extracted metals in surface sediments of the southwestern coastal Laizhou Bay, Bohai Sea: concentrations, spatial distributions and the indication of heavy metal pollution status. Mar Pollut Bull 76:128–138

Pathogenic Bacteria of Public Health Significance in Estuarine Mangrove Ecosystem

15

Krupali Poharkar, Swapnil Doijad, Savita Kerkar, and Sukhadeo Barbuddhe

Abstract

Mangroves provide a unique ecological niche to different microbes which play a significant role in nutrient recycling as well as various environmental activities. However, with the increase in urbanization, estuarine mangrove ecosystems are getting affected by different ways. Several 'non-innate' matters are getting introduced into these environments. Since last decade, increased incidences of pathogens in mangroves have been documented. Despite of their atypical environment, these pathogens can adapt, survive and persist in the mangrove. Several pathogenic bacteria of public health significance and their abundance have been reported. Evidences suggest that the mangrove biota being used as food plays a major role in the transmission of pathogens. Apparently, mangrove ecosystem is acting as a reservoir for many pathogens. This chapter describes the occurrence of different pathogens of public health significance in mangroves, the potential of mangroves as a reservoir of pathogens and the role of associated biota in transferring these pathogens to humans.

K. Poharkar (✉)
Departments of Biotechnology, Goa University, Goa 403206, India

National Institutes of Biotic Stress Management, Baronda, Raipur 493225, India
e-mail: krupalipoharkar.biotech@gmail.com

S. Doijad
Institute of Medical Microbiology, Justus-Liebig University, Giessen 35392, Germany

S. Kerkar
Department of Biotechnology, Goa University, Goa 403206, India

S. Barbuddhe
National Institutes of Biotic Stress Management, Baronda, Raipur 493225, India

15.1 Introduction

15.1.1 Mangrove Ecosystem

Mangroves are unique, highly productive, dynamic ecosystems found mainly in tropical and subtropical intertidal regions of the world, representing the most productive and valuable ecosystems. Mangrove ecosystem refers to groups of trees and shrubs that grow in saline and swampy coastal habitats. They cover an area of about 20 million hectares worldwide (English et al. 1997), of which 60–75 % is around the

M.M. Naik, S.K. Dubey (eds.), *Marine Pollution and Microbial Remediation*, DOI 10.1007/978-981-10-1044-6_15

tropical coastline. Mangrove ecosystems are rich in organic matter (Holguin et al. 2001; Zhou et al. 2009) and also efficient in biological nutrient recycling. These ecosystems potentially nourish a range of marine and terrestrial ecosystems through the transfer of nutrients and energy (Jennerjahn and Ittekott 2002; Vannucci 2000; Hyndes et al. 2014). These zones also play an important role in the oxidation, storage and release of terrestrial carbon, thereby affecting global carbon budgets (Cole et al. 2007; Downing et al. 2006; Downing et al. 2008). Mangrove continuously shed their leaves that are broken down by bacteria and fungi and release a significant amount of nutrients to nearby coastal areas. Mangroves act as a sink for nutrients and provide large quantities of detritus organic matter to nearby coastal waters (Krishna Prasad and Ramanathan 2008). When the nutrient-enriched mangrove water mixes with comparatively nutrient-poor neritic water by means of flow and tidal ebb, the productivity of coastal ecosystems improves. Thus mangroves play a significant role in maintaining the high productivity and biotic diversity of coastal waters (Kar and Satpathy 1995). Mangrove forest can truly be considered as evolutionarsy hotspots where marine organisms have undergone the transition to terrestrial species and terrestrial organisms have readapted to marine life (Saenger 2002; Yeragi and Yeragi 2014). Mangroves perform a wide range of ecological and economic functions such as stabilization of coastlines, enrichment of coastal waters, yielding commercial forest products and supporting coastal fisheries (Kathiresan and Bingham 2001). Mangrove leaf litter provides an essential nutrient base for food webs and, therefore, plays a crucial role in coastal and deep-sea fisheries. It serves as a nursery ground for fish and crustaceans and also supports a diversity of living organisms (Bacosa et al. 2013; Vannucci 2000). Millions of people all over the world depend directly or indirectly on the mangrove for their sustenance. They rely on the provision of a variety of food, timber, tannin, chemicals and medicines derived from mangrove forests or associated plants (Ewel et al. 1998; Glaser 2003; Stone 2006; Singh et al. 2012). Besides being a source for commercial products and fishery resource, it is developing as a site for eco-tourism (Kathiresan and Bingham 2001). These coastal areas are being subjected to high human pressures, as mass movement of people has been observed from the hinterlands towards the coastal areas during the late twentieth and twenty-first centuries. In spite of the huge social, economic and ecological importance of mangroves in tropical ecosystems (Ronback et al. 2007; Nagelkerken et al. 2008; Walters et al. 2008), since last decade, mangrove ecosystem is reduced at a rate of 1–2 % due to anthropogenic influence through deforestation and dumping activities (Duke et al. 2007; Kruitwagen et al. 2008). Continuous discharge of sewage from various sources affects the health and state of this ecosystem. Therefore, the environmental impact on these ecosystems needs to be monitored continuously to protect mangroves.

15.1.2 Physicochemical Characteristics of Mangroves

Estuarine and coastal areas are vulnerable to anthropogenic activities, which in turn affect the water quality of mangrove ecosystems. Physicochemical analyses help to determine water quality (Hamaidi-Chergui et al. 2013). Changes in the physicochemical concentration parameters such as pH, temperature, salinity, total dissolved solids and dissolved oxygen indicate changes in the condition of the water systems (Hacioglu and Dulger 2009). The water temperature is one of the most significant parameters that controls inborn physical qualities of water and plays a significant role in the solubility of salts and gases (Hamaidi-Chergui et al. 2013). Mangrove is coastal habitat where the water temperature does not go below 20 °C. A high organic content tends to decrease the pH in mangrove environment, while having pH greater than seven indicates increased salinity. Variation in coastal salinity is due to the effect of unpredictable rainfall, evaporation, precipitation, etc. Evaporation of water during the dry season leads to increase in salinity, while, during the wet season, due to rainfall and flood from rivers, dilution of water results in a decrease in salinity (Olatayo 2014). Dissolved oxygen (DO) is one of the important parameters that influences the aquatic life which further contributes to the biological and physical processes prevalent in the water (Srilatha et al. 2012). Optimal range of dissolved oxygen is 4–9 mg/L, while DO above 5 mg/L is supportive for marine life, whereas concentrations below this are considered potentially harmful (Olatayo 2014). Total solids may affect the water quality. Increased discharge of sewage into water bodies results in a high quantity of total dissolved solids that in turn influence the portability of water (Dhanlakshmi et al. 2013). Overall these physicochemical parameters indirectly control the microbial load present in mangroves.

15.2 Pathogens in Mangroves

15.2.1 Abundance of Pathogens

Mangroves are influenced by an input from terrestrial sources, iron-ore transporting barges, effluents from anchored casino boats, river runoff and various other anthropogenic factors. The mangrove marshland is extremely sensitive to environmental changes (Jiang et al. 2013). Sustained human activity and pollution continuously affect the diversity of the inhabiting microbes and may add or deplete the type of microbial flora in these ecosystems (Ristori et al. 2007). Anthropogenic activities increase a load of faecal bacteria and pathogens in this ecosystem (Malham et al. 2014). The level of faecal indicator bacteria and enteric pathogens is influenced by the discharge of domestic and industrial waste into the estuarine habitat (Touron et al. 2007); therefore, this environment becomes unfit for various activities such as recreation and fishing (Abbu and Lyimo 2007). These human and animal pathogens may tolerate variations in salinity, pH and environmental stress and may survive in atypical mangrove reservoir. Several lives depend on food from mangrove swamps; however, this ecosystem has not been delved extensively for the occurrence of foodborne pathogens. In turn, the water of the bays and estuaries usually contains vast microbial population composed of an indigenous group of organisms, as well as microorganisms introduced to these areas with the discharge of domestic and industrial wastes. Pathogens like *Escherichia coli*, *Vibrio* spp., *Salmonella* spp. and *Staphylococcus aureus* easily get added to the estuarine zone through domestic sewage discharge, land drainages and other discharges (Nagvenkar and Ramaiah 2009; Grisi and Lira 2010). Faecal pollution in the aquatic environment may lead to diseases in humans when foods harvested from these areas get consumed by people, through drinking water and during recreational activities (Atieno et al. 2013). Studies have reported the occurrence of pathogenic microorganisms, namely, *Vibrio cholerae*, *S. aureus*, *Salmonella*, *Shigella* and *E. coli* in mangrove ecosystems (Grisi and Lira 2010; Rodrigues et al. 2011; Poharkar et al. 2014). Indigenous bacterial flora (Desai et al. 2004; De Sousa and Bhosle 2012; Khandeparker et al. 2011) and pathogenic bacteria (Rodrigues et al. 2011; Ramaiah et al. 2007; Nagvenkar and Ramaiah 2009) have been isolated from mangrove ecosystems. The organic/inorganic content has also been determined (Attri et al. 2011; Krishnan and LokaBharathi 2009; Paula et al. 2009; Krishnan et al. 2007). But intensive environmental impact monitoring and assessment of these systems are still lacking (Peters et al. 1997; Penha-Lopes et. al. 2011), and the potential effects on the local population are not known. Impairment of mangrove environment due to the presence of pathogens is of great concern because of the consequences associated with public health, impacts to mangrove-originated biota and degradation of the overall usefulness of the water resource.

15.2.2 Persistent Pathogens

Some pathogens naturally present in the marine environment. *Aeromonas hydrophila*, *V. cholerae*, *V. vulnificus*, *V. parahaemolyticus* and *V. alginolyticus* are indigenous to the marine environment. Pathogens usually present in the marine environment can be transmitted to humans via inhalation, contact or ingestion of water or contaminated food. *Aeromonas* spp. are ubiquitously present in the marine environment and are increasingly reported for seafood and waterborne infections (Joseph et al. 2013). Previous reports have documented the occurrence of *Aeromonas* species in squid (Baldria and Alvero 1999), prawn (Thayumanavan et al. 2003) and mussel (Ottaviani et al. 2006) and their potential as health hazards (Austin and Austin 1993; Ghenghesh et al. 2008). *Vibrio* spp. are innate to both marine and estuarine environments (Malham et al. 2014). There are at least 12 *Vibrio* species that are recognized as human pathogens, although *V. cholerae*, *V. parahaemolyticus* and *V. vulnificus* account for the majority of *Vibrio*-related infections worldwide (Morris 2003). *Vibrio* spp. that are commonly encountered and having epidemic potentials in causing severe gastroenteritis are *V. cholerae* and *V. parahaemolyticus* (Daniels and Shafaie 2000; Ceccarelli et al. 2013). In addition, *V. vulnificus* can also cause severe infections in individuals with some underlying health conditions. Approximately 8000 people get ill each year due to *Vibrio* infection in the USA (Dechet et al. 2008) of which 5200 infections are of foodborne origin and about 2800 are from other sources (Dechet et al. 2008). Between 1995 and 2004, the World Health Organization (WHO) reported 100,000–300,000 cases of cholera. Approximately 99 % of cholera cases remain unreported each year. WHO estimates that actual case rates come up to 3–5 million with more than 120,000 cases leading to mortality (Zuckerman et al. 2007). Oysters collected during 2006–2007 from beaches, supermarket and restaurants were found to be contaminated with *V. parahaemolyticus* in Sao Paulo, Brazil (Sobrinho et al. 2011). Massive flooding in the US Gulf Coast caused 22 cases of *Vibrio* wound infection and five deaths in 2005 (CDC 2005). *Vibrio vulnificus* was found to infect 36 people including 10 deaths in 2013 at Florida, USA (Ross 2013).

15.2.3 Transient Pathogens

Pathogens of public health significance present in estuarine environments are mainly derived from human or animal faeces. Disposal of human faecal waste has been shown to be related to the occurrence of faecal coliforms in mangroves (Abbu and Lyimo 2007). A wide range of bacterial pathogens such as *Salmonella* spp., *Yersinia enterocolitica*, *Shigella* spp., *Enterococcus* spp., *Escherichia coli*, *Clostridium perfringens*, *Staphylococcus* spp. and *Campylobacter* spp. enter in mangrove environment through sewage discharge (Scott 2002). The source of *Salmonella* in the environment is increasingly being investigated as a potentially significant reservoir of *Salmonella* transmission (Schutze et al. 1999). *Salmonella* spp. are implicated in several foodborne outbreaks. Majority of the salmonellosis cases reported are caused by *S. enterica* that is generally associated with food products. Previously contamination of crab meat and other associated fishes has been linked with the presence of *Salmonella* spp. in the mangrove area (Grisi and Lira 2010; Lotfy et al. 2011). *E. coli* is a dominant bacterium in sewage, which can compete with the native microflora (Ramaiah et al. 2007). *E. coli* have been reported from polluted freshwater beaches (Walk et al. 2007), the tropical estuary (Chandran et al. 2008) and saltwater lakes (Chandran et al. 2013). *Shigella* species are found in the gastrointestinal tract of humans. Disposal of inadequately treated human sewage into recreational waters and lack of properly treated water supply were previously associated with outbreaks of *Shigella* in the USA (Leclerc et al. 2002). In an oyster-related outbreak of *S. sonnei*, 24 individuals were ill in Texas and the incident was associated with the disposal of faecal waste overboard from an oyster harvesting boat (Reeve et al. 1989). During the period 1992–2003, 14 % of all waterborne outbreaks occurred due to *Campylobacter* spp. (Malham et al. 2014). However, *Campylobacter jejuni* has been frequently associated with wastewaters and was

Table 15.1 The occurrence of pathogens in mangrove ecosystems

Transient pathogens
E. coli
Listeria monocytogenes
Salmonella spp.
Campylobacter spp.
Shigella spp.
Staphylococcus spp.
Enterococcus spp.
Yersinia enterocolitica
Persistent pathogens
Aeromonas spp.
Vibrio spp.

isolated from environmental water (Jokinen et al. 2011; Rodrıguez-Martınez et al. 2013). There are some reports of isolation of *Listeria* spp. from marine environment and associated food (Bou-m'handi et al. 2007; Colburn et al. 1990). In a previous report, Bou-m'handi et al. (2007) isolated *L. monocytogenes* from marine water, sediment and shellfishes harvested from the same environment in Morocco. Momtaz and Yadollahi (2013) isolated *L. monocytogenes* from marine foods such as fish and shrimp in Iran (Table 15.1).

15.3 Sources of Pathogens

15.3.1 Urbanization and Sewage Disposal

Migration of people to coastal areas has led to a rapid increase in urban wastewater production. These waste waters get disposed directly in the water bodies without any treatment and may cause severe damage on mangrove communities. Waste water contains diverse types of chemical as well as biological pollutants, including pathogens. It had been reported that every year approximately 25–30 million litres of untreated faecal matter was pumped into the brackish water lagoon of Lagos in Nigeria (Mastaller 1996). Discharge of faecal waste from the open toilets of fishing villages located by the mangrove estuaries is a major source of pathogens in the Kuala Sepetang mangroves of Malaysia (Ghaderpour et al. 2014). Along with domestic waste discharge insecticides, pesticides from adjacent agriculture and animal waste contribute to the contamination of mangroves. Domestic and municipal sewage is found to be responsible for the occurrence of abundant pathogenic microorganisms in mangrove areas of Paraíba do Norte River, state of Paraíba, Northeastern Brazil (Grisi and Lira 2010). Partially treated or untreated sewage combined with sewer outflows contains many pathogenic bacteria through sewage discharge which gain entry into mangrove ecosystems.

15.3.2 Industrialization

Several industries that are located in the vicinity of the mangroves add organic or inorganic matter into mangroves. Such organic/inorganic matter serves as food and positively influences the growth of bacteria including pathogens. Mainly wastes discharged from food industries like alcohol and sugar industries (Grisi and Lira 2010), shrimp aquaculture, fish industries, fertilizer industries and charcoal and timber industries (Marinebio 2015) carry an abundance of pathogenic microorganisms to mangrove habitat.

15.3.3 Miscellaneous Sources

Mangroves are influenced with inputs from terrestrial sources, river runoff and various other anthropogenic factors. Pathogenic microorganisms present in mangrove environments are generally derived from human or animal faeces. These pathogenic microorganisms enter into estuarine and coastal waters, generally during heavy rain or flood, and their actions may then have an impact on recreational and shellfish growing waters. Recreational activities such as swimming, boating and fishing in waters contaminated with faeces can pose a risk to human health (Fleisher et al. 1998). However, bather shedding can also be a source of pathogenic microorganisms in mangrove areas.

15.4 Survival of Pathogens in Atypical Habitat

15.4.1 Fate of Introduced Pathogens

When enteric pathogens are discharged into the marine environment, they come into an environment that is distinct from their typical pathogen–host habitat. However, numerous factors such as chemical, biological and physical dictate the fate of pathogens in estuarine ecosystems (Rhodes and Kator 1988). Environmental conditions in nature are rarely stable but fluctuate often between optimum and adverse. In mangrove, the pathogens encounter extremely different environment conditions such as high salinity, varying nutrient, various osmotic pressure and solar radiations (Kassen and Bell 1998; Leroi et al. 1994; Hughes et al. 2007; Ketola et al. 2004, 2013). These environmental conditions may affect the overall survival and proliferation rate. However, several pathogens that are capable of tolerating these fluctuating environmental conditions may overcome the environmental stresses and persist. Different pathogen types and even strains of the same pathogen possess different abilities to survive and persist (Anderson et al. 2005; Wanjugi and Harwood 2013). Under particular stressed conditions, bacteria may undergo several mutations, recombination of genes, horizontal uptake and disposal of genes which helps to survive bacteria in stress condition (Ishii and Sadowsky 2008).

15.4.2 Tolerance and Adaptations

Bacteria coordinately control gene expression to adapt and survive in fluctuating environmental conditions. When enteric bacteria are released into the marine environment, subjected to an immediate osmotic upshock, their ability to overcome this by means of several osmoregulatory systems could largely influence their subsequent survival in the marine environment (Gauthier et al. 1987). Upon an osmotic upshift, bacterial cells accumulate or synthesize particular osmoprotectant molecules, in order to balance osmotic pressure and avoid drastic loss of water from the cytoplasm (Csonka and Epstein 1996). Enteric bacteria are found to accumulate or synthesize molecules such as trehalose, glycine betaine and glutamic acid in order to regulate osmotic pressure (Rozen and Belkin 2001). Several studies have demonstrated the essential role of osmoregulatory mechanisms of enteric bacteria for survival in the marine environment.

Cells preadapted to high osmolarity are highly resistant to seawater (Munro et al. 1994, 1995; Gauthier et al. 1987). The molecular data available till date indicate that the most significant among adaptive systems, the rpoS regulon, plays an important role in the survival of enteric bacteria in the marine environment. The rpoS (σ^s) transcription factor controls the expression of a large number of genes involved in cellular responses to a diverse number of stresses, including osmotic stress, starvation, acid shock, heat shock, cold shock and oxidative damage (Rozen and Belkin 2001). As a minimum, 50 different genes were found to be under rpoS control; they are induced by a shift to a stationary growth phase, as well as by diverse stresses. Survival of enteric bacteria in the marine environment is greatly affected by both UV and visible light. In fact, light is considered to be the single most important contributor to bacterial die-off in the marine environment (Gameson and Gould 1975; Chamberlin and Mitchell 1978; Fujioka et al. 1981; Rozen and Belkin 2001). The presence of the rpoS gene in enteric bacteria showed a protective effect against light (Rozen and Belkin 2001). The rpoS$^+$ strains show higher resistance to various stresses than in the rpoS mutant, suggesting that the cross-protection endowed by oxidative, acidic, thermal and nutritional stresses was rpoS dependent (Munro et al. 1994).

Also prior to their arrival into the marine environment, pathogens are exposed to a variety of environmental conditions. In some cases, they are discharged directly from boats or bathers; in others they remain in reservoirs and/or are carried out to the marine environment through natural rivers. In several studies, it was found that the survival ability of enteric bacteria in the marine environment depends on their previous history (Gauthier et al. 1987; Munro et al. 1994; Rozen and Belkin 2001).

The concentration of living and nonliving particulate organic matter (POM), commonly higher in coastal regions, is capable of selectively enriching heterotrophic bacteria (Huq et al. 1983; Heidelberg et al. 2002; Grossart et al. 2005). This organic matter is a nutrient-rich hotspot that positively influences the growth of several bacteria including pathogens in coastal water (Eiler et al. 2007). Pathogens in the marine environment are often found in association with the surfaces of phytoplankton, sediments and suspended detritus. Algal and zooplankton blooms are found to promote proliferation of associated bacterial communities by providing microenvironments favouring growth and by releasing nutrients into the water (Lipp et al. 2002). In an earlier study, *Clostridium botulinum* spores were found in marine sediments with high overlying fish abundance suggesting deposition (Huss 1980), while sediments underlying farmed mussels were observed to maintain an enriched presence of vibrios relative to surrounding environments, possibly due to stimulated *Vibrio* growth in an organic-enriched environment (La Rosa et al. 2001). Filter-feeding shellfish are efficient bioconcentrators of small particles and pathogenic contaminants in the marine environments. Thus, overall various factors are responsible for tolerance and adaptation of pathogens in the marine environment. Also environmental conditions dictate the expression of certain genes, and such expression may be inherited in subsequent generations. This may thus result in an alien species adapting and adopting the mangrove swamps as its natural habitat and in turn becoming an indigenous organism of the mangrove ecosystem.

15.4.3 Survival and Persistence

Bacterial pathogens enter estuaries through several points and diffuse sources that include wildlife excrement, agricultural runoff, septic tank and sewage discharges. When these pathogens enter in the estuarine environment, they subsequently accumulate in the sediments and on the suspended sediments (flocs) where the deposition of flocculated particles promotes reservoirs of potentially pathogenic bacteria in the sediment. Macronutrient fluxes play an essential role in sediment dynamics, floc formation and pathogen survival, as it is well known that nutrient accessibility and sediment association boost the survival and persistence of the pathogens (Malham et al. 2014). Environmental conditions may select strains possessing virulence factors (Tamplin et al. 1996; Jackson et al. 1997; Faruque et al. 1998; Chakraborty et al. 2000); such factors may include attachment mechanisms to organic matter, motility and the ability to grow rapidly under nutrient-rich conditions. Attachment and colonization to the host surface are the primary steps in bacterial pathogenesis. The same factors are generally required for colonization of the human intestine and for the colonization of abiotic and biotic surfaces in the marine environment (Watnick et al. 1999; Chiavelli et al. 2001; Meibom et al. 2004). A marine environment is characterized by thermal variability, pH fluctuations, low nutrient availability and high osmolarity suggesting that a stress response is activated in pathogens in such environment. Several genes are involved in the survival of pathogens in the marine environment. These genes encode products involved in nutrient acquisition and utilization, motility and transcriptional regulation and contribute to survival and persistence of pathogens in the marine environment (Winfield and Groisman 2003). To survive in a marine environment, some pathogens employ a particular strategy in response to environmental stresses to retain its viability in the environment. It might go into a physiological state where it remains viable but not culturable (VBNC) (Tanaka et al. 2000; Rozen and Belkin 2001). In VBNC state, pathogens remain in the dormant state and survive for a long time in the marine environment. Ability to withstand various stresses apparently causes the continued prevalence of different pathogens in the mangrove swamps. The continued acclimatization of pathogens to this alien environment may turn out to be an indigenous niche, resulting in the permanent persistence of such pathogenic strains in these pristine ecosystems.

15.5 Transfer of Pathogens from Mangrove to Humans

15.5.1 Mangrove Biota as a Food

Crustaceans (shrimps, prawns and crabs) are abundant in mangrove forests. The crabs comprise the most important and abundant crustaceans in mangrove forests and subsequently have a significant role in mangrove ecosystems. Around 60 different species of crabs inhabiting the mangroves have been identified (UNEP-WCMC 2007). One of the most conspicuous species, the fiddler crab, is used as food by most of the people. Mangroves are also known as an essential nursery habitat for the diverse communities of fish and shrimps which find protection and abundant food in these environments. Many studies have demonstrated a strong relationship between the presence of mangroves and fish catch (Lee 2004; Manson et al. 2005; Meynecke et al. 2007); fishery catch is influenced by the relative abundance of mangroves in a region. Correlations have also been found between the mangrove areas and the catches of prawns in the fisheries adjacent to the mangroves (Staples et al. 1985; Pauly and Ingles 1986). Therefore, these studies provided important information on the fisheries–mangrove relationship and thus showed the economic importance of mangroves (Barbier and Strand 1998; Grasso 1998; Barbier 2000). Also, commercially important bivalves such as oysters, clams and mussels are commonly found in and around mangrove roots. Thus mangrove habitat acts as an important food resource, explored by people for several fishes (*Etroplus suratensis*, *Caranx malabaricus*, *Sparus berda*), crabs (*Scylla serrata*, fiddler crab) and mud clam (*Polymesoda erosa*) as commercial food (Clemente 2008; MSI 2013).

15.5.2 Contamination of Biota

Mangrove habitat is rich in valuable important food resource. But extensive anthropogenic activities end many natural functions of mangroves. Domestic and industrial sewage entering in this estuarine habitat contains diverse pollutants including viral and bacterial pathogens such as *E. coli*, *Vibrio cholerae*, *Staphylococcus aureus*, *Salmonella* spp. *and Pseudomonas aeruginosa*, toxic chemicals and a variety of organic and inorganic wastes (Grisi and Lira 2010; Poharkar et al. 2014). Bivalve molluscan shellfish, fish and crustaceans accumulate pathogens from estuarine waters contaminated with faeces and may serve as a source of pathogens to humans as these foods are likely to get consumed by people. Due to ocean currents, tidal action and turbulence, these pollutants get dispersed in the estuary and then concentrated in the food chain (Alam and Zafar 2013).

15.5.3 Occurrence of Pathogens in Biota

Mangroves and seafood are prospective reservoirs of pathogenic bacteria and play an important role in ecological and epidemiological studies (Joseph et al. 2013). Seafood harbours pathogenic microorganisms due to the texture of their flesh and also their microbe-loaded habitat (Joseph et al. 2013). Several studies documented the occurrence of pathogens in a mangrove-originated biota. In an earlier study, contamination of crab meat and other associated fish has been linked with the presence of *Salmonella* spp. in the mangrove area (Grisi and Lira 2010; Lotfy et al. 2011). In an another study, Kumar et al. (2001) reported the presence of pathogenic strains of *E. coli* in fishes and clams marketed in Mangalore, India. Consumption of *Salmonella*-contaminated food (sushi) caused 316 people to be ill in the USA (FSN 2012). In a recent report, Poharkar et al. (2014) reported pathogenic strains of *E. coli* in mangrove-originated biota off Goa, India. Several foodborne outbreaks have been reported previously due to the consumption of shellfish grown in sewage-contaminated water (Daniels et al. 2000). An outbreak of diarrheal illness caused by eating tuna paste contaminated with *E. coli* was described in Japan (Mitsuda et al. 1998). In Mangalore (India), an outbreak of food poisoning caused after eating fish contaminated by *S. weltevreden* affected 34 persons (Antony et al. 2009). Thus, the occurrence of pathogens in mangrove-associated biota defines the risk associated with public health.

15.5.4 Transmission Cycle of Pathogens Formed Between Human and Mangrove

Many enteric bacteria and faecal coliforms naturally inhabit the intestines of humans.

Human pathogens present in estuarine environments are primarily derived from human or animal faeces. Due to heavy rain and flood events, these pathogenic microorganisms get mixed from upstream sources to estuarine and coastal waters that may then have an impact on recreational and shellfish growing waters. However, bather shedding can also be a potential source of pathogenic microorganisms in coastal waters. A wide range of pathogens such as *Salmonella* spp., *Yersinia enterocolitica*, *Shigella* spp., *Escherichia coli*, *Clostridium perfringens*, *Staphylococcus* spp. and *Campylobacter* spp. may be present in domestic and industrial effluents (Scott 2002). Once these allochthonous pathogens are introduced into mangrove ecosystems, they can disperse far and wide to other regions. Exposure to sewage-related pathogens in the marine and estuarine environments generally occurs through incidental ingestion of contaminated water during recreational or commercial activities, such as fishing, swimming, boating and consumption of raw or partially cooked foods originating from such environments or food harvested from contaminated water (Schutze et al. 1999; Martinez-Urtaza et al. 2004; Brands et al. 2005). Fishes and crustaceans that can accumulate pathogens from estuarine waters contaminated with faeces may present a health risk when consumed raw or only lightly cooked (Lees 2000). In an earlier study at mangroves in Brazil (Keller et al., 2013), *E. coli* strains were found in water as well as mangrove-associated food for over a 14-month period indicating a history of chronic contamination. In a recent study, Poharkar et al. (2014) reported the presence of clonal *E. coli* strains in mangrove environment, associated food and clinical cases. This suggested that the pathogens form a vicious chain by entering into the mangroves through domestic discharges, subsequently survive in the mangrove areas, contaminate the associated food and re-enter the humans completing the cycle (Fig. 15.1).

15.6 Conclusions and Perspectives

Human interference adds pathogens directly or indirectly to the mangrove swamps. Ability to tolerate various stresses could cause the continued prevalence of pathogens in the mangrove swamps. The continued acclimatization of pathogens to this alien environment may turn out to be an indigenous niche, resulting in the permanent persistence of such pathogenic strains in these pristine ecosystems. Mangroves may thus turn out to be major reservoirs for pathogenic microorganisms which are critical for human health. There is a need for general awareness about this microbial contamination. Monitoring systems need to be established for the food being harvested and sold locally. Effective measures to control the direct disposal of the domestic waste in the mangroves and associated estuaries need to be implemented and ascertained in order to protect these so-called pristine environments. Protection of this ecosystem from adding undesirable microbial populations, appropriate policies and regulations should be implemented.

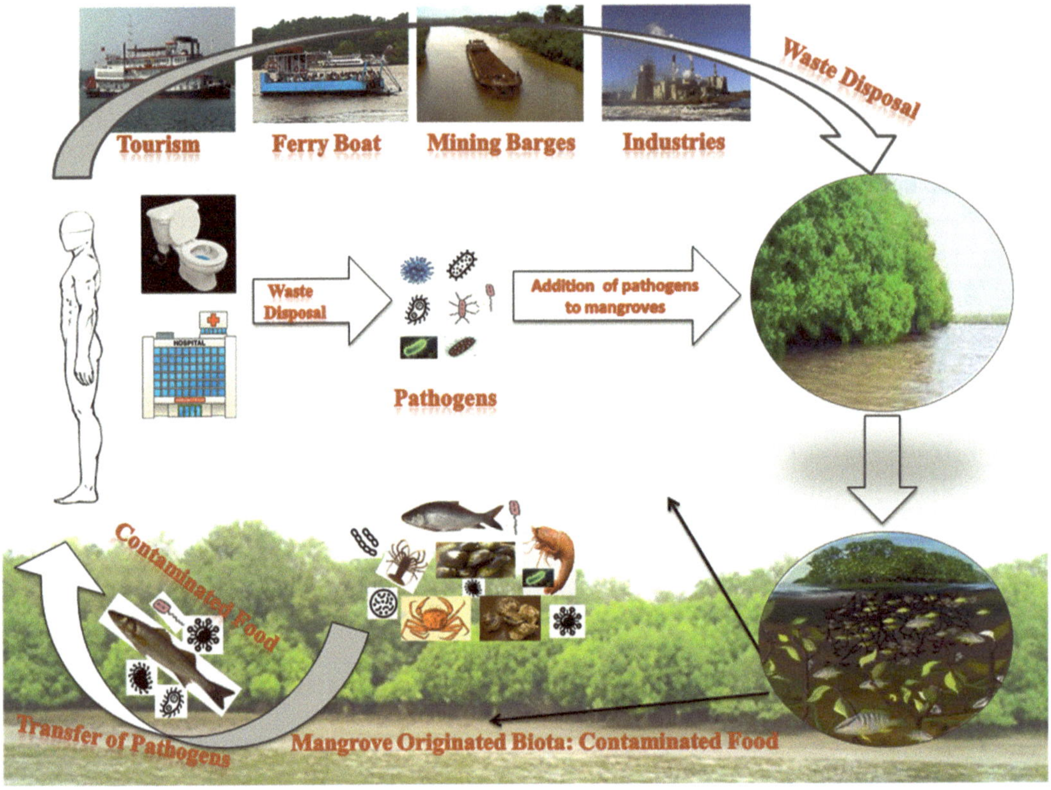

Fig. 15.1 Transmission cycle of pathogens formed between human and mangrove

References

Abbu AA, Lyimo TJ (2007) Assessment of fecal bacteria contamination in sewage and non-sewage impacted mangrove ecosystems along the coast of Dar Es Salaam. Tanzania J Sci 33:27–40

Alam W, Zafar M (2013) Spatial and temporal variation of *Escherichia coli* in water and soil with relation to water parameters at the estuary of Karnafuly River, Bangladesh. J Bacteriol Sci 5:1–8

Anderson KL, Whitlock JE, Harwood VJ (2005) Persistence and differential survival of fecal indicator bacteria in subtropical waters and sediments. Appl Environ Microbiol 71:3041–3048

Antony B, Dias M, Shetty AK, Rekha B (2009) Food poisoning due to *Salmonella* enterica serotype weltevreden in Mangalore. Indian J Med Microbiol 27:257–258

Atieno NR, Owuor OP, Omwoyo O (2013) Isolation of high antibiotic resistant fecal bacteria indicators, *Salmonella* and *Vibrio* species from raw abattoirs sewage in peri-urban locations of Nairobi, Kenya Greener. J Biol Sci 3:172–178

Attri K, Kerkar S, Loka Bharathi PA (2011) Ambient iron concentration regulates the sulfate reducing activity in the mangrove swamps of Diwar, Goa, India. Estuar Coast Shelf Sci 95:156–164

Austin B, Austin DA (1993) Bacterial fish pathogens. Ellis Horwood, Chichester

Bacosa HP, Suto K, Inoue C (2013) Degradation potential and microbial community structure of heavy oil-enriched microbial consortia from mangrove sediments in Okinawa, Japan. J Environ Sci Health 48:835–846

Baldria LR, Alvero MAC (1999) Aeromonas species from raw squid balls: a preliminary study of a potential food-borne pathogen and health hazard. J Environ Sci Health Manag 2:48–52

Barbier EB (2000) Valuing the environment as input: review of applications to mangrove-fishery linkages. Ecol Econ 35:47–61

Barbier EB, Strand I (1998) Valuing mangrove-fishery linkages—a case study of Campeche, Mexico. Environ Resour Econ 12:151–166

Bou-m'handi N, Jacquet C, Marrakchi AE, Martin P (2007) Phenotypic and molecular characterization of *Listeria monocytogenes* strains isolated from a marine environment in Morocco. Foodborne Pathog Dis 4:409–417

Brands DA, Inman AE, Gerba CP, Mare' CJ, Billington SJ, Saif LA, Levine JF, Joens LA (2005) Prevalence of *Salmonella* spp. in oysters in the United States. Appl Environ Microbiol 71:893–897

CDC (Centers for Disease Control and Prevention), 2005 *Vibrio* illnesses after hurricane Katrina. http://www.cdc.gov. Accessed 3 Mar 2014

Ceccarelli D, Hasan NA, Colwell RR (2013) Distribution and dynamics of epidemic and pandemic *Vibrio parahaemolyticus* virulence factors. Front Cell Infect Microbiol 3:97. doi:10.3389/fcimb.2013.00097

Chakraborty S, Mukhopadhyay AK, Bhadra RK, Ghosh AN, Mitra R, Shimada T, Yamasaki S, Faruque SM, Takeda Y, Colwell RR, Nair GB (2000) Virulence genes in environmental strains of *Vibrio cholerae*. Appl Environ Microbiol 66:4022–4028

Chamberlin CE, Mitchell R (1978) A decay model for enteric bacteria in natural waters. In: Mitchell R (ed) Water pollution microbiology, vol 2. Wiley, New York, pp 325–348

Chandran A, Hatha A, Varghese S, Sheeja KM (2008) Prevalence of multiple drug-resistant *Escherichia coli* serotypes in a tropical estuary, India. Microbes Environ 23:153–158

Chandran A, Suson PS, Thomas AP, Hatha M, Mazumder A (2013) Survival of multi-drug resistant enteropathogenic *Escherichia coli* and *Salmonella* paratyphi in Vembanadu lake as a function of saltwater barrier along southwest coast of India. J Water Health 11:324–332

Chiavelli DA, Marsh JW, Taylor RK (2001) The mannose-sensitive hemagglutinin of *Vibrio cholerae* promotes adherence to zooplankton. Appl Environ Microbiol 67:3220–3225

Clemente S (2008) Ecology and population dynamics of the mangrove clam *Polymesoda Erosa* (Solander 1876) in the mangrove ecosystem. http://library.unigoa.ac.in:8081/xmlui/handle/123456789/395. Accessed 14 Jan 2014

Colburn KG, Kaysner CA, Abeyta C, Wekell C, Wekell MM (1990) *Listeria monocytogenes* in California coast estuarine environment. Appl Environ Microbiol 56:2007–2011

Cole JJ, Prairie YT, Caraco NF, McDowell WH, Tranvik LJ (2007) Plumbing the global carbon cycle: integrating inland waters into the terrestrial carbon budget. Ecosystems 10:171–184

Csonka LN, Epstein W (1996) Osmoregulation. In: Neidhardt FC, Curtiss R, Ingraham JL, Lin ECC, Low KB, Magasanik B, Reznikoj WS, Riley M, Schaechter M, Umbarger HE (eds) *Escherichia coli* and *Salmonella*: cellular and molecular biology, vol 1. ASM Press, Washington, DC, pp 1210–1223

Daniels NA, Shafaie A (2000) A review of pathogenic *Vibrio* infections for clinicians. Infect Med 17:665–685

Daniels NA, Ray B, Easton A, Marano N, Kahn E, McShan AL, Del Rosario L, Baldwin T, Kingsley MA, Puhr ND, Wells JG, Angulo FJ (2000) Emergence of a new *Vibrio parahaemolyticus* serotype in raw oysters: a prevention quandary. J Am Med Assoc 284:1541–1545

Dechet AM, Patricia AY, Koram N, Painter J (2008) Non-foodborne *Vibrio* infections: an important cause of morbidity and mortality in the United States, 1997–2006. Clin Infect Dis 46:970–976

Desai RS, Krishnamurthy NK, Mavinkurve S, Bhosle S (2004) Alkaliphiles in estuarine mangrove regions of Goa. Indian J Mar Sci 33:177–180

DeSousa T, Bhosle S (2012) Isolation and characterization of a lipopeptide bioemulsifier produced by *Pseudomonas nitroreducens* TSB.MJ10 isolated from a mangrove ecosystem. Bioresour Technol 123:256–262

Dhanlakshmi V, Shanthi K, Remia KM (2013) Physiocochemical study of eutrophic pond in Pollachi town Tamilnadu, India. Int J Curr Microbiol Appl Sci 2:219–227

Downing JA, Prairie YT, Cole JJ, Duarte CM, Tranvik LJ (2006) The global abundance and size distribution of lakes, ponds, and impoundments. Limnol Oceanogr 51:2388–2397

Downing JA, Cole JJ, Middelburg JJ, Striegl RG, Duarte CM (2008) Sediment organic carbon burial in agriculturally eutrophic impoundments over the last century. Glob Biogeochem Cycles 22:1–10

Duke NC, Meynecke JO, Dittmann S, Ellison AM, Anger K, Berger U, Cannicci S, Diele K, Ewel KC, Field CD, Koedam N, Lee SY, Marchand C, Nordhaus I, Dahdouh-Guebas F (2007) A world without mangroves? Science 317:41–42

Eiler A, Gonzalez-Rey C, Allen S, Bertilsson S (2007) Growth response of *Vibrio cholerae* and other *Vibrio* spp. to cyanobacterial dissolved organic matter and temperature in brackish water. FEMS Microbiol Ecol 60:411–418

English S, Wilkinson C, Baker V (1997) Survey manual for tropical marine resources, 2nd edn. Australian Institute of Marine Science, Townsville, p 390

Ewel KC, Twilley RR, Ong JE (1998) Different kinds of mangrove forests provide different goods and services. Glob Ecol Biogeogr Lett 7:83–94

Faruque SM, Albert MJ, Mekalanos JJ (1998) Epidemiology, genetics, and ecology of toxigenic *Vibrio cholerae*. Microbiol Mol Biol Rev 62:1301–1314

Fleisher JM, Kay D, Wyer MD, Godfree AF (1998) Estimates of the severity of illnesses associated with bathing in marine recreational waters contaminated with domestic sewage. Int J Epidemiol 27:722–726

Fujioka RS, Hashimoto HH, Siwak EB, Young RH (1981) Eject of sunlight on survival of indicator bacteria in seawater. Appl Environ Microbiol 41:690–696

Gameson ALH, Gould DJ (1975) Ejects of solar radiation on the mortality of some terrestrial bacteria in sea water. In: Gameson ALH (ed) Proceedings of the international symposium on discharge of sewage from sea outfalls. Pergamon Press, London, pp 209–219

Gauthier MJ, Munro PM, Mohajer S (1987) Influence of salts and sodium chloride on the recovery of *Escherichia coli* from seawater. Curr Microbiol 15:5–10
Ghaderpour A, Nasori KNM, Chew LL, Chong VC, Thong KL, Chai LC (2014) Detection of multiple potentially pathogenic bacteria in Matang mangrove estuaries, Malaysia. Mar Pollut Bull 83:324–330
Ghenghesh KS, Ahmed SF, El-Khalek RA, Al-Gendy A, Klena J (2008) Aeromonas-associated infections in developing countries. J Infect Dev Ctries 2:81–98
Glaser M (2003) Interrelations between mangrove ecosystems, local economy and social sustainability in Caeté estuary, North Brazil. Wetl Ecol Manag 11:265–272
Grasso M (1998) Ecological-economic model for optimal mangrove trade off between forestry and fishery production: comparing a dynamic optimization and a simulation model. Ecol Model 112:131–150
Grisi LTCS, Lira KG (2010) The abundance of some pathogenic bacteria in mangrove habitats of paraiba do norte estuary and crabmeat contamination of mangrove crab *Ucides cordatus*. Braz Arch Biol Technol 53:227–234
Grossart HP, Levold F, Allgaier M, Simon M, Brinkhoff T (2005) Marine diatom species harbour distinct bacterial communities. Environ Microbiol 7:860–873
Hacioglu N, Dulger B (2009) Monthly variation of some physicochemical and microbiological parameters in Biga stream (Biga, Canakkale, Turkey). Afr J Biotechnol 8:1927–1937
Hamaidi-Chergui F, Errahmani MB, Benouaklil F, Hamaidi MS (2013) Preliminary study on physicochemical parameters and phytoplankton of chiffa river (Blida, Algeria). J Ecosyst 2013:1–9, http://dx.doi.org/10.1155/2013/148793
Heidelberg JF, Heidelberg KB, Colwell RR (2002) Bacteria of the γ-subclass proteobacteria associated with zooplankton in Chesapeake bay. Appl Environ Microbiol 68:5498–5507
Holguin G, Bashan Y, Vazquez P (2001) The role of sediment microorganisms in the productivity, conservation, and rehabilitation of mangrove ecosystem: an overview. Biol Fertil Soils 33:265–278
Hughes BS, Cullum AJ, Bennett AF (2007) An experimental evolutionary study on adaptation to temporally fluctuating pH in *Escherichia coli*. Physiol Biochem Zool 80:406–421
Huq A, Small EB, West PA, Huq MI, Rahman R, Colwell RR (1983) Ecological relationships between *Vibrio cholerae* and planktonic crustacean copepods. Appl Environ Microbiol 45:275–283
Huss HH (1980) Distribution of *Clostridium botulinum*. Appl Environ Microbiol 39:764–769
Hyndes GA, Nagelkerken I, McLeo RJ, Connolly RM, Lavery PS, Vanderklift MA (2014) Mechanisms and ecological role of carbon transfer within coastal seascapes. Biol Revives Camb Philos Soc 89:232–254
Ishii S, Sadowsky MJ (2008) *Escherichia coli* in the environment: implications for water quality and human health. Microbes Environ 23:101–108
Jackson JK, Murphree RL, Tamplin ML (1997) Evidence that mortality from *Vibrio vulnificus* infection results from single strains among heterogeneous populations in shellfish. J Clin Microbiol 35:2098–2101
Jennerjahn TC, Ittekkot V (2002) Relevance of mangroves for the production and deposition of organic matter along tropical continental margins. Naturwissenschaften 89:23–30
Jiang XT, Peng X, Deng GH, Sheng HF, Wang Y, Zhou HW, Tam NF (2013) Illumina sequencing of 16S rRNA tag revealed spatial variations of bacterial communities in a mangrove wetland. Microb Ecol 66:96–104
Jokinen C, Edge TA, Ho S, Koning W, Laing C, Mauro W, Medeiros D, Miller J, Robertson W, Taboada E, Thomas J, Topp E, Ziebell K, Gannon V (2011) Molecular subtypes of Campylobacter *spp*. Salmonella enterica, and Escherichia coli O157:H7 isolated from faecal and surface water samples in the Oldman River watershed, Alberta, Canada. Water Res 45:1247–1257
Joseph AV, Sasidharan RS, Nair HP, Bhat SG (2013) Occurrence of potential pathogenic *Aeromonas* species in tropical seafood, aquafarms and mangroves off Cochin coast in south India. Vet World 6(6):300–306
Kar CS, Satpathy GR (1995) Mangrove ecosystem and its biodiversity in Orissa coast with special reference to Bhitarkanika. In: PC Mishra, Behera N, Senapathi BK, Guru BC (eds) Advances in ecology and environmental sciences. Vedams Books 419–438
Kassen R, Bell G (1998) Experimental evolution in Chlamydomonas. IV. Selection in environments that vary through time at different scales. Heredity 80:732–741
Kathiresan K, Bingham BL (2001) Biology of mangroves and mangrove ecosystems. Adv Mar Biol 40:81–251
Keller R, Justino JF, Cassini ST (2013) Assessment of water and seafood microbiology quality in a mangrove region in Vitória. Braz J Water Health 11:573–580
Ketola T, Laakso J, Kaitala V, Airaksinen S (2004) Evolution of hsp90 expression in Tetrahymena thermophila (protozoa, ciliata) populations exposed to thermally variable environments. Evolution 58:741–748
Ketola T, Mikonranta L, Zhang J, Saarinen K, Ormala AM (2013) Fluctuating temperature leads to evolution of thermal generalism and preadaptation to novel environments. Evolution 67:2936–2944
Khandeparker R, Verma P, Meena RM, Deobagkar DD (2011) Phylogenetic diversity of carbohydrate degrading culturable bacteria from Mandovi and Zuari estuaries, Goa, west coast of India. Estuar Coast Shelf Sci 95:359–366
Krishna Prasad MB, Ramanathan A (2008) Distribution of rare earth elements in the pichavaram mangrove sediments of the southeast coast of India. J Coast Res 24:126–134
Krishnan KP, Loka Bharathi PA (2009) Organic carbon and iron modulate nitrification rates in mangrove swamps of Goa, south west coast of India. Estuar Coast Shelf Sci 84:419–426

Krishnan KP, Fernandes SO, Chandan GS, Loka Bharathi PA (2007) Bacterial contribution to mitigation of iron and manganese in mangrove sediments. Mar Pollut Bull 54:1427–1433

Kruitwagen G, Pratap HB, Covaci A, Wendelaar Bonga SE (2008) Status of pollution in mangrove ecosystems along the coast of Tanzania. Mar Pollut Bull 56:1022–1031

Kumar HS, Otta SK, Karunasagar I, Karunasagar I (2001) Detection of shiga-toxigenic *Escherichia coli* (STEC) in fresh seafood and meat marketed in Mangalore, India by pcr. Lett Appl Microbiol 33:334–338

La Rosa T, Mirto S, Marino A, Alonzo V, Maugeri TL, Mazzola A (2001) Heterotrophic bacteria community and pollution indicators of mussel farm impact in the Gulf of Gaeta (Tyrrhenian Sea). Mar Environ Res 52:301–321

Leclerc H, Schwartzbrod L, Dei-Cas E (2002) Microbial agents associated with water borne diseases. Crit Rev Microbiol 28(4):371–409

Lee SY (2004) Relationship between mangrove abundance and tropical prawn production: a re-evaluation. Mar Biol 145:943–949

Lees D (2000) Viruses and bivalve shellfish. Int J Food Microbiol 59:81–116

Leroi AM, Lenski RE, Bennett AF (1994) Evolutionary adaptation to temperature. III. Adaptation of *Escherichia coli* to a temporally varying environment. Evolution 48:1222–1229

Lipp EK, Huq A, Colwell RR (2002) Effects of global climate on infectious disease: the cholera model. Clin Microbiol Rev 15:757–770

Lotfy NM, Hassanein MA, Gawad GE, Taweel E, Bassem SM (2011) Detection of *Salmonella* spp. in aquatic insects, fish and water by MPN-PCR. World J Fish Mar Sci 1233:58–66

Malham SK, Rajko-Nenow P, Howlett E, Tuson KE, Perkins TL, Pallett DW, Wang H, Jago CF, Jonese DL, McDonaldb JE (2014) The interaction of human microbial pathogens, particulate material and nutrients in estuarine environments and their impacts on recreational and shellfish waters. Environ Sci Process Impacts 16:2145–2155

Manson FJ, Loneragan NR, Skilleter GA, Phinn SR (2005) An evaluation of the evidence for linkages between mangroves and fisheries: a synthesis of the literature and identification of research directions. Oceanogr Mar Biol Annu Rev 43:483–513

Marinebio (2015) Estuaries, salt marshes, and mangroves – MarineBio.org. MarineBio Conservation Society http://marinebio.org/oceans/estuaries-salt-marshes-mangroves. Accessed 24 Aug 2015

Martinez-Urtaza J, Liebana E, Garcia-Migura L, Perez-Pineiro P, Saco M (2004) Characterization of *Salmonella* enterica serovar typhimurium from marine environments in coastal waters of Galicia (Spain). Appl Environ Microbiol 70:4030–4034

Mastaller M (1996) Destruction of mangrove wetlands-causes and consequences. Nat Res Dev 43(44):37–57

Meibom KL, Li XB, Nielsen AT, Wu CY, Roseman S, Schoolnik GK (2004) The *Vibrio cholerae* chitin utilization program. Proc Natl Acad Sci 101:2524–2529

Meynecke JO, Lee SY, Duke NC, Warnken J (2007) Relationships between estuarine habitats and coastal fisheries in Queensland, Australia. Bull Mar Sci 80:773–793

Mitsuda T, Muto T, Yamada M, Kobayashi N, Toba N, Aihara Y, Ito A, Yokota S (1998) Epidemiological study of a food-borne outbreak of enterotoxigenic *Escherichia coli* O25:NM by plulsed-feild gel electrophoresis and randomly amplified polymorphic DNA analysis. J Clin Microbiol 36:652–656

Momtaz H, Yadollahi S (2013) Molecular characterization of *Listeria monocytogenes* isolated from fresh seafood samples in Iran. Diagn Pathol 8:149, http://www.diagnosticpathology.org/content/8/1/149

Morris JG (2003) Cholera and other types of vibriosis: a story of human pandemics and oysters on the half shell. Clin Infect Dis 37:272–280

MSI (2013) Mangrove ecosystem of Goa. Mangrove Fauna. http://mangrovesocietyofindia.in/mangrove_fauna.php. Accessed 25 Dec 2013

Munro PM, Clement RL, Flatau GN, Gauthier MJ (1994) Effect of thermal, oxidative, acidic, osmotic, or nutritional stresses on subsequent culturability of *Escherichia coli* in seawater. Microb Ecol 27:57–63

Munro PM, Flatau GN, Clement RL, Gauthier MJ (1995) Influence of the RpoS (KatF) sigma factor on maintenance of viability and culturability of *Escherichia coli* and *Salmonella typhimurium* in seawater. Appl Environ Microbiol 61:1853–1858

Nagelkerken I, Blaber SJM, Bouillon S, Green P, Haywood M, Kirton LG, Meynecke JO, Pawlik J, Penrose HM, Sasekumar A, Somerfield PJ (2008) The habitat function of mangroves for terrestrial and marine fauna: a review. Aquat Bot 89:155–185

Nagvenkar GS, Ramaiah N (2009) Abundance of sewage-pollution indicator and human pathogenic bacteria in a tropical estuarine complex. Environ Monit Assess 155:245–256

Olatayo AA (2014) Assessment of physico-chemical parameters of waters in Ilaje local government area of Ondo state, Nigeria. Int J Fish Aquat Stud 1:84–92

Ottaviani D, Santarelli S, Bacchiocchi S, Masini L, Ghittino C, Bacchiocchi I (2006) Occurrence and characterization of *Aeromonas* spp. in mussels from the Adriatic sea. Food Microbiol 23:418–422

Paula J, Schleyer MH, Krishnan KP, Loka Bharathi PA (2009) Organic carbon and iron modulate nitrification rates in mangrove swamps of Goa, south west coast of India. Estuar Coast Shelf Sci 84:419–426

Pauly D, Ingles J (1986) The relationship between shrimp yields and intertidal vegetation (Mangrove) areas: a reassessment. IOC/FAOWorkshop on Recruitment in Tropical Coastal Demersal Communities IOC, UNESCO, Paris, pp 227–284

Penha-Lopes G, Torres P, Cannicci S, Narciso L, Paula J (2011) Monitoring anthropogenic sewage pollution on mangrove creeks in southern Mozambique: a test of palaemon concinnus Dana, 1852 (palaemonidae) as a biological indicator. Environ Pollut 159:636–645

Peters EC, Gassman NJ, Firman JC, Richmond RH, Power EA (1997) Ecotoxicology of tropical marine ecosystems. Environ Toxicol Chem 16:12–40

Poharkar KV, Kerkar S, Doijad SP, Barbuddhe SB (2014) Prevalence and genetic profiles of Escherichia coli from mangroves and mangrove associated foods off Goa, India. Mar Pollut Bull 85:86–91

Ramaiah N, Rodrigues V, Alwares E, Rodrigues C, Baksh R, Jayan S, Mohandass C (2007) Sewage-pollution indicator bacteria. In: Shetye SR, DileepKumar M, Shankar D (eds) The Mandovi and Zuari estuaries, Ith edn. National Institute of Oceanography, Panjim, pp 115–120

Reeve G, Martin DL, Pappas J, Thompson RE, Greene KD (1989) An outbreak of shigellosis associated with the consumption of raw oysters. N Engl J Med 321:224–227

Rhodes MW, Kator H (1988) Survival of *Escherichia coli* and *Salmonella* spp. in Estuarine Environments. Appl Environ Microbiol 54:2902–2907

Ristori CA, Iaria ST, Gelli DS, Rivera IN (2007) Pathogenic bacteria 451 associated with oysters (Crassostrea brasiliana) and estuarine water along the south coast of Brazil. Int J Environ Health Res 17:259–269

Rodrigues V, Ramaiah N, Kakti S, Samant D (2011) Long-term variations in abundance and distribution of sewage pollution indicator and human pathogenic bacteria along the central west coast of India. Ecol Indic 11:318–327

Rodríguez-Martínez S, Cervero-Arag'o S, Gil-Martin I, Araujo R (2013) Multilocus sequence typing of Campylobacter jejuni and Campylobacter coli strains isolated from environmental waters in the Mediterranean area. Environ Res 127:56–62

Rönback P, Crona B, Ingwall L (2007) The return of ecosystem goods and services in replanted mangrove forests: perspectives from local communities in Kenya. Environ Conserv 34:313–324

Ross J (2013) Local man, 71, hospitalized with *Vibrio vulnificus* bacterial infection http://www.ocala.com dated 28th October 2013. Accessed 2 Dec 2013

Rozen Y, Belkin S (2001) Survival of enteric bacteria in seawater. FEMS Microbiol Rev 25:513–529

Saenger P (2002) Mangrove ecology, siviculture and conservation. Kluwer Academic Publishers, Dordrecht, pp 11–18

FSN (Food Safety News) (2012) CDC: 316 ill in multistate outbreak linked to Sushi tuna. *Salmonella* infections reported in 26 states and Washington, DC. http://www.foodsafetynews.com. Accessed 21 Sep 2015

Schutze GE, Sikes JD, Stefanova R, Cave MD (1999) The home environment and salmonellosis in children. Pediatrics 103(1):e1

Scott TM (2002) Detection of pathogens in the Venice lagoon: components for modeling and risk assessment. In: Proceedings of the bilateral Italy-U.S. workshop on mathematical models for water quality in an isolated environment. Venice, Italy. 14–16

Singh AK, Ansari A, Kumar D, Sarka UK (2012) Status, biodiversity and distribution of mangroves in India: an overview. Proceedings of international day for biodiversity, marine biodiversity, Uttarpradesh state board. pp 57–69

Sobrinho PS, Destro MT, Franco BD, Landgraf M (2011) Occurrence and distribution of *Vibrio parahaemolyticus* in retail oysters in Sao Paulo State. Braz J Food Microbiol 28:137–140

Srilatha G, Thilagavathi B, Varadharajan D (2012) Studies on the physico-chemical status of muthupet mangrove, south east coast of India. Adv Appl Sci Res 3:201–207

Staples DJ, Vance DJ, Heales DS (1985) Habitat requirements of juvenile penaeid prawns and their relationship to offshore fisheries. In: Rothlisberg PC, Hill BJ, Staples DJ (eds) Second Australian national prawn seminar, Kooralbyn, Queensland. CSIRO, Canberra, pp 47–54

Stone R (2006) A rescue effort for tsunami-ravaged mangrove forests. Science 314:404

Tamplin ML, Jackson JK, Buchrieser C, Murphree RL (1996) Pulsed-field gel electrophoresis and ribotype profiles of clinical and environmental *Vibrio vulnificus* isolates. Appl Environ Microbiol 62:3572–3580

Tanaka Y, Yamaguchi N, Nasu M (2000) Viability of *Escherichia coli* O157:H7 in natural river water determined by the use of flow cytometry. J Appl Microbiol 88:228–236

Thayumanavan T, Vivekhanandhan G, Savithakumari K, Subashkumar R, Lakshmanaperumalsamy P (2003) Incidence of haemolysin positive and drug resistant *Aeromonas hydrophila* in freshly caught finfish and prawn collected from major commercial fishes of coastal south India. FEMS Immunol Med Microbiol 36:41–45

Touron A, Berthe T, Gargala G, Fournier M, Ratajczak M, Servais P, Petit F (2007) Assessment of faecal contamination and the relationship between pathogens and faecal bacterial indicators in an estuarine environment (Seine, France). Mar Pollut Bull 54:1441–1450

UNEP-WCMC (2007) The mangroves of West-Central Africa. UNEP WCMC report pp 92. www.unep.org. Accessed 5 Aug 2015

Vannucci M (2000) What is so special about mangroves? Braz J Biol 61:599–603

Walk ST, Alm EW, Calhoun LM, Mladonicky JM, Whittam TS (2007) Genetic diversity and population structure of Escherichia coli isolated from freshwater beaches. Environ Microbiol 9:2274–2288

Walters BB, Rönnbäck P, Kovacs JM, Crona B, Hussain SA, Badola R, Primavera JH, Barbier E, Dahdouh-Guebas F (2008) Ethnobiology, socio-economics and management of mangrove forests: a review. Aquat Bot 89:220–236

Wanjugi P, Harwood VJ (2013) The influence of predation and competition on the survival of commensal and pathogenic fecal bacteria in aquatic habitats. Environ Microbiol 2013(15):517–526

Watnick PI, Fullner KJ, Kolter R (1999) A role for the mannose-sensitive hemagglutinin in biofilm formation by *Vibrio cholerae* El Tor. J Bacteriol 181:3606–3609

Winfield MD, Groisman EA (2003) Role of nonhost environments in the lifestyles of *Salmonella* and *Escherichia coli*. Appl Environ Microbiol 69:3687–3694

Yeragi SS, Yeragi SG (2014) Status, Biodiversity and distribution of mangroves in south konkan, Sindhudurg district, Maharashtra state India an overview. Int J Life Sci 2:67–69

Zhou X, Konkel ME, Call DR (2009) Type III secretion system 1 of *Vibrio parahaemolyticus* induces oncosis in both epithelial and monocytic cell lines. Microbiology 155:837–851

Zuckerman JN, Rombo L, Fisch A (2007) The true burden and risk of cholera: implications for prevention and control. Lancet Infect Dis 7:521–530

Global Concerns of Ship's Ballast Water Mediated Translocation of Bacteria

16

Lidita Khandeparker and A.C. Anil

Abstract

Shipping facilitates transportation of over 90 % of the world's commodities and eventually transfers ~10 billion tons of ballast water which is routinely taken aboard vessels to maintain their stability. The ballast water contains different nonindigenous species (NIS) that include bacteria, microscopic algae, virus, invertebrates, vertebrates, plants, etc. The movements of cargo ships between the continents and along the coastlines have facilitated the spread of these marine species to new localities where they have established themselves mostly in the ports and the coastal niches. Thus, ballast water is recognized as a potent vector of invasive coastal marine species that has threatened the biodiversity. The introduction of such alien organisms into the ecosystem is termed as "bioinvasion" and is among one of the greatest threats to the ocean health.

Microorganisms unlike many other organisms can be introduced into alien environments in bigger numbers as they are highly abundant, are capable of forming resting stages, and can withstand adverse environmental conditions. In ships, microorganisms are found either in ballast water, residual sediments, or biofilms formed on the interior tank surfaces and are also associated with the plankton which serves as microhabitat for bacteria. It is now well established that one of the primary vectors for the transport of vegetative and resting stages of aquatic microbes including disease-causing potentially pathogenic bacteria globally is the ballast water. The introduction of such pathogens has direct impact on the human health and thus has societal relevance. Thus, understanding microbiology of ballast water is of environmental importance as the discharged ballast water may contain infectious pathogens. The ballast water performance standards for different size classes of organisms are provided by the International Convention for the Control and Management of Ship's

L. Khandeparker (✉) • A.C. Anil
CSIR-National Institute of Oceanography,
Dona Paula, Goa 403004, India
e-mail: klidita@nio.org

M.M. Naik, S.K. Dubey (eds.), *Marine Pollution and Microbial Remediation*,
DOI 10.1007/978-981-10-1044-6_16

Ballast Water and Sediments (IMO 2004). As per these standards (D2 regulation of the convention), the planktonic forms need to be substantially reduced before discharge. In view of this, technologists are considering various options while designing the treatment technologies. The primary reason is the alteration in the microbial population as planktonic organisms are associated with the bacteria, and these bacteria are released in large numbers, while large planktonic organisms are destroyed.

The routine methods used for bacteriological assessment of seawater quality are lengthy. Recently, many scientists working in diverse fields have used flow cytometry (FCM), a technique that allows rapid and accurate counting of bacteria. A matrix-assisted laser desorption ionization-time of flight mass spectrometry (MALDI-TOF MS) using mass spectral libraries of different bacterial species has been also used to characterize different bacteria in ballast water. This method is cost-efficient, quicker, and reliable. However, most (~99 %) of prokaryotes in the environment cannot be cultured in the laboratory. The hurdle of exploring this segment of the microbiome has recently been solved by the development of different molecular tools which can accurately identify the unculturable microbiome and classify them further to species level, thereby overcoming taxonomic ambiguity. An overview of the ecology of microbes in ballast water and different analytical techniques that are used for monitoring bacteria in ballast water other than conventional methods is addressed.

16.1 Introduction

Ships move over 90 % of the world's commodities and in turn are responsible for the global transfer of approximately ten billion tons of ballast water which is carried by the ships to maintain their stability. While filling the ballast tanks, taxa of organisms are also boarded and carried from one region to another, the discharged ballast water release these organisms into new environment, creating a long distance inoculation and dispersion of various organisms. The introduction of such alien organisms into the ecosystem is termed as "bioinvasion" and is one among those listed as the greatest threats to the ocean health. Recently, ballast water has been identified as an important bioinvasion vector that can threaten the naturally evolved biodiversity (Anil et al. 2002; Khandeparker and Anil 2013). Other than ballast water, marine species are also transported as fouling communities attached to hulls, in sea chests, associated sediments, and other recesses in the hull structure (Carlton 1985, 1999; Coutts et al. 2003). When in their native locations, the invading organisms live in semblance and are controlled by a range of ecosystem interactions. However, when they are introduced into alien environments, they can either change the biodiversity of the food web or can directly impact the society and human health (Anil et al. 2002).

The ship's ballast tanks contain diverse groups of invertebrates, vertebrates, plants, microscopic algae, bacteria, viruses, etc. that are nonindigenous (Williams et al. 1988; Smith et al. 1996; Ruiz et al. 2000; Mimura et al. 2005; Drake et al. 2007). In the recent years, there has been a tremendous increase in the movements of vessels between the continents and along the coastlines that has facilitated the spread of NIS, especially in the estuarine and coastal water bodies. Since ballast water is thought to be the prime vector in the introduction of NIS of aquatic organisms to ports around the world, it is apparent that the coastal marine ecosystems may be the ones that are greatly invaded ecosystems worldwide (Grosholz 2002; Minton et al. 2005; Verling et al. 2005).

16.2 Pathogenic Bacteria in Ballast Water

Microorganisms are highly abundant in the natural environment with an ability to form spores, cysts, or physiological resting stages prolonging their survival and can withstand adverse environmental conditions owing to which they are introduced in large numbers into alien environments when compared to other organisms (Roszak et al. 1984; Hallegraeff and Bolch 1992). In a ship, microorganisms can be found in several locations such as within the ballast water, biofilms formed on the interior tank surfaces, residual sediments, etc. (Drake et al. 2007; Meyer et al. 2000). When compared to macro-organisms such as copepods and fish, aquatic microorganisms are orders of magnitude more abundant. The naturally occurring bacteria and viruses occur in the order of 10^6–10^{11} l^{-1} (Ducklow and Shiah 1993; Fuhrman 1999; Wommack and Colwell 2000). Microorganisms being smaller in size can be passively dispersed once released and have simple requirements for their survival owing to which they are well suited to be invasive when compared to metazoans (Deming 1997). Earlier studies have indicated that when compared to terrestrial pathogens, marine pathogens can spread locally much faster (Mc Callum et al. 2004). These pathogenic bacteria along with viruses, protists, and microalgae can be devastating on the economic resources and the ecosystems. Taking into consideration the relatively growing shipping industry, it is apparent that the risk of global dispersal of aquatic pathogens needs immediate attention.

The most extensively studied bacterium in ballast water has been *Vibrio cholerae*. *V. cholerae* is responsible for life-threatening diarrheal disease cholera that results in rapid dehydration and even death of infected persons (Finkelstein 1996). More than 200 O-antigen serogroups of *V. cholerae* are identified so far, but only two serogroups, O1 and O139, are known to cause epidemics and pandemics (Sack et al. 2003). Ballast water is one of the primary vectors responsible for the global transport of toxigenic *V. cholerae O1* and *O139* (Ruiz et al. 2000). *V. cholerae* O1 related epidemic in Latin America in 1991 raised concerns over the problem of ballast water related transfer of pathogenic microorganisms. Cholera was then placed as one of the "Ten Most Unwanted" in the Global Ballast Water Management Program (GloBallast 2002). Although few species of *V. cholerae* are pathogenic to humans, most of them are heterotrophic and metabolically versatile constituting part of the normal aquatic flora in ponds, lakes, riverine, estuarine environments, etc. (Paul et al. 2012). In India cholera epidemics crop up with seasonal regularity in the Gangetic delta (Colwell 1996), generally recur twice a year (Alam et al. 2007), with the maximum number of cases in the month of July–October (WHO 2010). Recently, the fresh outbreak that has occurred in Haiti highlights the harshness of this disease (Centers for Disease Control and Prevention 2010). Hasan et al. (2012) used a genomic approach to examine whether there is any difference among the Haitian *V. cholerae* O1 and non-O1/O139 strains. The comparative genome analyses of 76 genomes and the reference *V. cholerae* strains isolated from concurrent epidemics outside Haiti and other *V. cholerae* genomes that are available in the public database indicted that within the genome of *V. cholerae*, there are a considerable diversity and ongoing flux.

Plankton, especially zooplankton, serves as microhabitat for bacteria (Tang 2005). Ruiz et al. (2000) reported that the introduction of most of the pathogens to Chesapeake Bay was attributed to the bacteria associated with the plankton rather than the water column. The larval forms of marine invertebrates have an extensive surface area and complex body structures that can provide shelter to the bacteria. It is well known that copepod exoskeleton and their gut lining provide favorable surfaces for the bacterial attachment (Carman and Dobbs 1997). The bacteria associated with the copepods can be orders of magnitude higher than that in the ambient water (Tang 2005). Thus, plankton-associated microbes are of utmost importance in marine bioinvasion and in the implementation of the ship's ballast water treatment technologies. The carcasses of marine zooplankton when exposed to natural seawater are rapidly colonized and decomposed by the ambient bacteria (Tang et al. 2006), in turn release large numbers of bacteria and contribute to the bacterial production in the surrounding water. In fact in many aquatic ecosystems, decomposition of zooplankton carcasses by the microbes provides an alternative pathway for nutrient regeneration, elemental recycling, and microbial production (Lee and Fisher 1992).

The issue of NIS and ballast water is a cause of concern since the late 1980s and is under active consideration by the International Maritime Organization (IMO). The Ballast Water Management (BWM) Convention was adopted in February 2004 to address this issue. As per the regulations of the convention, the ship's ballast water needs to be managed to prescribe standards so that the risk of transferring harmful organisms is mitigated. This convention provides two standards (D1 and D2). Under regulation D1 of the convention, which is related to ballast water exchange standards, ballast water should be exchanged with an efficiency of at least 95 % volumetric exchange of the ballast water. This is not necessarily a full proof preventive method as this process can retain traces of coastal organisms. If these organisms happen to be euryhaline, they can still pose a threat. A similar conclusion was drawn by (Wonham et al. 2001; Taylor et al. 2007). According to regulation D2, which is ballast water performance standards, the treated ballast water that can be discharged should reduce the toxicogenic *V. cholerae* (O1 and O139) to <1 CFU 100 ml^{-1} or <1 CFU g^{-1} (wet weight) zooplankton in the sample. *Escherichia coli* numbers should be <250 CFU 100 ml^{-1} and intestinal *Enterococci* should be <100 CFU 100 ml^{-1}.

16.3 Treatment Technologies

A number of technologies (physical and chemical) or approaches have been evaluated for the treatment of ballast water (Jyoti and Pandit 2001; Lloyd's Register 2010). Recently, Tsolaki and Diamadopoulos (2010) summarized different available technologies used for ballast water treatment. Onboard treatment methods are categorized as physical separation, mechanical, or chemical methods. The biochemical composition and strength of the exoskeleton are different for different invertebrate larval forms; thus the energy required to macerate or destroy different size classes of larval forms is different (Holm et al. 2008). While treating macro-organisms, bacteria are released in large numbers; thus it is important not only to destroy the macro-organisms but also to kill the microbes harbored by them which is recently being realized as an additional concern in the ballast water treatment. It is evident that a combination of different methods can be more effective than one single method. Gavand et al. (2007) reported that combining sonication and advanced chemical oxidants could be more promising to eliminate aquatic algae and macroinvertebrates in ballast water. The cell membrane once ruptured (a physical consequence of cavitation), chemical oxidants can enter and attack the internal structures of the cell (Anand et al. 2007). Approximately 80 % of the zooplankton present in the seawater could be killed using hydrodynamic cavitation (Sawant et al. 2008). Jyoti and Pandit (2004) combined hydrodynamic cavitation, acoustic cavitation, and hydrogen peroxide and demonstrated that this combination can be better than anyone on its own for reducing the heterotrophic plate count bacteria as well as indicator microorganisms such as the total coliforms, fecal coliforms, and fecal *Streptococci*.

Recently, ultrasonic treatment was explored on the larvae of a dominant fouling organism, *Balanus amphitrite*, and the energy required for their destruction was quantified (Seth et al. 2010). It was observed that subsequent to pulverization of the barnacle larvae, bacterial abundance increased and the rate at which the bacteria were released was dependent on the power level and the treatment time (Seth et al. 2010). In another study by Khandeparker and Anil (2013), the epibiotic and endobiotic bacteria associated with barnacle nauplii, veliger larvae, and adults of the copepod *Oithona* sp. were characterized and quantified. According to this study, the smallest larval form examined, the veliger larva, harbored maximum numbers of bacteria, whereas the barnacle nauplius with the largest biovolume harbored the least. This indicates that size should not be overlooked during the treatment. The same study also reported that the abundance of *E. coli* and potential pathogenic bacteria such as *Streptococcus faecalis* and *V. cholerae* increased substantially subsequent to pulverization of zooplankton. When this pulverized zooplankton was aged in the dark to assess the contribution of bacteria from decaying debris, *Chromobacterium violaceum* emerged which is an opportunistic pathogen in animals and humans (Khandeparker and Anil 2013). Thus, microbiology of ballast water is of great relevance to the environment and ecosystem functioning as the discharged ballast water may harbor infectious pathogens, leading to their global distribution (Ruiz et al. 2000).

16.4 Methods for Quantification

The different conventional methods routinely used for bacteriological analysis of seawater are time consuming. The culturable pathogenic bacteria as specified in D2 guidelines for performance of ballast water treatment technologies (2004) are routinely estimated by the plating methods using specific media (Khandeparker et al. 2015). However, the plating method cannot account for the cells that are in a dormant state. It has been demonstrated that viable but non-culturable bacteria are active in metabolism and can still be infectious.

Recently, flow cytometry is being extensively used for counting microorganisms (Marie et al. 1997; Davey et al. 1999; Rattanasomboon et al. 1999; Ivanov 2000; Shvalov et al. 2000) and assessing their viability (Lopez-Amoros et al. 1997). It has also been used to evaluate the abundance of selected bacterial species in ballast water samples (Joachimsthal et al. 2004). Cells are usually labeled with fluorescent tags which allow them to be electronically identified while passing through a beam of laser light. This technique is comparatively faster and accurate (the most common instrument can detect more than 1000 cells s^{-1}). The instrument can analyze single cells, and it is possible to separate the heterogenous populations based upon differences in their size, internal complexity, and fluorescence which can be measured on each particle or cell. Flow cytometry can detect both nonviable and viable but non-culturable microorganisms which can still pose health risks which are not possible using culturing methods (Joachimsthal et al. 2003).

The use of fluorescence in situ hybridization (FISH) to identify specific bacteria within environmental samples has become a powerful tool that helps in molecular phylogenetic discrimination. Flow cytometry combined with FISH is currently a popular method of enumerating specific cells in environmental samples (Thomas et al. 1997; Not et al. 2002). Recently, Tomaru et al. (2010, 2014) assessed bacterial community in ballast water before and after a mid-ocean exchange using denaturing gradient gel electrophoresis (DGGE) and demonstrated changes in microbial communities over the course of the voyage. Recently, real-time PCR and NASBA methods have also been evaluated to specifically detect 1 CFU/100 ml of *V. cholerae* in ballast water. For this, ballast water samples were spiked with *V. cholerae* cells followed by enriching in alkaline peptone water before PCR or NASBA detection. This method is quick, can be performed within 7 h, and has the potential to be used for inspection of ballast water and enforcement control (Fykse et al. 2012). Emami et al. (2012) also evaluated a rapid and cost-effective matrix-assisted laser desorption ionization-time of flight mass spectrometry (MALDI-TOF MS) method for monitoring culturable bacteria in ballast water. Several marine bacterial species were characterized using this method. However, only those bacteria which can be cultured can be identified and only 0.001–0.1 % of marine bacteria are culturable (Oren 2004).

In order to address the genetics of uncultured organisms, metagenomics, the genomic analysis of microorganisms, is recently emerging as a powerful tool. The capability of high-throughput sequencing of 16S rRNA gene sequences using next-generation sequencing (NGS) technologies has been crucial in facilitating the discovery of microbiota biodiversity (Whiteley et al. 2012). Since the microbial world is the largest unexplored reservoir of biodiversity on the earth and many of the microorganisms have major impacts on oceanic processes, exploring and discovering microbiota invading different ecosystems through ballast water using metagenomics approach is a step ahead.

Acknowledgments We are grateful to the Director National Institute of Oceanography, Goa, for his support and encouragement. This is a CSIR-NIO contribution.

References

Alam M, Sultana M, Nair GB, Siddique AK, Hasan NA, Sack RB, Sack DA, Ahmad KU, Sadique A, Watanabe H, Grim CJ, Huq A, Colwell RR (2007) Viable but nonculturable *Vibrio cholerae* O1 in biofilms in the aquatic environment and their role in cholera transmission. Proc Natl Acad Sci 104(45):17801–17806

Anand H, Balasundaram B, Pandit AB, Harrison STL (2007) The effect of chemical pretreatment combined with mechanical disruption on the extent of disruption and release of intracellular protein from *E. coli*. Biochem Eng J 35(2):166–173

Anil AC, Venkat K, Sawant SS, Dileepkumar M, Dhargalkar VK, Ramaiah N, Harkantra SN, Ansari ZA (2002) Marine bioinvasion: concern for ecology and shipping. Curr Sci 83(3):214–219

Carlton JT (1985) Transoceanic and interoceanic dispersal of coastal marine organisms: the biology of ballast water. Oceanogr Mar Biol 23:313–371

Carlton JT (1999) Molluscan invasions in marine and estuarine communities. Malacologia 41(2):439–454

Carman KR, Dobbs FC (1997) Epibiotic microorganisms on copepods and other marine crustaceans. Microsc Res Tech 37(2):116–135

Centers for Disease Control and Prevention (2010) Update: cholera outbreak—Haiti, 2010. Morb Mortal Wkly Rep 59:1473–1479

Colwell RR (1996) Global climate and infectious disease: the cholera paradigm. Science 274(5295):2025–2031

Coutts AD, Moore KM, Hewitt CL (2003) Ships' sea-chests: an overlooked transfer mechanism for non-indigenous marine species? Mar Pollut Bull 46(11):1510–1513

Davey HM, Jones A, Shaw AD, Kell DB (1999) Variable selection and multivariate methods for the identification of microorganisms by flow cytometry. Cytometry 35(2):162–168

Deming JW (1997) Unusual or extreme high-pressure marine environments. In: Hurst CJ, Knudsen GR, McInerney MJ, Stetzenback LD, Walter MV (eds) ASM manual of environmental microbiology. ASM Press, Washington DC, pp 366–376

Drake LA, Doblin MA, Dobbs FC (2007) Potential microbial bioinvasions via ships' ballast water, sediment, and biofilm. Mar Pollut Bull 55(7):333–341

Ducklow HW, Shiah FK (1993) Bacterial production in estuaries. In: Ford TE (ed) Aquatic microbiology: an ecological approach. Blackwell Scientific Publications, Boston, pp 261–287

Emami K, Askari V, Ullrich M, Mohinudeen K, Anil AC, Khandeparker L, Burgess JG, Mesbahi E (2012) Characterization of bacteria in ballast water using MALDI-TOF mass spectrometry. PLoS One 7(6):38515

Finkelstein RA (1996) Cholera, *Vibrio cholerae* O1 and O139, and other pathogenic Vibrios. Medical Microbiology, 4th edn. University of Texas Medical Branch at Galveston, Galveston, Chapter 24

Fuhrman JA (1999) Marine viruses and their biogeochemical and ecological effects. Nature 399(6736):541–548

Fykse EM, Nilsen T, Nielsen AD, Tryland I, Delacroix S, Blatny JM (2012) Real-time PCR and NASBA for rapid and sensitive detection of *Vibrio cholerae* in ballast water. Mar Pollut Bull 64(2):200–206

Gavand MR, McClintock JB, Amsler CD, Peters RW, Angus RA (2007) Effects of sonication and advanced chemical oxidants on the unicellular green alga Dunaliella tertiolecta and cysts, larvae and adults of the brine shrimp Artemia salina: a prospective treatment to eradicate invasive organisms from ballast water. Mar Pollut Bull 54(11):1777–1788

Grosholz E (2002) Ecological and evolutionary consequences of coastal invasions. Trends Ecol Evol 17(1):22–27

Hallegraeff GM, Bolch CJ (1992) Transport of diatom and dinoflagellate resting spores in ships' ballast water: implications for plankton biogeography and aquaculture. J Plankton Res 14(8):1067–1084

Hasan NA, Choi SY, Eppinger M, Clark PW, Chen A, Alam M, Haley BJ, Taviani E, Hine E, Su Q, Tallon LJ, Prosper JB, Furth K, Hoq MM, Li H, Fraser-Liggett CM, Cravioto A, Huq A, Ravel J, Cebula TA, Colwell RR (2012) Genomic diversity of 2010 Haitian cholera outbreak strains. Proc Natl Acad Sci U S A 109(29):E2010–E2017

Holm ER, Stamper DM, Brizzolara RA, Barnes L, Deamer N, Burkholder JM (2008) Sonication of bacteria, phytoplankton and zooplankton: application to treatment of ballast water. Mar Pollut Bull 56(6):1201–1208

International Maritime Organization (2004) International convention for the control and management of ships' ballast water and sediments. International Maritime Organization, London

Ivanov V (2000) Monitoring the bacterial neuston. In: Rapid methods for analysis of biological materials in the environment. Springer, Dordrecht, pp 67–72

Joachimsthal EL, Ivanov V, Tay JH, Tay STL (2003) Flow cytometry and conventional enumeration of microorganisms in ships' ballast water and marine samples. Mar Pollut Bull 46(3):308–313

Joachimsthal EL, Ivanov V, Tay SL, Tay JH (2004) Bacteriological examination of ballast water in Singapore Harbour by flow cytometry with FISH. Mar Pollut Bull 49(4):334–343

Jyoti KK, Pandit AB (2001) Water disinfection by acoustic and hydrodynamic cavitation. Biochem Eng J 7(3):201–212

Jyoti KK, Pandit AB (2004) Effect of cavitation on chemical disinfection efficiency. Water Res 38(9):2249–2258

Khandeparker L, Anil AC (2013) Association of bacteria with marine invertebrates: implications for ballast water management. EcoHealth 10(3):268–276

Khandeparker L, Anil AC, Naik SD, Gaonkar CC (2015) Daily variations in pathogenic bacterial populations in a monsoon influenced tropical environment. Mar Pollut Bull 96:337–343

Lee BG, Fisher NS (1992) Decomposition and release of elements from zooplankton debris. Mar Ecol Prog Ser 88:117

Lloyd's Register Report (2010) Ballast water treatment technology, current status. Feb 2010. 3rd edn, London, UK, pp 7–35

López-Amorós R, Castel S, Comas-Riu J, Vives-Rego J (1997) Assessment of *E. coli* and *Salmonella* viability and starvation by confocal laser microscopy and flow cytometry using rhodamine 123, DiBAC4 (3), propidium iodide, and CTC. Cytometry 29(4):298–305

Marie D, Partensky F, Jacquet S, Vaulot D (1997) Enumeration and cell cycle analysis of natural populations of marine picoplankton by flow cytometry using the nucleic acid stain SYBR Green I. Appl Environ Microbiol 63(1):186–193

McCallum HI, Kuris A, Harvell CD, Lafferty KD, Smith GW, Porter J (2004) Does terrestrial epidemiology apply to marine systems? Trends Ecol Evol 19(11):585–591

McConnell M (2002) GloBallast legislative review – final report. GloBallast Monograph Series No. 1. IMO London

Meyer AE, Baier R, Hulsmann N, Galil B, Friedmann D, Forsberg R (2000) Risk assessment, prediction, and limitation of transport of bioinvaders in biofilms. Abstracts book, American Society of Limnology and Oceanography, Aquatic Sciences Meeting, Copenhagen, Denmark

Mimura H, Katakura R, Ishida H (2005) Changes of microbial populations in a ship's ballast water and sediments on a voyage from Japan to Qatar. Mar Pollut Bull 50(7):751–757

Minton MS, Verling E, Miller AW, Ruiz GM (2005) Reducing propagule supply and coastal invasions via ships: effects of emerging strategies. Front Ecol Environ 3(6):304–308

Not F, Simon N, Biegala IC, Vaulot D (2002) Application of fluorescent in situ hybridization coupled with tyramide signal amplification (FISH-TSA) to assess eukaryotic picoplankton composition. Aquat Microb Ecol 28(2):157–166

Oren A (2004) Prokaryote diversity and taxonomy: current status and future challenges. Philos Trans R Soc Lond 359:623–638

Paul D, Mahmud ZH, Islam MS, Neogi SB, Islam MS, Jahid IK (2012) Physiochemical conditions and contamination with Vibrios of surface water at Matlab, Bangladesh. J Microbiol Biotechnol Food Sci 2(2):713–729

Rattanasomboon N, Bellara SR, Harding CL, Fryer PJ, Thomas CR, Al-Rubeai M, McFarlane CM (1999) Growth and enumeration of the meat spoilage bacterium Brochothrix thermosphacta. Int J Food Microbiol 51(2):145–158

Roszak DB, Grimes DJ, Colwell RR (1984) Viable but nonrecoverable stage of *Salmonella enteritidis* in aquatic systems. Can J Microbiol 30(3):334–338

Ruiz GM, Rawlings TK, Dobbs FC, Drake LA, Mullady T, Huq A, Colwell RR (2000) Global spread of microorganisms by ships. Nature 408(6808):49–50

Sack RB, Siddique AK, Longini IM, Nizam A, Yunus M, Islam MS, Morris JG, Ali A, Huq A, Nair B, Qadri F, Faruque SH, Sack DA, Colwell RR (2003) A 4-year study of the epidemiology of *Vibrio cholerae* in four rural areas of Bangladesh. J Infect Dis 187(1):96–101

Sawant SS, Anil AC, Krishnamurthy V, Gaonkar C, Kolwalkar J, Khandeparker L, Desai DV, Mahulkar AV, Pandit AB (2008) Effect of hydrodynamic cavitation on zooplankton: a tool for disinfection. Biochem Eng J 42(3):320–328

Seth N, Chakravarty P, Khandeparker L, Anil AC, Pandit AB (2010) Quantification of the energy required for the destruction of *Balanus amphitrite* larva by ultrasonic treatment. J Mar Biol Assoc U K 90(07):1475–1482

Shvalov AN, Soini JT, Surovtsev IV, Kochneva GV, Sivolobova GF, Petrov AK, Maltsev VP (2000) Individual *Escherichia coli* cells studied from light scattering with the scanning flow cytometer. Cytometry 41(1):41–45

Smith LD, Wonham MJ, McCann LD, Reid DM, Carlton JT, Ruiz GM (1996) Biological invasions by nonindigenous species in United States waters: quantifying the role of ballast water and sediments. Parts I and II. The National Biological Invasions Shipping Study II (NABISS II) US Coast Guard and the US Department of Transportaion, Washington DC (Rep. No. CG-D-02-97, Gov. Access. No. ADA321543)

Tang KW (2005) Copepods as microbial hotspots in the ocean: effects of host feeding activities on attached bacteria. Aquat Microb Ecol 38(1):31–40

Tang KW, Hutalle KML, Grossart HP (2006) Microbial abundance, composition and enzymatic activity during decomposition of copepod carcasses. Aquat Microb Ecol 45(3):219–227

Taylor MD, MacKenzie LM, Dodgshun TJ, Hopkins GA, de Zwart EJ, Hunt CD (2007) Trans-Pacific shipboard trials on planktonic communities as indicators of open ocean ballast water exchange. Mar Ecol Prog Ser 350:41–54

Thomas JC, Desrosiers M, Pierre YS, Lirette P, Bisaillon JG, Beaudet R, Villemur R (1997) Quantitative flow cytometric detection of specific microorganisms in soil samples using rRNA targeted fluorescent probes and ethidium bromide. Cytometry 27(3):224–232

Tomaru A, Kawachi M, Demura M, Fukuyo Y (2010) Denaturing gradient gel electrophoresis shows that bacterial communities change with mid-ocean ballast water exchange. Mar Pollut Bull 60(2):299–302

Tomaru A, Kawachi M, Demura M, Fukuyo Y (2014) Changes in microbial communities, including both uncultured and culturable bacteria, with Mid-ocean Ballast-water exchange during a Voyage from Japan to Australia. PLoS One 9(5):e96274

Tsolaki E, Diamadopoulos E (2010) Technologies for ballast water treatment: a review. J Chem Technol Biotechnol 85(1):19–32

Verling E, Ruiz GM, Smith LD, Galil B, Miller AW, Murphy KR (2005) Supply-side invasion ecology: characterizing propagule pressure in coastal ecosystems. Proc R Soc Lond B Biol Sci 272(1569):1249–1257

Whiteley AS, Jenkins S, Waite I, Kresoje N, Payne H, Mullan B, Allcock R, O'Donnell A (2012) Microbial 16S rRNA Ion Tag and community metagenome sequencing using the Ion Torrent (PGM) Platform. J Microbiol Methods 91:80–88

Williams RJ, Griffiths FB, Van der Wal EJ, Kelly J (1988) Cargo vessel ballast water as a vector for the transport of non-indigenous marine species. Estuar Coast Shelf Sci 26(4):409–420

Wommack KE, Colwell RR (2000) Virioplankton: viruses in aquatic ecosystems. Microbiol Mol Biol Rev 64(1):69–114

Wonham MJ, Walton WC, Ruiz GM, Frese AM, Galil BS (2001) Going to the source: role of the invasion pathway in determining potential invaders. Mar Ecol Prog Ser 215:1–12

World Health Organization Geneva (2010) Weekly epidemiological record. 85(31): 293–308

Genotoxic Biomarkers as Indicators of Marine Pollution

17

M.K. Praveen Kumar, Avelyno D'costa, and S.K. Shyama

Abstract

The marine environment is the recipient of a wide range of pollutants including sewage, heavy metals, nanoparticles, petroleum hydrocarbons, and radiation which will ultimately affect the natural populations of aquatic organisms. These pollutants affect the integrity of the genome and lead to DNA damage. Genotoxicity tests such as the micronucleus test and comet assay are sensitive, reliable, and powerful techniques which are employed to assess the DNA damage. These tests can be designed on a wide variety of cells from any marine organism and can be performed either in the field or in the laboratory. The micronucleus test and comet assay can also be used as biomarkers along with other biochemical tests to evaluate the toxicity status of a particular marine water body. This review explores the use of these tests in different organisms under varying degrees of marine pollution. Further, it is recommended to use these two parameters together in order to confirm the holistic mutagenic/genotoxic effect of the test agents.

17.1 Introduction

Marine fauna, such as fish and bivalves, play a very crucial role in an aquatic ecosystem as consumers. Pollutants often get accumulated in sediment as well as the organisms of different trophic levels in aquatic ecosystems, including the benthic and pelagic animals. Fishes are the most predominant pelagic fauna in an aquatic environment representing the primary/secondary/tertiary consumers; whereas, bivalves are mostly sedentary in nature and are filter feeders. Larger fishes being positioned at the higher trophic level in our food chain can readily accumulate a variety of contaminants by ingestion of smaller species. Man, being at the apex of this food chain consumes varieties of fishes and bivalves as sea food and becomes the final recipient of these pollutants.

M.K. Praveen Kumar • A. D'costa
S.K. Shyama (✉)
Department of Zoology, Goa University,
Taleigao Plateau, Panaji 403206, Goa, India
e-mail: skshyama@gmail.com

M.M. Naik, S.K. Dubey (eds.), *Marine Pollution and Microbial Remediation*,
DOI 10.1007/978-981-10-1044-6_17

Biomonitoring of water bodies employing its natural populations of flora/fauna is of utmost importance to understand its pollution status. Ample reports are available on environmental monitoring/biomonitoring which are routinely carried out to assess the genotoxicity status of probable polluted water bodies and are compared with reference/pristine sites (Osman 2014). Exposure of fishes to toxic substances may damage their genome or in other words, exhibit genotoxic effects. Genotoxicity induces DNA changes leading to various diseases including cancer, reduced reproductive competence, teratogenicity, and may be even fatal, resulting in low fish catch. Further this may affect the normal food chain and result in a deleterious shift in the biotic community of this coastal ecosystem. These will also result in a drastic reduction in the quality/palatability of sea food (Chen and White 2004). Sea food is a major route of exposure of human populations to toxic chemicals. Fish and shellfish have been recognized as major mediators for contaminant transfer to humans and cause of innumerable sufferings in man (Al-Sabati and Metcalfe 1995). Due to pollution of the coastal environment with genotoxic substances, relevant genotoxicity tests are quickly gaining significance and several techniques are being developed to detect DNA damage and to identify these pollutants (Osman 2014).

17.2 Environmental Pollution

Marine pollution includes a range of threats such as runoffs, oil spills, untreated sewage discharge, heavy siltation, persistent organic pollutants (POPs), heavy metals from mining by-products, radioactive substances, and dumping of garbage. Many of these pollutants get collected at various depths, where they are consumed by marine organisms and are introduced into the global food chain. These pollutants may then ultimately get accumulated in the sediment and can pose a reverse threat to the benthic fauna of the aquatic ecosystem (Chen and White 2004). Polycyclic aromatic hydrocarbons, polychlorinated biphenyls, and polybrominated diphenyl ethers, when taken up by the body, form active metabolites which then cause DNA adducts and subsequent DNA damage (Wang et al. 2008). Barsiene et al. (2006a) observed a high incidence of micronuclei in the hemocytes of flounder (*Platichthys flesus*) which were contaminated with high concentrations of organic pollutants, evident by the presence of PAH metabolites in the bile.

Heavy metals often get accumulated in the marine environment at concentrations below the normal permissible levels by inputs from various sources such as industrial and domestic wastes. Most of the metals get readily concentrated and accumulated by aquatic organisms are relatively toxic even at fairly low concentrations. Discharge of heavy metals into river or any marine environment can change both marine species diversity and ecosystems, due to their toxicity and accumulative behavior (Bakan and Böke Özkoç 2007; Bat et al. 2009). Marine organisms such as fish tend to accumulate heavy metals many times higher than the concentration of that in water or sediment (Bat et al. 2009).

The uptake of xenobiotic contaminants, such as PCBs and PAHs, is highly detrimental to the biological integrity as well as the physiological functions of marine organisms. Many of these pollutants are chemical carcinogens and mutagens with the capacity to cause various types of DNA damage. Benzo[α]pyrene, a representative PAH, is reported to be converted at cellular level to the ROS, diol-epoxide (BaPDE), which can form stable adduct with DNA resulting into DNA strand breaks (Pisoni et al. 2004; Bihari and Fafandel 2004).

Nanogenotoxicology, which deals with the studies on the genotoxic effects of nanomaterials and nanodevices, is emerging as an important discipline along with the rapid expansion of the nanotechnology industry. Nanoparticles (NPs) are being used as a part of everyday products; thus, it is necessary to know and assess the possible toxic effects that are associated with them. The extremely small size of nanomaterials also means that they much more readily gain entry into the human body than larger-sized particles. Particles may gain access to the body by inhalation, ingestion, or through dermal contact, either deliberately or accidentally. Many metal NPs induce DNA strand breaks, oxidative DNA damage, mutations, and chromosomal aberrations. Particulate heavy metals (nano-sized) may interact with fish or other organisms as follows: (1) adsorption to the surface, i.e., cell, organ, or whole body, (2) cellular internalization, (3) dissolution of ions from the NP, and (4) mechanistic nano-effects such as formation of reactive oxygen species (ROS) (Baker et al. 2014). Gomes et al. (2013) exposed mussels to 10 μg/L of copper oxide (CuO) and silver (Ag) NPs for 15 days and reported DNA damage in hemocytes in the form of DNA strand breaks by the comet assay.

Radiations of various kinds widely used across the globe for diverse purposes are one of the major pollutants of our water bodies. Organisms get exposed to a wide range of radiation from these sources including radioactive effluents. Water bodies may get contaminated from industrial effluents intentionally or from nuclear facilities by accident. With most of the nuclear reactors being located in and around coastal areas, seas are often the ultimate recipient of radioactive contaminants (Praveen Kumar et al. 2014). Besides this, the nuclear accidents such as Chernobyl nuclear accident or the recent Fukushima disaster are the wake-up calls for employing radiation protection/safety of humans and the environment (Dallas et al. 2012). Since the risks of ionizing radiation to nonhuman biota being of considerable current interest to both the International Commission on Radiological Protection (ICRP) and the International Atomic Energy Agency (IAEA), they strongly recommend the impact assessment of radiation on those organisms (IAEA 1992; ICRP 2007). The major objective of the two international agencies, viz., FASSET (Framework for Assessment of Environmental Impact) and EPIC (Environmental Protection from Ionizing Contaminants in the Arctic), is to develop a methodology for protecting nonhuman organisms from ionizing contaminants (EPIC Project 2001; FASSET Project 2001).

17.3 Uses of Genotoxicity Parameters in Evaluation of Marine Pollution

The genotoxic effects of physical and chemical agents/pollutants can be monitored using a broad range of both in vivo and in vitro biomarker assays. The use of "biomarkers" in marine species has become a major tool for biomonitoring and assessing the environmental quality. This is based on the responses at the molecular and cellular level which represent the earliest signals of environmental disturbance (Depledge 1994; Serafim et al. 2011). Cytogenetic endpoints are proven as sensitive early biomarkers (Dallas and Jha 2015). Micronucleus (MN) test and comet assays are well-known genotoxicity tests often employed to predict mutagenicity/carcinogenicity of chemicals, including pharmaceuticals under in vivo and in vitro conditions (Shyama et al. 1991; Tice et al. 2000; Kirkland et al. 2005; Anbumani and Mohankumar 2012; Kadam et al. 2013; Dallas et al. 2013; Dasgupta et al. 2014; Guilherme et al. 2014; Pandey et al. 2014; Osman 2014; Praveen Kumar et al. 2015). These parameters are routinely used as biomarkers for monitoring marine pollution by genotoxic contaminants and can be combined with other physiological and biochemical biomarkers to fully assess the pollution status of various water bodies (Bolognesi and Cirillo 2014).

17.4 Micronucleus Test

Micronucleus test is an easy and useful test for the in vitro and in vivo genotoxicity assessment in marine organisms. Micronuclei result from acentric chromosome fragments or whole chromosomes lagging behind during metaphase to anaphase transition induced by clastogens or by spindle dysfunctions, respectively. Micronucleus test is routinely used for assessing the genotoxicity of compounds released from industries into aquatic environments (Barsiene et al. 2004; Cavaş and Ergene-Gozukara 2005a, b; Da Silva Souza and Fontanetti 2006). Fishes and bivalves are generally used to screen the potential genotoxicity of polluted water bodies (Bolognesi and Hayashi 2011). Natural populations of fishes collected from two different radiation-polluted sites showed high levels of MN (Ilyinskikh et al. 1998). Increased frequency of micronuclei was also reported in three fish species (*Platichthys flesus*, *Clupea harengus*, and *Zoarces viviparus*) collected from different regions of the Baltic Sea over a period of 3 years suggesting the periodical increase of pollution in the study site (Barsiene et al. 2012). X rays induced significant number of micronuclei in the erythrocytes of *Cyprinus carpio* (Gustavino et al. 2001) and also in the gill cells of *Oryzias latipes* (Takai et al. 2004a, b). Low-dose gamma radiation induced significant increase of micronuclei in fish cell lines, in vitro (Cassidy et al. 2007). A significant number of micronuclei were observed in the erythrocytes of the gamma-irradiated (5 Gy) fish, *Catla catla* (Hamilton) (Anbumani and Mohankumar 2011, 2012). An increase in the frequency of micronuclei in caged mussels (*Mytilus galloprovincialis*) transplanted at an offshore oil platform in the Adriatic Sea (Gorbi et al. 2008) supports the potential of these organisms to be used as biomonitoring species. Tritiated water (HTO) and tritiated glycine (T-Gly) induced a significant number of micronuclei in the hemocytes of *Mytilus edulis* (Jaeschke et al. 2010). Petroleum hydrocarbons of oil spills too induce genotoxicity as evident by the formation of micronuclei in the erythrocytes of flounder (*Platichthys flesus*) and gills of blue mussel (*Mytilus edulis*) (Barsiene et al. 2006b). Sacchi et al. (2013) reported the induction of micronuclei in the bivalve *Ruditapes philippinarum* at an estuarine coastal site polluted with polycyclic aromatic hydrocarbons and trace metals. MN levels in mussels (*Mytilus galloprovincialis*) collected from 17 sites along the Spanish Mediterranean coast were found to be the highest from metal-polluted sites (Fernandez et al. 2011). They also reported that environmental factors such as water temperature were able to influence the levels of metal-induced MN. Fishes collected from the Paraguay river within the Pantanal, Brazil, which is the world's largest wetland area and a region rich in biodiversity, revealed significantly high frequencies of MN in the erythrocytes probably due to the discharge of tannery effluents (Pimenta et al. 2013).

17.5 Comet Assay

The comet assay or single-cell gel electrophoresis (SCGE) is a rapid and sensitive technique that detects DNA single- or double-strand breaks and measures the level of migration of DNA from individual cell nuclei. The amount of the DNA in the tail region (tail DNA) of comet assay is commonly used for quantifying DNA strand breakage and represents the most reliable parameter (Mitchelmore and Chipman 1998). This test is also widely used for in vivo assessment in marine organisms (Fernández-Tajes et al. 2011; Frenzilli and Lyons 2013). Mussels are globally used as bioindicators for pollution of coastal and estuarine environments by metals and radionuclides (Lonsdale et al. 2009). Hagger et al. (2005) have reported the β- radiation induced genotoxic effects of tritiated water in the early life stages of the marine mollusk, *Mytilus edulis*. Further, the genotoxic effects of tritium (^3H) in the adult life stage of *Mytilus edulis* have been evaluated employing comet assay (indicating DNA single-strand breaks/alkali labile sites) in the hemocytes of radiation-exposed individuals (Jha et al. 2005, 2006). External and internal dose rates of ionizing radiation altered the DNA strand breakage in marine mussel *Mytilus edulis* (Alamri et al. 2012).

Polycyclic aromatic hydrocarbons were found to induce DNA strand breaks in the hemocytes and gill cells of three bivalve species (*Venerupis pullastra*, *Cerastoderma edule*, and *Mytilus galloprovincialis*) collected from an estuary contaminated with petroleum products (Fernandez-Tejes et al. 2011). Bottom feeders such as soles (*Solea senegalensis*) exhibited increased DNA strand breaks when they were exposed to sediment contaminated with heavy metals and PAHs and PCBs (Costa et al. 2008). Snails (*Littorina littorea*) and mussels (*Mytilus edulis*) collected from a contaminated harbor showed high levels of DNA damage due to presence of heavy metals and butyltin compounds (Rank 2009). Binelli et al. (2007) correlated the increase in DNA strand breaks in soft tissues of zebra mussels (*Dreissena polymorpha*) with chemicals such as polycyclic aromatic hydrocarbons, polybrominated diphenyl ethers, Dichlorodiphenyltrichloroethanes (DDTs), and hexachlorobenzene in the Lake Maggiore, Italy.

The most severe DNA lesions induced by ionizing toxicants are single- and double-strand breaks (Collins 2004). The alkaline single-cell gel electrophoresis assay allows the early detection of single-strand DNA breaks which may be induced by a range of genotoxic agents including radiation (Mayalpa et al. 1998; Kadam et al. 2013). Gamma irradiation is reported to induce DNA damage and cell cycle perturbations in the fish, *Catla catla* (Anbumani et al. 2012; Anbumani and Mohankumar 2015). A dose-dependent DNA damage was observed in zebrafsh exposed to ionizing radiation (Knowles 2002; Jarvis and Knowles 2003; Lemos et al. 2014). An increase in % tail DNA was observed in caged as well as feral European chubs (*Leuciscus cephalus*) from rivers polluted with a combination of PAHs, PCBs, organochlorine pesticides (OCPs), and heavy metals around Birmingham, UK (Winter et al. 2004). De Andrade et al. (2004) also reported increased DNA damage in mullet and sea catfish from two rivers in Brazil that were possibly contaminated with heavy metals and hydrocarbons. Dabs (*Limanda limanda*) collected from the mouth of the Seine estuary in France exhibited high DNA damage (% tail DNA) and were found to be positively correlated with the concentration of biliary PAH metabolites (Devier et al. 2013). Cockles (*Cerastoderma edule*) and blue mussels (*Mytilus edulis*) collected from an estuary in South West England exhibited high DNA damage due to the presence of heavy metals discharged by industries (Dallas et al. 2012). Almeida et al. (2013) reported low levels of DNA damage from a coastal lagoon in Portugal and has also demonstrated the use of the comet assay to evaluate genotoxicity from places of low contamination.

17.6 Conclusions

The presence of large quantities of pollutants such as heavy metals, petroleum hydrocarbons, radiation, and nanoparticles causes several adverse effects in marine fauna and may affect the fish catch and palatability. Further, these pollutants may enter the food chain and ultimately affect humans. The micronucleus test detects chromosomal and genomic mutations (chromosomal damage and/or alteration of mitotic spindles), whereas the alkaline comet assay detects primary DNA damage, expressed as single-strand breaks, including the ones which are associated with incompletely repaired excision. A combined use of MN test and alkaline single-cell gel electrophoresis confirms the possible single as well as double-strand breaks induced by genotoxic agents. Accordingly, it is recommended to use these two parameters together in order to confirm the holistic mutagenic/genotoxic effect of the test agents (Figs. 17.1 and 17.2).

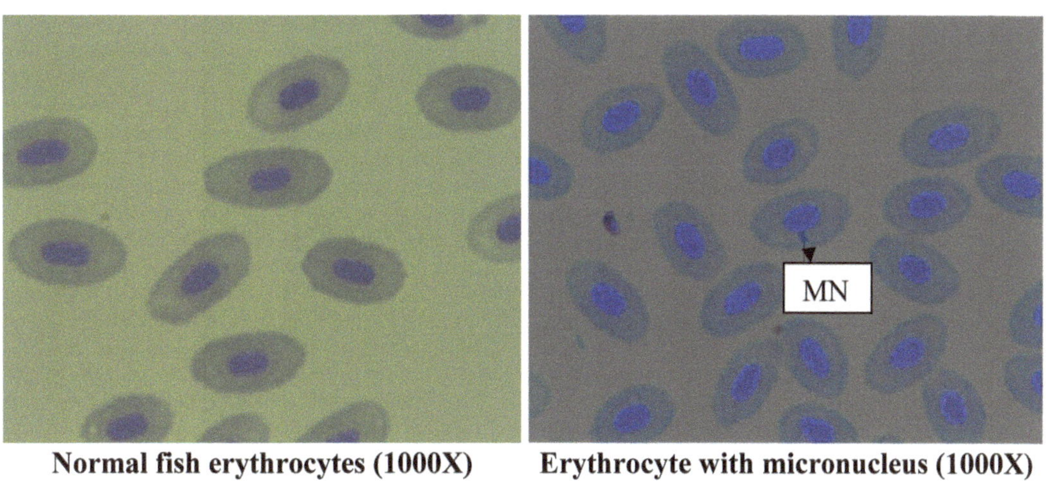

Fig. 17.1 Normal erythrocytes and erythrocyte with a micronucleus in zebra fish

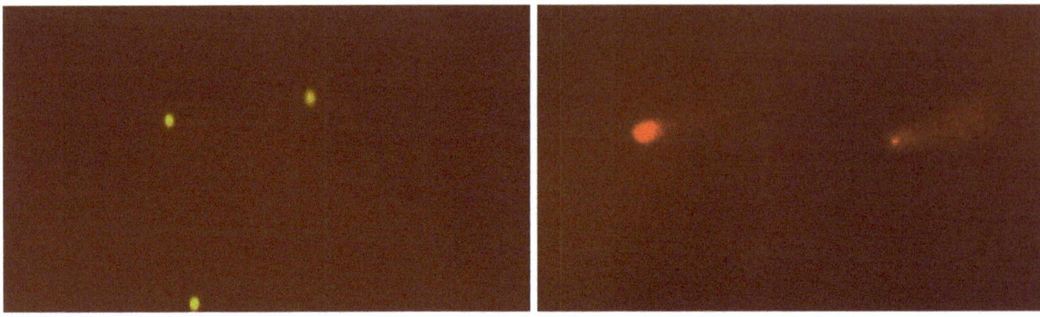

Fig. 17.2 Comet assay in erythrocytes of zebra fish

References

Alamri OD, Cundy AB, Di Y, Jha AN, Rotchell JM (2012) Ionizing radiation-induced DNA damage response identified in marine mussels, *Mytilus sp*. Environ Pollut 168:107–112

Almeida C, Pereira CG, Gomes T, Cardoso C, Bebianno MJ, Cravo A (2013) Genotoxicity in two bivalve species from a coastal lagoon in the south of Portugal. Mar Environ Res 89:29–38

Al-Sabti K, Metcalfe CD (1995) Fish micronuclei for assessing genotoxicity in water. Mutat Res 343:2–3

Anbumani S, Mohankumar MN (2011) Nuclear and cytoplasmic abnormalities in the fish *Catla catla* (Hamilton) exposed to chemicals and ionizing radiation. Res J Environ Sci 5:867–877

Anbumani S, Mohankumar MN (2012) Gamma radiation induced micronuclei and erythrocyte cellular abnormalities in the fish *Catla catla*. Aquat Toxicol 122–123:125–132

Anbumani S, Mohankumar MN (2015) Gamma radiation induced cell cycle perturbations and DNA damage in *Catla Catla* as measured by flow cytometry. Ecotox Environ Safe 113:18–22

Anbumani S, Mohankumar MN, Selvanayagam M (2012) Flow cytometric assessment of DNA damage in the fish *Catla catla* (Ham.) exposed to gamma radiation. J Adv Biotechnol 11(8):36–40

Bakan G, Böke Özkoç H (2007) An ecological risk assessment of the impact of heavy metals in surface sediments on biota from the mid-Black Sea coast of Turkey. Int J Environ Stud 64(1):45–57

Baker TJ, Tyler CR, Galloway TS (2014) Impacts of metal and metal oxide nanoparticles on marine organisms. Environ Pollut 186:257–271

Baršiene J, Lazutka J, Šyvokiene J, Dedonyte V, Rybakovas A, Bjornstad A, Andersen OK (2004) Analysis of micronuclei in blue mussels and fish from the Baltic and the North Seas. Environ Toxicol 19:365–371

Baršiene J, Lehtonen KK, Koehler A, Broeg K, Vuorinen PJ, Lang T, Pempkowiak J, Syvokien J, Dedonyte V, Rybakovas A, Repecka R, Vuontisjarvi H, Kopecka J (2006a) Biomarker responses in flounder (*Platichthys flesus*) and mussel (*Mytilus edulis*) in the Klaipeda-Butinge area (Baltic Sea). Mar Pollut Bull 53:422–436

Baršiene J, Dedonyte V, Rybakovas A, Andreikenait L, Andersen OK (2006b) Investigation of micronuclei and other nuclear abnormalities in peripheral blood and kidney of marine fish treated with crude oil. Aquat Toxicol 78S:S99–S104

Baršienė J, Rybakovas A, Lang T, Grygiel W, Andreikėnaitė L, Michailovas A (2012) Risk of environmental genotoxicity in the Baltic Sea over the period of 2009–2011 assessed by micronuclei frequencies in blood erythrocytes of flounder (*Platichthys flesus*), herring (*Clupea harengus*) and eelpout (*Zoarces viviparus*). Mar Environ Res 77:35–42

Bat L, Gökkurt O, Sezgin M, Üstün F, Sahin F (2009) Evaluation of the Black Sea Land based sources of pollution the coastal region of Turkey. Open Mar Biol J 3:112–124

Bihari N, Fafandel M (2004) Interspecies differences in DNA single strand breaks caused by benzo(α)pyrene and marine environment. Mutat Res/Fundam Mol Mech Mutagen 552(1/2):209–217

Binelli A, Riva C, Provini A (2007) Biomarkers in Zebra mussel for monitoring and quality assessment of Lake Maggiore (Italy). Biomarkers 12(4):349–368

Bolognesi C, Cirillo S (2014) Genotoxicity biomarkers in aquatic bioindicators. Curr Zool 60(2):273–284

Bolognesi C, Hayashi M (2011) Review: micronucleus assay in aquatic animals. Mutagenesis 26(1):205–213

Cassidy CL, Lemon JA, Boreham DR (2007) Impacts of low-dose gamma-radiation on genotoxic risk in aquatic ecosystems. Dose-Response 5:323–332

Cavaş T, Ergene-Gozukara S (2005a) Induction of micronuclei and nuclear abnormalities in *Oreochromis niloticus* following exposure to petroleum refinery and chromium processing plant effluents. Aquat Toxicol 74:264–271

Cavaş T, Ergene-Gozukara S (2005b) Micronucleus test in fish cells: a bioassay for *in situ* monitoring of genotoxic pollution in the marine environment. Environ Mol Mutagen 46:64–70

Chen G, White PA (2004) The mutagenic hazards of aquatic sediments: a review. Mutat Res 567(2–3):151–225

Collins AR (2004) The comet assay for DNA damage and repair: principles, applications and limitations. Mol Biotechnol 26:249–261

Costa PM, Lobo J, Caeiro S, Martins M, Ferreira AM, Caetano M, Vale C, Delvalls TA, Costa MH (2008) Genotoxic damage in *Solea senegalensis* exposed to sediments from the Sado Estuary (Portugal): effects of metallic and organic contaminants. Mutat Res 654(1):29–37

Da Silva Souza T, Fontanetti CS (2006) Micronucleus test and observation of nuclear alterations in erythrocytes of Nile tilapia exposed to waters affected by refinery effluent. Ecotox Environ Safe 605:87–93

Dallas LJ, Jha AN (2015) Applications of biological tools or biomarkers in aquatic biota: a case study of the Tamar Estuary, South West England. Mar Pollut Bull 95(2):618–633

Dallas LJ, Keith-Roach M, Lyons BP, Jha AN (2012) Assessing the impact of ionizing radiation on aquatic invertebrates: a critical review. Radiat Res 177:693–716

Dallas LJ, Cheung VV, Fisher AS, Jha AN (2013) Relative sensitivity of two marine bivalves for detection of genotoxic and cytotoxic effects: a field assessment in the Tamar Estuary, South West England. Environ Monit Assess 185(4):3397–3412

Dasgupta S, Cao A, Mauer B, Yan B, Uno S, McElroy A (2014) Genotoxicity of oxy-PAHs to Japanese medaka (*Oryzias latipes*) embryos assessed using the comet assay. Environ Sci Pollut Res Int 21:13867–13876

de Andrade VM, da Silva J, Silva FR, Heuser VD, Dias JF, Yoneama ML, Freitas TR (2004) Fish as bioindicators to assess the effects of pollution in two southern Brazilian rivers using the Comet assay and micronucleus test. Environ Mol Mutagen 44(5):459–468

Depledge MH (1994) The rational basis for the use of biomarkers as ecological tools. In: Fossi MC, Leonzio C (eds) Nondestructive biomarkers in vertebrates. Lewis Publishers, Boca Raton, pp 271–295

Dévier MH, Dû-Lacoste L, Akcha F, Morin B, Peluhet L, Le Menach K, Burgeot T, Budzinski H (2013) Biliary PAH metabolites, EROD activity and DNA damage in dab (*Limanda limanda*) from Seine Estuary (France). Environ Sci Pollut Res 20:708–722

EPIC Project (2001) Environmental contamination from ionising contaminants in the Arctic. First Annual Report (01.11.2000–31.10.2001). Project ICA2-CT-2000-10032

FASSET Project (2001) Framework for assessment of environmental impact. Progress report N 1 (1.11.2000–31.10.2001). A project within the EC 5th framework programme. http://www.fasset.org

Fernández B, Campillo JA, Martínez-Gómez C, Benedicto J (2011) Micronuclei and other nuclear abnormalities in mussels (*Mytilus galloprovincialis*) as biomarkers of cyto-genotoxic pollution in mediterranean waters. Environ Mol Mutagen 52(6):479–491

Fernández-Tajes J, Flórez F, Pereira S, Rábade T, Laffon B, Méndez J (2011) Use of three bivalve species for biomonitoring a polluted estuarine environment. Environ Monit Assess 177(1–4):289–300

Frenzilli G, Lyons BP (2013) The comet assay in marine animals. Methods Mol Biol 1044:363–372

Gomes T, Araújo O, Pereira R, Almeida AC, Cravo A, Bebianno MJ (2013) Genotoxicity of copper oxide and silver nanoparticles in the mussel *Mytilus galloprovincialis*. Mar Environ Res 84:51–59

Gorbi S, Virno Lamberti C, Notti A, Benedetti M, Fattorini D, Moltedo G, Regoli F (2008) An ecotoxicological protocol with caged mussels, *Mytilus galloprovincialis*, for monitoring the impact of an offshore platform in the Adriatic Sea. Mar Environ Res 65(1):34–49

Guilherme S, Santos MA, Gaivao I, Pacheco M (2014) Genotoxicity evaluation of the herbicide Garlon®and its active ingredient (triclopyr) in fish (*Anguilla anguilla L.*) using the comet assay. Environ Toxicol. Epub http://dx.doi.org/10.1002/tox.21980

Gustavino B, Scornajenghi KA, Minissi S, Ciccotti E (2001) Micronuclei induced in erythrocytes of *Cyprinus carpio* (teleostei, Pisces) by X-rays and colchicines. Mutat Res 494:151–159

Hagger JA, Atienzar FA, Jha AN (2005) Genotoxic, cytotoxic, developmental and survival effects of tritiated water in the early life stages of the marine mollusc, *Mytilus edulis*. Aquat Toxicol 74:205–217

IAEA (1992) Effects of ionizing radiation on plants and animals at levels implied by current radiation protection standards, vol 332, Technical reports series. International Atomic Energy Agency, Vienna

ICRP (2007) The 2007 recommendations of the international commission on radiological protection. ICRP publication 103. Ann ICRP 37:2–4

Ilyinskikh NN, Ilyinskikh EN, Ilyinskikh IN (1998) Micronucleated erythrocytes frequency and radiocesium bioconcentration in pikes (*Esox lucius*) caught in the Tom River near the nuclear facilities of the Siberian Chemical Complex (Tomsk-7). Fundam Mol Mech Mutagen 421:197–203

Jaeschke BC, Millward GE, Moody AJ, Jha AN (2010) Tissue-specific incorporation and genotoxicity of different forms of tritium in the marine mussel, Mytilus edulis. Environ Pollut 159(1):274–280

Jarvis RB, Knowles JF (2003) DNA damage in zebrafish larvae induced by exposure to low-dose rate gamma-radiation: detection by the alkaline comet assay. Mutat Res-Genet Toxicol Environ Mutagen 541:63–69

Jha AN, Dogra Y, Turner A, Millward GE (2005) Impact of low doses of tritium on the marine mussel, Mytilus edulis: genotoxic effects and tissue-specific bioconcentration. Mutat Res 586:47–57

Jha AN, Dogra Y, Turner A, Millward GE (2006) Are low doses of tritium genotoxic to *Mytilus edulis*? Mar Environ Res 62:297–300

Kadam SB, Shyama SK, Almeida VG (2013) Evaluation of the in vivo genotoxic effects of gamma radiation on the peripheral blood leukocytes of head and neck cancer patients undergoing radiotherapy. Mutat Res Genet Toxicol Environ Mutagen 752:42–46

Kirkland D, Aardema M, Henderson L, Mueller L (2005) Evaluation of the ability of a battery of 3 in vitro genotoxicity tests to discriminate rodent carcinogens and non- carcinogens. I. Sensitivity, specificity and relative predictivity. Mutat Res 584:1–256

Knowles JF (2002) An investigation into the effects of chronic radiation on fish. In: R&D Technical Reports. Environment Agency, Bristol. P3-053/TR

Lemos J, Costa P, Cunha L, Carvalho AP, Neuparth T, Genésio P, Ponte F, Costa P, Crespo P, Metello L (2014) Assessing radiobiological effects of diagnostic level doses of ionizing radiation on zebrafish using comet assay. J Nucl Med 55:1426

Lonsdale DJ, Cerrato RM, Holland R, Mass A, Holt L, Schaffner RA, Pan J, Caron DA (2009) Influence of suspension-feeding bivalves on the pelagic food webs of shallow, coastal embayments. Aquat Biol 6:263–279

Malyapa SR, Bi C, Ahern EW, Roti Roti JL (1998) Detection of DNA damage by the alkaline comet assay after exposure to low-dose gamma radiation. Radiat Res 149:396–400

Mitchelmore CL, Chipman JK (1998) DNA strand breakage in aquatic organisms and the potential value of the comet assay in environmental monitoring. Mutat Res 399(2):135–147

Osman AGM (2014) Genotoxicity tests and their contributions in aquatic environmental research. J Environ Prot 5:1391–1399

Pandey AK, Nagpure NS, Trivedi SP (2014) Evaluation of genotoxicity of profenofos to freshwater fish *Channa punctatus* (Bloch) using the micronucleus assay. Afr J Biotechnol 13(39):3985–3988

Pimenta VM, da Silva JM, Nepomuceno JC, Pavanin LA (2013) In situ assessment of the Paraguay River water, in Brazilian Pantanal, by means of micronucleus assay with fish and chemical analysis. Bull Environ Contam Toxicol 90:427–433

Pisoni M, Cogotzi L, Frigeri A, Corsi I, Bonacci S, Iacocca A, Lancini L, Mastrototaro F, Focardi S, Svelti M (2004) DNA adducts, benzo(α)pyrene monooxygenase activity and lysosomal membrane stability in *Mytilus galloprovincialis* from different areas in Taranto coastal waters (Italy). Environ Res 96:163–175

Praveen Kumar MK, Shyama SK, Sonaye BS, Naik RU, Kadam SB, Bipin PD, D'costa A, Chaubey RC (2014) Evaluation of γ- radiation induced DNA damage in two species of bivalves and their relative sensitivity using comet assay. Aquat Toxicol 150:1–8

Praveen Kumar MK, Shyama SK, Sonaye BS, D'costa A, Chaubey RC (2015) The effect of gamma radiation on the Common carp (Cyprinus carpio): in vivo genotoxicity assessment with the micronucleus and comet assay. Mutat Res 792:19–25

Rank J (2009) Intersex in Littorina littorea and DNA damage in Mytilus edulis as indicators of harbour pollution. Ecotoxicol Environ Saf 72(4):1271–1277

Sacchi A, Mouneyrac C, Bolognesi C, Sciutto A, Roggieri P, Fusi M, Beone GM, Capri E (2013) Biomonitoring study of an estuarine coastal ecosystem, the Sacca di Goro lagoon, using *Ruditapes philippinarum* (Mollusca: Bivalvia). Environ Pollut 177:82–89

Serafim A, Lopes B, Company R, Cravo A, Gomes T, Sousa V, Bebianno MJ (2011) A multi-biomarker approach in cross-transplanted mussels *Mytilus galloprovincialis*. Ecotoxicology 20(8):1959–1974

Shyama SK, Rahiman MA, Vijayalaxmi KK (1991) Genotoxic effect of Anovlar 21, an oral contraceptive, on mouse bone marrow. Mutat Res 260:47–53

Takai A, Kagawa N, Fujikawa K (2004a) Dose- and time-dependent responses for micronucleus induction by X-rays and fast neutrons in gill cells of medaka (*Oryzias latipes*). Environ Mol Mutagen 44:108–112

Takai A, Kagawa N, Fujikawa K (2004b) Susceptibility of male and female medaka (*Oryzias latipes*) fish to spontaneous and X-ray induced micronucleus formation in gill cells. Mutat Res 558:131–136

Tice RR, Agurell E, Anderson D, Burlinson B, Hartman A, Kobayashi H, Miyamae Y, Royas E, Ryu JC, Sasaki YF (2000) Single cell gel/comet assay: guidelines for in vitro and in vivo genetic toxicology testing. Environ Mol Mutagen 35:206–221

Wang XH, Hong HS, Mu JL, Lin JQ, Wang SH (2008) Polycyclic aromatic hydrocarbon (PAH) metabolites in marine fishes as a specific biomarker to indicate PAH pollution in the marine coastal environment. J Environ Sci Health A Tox Hazard Subst Environ Eng 43(3):219–226

Winter MJ, Day N, Hayes RA, Taylor EW, Butler PJ, Chipman JK (2004) DNA strand breaks and adducts determined in feral and caged chub (Leuciscus cephalus) exposed to rivers exhibiting variable water quality around Birmingham, UK. Mutat Res 552(1–2):163–175

Kolibri 360 GmbH
Ferdinand-Jühlke-Straße 7
99095 Erfurt, Deutschland
produktsicherheit@kolibri360.de